Contents

Chapter 1..2
Chapter 2...12
Chapter 3...59
Chapter 4 87
Chapter 5..107
Chapter 6..161
Chapter 7..183
Chapter 8..227
Chapter 9..271
Chapter 10...316
Chapter 11...339
Chapter 12...335
Chapter 13...339

Instructor's Solutions Manual

Milton Loyer

Penn State University

to accompany

Elementary Statistics

Ninth Edition

Mario F. Triola

Dutchess Community College

Boston San Francisco New York
London Toronto Sydney Tokyo Singapore Madrid
Mexico City Munich Paris Cape Town Hong Kong Montreal

Reproduced by Pearson Addison-Wesley from electronic files supplied by the author.

Copyright © 2004 Pearson Education, Inc.
Publishing as Pearson Addison-Wesley, 75 Arlington Street, Boston, MA 02116

All rights reserved. No part of this publication may be reproduced, stored in a retrieval system, or transmitted, in any form or by any means, electronic, mechanical, photocopying, recording, or otherwise, without the prior written permission of the publisher. Printed in the United States of America.

ISBN 0-321-12212-7

1 2 3 4 5 VHG 06 05 04 03

INTRODUCTION

by Milton Loyer

This *Instructor's Solutions Manual* contains solutions to all the exercises in the text. For convenience, we include herein all the information given in the ***Student's Solutions Manual*** as it appears in that volume – except that the student solutions are sometimes given in an expanded format made user-friendlier by the inclusion of additional white space. Boldface type identifies the exercise numbers and NOTE designations for material appearing only in the instructor's manual. Accordingly, even numbers not in boldface type identify solutions appearing for one reason or another in the student's manual.

This manual includes six sample syllabi to help instructors plan a course outline. These syllabi appear at the back of the text, after page 413. Also included at the back of this manual is a guide to the videos, designed to supplement most sections in the book. These videos are an excellent resource for instructors involved with distance learning, individual study, and self-paced learning programs.

STUDENT'S SOLUTIONS MANUAL PREFACE

This manual contains the solutions to the odd-numbered exercises for each section of the textbook *Elementary Statistics*, Ninth Edition, by Mario Triola, and the solutions for all of the end-of-chapter review and cumulative review exercises of that text. To aid in the comprehension of calculations, worked problems typically include intermediate steps of algebraic and/or computer/calculator notation. When appropriate, additional hints and comments are included and prefaced by NOTE.

Many statistical problems are best solved using particular formats. Recognizing and following these patterns promote understanding and develop the capacity to apply the concepts to other problems. This manual identifies and employs such formats whenever practicable.

For best results, read the text carefully *before* attempting the exercises, and attempt the exercises *before* consulting the solutions. This manual has been prepared to provide a check and extra insights for exercises that have already been completed and to provide guidance for solving exercises that have already been attempted but have not been successfully completed.

I would like to thank Mario Triola for writing an excellent elementary statistics book and for inviting me to prepare this solutions manual.

# Chapter 1

Introduction to Statistics

1-2 Types of Data

1. Parameters, since 87 and 13 refer to the entire population.
2. Statistic, since 4.2 refers to the selected sample.
3. Statistic, since 0.65 refers to the selected sample.
4. Parameter, since 706 refers to the entire population.
5. Discrete, since annual salary cannot assume any value over a continuous span – as an annual salary of $98,765.4321, for example, is not a physical possibility. The annual presidential salary is typically stated in multiples of $1000 - and it cannot be paid in US currency in units smaller than a cent.
6. Continuous, since weight can be any value over a continuous span.
7. Discrete, since the calculated percent cannot be any value over a continuous span. Because the number having guns in their homes must be an integer, the possible values for the percent are limited to multiples of 100/1059.
8. Discrete, since the number of defective masks must be an integer.
9. Ratio, since differences are meaningful and zero height has a natural meaning.
10. Ordinal, since the ratings give relative position in a hierarchy.
11. Interval, since differences are meaningful but ratios are not. Refer to exercise #21.
12. Nominal, since the number are for identification only. The numbers on the jerseys are merely numerical names for the players.
13. Ordinal, since the ratings give relative position in a hierarchy.
14. Nominal, since the numbers are used for identification only. Even though SS numbers are assigned chronologically within regions and can be placed in numerical order, there is no meaningful way to compare 208-34-3337 and 517-94-1439. If all the numbers had been assigned chronologically beginning with 000-00-0001, like the order of finishers in a race, then SS numbers would illustrate the ordinal level of measurement.
15. Ratio, since differences are meaningful and zero "yes" responses has a natural meaning.
16. Nominal, since the numbers are used for identification only. Even though they are assigned alphabetically within regions, zip codes are merely numerical names for post offices. If the numbers had resulted from placing all the post offices in one large list alphabetically (or by city size, or by mail volume, etc.), so that 17356 [Red Lion, PA] was the 17,356th post office in the list, like the order of finishers in a race, then zip codes would illustrate the ordinal level of measurement.
17. a. The sample is the ten adults who were asked for an opinion.
    b. There is more than one possible answer. The population the reporter had in mind may have been all the adults living in that particular city.
    No. People living in that city but who do not pass by that corner (shut-ins, those with no business in that part of town, etc.) will not be represented in the sample. Furthermore, it was not specified that the reported selected the ten adults in an unbiased manner - e.g., he may have (consciously or subconsciously) selected ten friendly-looking people.

INSTRUCTOR'S SOLUTIONS Chapter 1   S-3

18. **NOTE:** This is a non-trivial problem. By definition, the sample must be a subset of the population – i.e., the sample and the population must have the same units. In this particular problem, the researchers appear to be interested in the TV sets that are in use at a particular time – and, of course, the number of TV sets in use at a selected household may be 0, 1, or more than 1.
    a. The sample is the TV's that are in use at that time the randomly selected households.
    b. The population is all the TV's in use at that time in the entire country.
    Yes. If the households are selected at random, then each TV in use at the time has the same chance of being included - which means that there are no biases that would prevent the sample from being representative.

19. a. The sample is the 1059 randomly selected adults.
    b. The population is all the adults in the entire country. NOTE: While it is not specified that the adults were selected from within USA, this is the logical assumption.
    Yes. If the adults in the sample were randomly selected, there are no biases that would prevent the sample from being representative. NOTE: This is a sample of adults, not of homes. The appropriate inference is that about 39% of the adults live in a home with a gun, not that there is a gun in about 39% of the homes. Since adults were selected at random, homes with more adults are more likely to be represented in the survey – and that bias would prevent the survey from being representative of all homes.

20. a. The sample is the 65 adults who responded to the survey.
    b. The population for this portion of the project is all the adults that the graduate student knows.
    No. The sample includes only those who cared enough about the graduate student and/or the issue in question to offer a response. Those who did not care enough to respond will not be represented in the sample.

21. Temperature ratios are not meaningful because a temperature of 0° does not represent the absence of temperature in the same sense that $0 represents the absence of money. The zero temperature in the example (whether Fahrenheit or Centigrade) was determined by a criterion other than "the absence of temperature."

22. That value has no meaning.
    NOTE: When there are only two categories coded 0 and 1, the average gives the proportion of cases in the category coded 1. When there are more than two categories, such calculations with nominal data have no meaning.

23. This is an example of ordinal data. It is not interval data because differences are not meaningful – i.e., the difference between the ratings +4 and +5 does not necessarily represent the same differential in the quality of the food as the difference between the ratings 0 and +1.

1-3 Critical Thinking

NOTE: The alternatives given in exercises #1-4 are not the only possible answers.

1. Males weigh more than females – and truck drivers tend to be male, while about half the adults who do not drive trucks are female.

2. Homeowners tend to be more affluent than renters – and thus they can afford better medical care.

3. There are more minority drivers than white drivers in Orange County – if the rates of speeding were the same for both types of drivers, you would expect the more numerous type of driver to be issued more tickets.

4. Colds generally do not last more than two weeks – you would expect all the subjects to be improved in two weeks no matter what medicine they took or didn't take.

5. The study was financed by those with a vested interest in promoting chocolate, who would be more likely to report only the favorable findings. While chocolate contains an antioxidant associated with decreased risk of heart disease and stroke, it may also contain other ingredients associated with an increased risk of those or other health problems.

6. The census counting process is not 100% accurate, especially among those (such as illegal aliens) who may not wish to be counted. Giving such a precise figure suggests a degree of accuracy not actually warranted.

7. No. The sample was self-selected and therefore not necessarily representative of the general population (or even of the constituencies of the various women's groups contacted).

8. Because the 186,000 respondents were self-selected and not randomly chosen, they are not necessarily representative of the general population and provide no usable information about the general population. In addition, the respondents were self-selected from a particular portion of the general population – viz., persons watching "Nightline" and able to spend the time and money to respond.

9. Telephone directories do not contain unlisted numbers, and consumer types electing not be listed would not be represented in the survey. In addition, sometimes numbers appear in directories more than once (e.g., under the husband's name and under the wife's name), and consumer types electing multiple listings would be over-represented in the survey. Also excluded from the survey would be those without telephones (for economic, religious, or other reasons), transient persons (both frequent movers and those who just happened to have moved) not in the current directory, and persons whose living arrangements (group housing, extended families, etc.) do not include a phone in their own name.

10. A person could make the same claim about anything found in cities but not in rural areas. Consider, for example, the claim that fire hydrants cause crime – because crime rates are higher in cities with fire hydrants than in rural areas that have no fire hydrants.

11. Motorcyclists that died in crashes in which helmets may have saved their lives could not be present to testify.

12. They survey may not produce honest responses. A financial consultant, or any person dealing with the public, will probably give better service to those he likes – and so a respondent who knows that his name will be linked with the answers he supplies might hesitate to give responses that would upset his financial consultant.

13. No. While 0.94 is the average of the 29 different brands in the survey, it is not a good estimate of the average of all the cigarettes smoked for at least two significant reasons. First, the 29 brands in the survey are not representative of all brands on the market – they include only 100mm filtered brands that are not menthol or light. Secondly, some brands are much more popular than others. Even if every brand on the market were in the survey, the average would not be a good estimate of all the cigarettes smoked – the values of the popular brands will carry more weight when the amounts of each brand that are actually smoked are taken into consideration.

14. There are several possible answers. (1) Since tallness is perceived to be a favorable attribute, people tend to overstate their heights; at the very least, people would tend to round to the next highest inch and not to the nearest inch. (2) Many people do not really accurately know their height. (3) Because Americans tend to express height in feet and inches, errors might occur either in converting heights to all inches or in misstatements like 52" for 5'2". (4) Because many cultures express height in centimeters, some people might not know or be able to readily calculate their heights in inches.

15. Most people born after 1945 are still living. This means the sample would include only those who did not reach their normal life expectancy and would not be representative.

16. The decreased rate that occurred after the recommendation to use the supine position was not necessarily *caused* by the use of that position. It could be that general medical

advances were lowering the rates of all infant mortality, including death by SIDS.

NOTE for exercises #17-20: Multiplying or dividing by 1 will not change a value. Since 100% is the same as 1, multiplying or dividing by 100% will not change a value – but remember that the percent sign is a necessary unit (just like inches or pounds) that cannot be ignored.

17. a. $17/25 = .68 = .68 \times 100\% = 68\%$
    b. $35.2\% = 35.2\%/100\% = .352$
    c. $57\% = 57\%/100\% = .57; .57 \times 1500 = 855$
    d. $.486 = .486 \times 100\% = 48.6\%$

18. a. $26\% = 26\%/100\% = .26; .26 \times 950 = 247$
    b. $5\% = 5\%/100\% = .05$
    c. $.01 = .01 \times 100\% = 1\%$
    d. $527/1200 = .439 = .439 \times 100\% = 43.9\%$

19. a. There is not enough accuracy in the reported percent to determine an exact answer. While $.52 \times 1038 = 539.76$ suggests an answer of 540, x/1038 rounds to 52% for $535 \le x \le 544$. The actual number could be anywhere from 535 to 544 inclusive.
    b. $52/1038 = .05 = 5\%$

20. a. $19/270 = .07 = 7\%$
    b. $270 \times .030 = 8$

21. There is not enough accuracy in the reported percents to determine an exact answer.
    It is tempting to conclude that
    $.08 \cdot (1875) = 150$ admit to committing a campus crime
    $.62 \cdot (150) = 93$ say they did so under the influence of alcohol or drugs.
    But x/1875 rounds to 8% for $141 \le x \le 159$, and so the number admitting to a crime could be anywhere from 141 to 159 inclusive. Similarly, the number that say they committed the crime under the influence of alcohol or drugs could be anywhere from 87 to 99 inclusive – since 87/141 and 99/159 both round to 62%.

22. Since 100% is the totality of whatever is being measured, a decrease of 100% would mean that nothing is left. A decrease of more than 100% would be possible only in situations (e.g., those involving money) where negative amounts have meaning. An increase of more than 100%, however, presents no problems – as an increase of 200%, for example, means an addition of twice the original amount to produce three times the starting quantity.
    a. Refer to the explanation above. Since negative plaque has no meaning, the amount of plaque cannot be reduced by more than 100%.
    b. Refer to the explanation above. Since negative investment generally has no meaning, the amount of foreign investment cannot fall more than 100%. The only possible literal interpretation is that the net flow of foreign money is now negative, and four times what it used to be when it was positive.

23. Assuming that each of the 20 individual subjects is ultimately counted as a success or not (i.e., that there are no "dropouts" or "partial successes"), the success rates in fraction form must be one of 0/20, 1/20, 2/20,..., 19/20, 20/20. In percentages, these rates are multiples of 5 (0%, 5%, 10%,..., 95%, 100%), and values such as 53% and 58% are not mathematical possibilities.

24. a. Since 100% is the totality of whatever is being measured, removing 100% of some quantity means that none of it is left.
    b. Reducing plaque by over 300% would mean removing three times as much plaque as is there, and then removing even more!

25. The answers will vary according to student interests. Here is one example.
    a. Should our school consider a dress code?
    b. Should our school consider a dress code outlawing lewd or unsanitary practices?
    c. Should our school consider a dress code limiting personal freedom and diverting attention from more important educational issues?

S-6 INSTRUCTOR'S SOLUTIONS Chapter 1

26. The following figures meet the stated criteria.
   a. The graph on the left is an objective presentation of the data, for the height (and area) of the female bar is 74% of the height (and area) of the male bar.
   b. The graph on the right exaggerates the differences by "cutting off" the bottom of the figure and starting at 70. Even though the vertical labeling gives the correct values, the graph is misleading since the height (and area) of the female bar is only 4/30 = 13% of the height (and area) of the male bar – and so the visual impression is that the female salary is only 13% of the male salary.

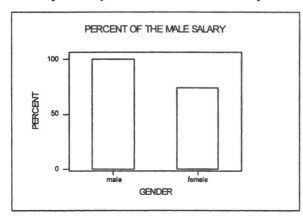
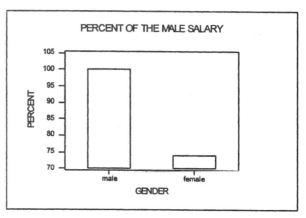

1-4 Design of Experiments

1. Experiment, since the effect of an applied treatment is measured.
2. Observational study, since specific characteristics are measured on unmodified subjects
3. Observational study, since specific characteristics are measured on unmodified subjects.
4. Experiment, since the effect of an applied treatment is measured.
5. Retrospective, since the data are collected by going back in time.
6. Prospective, since the data are to be collected in the future from a group sharing a common factor.
7. Cross-sectional, since the data are collected at one point in time.
8. Retrospective, since the data are collected by going back in time.
9. Convenience, since the sample was simply those who happen to pass by.
10. Systematic, since every 100th element in the population was chosen.
11. Random, since each adult has an equal chance of being selected.
    NOTE: This is a complex situation. The above answer ignores the fact that there is not a 1 to 1 correspondence between adults and phone numbers. An adult with more than one number will have a higher chance of being selected. An adult who shares a phone number with other adults will have a lower chance of being selected. And, of course, adults with no phone numbers have no chance of being selected at all.
12. Stratified, since the population of interest (assumed to be all car owners) was divided into 5 subpopulations from which the actual sampling was done.
13. Cluster, since the population of interest (assumed to be all students at The College of Newport) was divided into classes which were randomly selected in order to interview all the students in each selected class.
    NOTE: Ideally the division into classes should place each student into one and only one

class (e.g., if every student must take exactly one PE class each semester, select the PE classes at random). In practice such divisions are often made in ways that place some students in none of the classes (e.g., by selecting from all 2 pm M-W-F classes) or in more than one of the classes (e.g., by selecting from all the classes offered in the college). With careful handling, imperfect divisions should not significantly affect the results.

14. Random, since each Corvette owner has an equal chance of being selected.
    NOTE: this is also an example of simple random sampling, since every possible group of 50 Corvette owners has the same chance of being chosen.

15. Systematic, since every fifth element in the population (assumed to be all drivers passing the checkpoint during its operation) was sampled.

16. Cluster, since the population (assumed to be all voters) will be divided into polling stations which will be randomly selected in order to interview all the voters at each station.

17. Stratified, since the population of interest (assumed to be all workers) was divided into 3 subpopulations from which the actual sampling was done.

18. Convenience, since the sample is those who happen to be in his family.

19. Cluster, since the population of interest (assumed to be all cardiac patients) were divided into hospitals which were randomly selected in order to survey all such patients at each hospital.

20. Stratified, since the population of interest will be divided into subpopulations from which the actual sampling will be done.

21. Yes, this is a random sample because each of the 1000 tablets (assumed to be the population) has an equal chance of being selected. Yes, this is a simple random sample because every possible group of 50 tablets has the same chance of being chosen.

22. Yes, this is a random sample of the class because each of the 30 students has an equal chance (viz., 1/5) of being selected. No, it is not a simple random sample because every possible group of 6 students does not have the same chance of being chosen – groups of 6 with students from different rows have no chance of being chosen.

23. No, this is not a random sample because each element in the population does not have an equal chance of being selected. If the population of interest is all the residents of the city, then those who do not pass that street corner have no chance of being selected. Even if the population of interest is all those who pass that street corner, the reporter's personal preferences would likely prevent each person from having an equal chance of being selected – e.g., the reporter might not want to stop a woman with three noisy children, or a man walking very fast and obviously in a hurry to get somewhere. Since the sample is not random, it cannot be a simple random sample.

24. Yes, this is a random sample because each unit on the conveyor belt has an equal chance (viz., 1/100) of being selected. No, this is not a simple random sample because all possible samples do not have the same chance of being chosen – samples containing two consecutive units, for example, cannot occur.
    NOTE: The above answer assumes the units are placed on the conveyor belt in random order, or that the engineer begins his sampling at a random moment. If the units are placed on the conveyor belt in some order and the engineer wants to begin sampling as soon as the first units are placed on the belt, then he should randomly choose a number from 1 to 100 to identify the first unit to be selected.

25. Yes, this is a random sample because each person in Orange County has an equal chance of being selected. No, this is not a simple random sample because every possible sample of 200 Orange County residents does not have the same chance of being chosen – a sample with 75 men and 45 women, for example, is not possible. The survey will include a simple random sample of Orange County men and a simple random sample of Orange County women, but not a simple random sample of Orange County adults.

26. Yes, this is a random sample because each adult resident of Newport has an equal chance (viz, $10/n$ – where n is the total number of blocks in Newport) of being selected. No, this is not a simple random sample because every possible sample does not have the same chance of being chosen – a sample with at least one adult from every block, for example, cannot occur.

27. Assume that "students who use this book" refers to a particular semester (or other point in time). Before attempting any sampling procedure below, two preliminary steps are necessary: (1) obtain from the publisher a list of each school that uses the book and (2) obtain from each school the number of students enrolled in courses requiring the text. Note that using only the number of books sold to the school by the publisher will miss those students purchasing used copies turned in from previous semesters. "Students" in parts (a)-(e) below is understood to mean students using this book.
  a. Random - conceptually (i.e., without actually doing so) make a list of all N students by assigning to each school one place on the list for each of its students; pick 100 random numbers from 1 to N -- each number identifies a particular school and a conceptual student; for each identified school select a student at random (or more than one student, if that school was identified more than once). In this manner every student has the same chance of being selected -- in fact, every group of 100 has the same chance of being selected, and the sample is a simple random sample.
  b. Systematic - place the n schools in a list, pick a random number between 1 and $(n/10)$ to determine a starting selection, select every $(n/10)^{th}$ school; at each of the 10 systematically selected schools, randomly (or systematically) select and survey 10 students.
  c. Convenience - randomly (or conveniently) select and survey 100 students from your school.
  d. Stratified - divide the schools into categories (e.g., 2 year, 4 year private, 4 year public, proprietary) and select 1 school from each category; at each school randomly (or by a stratified procedure) select and survey the number of students equal to the percent of the total students that are in that category.
  e. Cluster - select a school at random and survey all the students (or cluster sample again by grouping the students in classes and selecting a class at random in which to survey all the students); repeat (i.e., select another school at random) as necessary until the desired sample of 100 is obtained.

28. Confounding occurs when the researcher is not able to determine which factor (often one planned and one unplanned) produced an observed effect. If a restaurant tries adding an evening buffet for one week and it happens to be the same week that a nearby theater shows a real blockbuster that attracts unusual crowds to the neighborhood, the restaurant cannot know whether its increased business is due to the new buffet or to the extra traffic created by the theater.

29. No. Even though every state has the same chance of being selected, not every voter has the same chance of being selected. Once a state is selected, any particular registered voter in a small state is competing against fewer peers and has a higher chance of being selected than any particular registered voter in a large state.

30. The study may be described as
  a. prospective because the data are to be collected in the future, after the patients are assigned to different treatment groups.
  b. randomized because the patients were assigned to one of the three treatment groups at random.
  c. double-blind because neither the patient nor the evaluating physician knew which treatment had been administered.
  d. placebo-controlled because there was a treatment with no drug that allowed the researchers to distinguish between real effects of the drugs in question and (1) the psychological effects of "being treated" and/or (2) the results of not receiving any drugs.

INSTRUCTOR'S SOLUTIONS Chapter 1   S-9

31. a. The two samples are self-selected, and the participants were not randomly assigned to their respective groups. It appears that the researchers asked a group of drivers without cell phones to participate, and then approached a group of drivers who already had cell phones. It could be that drivers who self-select to have cell phones are anxious types that would have more accidents anyway.
   b. Those with cell phones were specifically asked to use the phones while driving. This is not likely to produce the same results as "normal" usage – not to mention the fact that knowing they are part of a study will likely put them on their best behavior and not represent their typical conduct in such a situation.

Review Exercises

1. No, the responses cannot be considered representative of the population of the United States for at least the following reasons.
   a. Only AOL subscribers were contacted. Those who are not AOL subscribers, not to mention those with no Internet connections at all, will not be represented. This will introduce a significant bias – since non-Internet users might to be the less affluent, less educated and less business-oriented.
   b. The sample was self-selected. The responses represent only those who had enough time and/or interest in the question. Typically, those with strong feelings one way or the other are more likely to respond and will be over-represented in the results.

2. Let N be the total number of full-time students and n be the desired sample size.
   a. Random. Obtain a list of all N full-time students, number the students from 1 to N, select n random numbers from 1 to N, and poll each student whose number on the list is one of the random numbers selected.
   b. Systematic. Obtain a list of all N full-time students, number the students from 1 to N, let m be the largest integer less than the fraction N/n, select a random number between 1 and m, begin with the student whose number is the random number selected, and poll that student and every mth student thereafter.
   c. Convenience. Select a location (e.g., the intersection of major campus walkways) by which most of the students usually pass, and poll the first n full-time students that pass.
   d. Stratified. Obtain a list of all N full-time students and the gender of each, divide the list by gender, and randomly select and poll n/2 students from each gender.
   e. Cluster. Obtain a list of all the classes meeting at a popular time (e.g., 10 am Monday), estimate how many of the classes would be necessary to include n students, select that many of the classes at random, and poll all of the students in each selected class.

3. a. Ratio, since differences are valid and there is a meaningful 0.
   b. Ordinal, since there is a hierarchy but differences are not valid.
   c. Nominal, since the categories have no meaningful inherent order.
   d. Interval, since differences are valid but the 0 is arbitrary.

4. a. Discrete, since the number of shares held must be an integer.
      NOTE: Even if partial shares are allowed (e.g., 5½ shares), the number of shares must be some fractional value and not any value on a continuum -- e.g., a person could not own π shares.
   b. Ratio, since differences between values are consistent and there is a natural zero.
   c. Stratified, since the set of interest (all stockholders) was divided into subpopulations (by states) from which the actual sampling was done.
   d. Statistic, since the value is determined from a sample and not the entire population.
   e. There is no unique correct answer, but the following are reasonable possibilities.
      (1) The proportion of stockholders holding above that certain number of shares (which would vary from company to company) that would make them "influential." (2) The proportion of stockholders holding below that certain number of shares (which would vary from company to company) that would make them "insignificant." (3) The numbers of shares (and hence the degree of influence) held by the largest stockholders.

f. There are several possible valid answers. (1) The results would be from a self-selected group (i.e., those who chose to respond) and not necessarily a representative group. (2) If the questionnaire did not include information on the numbers of shares owned, the views of small stockholders (who are probably less knowledgeable about business and stocks) could not be distinguished from those of large stockholders (whose views should carry more weight).

5. a. Systematic, since the selections are made at regular intervals.
    b. Convenience, since those selected were the ones who happened to attend.
    c. Cluster, since the stockholders were organized into groups (by stockbroker) and all the stockholders in the selected groups were chosen.
    d. Random, since each stockholder has the same chance of being selected.
    e. Stratified, since the stockholders were divided into subpopulations from which the actual sampling was done.

6. a. Blinding occurs when those involved in an experiment (either as subjects or evaluators) do not know whether they are dealing with a treatment or a placebo. It might be used in this experiment by (a) not telling the subjects whether they are receiving Sleepeze or the placebo and/or (b) not telling any post-experiment interviewers or evaluators which subjects received Sleepeze and which ones received the placebo. Double-blinding occurs when neither the subjects nor the evaluators know whether they are dealing with a treatment or a placebo.
    b. The data reported will probably involve subjective assessments (e.g., "On a scale of 1 to 10, how well did it work?") that may be subconsciously influenced by whether the subject was known to have received Sleepeze or the placebo.
    c. In a completely randomized block design, subjects are assigned to the groups (in this case to receive Sleepeze or the placebo) at random.
    d. In a rigorously controlled block design, subjects are assigned to the groups (in this case to receive Sleepeze or the placebo) in such a way that the groups are similar with respect to extraneous variables that might affect the outcome. In this experiment it may be important to make certain each group has approximately the same age distribution, degree of insomnia, number of males, number users of alcohol and/or tobacco, etc.
    e. Replication involves repeating the experiment on a sample of subjects large enough to ensure that atypical responses of a few subjects will not give a distorted view of the true situation.

Cumulative Review Exercises

NOTE: Throughout the text intermediate mathematical steps will be shown as an aid to those who may be having difficulty with the calculations. In practice, most of the work can be done continuously on calculators and the intermediate values are unnecessary. Even when the calculations cannot be done continuously, DO NOT WRITE AN INTERMEDIATE VALUE ON YOUR PAPER AND THEN RE-ENTER IT IN THE CALCULATOR. That practice can introduce round-off errors and copying errors. Store any intermediate values in the calculator so that you can recall them with infinite accuracy and without copying errors.

1. $\frac{169.1+144.2+179.3+178.5+152.6+166.8+135.0+201.5+175.2+139.0}{10} = \frac{1638.5}{10}$
$= 163.85$

2. $\frac{98.20 - 98.60}{.62} = \frac{-.40}{.62} = -.645$

3. $\dfrac{98.20 - 98.60}{0.62/\sqrt{106}} = \dfrac{-.40}{.0602} = -6.642$

4. $\left[\dfrac{(1.96)(15)}{2}\right]^2 = \left[\dfrac{29.4}{2}\right]^2 = [14.7]^2 = 216.09$

5. $\sqrt{\dfrac{(5-7)^2 + (12-7)^2 + (4-7)^2}{3-1}} = \sqrt{\dfrac{(-2)^2 + (5)^2 + (-3)^2}{2}} = \sqrt{\dfrac{4 + 25 + 9}{2}} = \sqrt{\dfrac{38}{2}}$
$= \sqrt{19} = 4.359$

6. $\dfrac{(183 - 137.09)^2}{137.09} + \dfrac{(30 - 41.68)^2}{41.68} = \dfrac{(45.91)^2}{137.09} + \dfrac{(-11.68)^2}{41.68} = \dfrac{2107.7281}{137.09} + \dfrac{136.4224}{41.68}$

$= 15.375 + 3.273 = 18.647$

7. $\sqrt{\dfrac{10(513.27) - 71.5^2}{10(9)}} = \sqrt{\dfrac{5132.7 - 5112.25}{90}} = \sqrt{\dfrac{20.45}{90}} = \sqrt{.2272} = .477$

8. $\dfrac{8(151{,}879) - (516.5)(2176)}{\sqrt{8(34{,}525.75) - 516.5^2}\,\sqrt{8(728{,}520) - 2176^2}} = \dfrac{1215032 - 1123904}{\sqrt{9433.75}\,\sqrt{1093184}} = \dfrac{91128}{1015522} = .897$

9. $0.95^{500} = 7.27\text{E-}12 = 7.27 \cdot 10^{-12}$
$= .00000000000727$; moving the decimal point left 12 places

NOTE: Calculators and computers vary in their representation of such numbers. This manual assumes they will be given in scientific notation as a two-decimal number between 1.00 and 9.99 inclusive followed by an indication (usually E for *exponent* of the multiplying power of ten) of how to adjust the decimal point to obtain a number in the usual notation (rounded to three significant digits).

10. $8^{14} = 4.40\text{E+}12 = 4.40 \cdot 10^{12}$
$= 4{,}400{,}000{,}000{,}000$; moving the decimal point right 12 places

11. $9^{12} = 2.82\text{E+}11 = 2.82 \cdot 10^{11}$
$= 282{,}000{,}000{,}000$; moving the decimal point right 11 places

12. $.25^{17} = 5.82\text{E-}11 = 5.82 \cdot 10^{-11}$
$= .0000000000582$; moving the decimal point left 11 places

Chapter 2

Describing, Exploring, and Comparing Data

2-2 Frequency Distributions

1. Subtracting the first two consecutive lower class limits indicates that the class width is 100 - 90 = 10. Since there is a gap of 1.0 between the upper class limit of one class and the lower class limit of the next, class boundaries are determined by increasing or decreasing the appropriate class limits by (1.0)/2 = 0.5. The class boundaries and class midpoints are given in the table below.

pressure	class boundaries	class midpoint	frequency
90 - 99	89.5 - 99.5	94.5	1
100 - 109	99.5 - 109.5	104.5	4
110 - 119	109.5 - 119.5	114.5	17
120 - 129	119.5 - 129.5	124.5	12
130 - 139	129.5 - 139.5	134.5	5
140 - 149	139.5 - 149.5	144.5	0
150 - 159	149.5 - 159.5	154.5	1
			40

   NOTE: Although they often contain extra decimal points and may involve consideration of how the data were obtained, class boundaries are the key to tabular and pictorial data summaries. Once the class boundaries are obtained, everything else falls into place. Here the first class width is readily seen to be 99.5 - 89.5 = 10.0 and the first midpoint is (89.5 + 99.5)/2 = 94.5. In this manual, class boundaries will typically be calculated first and then used to determine other values. In addition, the sum of the frequencies is an informative number used in many subsequent calculations and will be shown as an integral part of each table.

2. Since the gap between classes as presented is 1.0, the appropriate class limits are increased or decreased by (1.0)/2 = 0.5 to obtain the class boundaries and the following table.

pressure	class boundaries	class midpoint	frequency
80 - 99	79.5 - 99.5	89.5	9
100 - 119	99.5 - 119.5	109.5	24
120 - 139	119.5 - 139.5	129.5	5
140 - 159	139.5 - 159.5	149.5	1
160 - 179	159.5 - 179.5	169.5	0
180 - 199	179.5 - 199.5	189.5	1
			40

   The class width is 99.5 - 79.5 = 20; the first midpoint is (79.5 + 99.5)/2 = 89.5.

3. Since the gap between classes as presented is 1.0 the appropriate class limits are increased or decreased by (1.0)/2 = .05 to obtain the class boundaries and the following table.

cholesterol	class boundaries	class midpoint	frequency
0 - 199	-0.5 - 199.5	99.5	13
200 - 399	199.5 - 399.5	299.5	11
400 - 599	399.5 - 599.5	499.5	5
600 - 799	599.5 - 799.5	699.5	8
800 - 999	799.5 - 999.5	899.5	2
1000 - 1199	999.5 - 1199.5	1099.5	0
1200 - 1399	1199.5 - 1399.5	1299.5	1
			40

   The class width is 199.5 - (-0.5) = 200; the first midpoint is (-0.5 + 199.5)/2 = 99.5.

4. Since the gap between classes as presented is 0.1, the appropriate class limits are increased or decreased by $(0.1)/2 = 0.05$ to obtain the class boundaries and the following table.

mass index	class boundaries	class midpoint	frequency
15.0 - 20.9	14.95 - 20.95	17.95	10
21.0 - 26.9	20.95 - 26.95	23.95	15
27.0 - 32.9	26.95 - 32.95	29.95	11
33.0 - 38.9	32.95 - 38.95	35.95	2
39.0 - 44.9	38.95 - 44.95	41.95	2
			40

The class width is $20.95 - 14.95 = 6$; the first midpoint is $(14.95 + 20.95)/2 = 17.95$.

5. The relative frequency for each class is found by dividing its frequency by 40, the sum of the frequencies. **NOTE:** As before, the sum is included as an integral part of the table. For relative frequencies, this should always be 1.000 (i.e., 100%) and serves as a check for the calculations.

pressure	relative frequency
90 - 99	.025
100 - 109	.100
110 - 119	.425
120 - 129	.300
130 - 139	.125
140 - 149	.000
150 - 159	.025
	1.000

6. The relative frequency for each class is found by dividing its frequency by 40, the sum of the frequencies. **NOTE:** As before, the sum is as an integral part of the table. For relative frequencies, this should always be 1.000 (i.e., 100%) and serves as a check for the calculations.

pressure	relative frequency
80 - 99	.225
100 - 119	.600
120 - 139	.125
140 - 159	.025
160 - 179	.000
180 - 199	.025
	1.000

7. The relative frequency for each class is found by dividing its frequency by 40, the sum of the frequencies. **NOTE:** In #5, the relative frequencies were expressed as decimals; here they are expressed as percents. The choice is arbitrary.

cholesterol	relative frequency
0 - 199	.325
200 - 399	.275
400 - 599	.125
600 - 799	.200
800 - 999	.050
1000 -1199	.000
1200 -1399	.025
	1.000

8. The relative frequency for each class is found by dividing its frequency by 40, the sum of the frequencies. **NOTE:** In #6, the relative frequencies were expressed as decimals; here they are expressed as percents. The choice is arbitrary.

mass index	relative frequency
15.0 - 20.9	.250
21.0 - 26.9	.375
27.0 - 32.9	.275
33.0 - 38.9	.050
39.0 - 44.9	.050
	1.000

9. The cumulative frequencies are determined by repeated addition of successive frequencies to obtain the combined number in each class and all previous classes. **NOTE:** Consistent with the emphasis that has been placed on class boundaries, we choose to use upper class boundaries in the "less than" column. Conceptually, pressures occur on a continuum and the integer values reported are assumed to be the nearest whole number representation of the precise measure. An exact pressure of 99.7, for example, would be reported as 100 and fall in the second class. The values in the first class, therefore, are better described as being "less than 99.5" (using the upper class boundary) than as being "less than 100."

This distinction becomes crucial in the construction of pictorial representations in the next section. In addition, the fact that the final cumulative frequency must equal the total number (i.e, the sum of the frequency column) serves as a check for calculations. The sum of cumulative frequencies, however, has absolutely no meaning and is not included.

(#9) pressure	cumulative frequency	(#10) pressure	cumulative frequency
less than 99.5	1	less than 99.5	9
less than 109.5	5	less than 119.5	33
less than 119.9	22	less than 139.5	38
less than 129.5	34	less than 159.5	39
less than 139.5	39	less than 179.5	39
less than 149.5	39	less than 199.5	40
less than 159.5	40		

10. The cumulative frequencies are determined by repeated addition of successive frequencies to obtain the combined number in each class and all previous classes. **NOTE:** See NOTE for #9.

11. The cumulative frequencies are determined by repeated addition of successive frequencies to obtain the combined number in each class and all previous classes. NOTE: Consistent with the emphasis that has been placed on class boundaries, we choose to use upper class boundaries in the "less than" column.

(#11) cholesterol	cumulative frequency	(#12) mass index	cumulative frequency
less than 199.5	13	less than 20.95	10
less than 399.5	24	less than 26.95	25
less than 599.5	29	less than 32.95	36
less than 799.5	37	less than 38.95	38
less than 999.5	39	less than 44.95	40
less than 1199.5	39		
less than 1399.5	40		

12. The cumulative frequencies are determined by repeated addition of successive frequencies to obtain the combined number in each class and all previous classes. **NOTE:** See NOTE for #11.

13. The relative frequencies are determined by dividing the given frequencies by 200, the sum of the given frequencies. The fact that the sum of the relative frequencies is 1.000 provides a check to the arithmetic. For a fair die, we expect each relative frequency to be close to 1/6 = .167. As these relative frequencies all fall between .135 and .210, they do not appear to differ significantly from the values expected for a fair die.

(#13) outcome	relative frequency	(#14) digit	relative frequency
1	.135	0	11.250%
2	.155	1	7.500%
3	.210	2	8.750%
4	.200	3	5.625%
5	.140	4	10.625%
6	.160	5	12.500%
	1.000	6	13.125%
		7	16.250%
		8	4.375%
		9	10.000%
			100.000%

14. The relative frequencies are determined by dividing the given frequencies by 160, the sum of the given frequencies. The fact that the sum of the relative frequencies is 1.000 provides a check to the arithmetic. For random selection with all digits equally likely, we expect each relative frequency to be close to 1/10 = 10%. As these relative frequencies range from 4% (less than ½ of expectation) to 16% (more than 1½ of the expectation,

15. For a lower class limit of 0 for the first class and a class width of 50, the frequency distribution is given at the right.
NOTE: The class limits for the first class are 0-49 and not 0-50.

weight (lbs)	frequency
0 - 49	6
50 - 99	10
100 - 149	10
150 - 199	7
200 - 249	8
250 - 299	2
300 - 349	4
350 - 399	3
400 - 449	3
450 - 499	0
500 - 549	1
	54

16. For a lower class limit of 96.5 for the first class and a class width of 0.4, the frequency distribution is given at the right.
NOTE: The classes may be identified using either class limits or class boundaries. While class boundaries typically introduce another decimal place in the presentation, they eliminate the impression that some data might have occurred in gaps between the classes.

temperature	frequency
96.45 - 96.85	1
96.85 - 97.25	8
97.25 - 97.65	14
97.65 - 98.05	22
98.05 - 98.45	19
98.45 - 98.85	32
98.85 - 99.25	6
99.25 - 99.65	4
	106

Two notable features of the frequency table are: [NOTE: Answers will vary]
(1) The class containing 98.6 has the greatest frequency, but the data do not center around that class.
(2) The data appear to have two peaks (near 97.85 and 98.65) with a slight valley between them – that could signify the placing together of two unimodal subpopulations (e.g., male/female, athletic/non-athletic, etc.).

17. The separate frequency distributions are given below.

male head circumference	frequency
34.0 - 35.9	2
36.0 - 37.9	0
38.0 - 39.9	5
40.0 - 41.9	29
42.0 - 43.9	14
	50

female head circumference	frequency
34.0 - 35.9	1
36.0 - 37.9	3
38.0 - 39.9	14
40.0 - 41.9	27
42.0 - 43.9	5
	50

It appears that the head circumferences tend to be larger for baby boys than for baby girls.

18. The separate frequency distributions are given below.

tobacco use (seconds)	frequency
0 - 99	39
100 - 199	6
200 - 299	4
300 - 399	0
400 - 499	0
500 - 599	1
	50

alcohol use (seconds)	frequency
0 - 99	46
100 - 199	3
200 - 299	0
300 - 399	0
400 - 499	1
	50

There does not appear to be a significant difference between the lengths of tobacco use and the lengths of alcohol use in animated movies for children.

19. The separate relative frequency distributions are given below, to the right of the figure giving the actual frequencies. The relative frequencies were obtained by dividing the actual frequencies for each gender by the total frequencies for that gender.

M	age	F	male ages	relative frequency	female ages	relative frequency
11	19-28	8	19-28	.099	19-28	.205
43	29-38	18	29-38	.387	29-38	.462
31	39-48	4	39-48	.279	39-48	.103
22	49-58	7	49-58	.198	49-58	.179
4	59-68	2	59-68	.036	59-68	.051
111		39		1.000		1.000

Both genders have more runners in the 29-38 age group than in any other ten year spread. But the second most populous category is the one above that for the males, and the one below that for the females. It appears that the male runners tend to be slightly older than the female runners.

20. Assuming that "start the first class at" refers to the first lower class limit produces the following frequency tables for Regular and Diet Coke. In each case the relative frequencies were obtained by dividing the observed frequencies by 36.

REGULAR COKE weight (lbs)	relative frequency	DIET COKE weight (lbs)	relative frequency
.7900 - .7949	.028	.7750 - .7799	.111
.7950 - .7999	.000	.7800 - .7849	.361
.8000 - .8049	.028	.7850 - .7899	.417
.8050 - .8099	.083	.7900 - .7949	.111
.8100 - .8149	.111		1.000
.8150 - .8199	.472		
.8200 - .8249	.167		
.8250 - .8299	.111		
	1.000		

There are two significant differences between the data sets: the weights for Regular Coke are considerably larger than those for Diet Coke, the weights for Regular Coke cover a much wider range than those for Diet Coke. The first difference might be explained by the weight of the sugar present in Regular Coke but not in Diet Coke. The second difference suggests that the sugar might not be evenly distributed throughout the large amounts of Regular Coke from which the cans are filled - assuming the cans are filled by volume, the lighter cans could be coming from a portion of the mixture containing less sugar.

NOTE: The Diet Coke frequency table has only 4 categories, which is usually not sufficient to give a picture of the nature of the distribution. This is allowable in this context, since the class width and class limits employed work well with the other cola data and permit meaningful comparisons across the data sets.

21. Assuming that "start the first class at 200 lb" refers to the first lower class limit produces the frequency table given at the right.

weight (lbs)	frequency
200 - 219	6
220 - 239	5
240 - 259	12
260 - 279	36
280 - 299	87
300 - 319	28
320 - 339	0
340 - 359	0
360 - 379	0
380 - 399	0
400 - 419	0
420 - 439	0
440 - 459	0
460 - 479	0
480 - 499	0
500 - 519	1
	175

In general, an outlier can add several rows to a frequency table. Even though most of the added rows have frequency zero, the table tends to suggest that these are possible valid values - thus distorting the reader's mental image of the distribution.

INSTRUCTOR'S SOLUTIONS Chapter 2 S-17

22. Let n = the number of data values
x = the number of classes.
Solve the given formula x = 1 + (log n)/(log 2)
for n to get n = 2^{x-1}.
Use the value x = 5.5, 6.5, 7.5,... to get the cut-ff values for n as follows.

x	n = 2^{x-1}	
5.5	22.63	(for n < 22.63, x rounds to 5; for n > 22.63, n rounds to 6)
6.5	45.25	(for n < 45.25, x rounds to 6; for n > 45.25, x rounds to 7)
7.5	90.51	(etc.)
8.5	181.02	
9.5	362.04	
10.5	724.04	
11.5	1448.15	
12.5	2896.31	

Assuming n ≥ 16, use the cut-off values to complete the table as follows.

n	ideal # of classes
16– 22	5
23– 45	6
46– 90	7
91– 181	8
182– 362	9
363– 724	10
725–1448	11
1448–2896	12

2-3 Visualizing Data

1. The answer depends upon what is meant by "the center." Since the ages seem to range from about 10 to about 70, 40 could be called the center value - in the sense that it is half way between the lowest and highest points on the horizontal axis that represent observed data. Since the ages are concentrated at the lower end of the scale, an age near 24 could be called the center value - in the sense that about ½ of the 131 ages appear to be below 24 and about ½ of the ages appear to be above 24. Or an age near 26 could be called the center value - in the sense that a fulcrum placed under 26 on the horizontal would appears to nearly "balance" the histogram.

2. The lowest and highest possible ages appear to be about 10 and 70 respectively.

3. The percentage younger than 30 is about (1+38+38+16)/131 = 93/131 = .710 = 71.0%.

4. The class width is 5 - the class at the left with frequency 1 appears to extend from about 10 to 15, and the adjacent class with frequency 38 appear to extend from about 15 to 20.

5. Since, the shading representing Group A covers about 40% of the "pie," the approximate percentage of people with Group A blood is 40%. If the chart is based on a sample of 500 people, approximately (.40)(500) = 200 people had Group A blood.

6. Since, the shading representing Group B covers about 10% of the "pie," the approximate percentage of people with Group B blood is 10%. If the chart is based on a sample of 500 people, approximately (.10)(500) = 50 people had Group B blood.

S-18 INSTRUCTOR'S SOLUTIONS Chapter 2

7. Obtain the relative frequencies by dividing each frequency by the total frequency for each sample. The two relative frequency histograms are given, using the same scale, below at the right. Each sample distribution spreads out in both directions from a "typical" (or most frequently occurring value), but the faculty data appears to be shifted to the left by about one interval. Since each interval represents 3 years, the faculty cars are about 3 years newer than the student cars. In addition, the majority of the student car ages fall below the class with the highest frequency, while faculty car ages tend to occur above their most populous class.

NOTE: The class boundaries in the histograms are -0.5, 2.5, 5.5, etc. Assuming the car ages were obtained by subtracting the model year from the calendar year, that boundary scheme is consistent with the new models being introduced ½ year "early" and sold at a steady rate during the calendar year. In truth, this data is difficult to present in a graph that accounts for the underlying continuity of the ages. The figures given below are probably the best for the level of this text.

STUDENTS

age	frequency	relative frequency
0- 2	23	.106
3- 5	33	.152
6- 8	63	.290
9-11	68	.313
12-14	19	.088
15-17	10	.046
18-20	1	.005
21-23	0	.000
	217	1.000

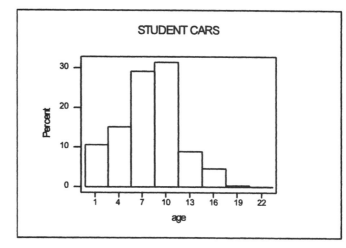

FACULTY

age	frequency	relative frequency
0- 2	30	.197
3- 5	47	.309
6- 8	36	.237
9-11	30	.197
12-14	8	.053
15-17	0	.000
18-20	0	.000
21-23	1	.007
	152	1.000

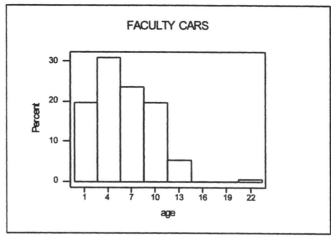

8. See the figure below. The bars extend from class boundary to class boundary. Each axis is labeled numerically <u>and</u> with the name of the quantity represented. Although the posted limit is 30 mph, it appears that the police ticket only those traveling at least 42 mph.

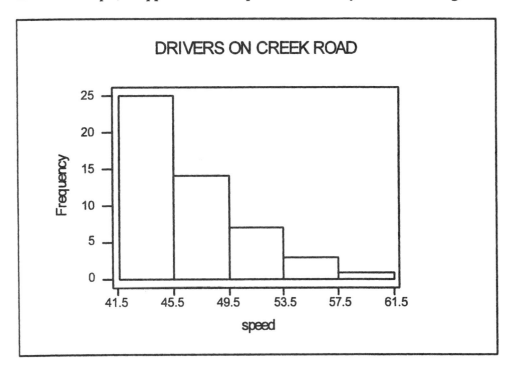

9. See the figure below. The class boundaries are -0.5, 49.5, 99.5, 149.5,...549.5. The bars extend from class boundary to class boundary. For visual simplicity, the horizontal axis has been labeled 0, 50, 150, etc. The "center" of the weights (i.e., the "balance point" of the histogram) appears to be about 190.

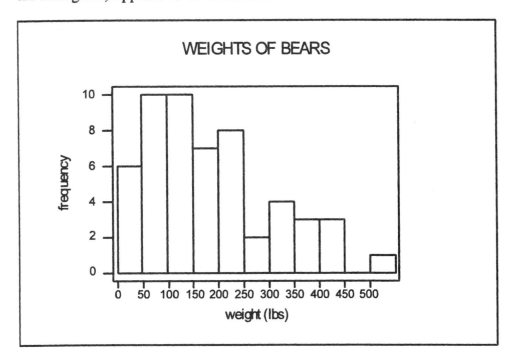

10. See the figure below. The distribution appears to be approximately bell-shaped and suggests that the average body temperature may be less than the commonly believed 98.6°F. The distribution of normal temperatures is important because it may be used to identify abnormal temperatures.

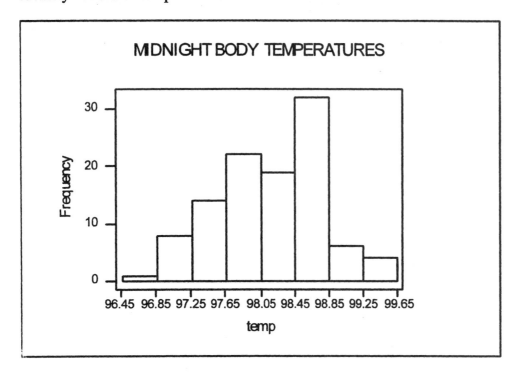

11. See the figure below. The frequencies are plotted above the class midpoints, and "extra" midpoints are added so that both polygons begin and end with a frequency of zero. The females are represented by the solid line, and the males by the dashes.

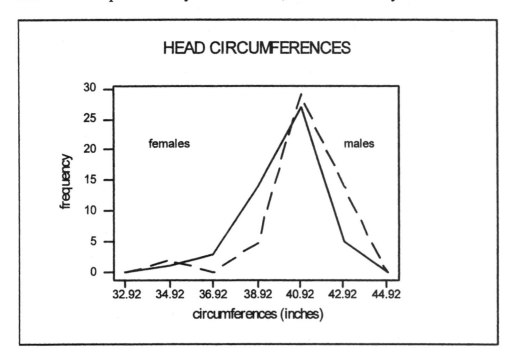

Both polygons have approximately the same shape, but the male circumferences appear to be slightly larger.

12. See the figure below. The frequencies are plotted above the class midpoints, and "extra midpoints are added so that both polygons begin and end with a frequency of zero. Tobacco is represented by the solid line, and alcohol by the dashes.

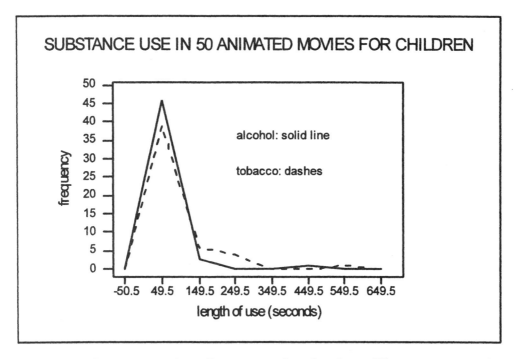

The two polygons are virtually on top of each other. There appears to be no significant difference in the lengths of time that alcohol and tobacco use are shown in the films.

13. See the figures below. Since ages are reported as of the last birthday, the class boundaries are 19,29,39,49,59,69 – i.e., the first class in each figure includes all ages from 19.000 to 28.999.

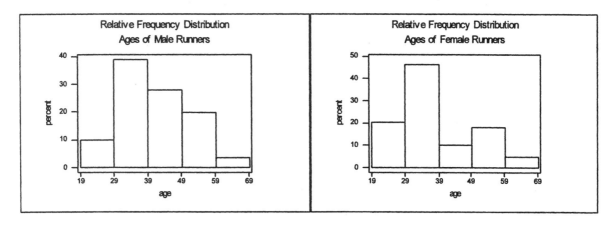

The two histograms are almost identical, except for the "dip" at age 39-49 for the females. It could be that physical or family conditions that typically occur about that age for females are not conducive to running marathons. Because of that dip, it appears that the male runners are slightly older than their female counterparts.

14. See the figure below. The frequencies are plotted above the class midpoints, and "extra" midpoints are added so that both polygons begin and end with a frequency of zero. The horizontal axis is labeled for convenience of presentation, but the frequencies are plotted beginning at .77245 and every .005 thereafter – .77245, .77745,...,.82745,.83245.

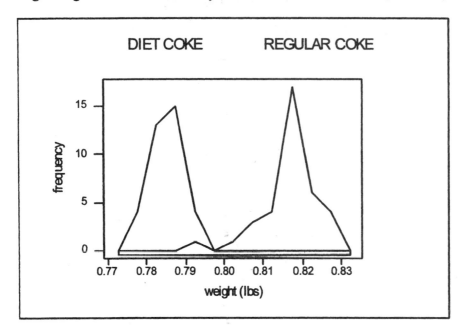

The two distributions have similar shapes, but the diet Coke weights are less than those of regular Coke - probably because of the weight of the sugar in regular Coke.

15. The original numbers are listed by the row in which they appear in the stem-and-leaf plot.

    ```
    stem | leaves      original numbers
      20 | 0005        200, 200, 200, 205
      21 | 69999       216, 219, 219, 219, 219
      22 | 2233333     222, 222, 223, 223, 223, 223, 223
      23 |
      24 | 1177        241, 241, 247, 247
    ```

16. The original numbers are listed by the row in which they appear in the stem-and-leaf plot.

    ```
    stem | leaves           original numbers
      50 | 12 12 12 55      5012, 5012, 5012, 5055
      51 |
      52 | 00 00 00 00      5200, 5200, 5200, 5200
      53 | 27 27 35         5327, 5327, 5335
      54 | 72               5472
    ```

17. The dotplot is constructed using the original scores as follows. Each space represents 1 unit.

    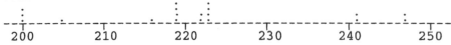

18. The dotplot is constructed using the original scores as follows.

    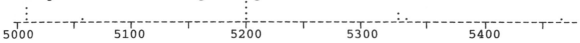

19. The expanded stem-and-leaf plot below on the left is one possibility. **NOTE:** The text claims that stem-and-leaf plots enable us to "see the distribution of data and yet keep all the information in the original list." Following the suggestion to round the nearest inch not only loses information but also uses subjectivity to round values exactly half way between. Since always rounding such values "up" creates a slight bias, many texts suggest rounding toward the even digit – so that 33.5 becomes 34, but 36.5 becomes 36. The technique below of using superscripts to indicate the occasional decimals is both mathematically clear and visually uncluttered.

    ```
    stem | leaves
       3 | 6 7
       4 | 0 0 1 3 3⁵
       4 | 6 6 7 8 8 9
       5 | 0 2 2⁵ 3 3 4
       5 | 7³ 7⁵ 8 9 9 9
       6 | 0 0⁵ 1 1 1⁵ 2 3 3 3 3⁵ 4 4 4
       6 | 5 5 6⁵ 7 7⁵ 8⁵
       7 | 0 0⁵ 2 2 2 2 3 3⁵
       7 | 5 6⁵
    ```

20. The stem-and-left plot below is one possibility. **NOTE:** The text claims that stem-and-leaf plots enable us to "see the distribution of data and yet keep all the information in the original list." Following the suggestion to round the nearest tenth not only loses information but also uses subjectivity to round values exactly half way between. Since always rounding such values "up" creates a slight bias, many texts suggest rounding toward the even digit – so that 2.35 becomes 2.4, but 3.05 becomes 3.0. The figure below is accurate and uncluttered.

    ```
    stem | leaves
     0.  | 15 27 38
     0.  | 58 63 65 72 74 80 85 89 92 92 93
     1.  | 14 15 28 36 40 41 41 44 44 45 45 48 49
     1.  | 53 53 58 61 68 74 81
     2.  | 00 04 10 13 13 17 19 19 30 31 35 44
     2.  | 66 68 83 83 87 88 93 96 97
     3.  | 05 36 42
     3.  | 53
     4.  | 37
     4.  | 69
     5.  | 28
    ```

21. See the figure below, with bars arranged in order of magnitude. Networking appears to be the most effective job-seeking approach.

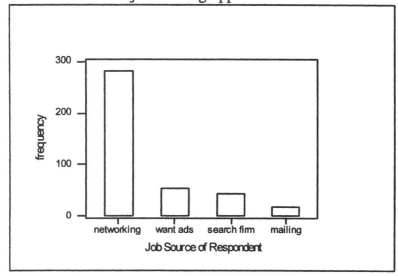

22. The pie chart is given below at the left. The Pareto chart appears to be more effective in showing the relative importance of job sources.

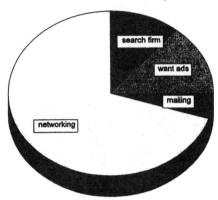
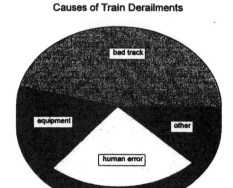

23. See the figure above at the right. The sum of the frequencies is 50; the relative frequencies are $23/50 = 46\%$, $9/50 = 18\%$, $12/50 = 24\%$, and $6/50 = 12\%$. The corresponding central angles are $(.46)360° = 165.6°$, $(.18)360° = 64.8°$, $(.24)360° = 86.4°$, and $(.12)360° = 43.2°$. To be complete, the figure needs to be titled with the name of the quantity being measured.

24. See the following figure, with bars arranged in order of magnitude. The Pareto chart appears to be more effective in showing the relative importance of the causes of train derailments.

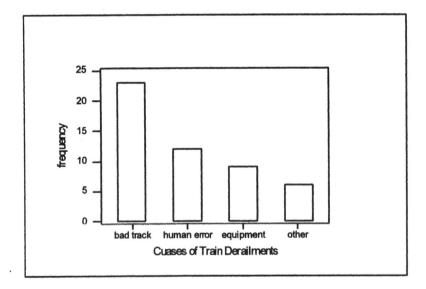

25. The scatter diagram is given below. The figure should have a title, and each axis should be labeled both numerically and with the name of the variable. An "x" marks a single occurrence, while numbers indicate multiple occurrences at a point. Cigarettes high in tar also tend to be high in CO. The points cluster about a straight line from (0,0) to (18,18), indicating that the mg of CO tends to be about equal to the mg of tar.

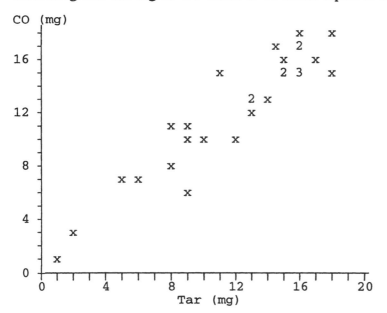

26. The scatter diagram is given below. The figure should have a title, and each axis should be labeled both numerically and with the name of the variable. An "x" marks a single occurrence, while numbers indicate the multiple occurrences at a point. The pattern indicates that bears with large necks also tend to have high weights.

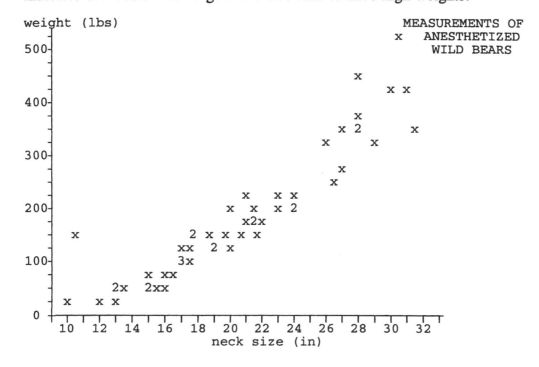

27. The time series graph is given below. An investor could use this graph to project where the market might be next year and invest accordingly. The flat portion at the right might be suggesting that the market has reached a point at which it will level off for a while – and that the rapid growth of the last several years may be at end..

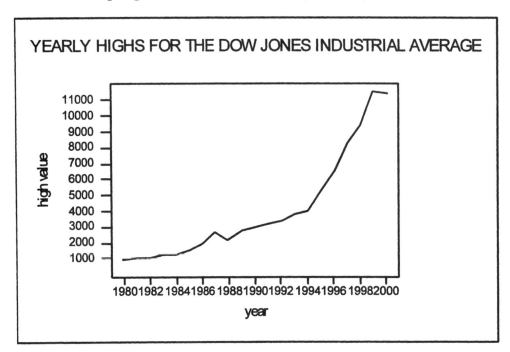

28. The time series graph is given below. There appears to be a downward trend – which may be due to increased seat belt use and/or safety improvements in both automobiles and the highways.

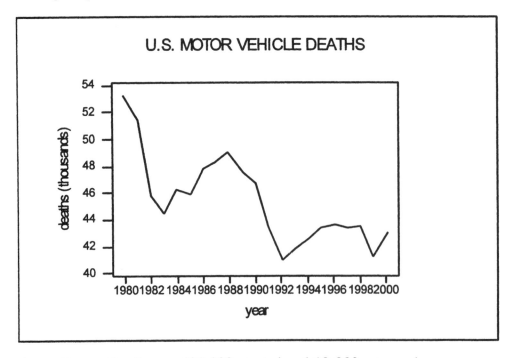

29. According to the figure, 422,000 started and 10,000 returned.
    10,000/422,000 = 2.37%

30. According to the figure, there were 50,000 men prior to crossing the Berezina River on the

return trip and 28,000 after crossing the river. The number who died was 22,000 or 22,000/50,000 = 44%.
NOTE: The figure is a summary that gives only periodic counts, and it is not necessarily true that 22,000 died while crossing the river.

31. The figure indicates the number of men had just dropped to 37,000 on November 9 when the temperature was 16°F (-9°C), and had just dropped to 24,000 on November 14 when the temperature was -6°F (-21°C). The number who died during that time, therefore, was 37,000 - 24,000 = 13,000.

32. The figure indicates that 100,000 men reached Moscow and that 20,000 of those survived to rejoin about 30,000 of their comrades at Botr. The number who died on that portion of the return trip, therefore, was 100,000 - 20,000 = 80,000.

33. NOTE: Exercise #20 of section 2-2 dealt with frequency table representations of this data and specified a class width of 20. Using a class width with an odd number of units of measure allows the class midpoint to have the same number of decimal places as the original data and often produces more appealing visual representations. Here we use a class width of 25 – with class midpoints of 200, 225, etc. – and so the figures will differ from ones made using the previous classes. The histogram bars extend from class boundary to class boundary (i.e., from 187.5 to 212.5 for the first class), but for convenience the labels have been placed at the class midpoints.

   a. The histogram with the outlier is as shown below.

   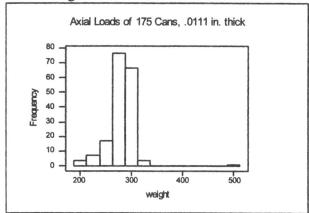

   b. The histogram without the outlier is as shown below.

   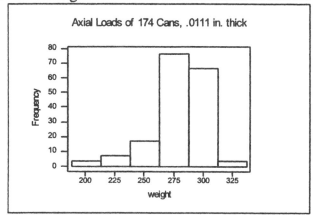

   c. The basic shape of the histogram does not change, except that a distant piece has been "broken off." NOTE: The redrawn histogram in part (b) should not be an exact copy of the one in part (a) with the distant bar erased. Since removing the distant bar reduces the effective width of the figure significantly, the rule that the height should be approximately 3/4 of the width requires either making the remaining bars wider (and keeping the figure's height) or making them shorter (and keeping the figure's reduced width). To do otherwise produces a figure too tall for its width – and one that tends to visually overstate the differences between classes.

34. a. The final form of the back-to-back stem-and-leaf plot is given below, an example of adapting a standard visual form in order to better communicate the data. While such decisions are arbitrary, we choose to display "outward" from the central stem but to keep the actors' ages in increasing order from left to right.

```
 actor's age|  |actress' age
            |2 |145666678
 1222356677899|3 |001133344445557789
0001223345566788|4 |111249
          13566|5 |0
           0012|6 |011
              6|7 |4
               |8 |0
```

b. Female Oscar winners tend to be younger than male Oscar winners. If one assumes that acting ability doesn't peak differently for females and males, the data may reveal a difference in the standards by which females and males are judged.

2-4 Measures of Center

NOTE: As it is common in mathematics and statistics to use symbols instead of words to represent quantities that are used often and/or that may appear in equations, this manual employs symbols for the measures of central tendency as follows:
 mean = $\bar{x}$ mode = M
 median = $\tilde{x}$ midrange = m.r.
Also, this manual generally follows the author's guideline of presenting means, medians and midranges accurate to one more decimal place than found in the original data. The mode, the only measure which must be one of the original pieces of data, is presented with the same accuracy as the original data. This manual will, however, recognize the following two exceptions to these guidelines.
(1) When there are an odd number of data, the median will be one of the original values and will not be reported to one more decimal place. And so the median of 1,2,3,4,5 would be reported as 3.
(2) When the mean falls <u>exactly</u> halfway between two values that meet the guidelines, an extra decimal place ending in 5 will be used. And so the mean of 1,2,3,3 would be reported as 9/4 = 2.25.

1. Arranged in order, the 6 scores are: 0 0 0 176 223 548
 a. $\bar{x} = (\Sigma x)/n = (947)/6 = 157.8$ c. M = 0
 b. $\tilde{x} = (0 + 176)/2 = 88.0$ d. m.r. = (0 + 548)/2 = 274.0
Whether or not there is a problem with scenes of tobacco use in animated children's films depends on the scene. If a likeable character uses tobacco without any criticism or ill effects, this would suggest to children that such usage is appropriate and acceptable. In most children's films, however, tobacco usage is typically associated with evil and undesirable characters and situations – suggesting that tobacco usage is not appropriate or acceptable for decent persons.

NOTE: The median is the middle score <u>when the scores are arranged in order</u>, and the midrange is halfway between the first and last score <u>when the scores are arranged in order</u>. It is therefore usually helpful to begin by placing the scores in order. This will not affect the mean, and it may

INSTRUCTOR'S SOLUTIONS Chapter 2   S-29

also aid in identifying the mode. In addition, no measure of central tendency can have a value lower than the smallest score or higher than the largest score – remembering this helps to protect against gross errors, which most commonly occur when calculating the mean.

2. Arranged in order, the 12 scores are:
    70.9 74.0 78.6 79.2 79.5 80.2 82.5 83.7 84.3 84.6 85.3 86.2
    a. $\bar{x} = (\Sigma x)/n = (969.0)/12 = 80.75$ c. M = [none]
    b. $\tilde{x} = (80.2 + 82.5)/2 = 81.35$ d. m.r. = (70.9 + 86.2)/2 = 78.55
    Yes; if the pages were randomly selected, their sample mean is likely to be a reasonable estimate of the mean level of the whole book.

3. Arranged in order, the 16 scores are:
    .03 .07 .09 .13 .13 .17 .24 .30 .39 .43 .43 .44 .45 .47 .47 .48
    a. $\bar{x} = (\Sigma x)/n = (4.72)/16 = .295$ c. M = .13,.43,.47 [tri-modal]
    b. $\tilde{x} = (.30 + .39)/2 = .345$ d. m.r. = (.03 + .48)/2 = .255
    No; this sample means weights each brand of cereal equally and does not take into account which of the cereals have higher (or lower) rates of consumption.

4. Arranged in order, the 15 scores are:
    17.7 19.9 19.6 20.6 21.4 22.0 23.8 24.0 25.2 27.5 28.9 29.1 29.9 33.5 37.7
    a. $\bar{x} = (\Sigma x)/n = (380.5)/15 = 25.37$ c. M = 19.6
    b. $\tilde{x} = 24.0$ d. m.r. = (17.7 + 37.7)/2 = 27.70
    Yes; the mean of this sample is reasonably close to 25.74, the overall mean of all 40 values.

5. Arranged in order, the 15 scores are:
    .12 .13 .14 .16 .16 .16 .17 .17 .17 .18 .21 .24 .24 .27 .29
    a. $\bar{x} = (\Sigma x)/n = (2.81)/15 = .187$ c. M = .16,.17 [bi-modal]
    b. $\tilde{x} = .17$ d. m.r. = (.12 + .29)/2 = .205
    Yes; these values appear to be significantly above the allowed maximum.

6. Arranged in order, the 18 scores are:
    14 16 17 18 20 21 23 24 25 27 28 30 31 34 37 38 40 42
    a. $\bar{x} = (\Sigma x)/n = (485)/18 = 26.9$ c. M = [none]
    b. $\tilde{x} = (25 + 27)/2 = 26.0$ d. m.r. = (14 + 42)/2 = 28.0
    Yes; these results support the common belief that those killed in motorcycle accidents come from among the younger drivers.

7. Arranged in order, the 20 scores are:
    15 17 17 17 17 17 17 18 18 18 18 18 19 19 19 19 20 21 21 21
    a. $\bar{x} = (\Sigma x)/n = (366)/20 = 18.3$ c. M = 17
    b. $\tilde{x} = (18 + 18)/2 = 18.0$ d. m.r. = (15 + 21)/2 = 18.0
    The sample results appear to be very consistent – i.e., there appears to be little variation from person to person. This suggests that there is little variation in the population, and that the sample mean should be a good estimate of the population mean – in the sense that other samples would likely produce similar results that vary little from this sample

8. Arranged in order, the 12 scores are:
    654.2 661.3 662.2 662.7 667.0 667.4 669.8 670.7 672.2 672.2 672.6 679.2
    a. $\bar{x} = (\Sigma x)/n = (8011.5)/12 = 667.625$ c. M = 672.2
    b. $\tilde{x} = (667.4 + 669.8)/2 = 668.60$ d. m.r. = (654.2 + 679.2)/2 = 666.70
    A serious consequence of having weights that vary too much would be the uncertainty about how much medication was actually being taken.

9. Arranged in order, the scores are as follows.
   JV: 6.5 6.6 6.7 6.8 7.1 7.3 7.4 7.7 7.7 7.7
   Pr: 4.2 5.4 5.8 6.2 6.7 7.7 7.7 8.5 9.3 10.0

   Jefferson Valley
   $n = 10$
   $\bar{x} = (\Sigma x)/n = (71.5)/10 = 7.15$
   $\tilde{x} = (7.1 + 7.3)/2 = 7.20$
   $M = 7.7$
   m.r. $= (6.5 + 7.7)/2 = 7.10$

   Providence
   $n = 10$
   $\bar{x} = (\Sigma x)/n = (71.5)/10 = 7.15$
   $\tilde{x} = (6.7 + 7.7)/2 = 7.20$
   $M = 7.7$
   m.r. $= (4.2 + 10.0)/2 = 7.10$

   Comparing only measures of central tendency, one might suspect the two sets are identical. The Jefferson Valley times, however, are considerably less variable.
   NOTE: This is the reason most banks have gone to the single waiting line. While it doesn't make service faster, it makes service times more equitable by eliminating the "luck of the draw" – i.e., ending up by pure chance in a fast or slow line and having unusually short or long waits.

10. Arranged in order, the scores are as follows.
    Regular: .8150 .8163 .8181 .8192 .8211 .8247
    Diet: .7758 .7773 .7844 .7861 .7868 .7896

    Regular
    $n = 6$
    $\bar{x} = (\Sigma x)/n = (4.9144)/6 = .81907$
    $\tilde{x} = (.8181 + .8192)/2 = .81865$
    $M = $ [none]
    m.r. $= (.8150 + .8247)/2 = .81985$

    Diet
    $n = 6$
    $\bar{x} = (\Sigma x)/n = (4.7000)/6 = .78333$
    $\tilde{x} = (.7844 + .7861)/2 = .78525$
    $M = $ [none]
    m.r. $= (.7758 + .7896)/2 = .78270$

    Diet Coke appears to weigh less than regular Coke, perhaps because it contains less sugar.

11. Arranged in order, the scores are as follows.
    McDonald's: 92 118 128 136 153 176 192 193 240 254 267 287
    Jack in the Box: 74 109 109 190 229 255 270 300 328 377 428 481

    McDonald's
    $n = 12$
    $\bar{x} = (\Sigma x)/n = (2236)/12 = 186.3$
    $\tilde{x} = (176 + 192)/2 = 184.0$
    $M = $ [none]
    m.r. $= (92 + 287)/2 = 189.5$

    Jack in the Box
    $n = 12$
    $\bar{x} = (\Sigma x)/n = (3150)/12 = 262.5$
    $\tilde{x} = (255 + 270)/2 = 262.5$
    $M = 109$
    m.r. $= (74 + 481)/2 = 277.5$

    McDonald's appears to be faster. Yes; the difference appears to be significant.

12. Arranged in order, the scores are as follows.
    BC: 119 125 126 126 128 128 129 131 131 131 132 138
    AD: 126 126 129 130 131 133 134 136 137 138 139 141

    4000 BC skulls
    $n = 12$
    $\bar{x} = (\Sigma x)/n = (1544)/12 = 128.7$
    $\tilde{x} = (128 + 129)/2 = 128.5$
    $M = 131$
    m.r. $= (119 + 138)/2 = 128.5$

    150 AD skulls
    $n = 12$
    $\bar{x} = (\Sigma x)/n = (1600)/12 = 133.3$
    $\tilde{x} = (133 + 134)/2 = 133.5$
    $M = 126$
    m.r. $= (126 + 141)/2 = 133.5$

    Yes; the head sizes appear to have changed, indicating possible interbreeding.
    NOTE: This example illustrates the difficulty with using the mode to measure what is typical. When the observed values are approximately evenly spread out across all the scores and do not "bunch up" around a most popular score, the mode could be any score (no matter how unrepresentative) that just happened to occur twice.

13. The following values were obtained for the head circumference data, where x_i indicates the ith score from the ordered list.

    males
    $n = 50$
    $\bar{x} = (\Sigma x)/n = 2054.9/50 = 41.10$
    $\tilde{x} = (x_{25}+x_{26})/2$
    $\ \ = (41.1+41.1)/2 = 41.10$

    females
    $n = 50$
    $\bar{x} = (\Sigma x)/n = 2002.4/50 = 40.05$
    $\tilde{x} = (x_{25}+x_{26})/2)/2$
    $\ \ = (40.2+40.2 = 40.20$

    No; on the basis of the means and medians alone, there does not appear to be a significant difference between the genders.

14. The following Flesch-Kincaid Grade Level Rating values were obtained for the three authors, where x_i indicates the ith score from the ordered list.

    Clancy
    $n = 12$
    $\bar{x} = (\Sigma x)/n$
    $\ \ = 78.0/12 = 6.50$
    $\tilde{x} = (x_6+x_7)/2$
    $\ \ = (5.4+6.1)/2 = 5.75$

    Rowling
    $n = 12$
    $\bar{x} = (\Sigma x)/n$
    $\ \ = 60.9/12 = 5.075$
    $\tilde{x} = (x_6+x_7)/2$
    $\ \ = (4.9+5.2)/12 = 5.05$

    Tolstoy
    $n = 12$
    $\bar{x} = (\Sigma x)/n$
    $\ \ = 101.2/12 = 8.43$
    $\tilde{x} = (x_6+x_7)/2$
    $\ \ = (8.2+8.3)/2 = 8.30$

    Yes; the ratings appear to be different.

15. The following values were obtained for Boston rainfall, where x_i indicates the ith score from the ordered list.

    Thursday
    $n = 52$
    $\bar{x} = (\Sigma x)/n = 3.57/52 = .069$
    $\tilde{x} = (x_{26}+x_{27})/2 = (.00 + .00)/2 = .000$

    Sunday
    $n = 52$
    $\bar{x} = (\Sigma x)/n = 3.52/52 = .068$
    $\tilde{x} = (x_{26}+x_{27})/2 = (.00 + .00)/2 = .000$

    If "it rains more on weekends" refers to the amount of rain, the data do not support the claim. The amount of rainfall appears to be virtually the same for Thursday and Sunday. If "it rains more on weekends" refers to the frequency of rain (regardless of the amount), then the proportions of days on which there was rain would have to be compared.

16. The following lengths (in seconds) of scenes showing tobacco and alcohol use were obtained for the animated children's movies, where x_i indicates the ith score from the ordered list.

    tobacco
    $n = 50$
    $\bar{x} = (\Sigma x)/n = 2872/50 = 57.4$
    $\tilde{x} = (x_{25}+x_{26})/2 = (5+6)/2 = 5.5$

    alcohol
    $n = 50$
    $\bar{x} = (\Sigma x)/n = 1623/50 = 32.5$
    $\tilde{x} = (x_{25}+x_{26})/2 = (0+3)/2 = 1.5$

    Yes; there does appear to be a difference in the times. Whether or not "more time" is synonymous with a "larger problem" depends on the nature of the scenes. If the scenes where the product is used present the product in a negative manner, then "more time" spent in a manner critical of the product corresponds to a "more wholesome" attitude being conveyed.

17. The x values below are the class midpoints from the given frequency table.

x	f	x·f
44.5	8	356.0
54.5	44	2398.0
64.5	23	1483.5
74.5	6	447.0
84.5	107	9041.5
94.5	11	1039.5
104.5	1	104.5
	200	14870.0

    $\bar{x} = (\Sigma x \cdot f)/n$
    $\ \ = (14870.0)/200$
    $\ \ = 74.35$

S-32 INSTRUCTOR'S SOLUTIONS Chapter 2

NOTE: The mean time was calculated to be 74.35 minutes. According to the rule given in the text, this value should be rounded to one decimal place. The text describes how many decimal places to present in an answer, but not the actual rounding process. When the figure to be rounded is <u>exactly</u> half-way between two values (i.e., the digit in the position to be discarded is a 5, and there are no further digits because the calculations have "come out even"), there is no universally accepted rounding rule. Some authors say to always round up such a value; others correctly note that always rounding up introduces a consistent bias, and that the value should actually be rounded up half the time and rounded down half the time. And so some authors suggest rounding toward the even value (e.g., .65 becomes .6 and .75 becomes .8), while others simply suggest flipping a coin. In this manual, answers <u>exactly</u> half-way between will be reported without rounding (i.e., stated to one more decimal than usual).

18. The individual x values may be used directly ion the calculations as follows.

x	f	x·f
1	27	27
2	31	62
3	42	126
4	40	160
5	28	140
6	32	192
	200	707

$\bar{x} = (\Sigma x \cdot f)/n$
$= (707)/200$
$= 3.5$

This sample mean of 3.5 is identical (to one decimal accuracy) to the true mean of the equally-weighted values 1,2,3,4,5,6. While loading the die might be expected to affect the relative frequency of occurrence of some of the values, it this case it appears not to have affected the overall mean value that occurs.

19. The x values below are the class midpoints from the given frequency table.

x	f	x·f
43.5	25	1087.5
47.5	14	665.0
51.5	7	360.5
55.5	3	166.5
59.5	1	59.5
	50	2339.0

$\bar{x} = (\Sigma x \cdot f)/n$
$= (2339.0)/50$
$= 46.8$

The mean speed of 46.8 mi/hr of those ticketed by the police is more than 1.5 times the posted speed limit of 30 mi/hr. NOTE: This indicates nothing about the mean speed of *all* drivers, a figure which may or may not be higher than the posted limit.

20. The x values below are the class midpoints from the given frequency table.

x	f	x·f
96.65	1	96.65
97.05	8	776.40
97.45	14	1634.30
97.85	22	2152.70
98.25	19	1866.75
98.65	32	3156.80
99.05	6	594.30
99.45	4	397.80
	106	10405.70

$\bar{x} = (\Sigma x \cdot f)/n$
$= (10405.70)/106$
$= 98.17$

This sample mean of 98.17°F appears to be significantly less than the commonly assumed population mean of 98.6°F.

21. a. Arranged in order, the original 54 scores are:
    26 29 34 40 46 48 60 62 64 65 76 79 80 86 90
    94 105 114 116 120 125 132 140 140 144 148 150 150 154 166
    166 180 182 202 202 204 204 212 220 220 236 262 270 316 332
    344 348 356 360 365 416 436 446 514
    $\bar{x} = (\Sigma x)/n = (9876)/54 = 182.9$

    b. Trimming the highest and lowest 10% (or 5.4 = 5 scores), the remaining 44 scores are:
    48 60 62 64 65 76 79 80 86 90 94 105 114 116 120
    125 132 140 140 144 148 150 150 154 166 166 180 182 202 202
    204 204 212 220 220 236 262 270 316 332 344 348 356 360
    $\bar{x} = (\Sigma x)/n = (7524)/44 = 171.0$

    c. Trimming the highest and lowest 20% (or 10.8 = 11 scores), the remaining 32 scores are:
    79 80 86 90 94 105 114 116 120 125 132 140 140 144 148
    150 150 154 166 166 180 182 202 202 204 204 212 220 220 236
    262 270
    $\bar{x} = (\Sigma x)/n = (5093)/32 = 159.2$

    In this case, the mean gets smaller as more scores are trimmed. In general, means can increase, decrease, or stay the same as more scores are trimmed. The mean decreased here because the higher scores were farther from the original mean than were the lower scores.

22. No, the result is not the national average teacher's salary. Since there are many more teachers in California, for example, than Nevada, those state averages should not be counted equally – a weighted average (where the weights are the numbers of teachers in each state) should be used.

23. If n=10 values have $\bar{x} = (\Sigma x)/n = 75.0$, then $\Sigma x = 750$.
    a. Since the sum of the 9 given values is 698, the 10th value must be 750-698 = 52.
    b. Generalizing the result in part (a), n-1 values can be freely assigned. The nth value must be whatever is needed (whether positive or negative) to reach the sum $(\Sigma x) = n \cdot \bar{x}$ determined by the given mean.

24. Treating the 5^+ values as 5.0, $\bar{x} = (\Sigma x)/n = 17.1/5 = 3.42$. Since continuing the experiment would only increase Σx, one can conclude that the actual mean battery life in the sample is greater than 3.42.

25. The w values below are the weights – which can sum to any value, since the weighted mean is found by dividing by Σw.

x	w	x·w
65	.15	9.75
83	.15	12.45
80	.23	12.00
90	.15	13.50
92	.40	36.80
	1.00	84.50

    $\bar{x} = (\Sigma x \cdot w)/(\Sigma w)$
    $= (84.50)/(1.00)$
    $= 84.50$

26. Let the original data, arranged in order, be: $x_1, x_2, ..., x_n$
    a. Arranged in order, the new data would be: $x_1+k, x_2+k, ..., x_n+k$

    In general, adding (or subtracting) a constant k from each score will add (or subtract) k from each measure center.
    new $\bar{x} = [\Sigma(x+k)]/n = [\Sigma x + nk]/n = (\Sigma x)/n + (nk)/n = \bar{x} + k$
    new $\tilde{x} = \tilde{x} + k$ [from the ordered list]
    new M = M+k [from the ordered list]
    new m.r. = $[(x_1+k) + (x_n+k)]/2 = [x_1+x_n+2k]/2 = (x_1+x_n)/2 + (2k)/2 = $ m.r. $+k$

b. Arranged in order, the new data would be: $k \cdot x_1, k \cdot x_2, \ldots, k \cdot x_n$
In general, multiplying (or dividing) each score by a constant k will multiply (or divide) each measure of center by k.
new $\bar{x} = [\Sigma(k \cdot x)]/n = [k \cdot (\Sigma x)]/n = k \cdot [(\Sigma x)/n] = k \cdot \bar{x}$
new $\tilde{x} = k \cdot \tilde{x}$ [from the ordered list]
new $M = k \cdot M$ [from the ordered list]
new m.r. $= (k \cdot x_1 + k \cdot x_n)/2 = k \cdot (x_1 + x_n)/2 = k \cdot$ m.r.
NOTE: Do not assume the principle in (a) and (b) extends automatically to other mathematical operations. A short example will show, for example, that the mean of the squares of a set of data is not necessarily the square of their original mean.

27. Let $\bar{x}_h$ stand for the harmonic mean: $\bar{x}_h = n/[\Sigma(1/x)] = 2/[1/40 + 1/60] = 2/[.0417]$
$= 48.0$

28. The geometric mean of the five values is the fifth root of their product.
$\sqrt[5]{(1.10)(1.08)(1.09)(1.12)(1.07)} = 1.092$

29. R.M.S. $= \sqrt{\Sigma x^2/n} = \sqrt{[(110)^2 + (0)^2 + (-60)^2 + (12)^2]/4} = \sqrt{15844/4} = \sqrt{3961} = 62.9$

30. There are 40 scores. The median $(x_{20} + x_{21})/2$ is in the 100-199 class. The relevant values are as follows.
   lower class limit of the median class = 100   n = 40 total scores
   class width = 100   m = 11 scores entering the median class
   frequency of the median class = 12
$\tilde{x}$ = (median class lower limit) + (class width) $\cdot$ [(n+1)/2 - (m+1)]/(median class frequency)
$= 100 + 100 \cdot [41/2 - 12]/(12)$
$= 100 + 100 \cdot [8.5]/(12) = 100 + 70.8 = 170.8$
This compares well with the value 170 obtained using the original list of data. Since frequency tables are summaries which do not contain all the information, values obtained from the original data should always be preferred to ones obtained from frequency tables.

2-5 Measures of Variation

NOTE: Although not given in the text, the symbol R will be used for the range throughout this manual. Remember that the range is the difference between the highest and the lowest scores, and not necessarily the difference between the last and the first values as they are listed. Since calculating the range involves only the subtraction of 2 original pieces of data, that measure of variation will be reported with the same accuracy as the original data.

1.

x	x - $\bar{x}$	$(x-\bar{x})^2$	x^2
0	-157.83	24910.3089	0
0	-157.83	24910.3089	0
0	-157.83	24910.3089	0
176	18.17	330.1489	30976
223	65.17	4247.1289	49729
548	390.17	152232.6289	300304
947	0.02	231540.8834	381009

$\bar{x} = (\Sigma x)/n = 947/6 = 157.83$
R = 5480 = 548

by formula 2-4,
$s^2 = \Sigma(x-\bar{x})^2/(n-1)$
$= 231540.8834/5$
$= 46308.17$
$= 46308.2$

by formula 2-5,
$s^2 = [n(\Sigma x^2) - (\Sigma x)^2]/[n(n-1)]$
$= [6(381009) - (947)^2]/[6(5)]$
$= [1388888889245]/[30]$
$= 46308.17$
$= 46308.2$

$s = \sqrt{46308.17} = 215.2$
The times appear to vary widely.

NOTE: When finding the square root of the variance to obtain the standard deviation, use all the decimal places of the variance, and not the rounded value reported as the answer. The best way to do this is either to keep the value on the calculator display or to place it in the memory. Do not copy down all the decimal places and then re-enter them to find the square root, as that could introduce round-off and/or copying errors.

When using formula 2-4, constructing a table having the first three columns shown above helps to organize the calculations and makes errors less likely. In addition, verify that $\Sigma(x-\bar{x}) = 0$ [except for a possible small discrepancy at the last decimal, due to using a rounded value for the mean] before proceeding – if such is not the case, there is an error and further calculation is fruitless. For completeness, and as a check, both formulas 2-4 and 2-5 were used above. In general, only formula 2-5 will be used throughout the remainder of this manual for the following reasons:
   (1) When the mean does not "come out even," formula 2-4 involves round-off error and/or many messy decimal calculations.
   (2) The quantities Σx and Σx^2 needed for formula 2-5 can be found directly and conveniently on the calculator from the original data without having to construct a table like the one above.

2. preliminary values: n = 12, Σx = 969.0, Σx^2 = 78487.82
   R = 86.2 - 70.9 = 15.3
   $s^2 = [n(\Sigma x^2) - (\Sigma x)^2]/[n(n-1)] = [12(78487.82) - (969.0)^2]/[12(11)]$
   = (2892.84)/132 = 21.92
   s = 4.68
   Yes; since the 12 pages were selected at random, s = 4.68 should be a reasonable estimate for the entire book.

3. preliminary values: n = 16, Σx = 4.72, Σx^2 = 1.8144
   R = .48 - .03 = .45
   $s^2 = [n(\Sigma x^2) - (\Sigma x)^2]/[n(n-1)] = [16(1.8144) - (4.72)^2]/[16(15)]$
   = (6.7520)/240 = .028
   s = .168
   No; this value is not likely to be a good estimate for the standard deviation among the amounts in each gram consumed by the population. This value gives equal weight to each cereal in the sample, no matter how popular (or unpopular) it is with consumers. The sample is not representative of the consumption patters.

NOTE: The quantity $[n(\Sigma x^2) - (\Sigma x)^2]$ cannot be less than zero. A negative value indicates that there is an error and that further calculation is fruitless. In addition, remember to find the value for s by taking the square root of the precise value of s^2 showing on the calculator display before it is rounded to one more decimal place than the original data.

4. preliminary values: n = 15, Σx = 380.5, Σx^2 = 10101.23
   R = 37.7 - 17.7 = 20.0
   $s^2 = [n(\Sigma x^2) - (\Sigma x)^2]/[n(n-1)] = [15(1010.23) - (380.5)^2]/[15(14)]$
   = (6738.20)/210 = 32.09
   s = 5.66
   Yes; this sample value is reasonably close to the true value for all women in the data set.

5. preliminary values: n = 15, Σx = 2.81, Σx^2 = .5631
   R = .29 - .12 = .17
   $s^2 = [n(\Sigma x^2) - (\Sigma x)^2]/[n(n-1)] = [15(.5631) - (2.81)^2]/[15(14)]$
   = (.5504)/210 = .00262
   s = .051
   No; the intent is to lower the mean amount. If all drivers had high concentrations (e.g., between .35 and .30), there would be a low s – but a high level of drunk driving.

NOTE: Following the usual round-off rule of giving answers with one more decimal than the original data produces a variance of .003, which has only one significant digit. This is an unusual situation occurring because values less than 1.0 become smaller when they are squared. Since the original data had 2 significant digits, we provide 3 significant digits for the variance.

6. preliminary values: $n = 18$, $\Sigma x = 485$, $\Sigma x^2 = 14343$
  $R = 42 - 14 = 28$
  $s^2 = [n(\Sigma x^2) - (\Sigma x)^2]/[n(n-1)] = [18(14343) - (485)^2]/[18(17)] = (22949)/306 = 75.0$
  $s = 8.7$
  Since motorcycle drivers tend to come from a particular age group (viz., the younger drivers), their ages would vary less than those in the general population of all licensed drivers.

7. preliminary values: $n = 20$, $\Sigma x = 366$, $\Sigma x^2 = 6746$
  $R = 21 - 15 = 6$
  $s^2 = [n(\Sigma x^2) - (\Sigma x)^2]/[n(n-1)] = 20(6746) - (366)^2]/[20(19)] = (964)/380 = 2.5$
  $s = 1.6$
  The fact that both R and s are small (relative to the recorded data values) means that reaction times tended to be consistent and showed little variation.

8. preliminary values: $n = 12$, $\Sigma x = 8011.5$, $\Sigma x^2 = 5349166.83$
  $R = 679.2 - 654.2 = 25.0$
  $s^2 = [n(\Sigma x^2) - (\Sigma x)^2]/[n(n-1)] = [12(5349166.83) - (8011.5)^2]/[12(11)]$
  $= (5869.71)/132 = 44.47$
  $s = 6.67$
  While there is always room for improvement, it appears that the variability from pill to pill is at an acceptable level. The standard deviation is about 1% of the typical weight, and the values of s and R are consistent with the Range Rule of Thumb – that $s \approx R/4$.

9. Jefferson Valley
  $n = 10$, $\Sigma x = 71.5$, $\Sigma x^2 = 513.27$
  $R = 7.7 - 6.5 = 1.2$
  $s^2 = [n(\Sigma x^2) - (\Sigma x)^2]/[n(n-1)]$
  $= [10(513.27) - (71.5)^2]/[10(9)]$
  $= 20.45/90 = 0.23$
  $s = 0.48$

  Providence
  $n = 10$, $\Sigma x = 71.5$, $\Sigma x^2 = 541.09$
  $R = 10.0 - 4.2 = 5.8$
  $s^2 = [n(\Sigma x^2) - (\Sigma x)^2]/[n(n-1)]$
  $= [10(541.09) - (71.5)^2]/[10(9)]$
  $= 298.65/90 = 3.32$
  $s = 1.82$

  Exercise #9 of section 2-4 indicated that the mean waiting time was 7.15 minutes at each bank. The Jefferson Valley waiting times, however, are considerably less variable. The range measures the difference between the extremes. The longest and shortest waits at Jefferson Valley differ by a little over 1 minute (R=1.2), while the longest and shortest waits at Providence differ by almost 6 minutes (R=5.8). The standard deviation measures the typical difference from the mean. A Jefferson Valley customer usually receives service within about ½ minute (s=0.48) of 7.15 minutes, while a Providence customer usually receives service within about 2 minutes (s=1.82) of the mean.

10. Regular
  $n = 6$, $\Sigma x = 4.9144$, $\Sigma x^2 = 4.02528224$
  $R = .8247 - .8150 = .0097$
  $s^2 = [n(\Sigma x^2) - (\Sigma x)^2]/[n(n-1)]$
  $= [6(4.02528224) - (4.9144)^2]/[6(5)]$
  $= .00036608/30 = .000012202$
  $s = .00349$

  Diet
  $n = 6$, $\Sigma x = 4.7000$, $\Sigma x^2 = 3.68181990$
  $R = .7896 - .7758 = .0138$
  $s^2 = [n(\Sigma x^2) - (\Sigma x)^2]/[n(n-1)]$
  $= [6(3.68181990) - (4.7000)^2]/[6(5)]$
  $= .00091940/30 = .000030633$
  $s = .00554$

  There appears to be slightly more variation among the Diet Coke weights than among those for Regular Coke.

INSTRUCTOR'S SOLUTIONS Chapter 2   S-37

11. McDonald's
     $n = 12$, $\Sigma x = 2236$, $\Sigma x^2 = 461540$
     $R = 287 - 92 = 195$
     $s^2 = [n(\Sigma x^2) - (\Sigma x)^2]/[n(n-1)]$
         $= [12(461540) - (2236)^2]/[12(11)]$
         $= 538784/132 = 4081.7$
     $s = 63.9$

     Jack in the Box
     $n = 12$, $\Sigma x = 3150$, $\Sigma x^2 = 1009962$
     $R = 481 - 74 = 407$
     $s^2 = [n(\Sigma x^2) - (\Sigma x)^2]/[n(n-1)]$
         $= [12(1009962) - (3150)^2]/[12(11)]$
         $= 2197044/132 = 16644.3$
     $s = 129.0$

     There appears to be about twice the variability among the times at Jack in the Box as there is at McDonald's. In other words, McDonald's seems to be doing better at producing a uniform system for dealing with all orders.

12. 4000 BC
     $n = 12$, $\Sigma x = 1544$, $\Sigma x^2 = 198898$
     $R = 138 - 119 = 19$
     $s^2 = [n(\Sigma x^2) - (\Sigma x)^2]/[n(n-1)]$
         $= [12(198898) - (1544)^2]/[12(11)]$
         $= 2840/132 = 21.5$
     $s = 4.6$

     150 AD
     $n = 12$, $\Sigma x = 1600$, $\Sigma x^2 = 213610$
     $R = 141 - 126 = 15$
     $s^2 = [n(\Sigma x^2) - (\Sigma x)^2]/[n(n-1)]$
         $= [12(213610) - (1600)^2]/[12(11)]$
         $= 3320/132 = 25.2$
     $s = 5.0$

     There is approximately the same amount of variability in skull breadths within each period.

13. males
     $n = 50$, $\Sigma x = 2054.9$, $\Sigma x^2 = 84562.23$
     $s^2 = [n(\Sigma x^2) - (\Sigma x)^2]/[n(n-1)]$
         $= [50(84562.23) - (2054.9)^2]/[50(49)]$
         $= 5497.49/2450 = 2.24$
     $s = 1.50$

     females
     $n = 50$, $\Sigma x = 2002.4$, $\Sigma x^2 = 80323.82$
     $s^2 = [n(\Sigma x^2) - (\Sigma x)^2]/[n(n-1)]$
         $= [50(80323.82) - (2002.4)^2]/[50(49)]$
         $= 6585.24/2450 = 2.69$
     $s = 1.64$

     While there was slightly more variability among the female values in this particular sample, there does not appear to be a significant difference between the genders.

14. Clancy
     $n = 12$, $\Sigma x = 78.0$
     $\Sigma x^2 = 572.94$
     $s^2 = [n(\Sigma x^2) - (\Sigma x)^2]/[n(n-1)]$
         $= [12(572.94) - (78.0)^2]/[12(11)]$
         $= 791.28/132 = 5.99$
     $s = 2.45$

     Rowling
     $n = 12$, $\Sigma x = 60.9$
     $\Sigma x^2 = 572.94$
     $s^2 = [n(\Sigma x^2) - (\Sigma x)^2]/[n(n-1)]$
         $= [12(324.07) - (60.9)^2]/[12(11)]$
         $= 180.03/132 = 1.36$
     $s = 1.17$

     Tolstoy
     $n = 12$, $\Sigma x = 101.2$
     $\Sigma x^2 = 572.94$
     $s^2 = [n(\Sigma x^2) - (\Sigma x)^2]/[n(n-1)]$
         $= [12(897.82) - (101.2)^2]/[12(11)]$
         $= 532.40/132 = 4.03$
     $s = 2.01$

     The reading difficulty level seems to vary from page to page most in the Clancy book and least in the Rowling book. As 2.45 is more than twice 1.17, it appears that the difference may be significant.

15. Thursday
     $n = 52$, $\Sigma x = 3.57$, $\Sigma x^2 = 1.6699$
     $s^2 = [n(\Sigma x^2) - (\Sigma x)^2]/[n(n-1)]$
         $= [52(1.6699) - (3.57)^2]/[52(51)]$
         $= 74.0899/2652 = .02793$
     $s = .167$

     Sunday
     $n = 52$, $\Sigma x = 3.52$, $\Sigma x^2 = 2.2790$
     $s^2 = [n(\Sigma x^2) - (\Sigma x)^2]/[n(n-1)]$
         $= [52(2.2790) - (3.52)^2]/[52(51)]$
         $= 106.1176/2652 = .04001$
     $s = .200$

     The values are close, although there may be slightly more variation among the amounts for Sundays than among the amounts for Thursdays.

16. tobacco
     $n = 50$, $\Sigma x = 2872$, $\Sigma x^2 = 694918$
     $s^2 = [n(\Sigma x^2) - (\Sigma x)^2]/[n(n-1)]$
         $= [50(694918) - (2872)^2]/[50(49)]$
         $= 26497516/2450 = 10815.3$
     $s = 104.0$

     alcohol
     $n = 50$, $\Sigma x = 1623$, $\Sigma x^2 = 268331$
     $s^2 = [n(\Sigma x^2) - (\Sigma x)^2]/[n(n-1)]$
         $= [50(268331) - (1623)^2]/[50(49)]$
         $= 10782421/2450 = 4401.0$
     $s = 66.3$

While there was slightly more variability among the lengths of tobacco usage in this particular sample, there does not appear to be a significant difference between the products.

17.

x	f	f·x	f·x²
44.5	8	356.0	15842.00
54.5	44	2398.0	130691.00
64.5	23	1483.5	95685.75
74.5	6	447.0	33301.50
84.5	107	9041.5	764006.75
94.5	11	1039.5	98232.75
104.5	1	104.5	10920.25
	200	14870.0	1148680.00

$s^2 = [n(\Sigma f \cdot x^2) - (\Sigma f \cdot x)^2]/[n(n-1)]$
$= [200(1148680.00) - (1487.0)^2]/[200(199)] = (8619100.00)/29800 = 216.56$
$s = 14.7$

18.

x	f	f·x	f·x²
1	27	27	27
2	31	62	124
3	42	126	378
4	40	160	640
5	28	140	700
6	32	192	1152
	200	707	3021

$s^2 = [n(\Sigma f \cdot x^2) - (\Sigma f \cdot x)^2]/[n(n-1)]$
$= [200(3021) - (707)^2]/[200(199)] = (104351)/39800 = 2.6$
$s = 1.6$

NOTE: This compares favorably with $\sigma = 1.708$, the true standard deviation among the values produced by a balanced die.

19.

x	f	f·x	f·x²
43.5	25	1087.5	47305.25
47.5	14	665.0	31587.50
51.5	7	360.5	18565.75
55.5	3	166.5	9240.75
59.5	1	59.5	3540.25
	50	2339.0	110239.50

$s^2 = [n(\Sigma f \cdot x^2) - (\Sigma f \cdot x)^2]/[n(n-1)]$
$= [50(110239.50) - (2339.0)^2]/[50(49)] = (41054.00)/2450 = 16.76$
$s = 4.1$

20. The x values below are the class midpoints from the given frequency table.

x	f	f·x	f·x²
96.65	1	96.65	9341.2225
97.05	8	776.40	75349.6200
97.45	14	1634.30	132951.0350
97.85	22	2152.70	210641.6950
98.25	19	1866.75	183408.1875
98.65	32	3156.80	311418.3200
99.05	6	594.30	58865.5150
99.45	4	397.80	39561.2100
	106	10405.70	1021536.7050

$s^2 = [n(\Sigma f \cdot x^2) - (\Sigma f \cdot x)^2]/[n(n-1)]$
$= [106(1021536.7050) - (10504.70)^2]/[106(105)] = (4297.71)/11130 = .3861$
$s = .62$

21. Answers will vary. Assuming that the ages range from 25 to 75, the Range Rule of Thumb suggests $s \approx$ range/4 = (75-25)/4 = 50/4 = 12.5.

INSTRUCTOR'S SOLUTIONS Chapter 2   S-39

22. Answers will vary. Assuming that the grades range from 50 to 100, the Range Rule of Thumb suggests s ≈ range/4 = (100-50)/4 = 50/4 = 12.5.

23. Given $\bar{x}$ = 38.86 and s = 3.78, the Range Rule of Thumb suggests
    minimum "usual" value = $\bar{x}$ − 2s = 38.86 − 2(3.78) = 38.86 − 7.56 = 31.30
    maximum "usual" value = $\bar{x}$ + 2s = 38.86 + 2(3.78) = 38.86 + 7.56 = 46.42
    Yes, in this context a length of 47.0 cm would be considered unusual.

24. Given $\bar{x}$ = 63.6 and s = 2.5, the Range Rule of Thumb suggests
    minimum "usual" value = $\bar{x}$ − 2s = 63.6 − 2(2.5) = 63.6 − 5.0 = 58.6
    maximum "usual" value = $\bar{x}$ + 2s = 63.6 + 2(2.5) = 63.6 + 5.0 = 68.6
    Yes, in this context a woman 72" tall would be considered unusually tall.

25. a. The limits 61.1 and 66.1 are 1 standard deviation from the mean. The Empirical Rule for Data with a Bell-shaped Distribution states that about 68% of the heights should fall within those limits.
    b. The limits 56.1 and 76.1 are 3 standard deviations from the mean. The Empirical Rule for Data with a Bell-shaped Distribution states that about 99.7% of the heights should fall within those limits.

26. a. The limits .80931 and .82433 are 1 standard deviation from the mean. The Empirical Rule for Data with a Bell-shaped Distribution states that about 68% of the weights should fall within those limits.
    b. The limits .80180 and .83184 are 2 standard deviations from the mean. The Empirical Rule for Data with a Bell-shaped Distribution states that about 95% of the weights should fall within those limits.

27. The limits 58.6 and 68.6 are 2 standard deviations from the mean. Chebyshev's Theorem states that there must be at least $1 - 1/k^2$ of the scores within k standard deviations of the mean. Here k = 2, and so the proportion of the heights within those limits is at least $1 - 1/2^2 = 1 - 1/4 = 3/4 = 75\%$.

28. The limits .79429 and .83935 are 3 standard deviations from the mean. Chebyshev's Theorem states that there must be at least $1 - 1/k^2$ of the scores within k standard deviations of the mean. Here k = 3, and so the proportion of the heights within those limits is at least $1 - 1/3^2 = 1 - 1/9 = 8/9 = 89\%$.

29. The following values were obtained using computer software. Using the accuracy for $\bar{x}$ and s given by the computer, we calculate the coefficient of variation to the nearest .01%.

    calories
    n = 16
    $\bar{x}$ = 3.7625
    s = .2217
    CV = s/$\bar{x}$
       = .2217/3.7625
       = .0589 = 5.89%

    grams of sugar
    n = 16
    $\bar{x}$ = .2950
    s = .1677
    CV = s/$\bar{x}$
       = .1677/.2950
       = .5685 = 56.85%

    Relative to their mean amounts, there is considerably more variability in the grams of sugar than in the calories. There are two possible explanations: (1) The amount of calories may be easier to control than the weight of sugar. (2) The weights of sugar are so small that any slight deviation seems large by comparison.

30. The following values were obtained using computer software. Using the accuracy for $\bar{x}$ and s given by the computer, we calculate the coefficient of variation to the nearest .01%.

Regular Coke
n = 36
$\bar{x}$ = .81682
s = .00751
CV = s/$\bar{x}$
    = .00751/.81682
    = .0092 = 0.92%

Regular Pepsi
n = 36
$\bar{x}$ = .82410
s = .00570
CV = s/$\bar{x}$
    = .00570/.82410
    = .0069 = 0.69%

As each of the coefficients falls between ½% and 1%, they do not appear to differ significantly.

31. A standard deviation of s = 0 is possible only when s^2 = 0, and $s^2 = \Sigma(x-\bar{x})^2/(n-1)$ = 0 only when $\Sigma(x-\bar{x})^2$ = 0. Since each $(x-\bar{x})^2$ is non-negative, $\Sigma(x-\bar{x})^2$ = 0 only when every $(x-\bar{x})^2$ = 0 – i.e., only when every x is equal to $\bar{x}$. In simple terms, zero variation occurs only when all the scores are identical.

32. a. dollars; the units for the standard deviation are always the same as the original units
b. dollars squared; the units for the variance are always the square of the original units

33. The Everlast brand is the better choice. In general, a smaller standard deviation of lifetimes indicates more consistency from battery to battery – signaling a more dependable production process and a more dependable final product. Assuming a bell-shaped distribution of lifetimes, for example, that empirical rule states that about 68% of the lifetimes will fall within one standard deviation of the mean. Here, those limits would be
   for Everlast: 50 ± 2 or 48 months to 52 months
   for Endurance: 50 ± 6 or 44 months to 56 months
While a person might be lucky and purchase a long-lasting Endurance battery, an Everlast battery is much more likely to last for the advertised 48 months.

34. A large effect; the new standard deviation will be much higher.
NOTE: Very roughly, standard deviation measures "typical spread." If 20 scores have s=1, then the "total spread" is about 20. If the new score is about 20 away from the others (either higher or lower), the new total spread will be about 40 and the new typical spread will be about 2.

35.
section 1
n = 11, Σx = 201, Σx^2 = 4001
R = 20 - 1 = 19
$s^2 = [n(\Sigma x^2) - (\Sigma x)^2]/[n(n-1)]$
    = $[11(4001) - (201)^2]/[11(10)]$
    = 3610/110 = 32.92
s = 5.7

section 2
n = 11, Σx = 119, Σx^2 = 1741
R = 19 - 2 = 17
$s^2 = [n(\Sigma x^2) - (\Sigma x)^2]/[n(n-1)]$
    = $[11(1741) - (119)^2]/[11(10)]$
    = 1990/110 = 45.36
s = 6.7

The range values give the impression that section 1 had more variability than section 2.
The range can be misleading because it is based only on the extreme scores. In this case, the lowest score in section 1 was so distinctly different from the others that to include it in any measure trying to give a summary about the section as a whole would skew the results. For the mean, where the value is only one of 11 used in the calculation, the effect is minimal; for the range, where the value is one of only 2 used in the calculation, the effect is dramatic. The standard deviation values give the impression that section 2 had slightly more variability.
NOTE: In this case, section 2 seems considerably more variable (or diverse), and even the standard deviation by itself fails to accurately distinguish between the sections.

INSTRUCTOR'S SOLUTIONS Chapter 2   S-41

36. a. for each value, let $y_i = x_i + k$
    part (a) of exercise #26 of section 2-4 determined $\bar{y} = \bar{x} + k$
    $R_y = (x_n + k) - (x_1 + k) = x_n + k - x_1 - k = x_n - x_1 = R_x$
    $s_y^2 = \Sigma[y - \bar{y}]^2/(n-1) = \Sigma[(x+k) - (\bar{x}+k)]^2/(n-1) = \Sigma[x - \bar{x}]^2/(n-1) = s_x^2$
    $s_y = s_x$
    Adding a constant value k to each score does not affect the values of the measures of dispersion. In non-statistical terms, shifting everything by k units does not affect the spread of the scores.

    b. for each value, let $y_i = k \cdot x_i$
    part (b) of exercise #26 of section 2-4 determined $\bar{y} = k \cdot \bar{x}$
    $R_y = k \cdot x_n - k \cdot x_1 = k \cdot (x_n - x_1) = k \cdot R_x$
    $s_y^2 = \Sigma[y - \bar{y}]^2/(n-1) = \Sigma[k \cdot x - k \cdot \bar{x}]^2/(n-1) = k^2 \cdot \Sigma[x - \bar{x}]^2/(n-1) = k^2 \cdot s_x^2$
    $s_y = k \cdot s_x$
    Multiplying each score by the value k multiplies both the range and the standard deviation by k. The variance, whose units are the square of the units in the problem, is multiplied by k^2.

    c. Summarizing the results of this exercise [and exercise #26 of section 2-4],

    ▸ Adding (or subtracting) a constant to each score will add (or subtract) that constant to the mean but will not change the standard deviation.
    ▸ Multiplying (or dividing) each score by a constant will multiply (or divide) both the mean and the standard deviation by that constant.

    Mathematically, if $y = a \cdot x + b$
    then $\bar{y} = a \cdot \bar{x} + b$
    and $s_y = a \cdot s_x$
    Applying this to C and F, the body temperatures in degrees Celsius and degrees Fahrenheit, where $C = 5(F-32)/9 = (5/9) \cdot F - 160/9$
    $\bar{C} = (5/9) \cdot \bar{F} - 160/9 = (5/9)(98.20) - 160/9 = 36.78°C$
    $s_C = (5/9) \cdot s_F = (5/9)(0.62) = 0.34°C$

37. The following values were obtained for the smokers in Table 2-1 using computer software. All subsequent calculations use the accuracy of these values.
    n = 40 $\bar{x}$ = 172.5   s = 119.5
    The signal-to-noise ratio is
    $\bar{x}/s = 172.5/119.5 = 1.44$.

38. The following values were obtained for the smokers in Table 2-1 using computer software. All subsequent calculations use the accuracy of these values.
    n = 40 $\bar{x}$ = 172.5 $\tilde{x}$ = 170.0   s = 119.5
    Pearson's index is skewness is
    $I = 3(\bar{x} - \tilde{x})/s = 3(172.5 - 170.0)/119.5 = 3(2.5)/119.5 = .063$
    Since $-1.00 < I < 1.00$ for these data, there is not significant skewness.

39. The most spread, and hence the largest standard deviation, occurs when the scores are evenly divided between the minimum and maximum values – i.e., for ½ the values occurring at the minimum and the other ½ occurring at the maximum. For n=10 scores between 70 and 100 inclusive, the largest possible standard deviation occurs for the values:
    70 70 70 70 70 100 100 100 100 100
    preliminary values: n = 10, Σx = 850, Σx^2 = 74500
    $s^2 = [n(\Sigma x^2) - (\Sigma x)^2]/[n(n-1)] = [10(74500) - (850)^2]/[10(9)]$
    $= (22500)/90 = 250$
    s = 15.8
    NOTE: This agrees with the notion that the "standard deviation" is describing the "typical spread" of the data. Since the mean of the values is $\bar{x} = (\Sigma x)/n = 850/10 = 85$, each of

the values is exactly 15 away from the center and any reasonable measure of typical spread should give a value close to 15.

40. Every score must be within $s \cdot \sqrt{n-1} = (5.0) \cdot \sqrt{16} = 20.0$ of the mean. Since $\bar{x} = 75.0$, every score must be within 75.0 ± 20.0, or from 55.0 to 95.0. Kelly is not telling the truth. If she really had a 97, either the mean would have to be higher than 75.0 or the standard deviation would have to be greater than 5.0.

41. For greater accuracy and understanding, we use 3 decimal places and formula 2-4.
  a. the original population

x	x-μ	(x-μ)²
3	-3	9
6	0	0
9	3	9
18	0	18

  $\mu = (\Sigma x)/N = 18/3 = 6$

  $\sigma^2 = \Sigma(x-\mu)^2/N = 18/3 = 6$
  $[\sigma = 2.449]$

  b. the nine samples: using $s^2 = \Sigma(x-\bar{x})^2/(n-1)$ [for each sample, n = 2]

sample	$\bar{x}$	s^2	s
3,3	3.0	0	0
3,6	4.5	4.5	2.121
3,9	6.0	18.0	4.245
6,3	4.5	4.5	2.121
6,6	6.0	0	0
6,9	7.5	4.5	2.121
9,3	6.0	18.0	4.245
9,6	7.5	4.5	2.121
9,9	9.0	0	0
	54.0	54.0	16.974

  mean of the 9 s^2 values
  $(\Sigma s^2)/9 = 54.0/9 = 6$

  c. the nine samples: using $\sigma^2 = \Sigma(x-\mu)^2/N$ [for each sample, N = 2]

sample	μ	σ^2
3,3	3.0	0
3,6	4.5	2.25
3,9	6.0	9.00
6,3	4.5	2.25
6,6	6.0	0
6,9	7.5	2.25
9,3	6.0	9.00
9,6	7.5	2.25
9,9	9.0	0
	54.0	27.00

  mean of the 9 σ^2 values
  $(\Sigma \sigma^2)/9 = 27.00/9 = 3$

  d. The approach in (b) of dividing by n-1 when calculating the sample variance gives a better estimate of the population variance. On the average, the approach in (b) gave the correct population variance of 6. The approach in (c) of dividing by n underestimated the correct population variance. When computing sample variances, divide by n-1 and not by n.

  e. No. An unbiased estimator is one that gives the correct answer on the average. Since the average value of s^2 in part (b) was 6, which was the correct value calculated for σ^2 in part (a), s^2 is an unbiased estimator of σ^2. Since the average value of s in part (b) is $(\Sigma s)/9 = 16.974/9 = 1.886$, which is not the correct value of 2.449 calculated for σ in part (a), s is not an unbiased estimator of σ.
  NOTE: Since the average value of $\bar{x}$ in part (b) is $(\Sigma \bar{x})/9 = 54.0/9 = 6.0$, which is the correct value calculated for μ in part (a), $\bar{x}$ is an unbiased estimator of μ.

42. a. the original population

x	x-μ	|x-μ|
3	-3	3
6	0	0
9	3	3
18	0	6

$\mu = (\Sigma x)/N = 18/3 = 6$

$MAD = \Sigma|x-\mu|/N = 6/3 = 2$

b. the nine samples: using $MAD = \Sigma|x-\bar{x}|/n$ [for each sample, n = 2]

sample	$\bar{x}$	MAD
3,3	3.0	0
3,6	4.5	1.5
3,9	6.0	3.0
6,3	4.5	1.5
6,6	6.0	0
6,9	7.5	1.5
9,3	6.0	3.0
9,6	7.5	1.5
9,9	9.0	0
	54.0	12.0

mean of the 9 MAD values

[Σ(MAD)]/9 = 12.0/9 = 1.33

c. Since the average value of the sample MAD's in part (b), does not equal the true population MAD in part (a), the sample MAD is not an unbiased estimator of the population MAD. In the given example, dividing by (n-1) instead of n would be dividing by 1 instead of 2. That would double each of the sample MAD's and make the average MAD 24/9 = 2.67, which is still not equal to the value in part (a). Therefore division by n-1 instead of n does not produce an unbiased estimator for the mean absolute deviation as it did for the variance.

2-6 Measures of Relative Standing

1. a. $x - \mu = 160 - 100 = 60$
   b. $60/\sigma = 60/16 = 3.75$
   c. $z = (x - \mu)/\sigma = (160 - 100)/16 = 60/16 = 3.75$
   d. Since Einstein's IQ converts to a z score of $3.75 > 2$, it is considered unusually high.

2. a. $x - \mu = 48 - 72.9 = -24.9$
   b. $-24.9/\sigma = -24.9/12.3 = -2.02$
   c. $z = (x - \mu)/\sigma = (48 - 72.9)/12.3 = -24.9/12.3 = -2.02$
   d. Since the author's pulse rate converts to a z score of $-2.02 < -2$, it would be considered unusually low. Athletes and people who exercise regularly tend to have low pulse rates; it could be that the author falls into that category.

   NOTE: It is also acceptable to use $|x - \mu| = 24.9$ in part (a). This means that the answer in (b) is +2.02. The z score is the *signed* number of standard deviations from the mean.

3. In general, $z = (x - \mu)/\sigma$.
   a. $z_{60} = (60 - 69.0)/2.8 = -3.21$
   b. $z_{85} = (85 - 69.0)/2.8 = 5.71$
   c. $z_{69.72} = (69.72 - 69.0)/2.8 = 0.26$

4. In general, $z = (x - \mu)/\sigma$.
   a. $z_{100} = (100 - 98.20)/.62 = 2.90$
   b. $z_{96.96} = (96.96 - 98.20)/.62 = -2.00$
   c. $z_{98.20} = (98.20 - 98.20)/.62 = 0.00$

5. $z = (x - \mu)/\sigma$
   $z_{70} = (70 - 63.6)/2.5 = 2.56$
   Yes, that height is considered unusual since $2.56 > 2.00$.

S-44 INSTRUCTOR'S SOLUTIONS Chapter 2

6. $z = (x - \mu)/\sigma$
    $z_{308} = (308 - 268)/15 = 2.67$
    Yes, the length is considered unusual since $2.67 > 2.00$. The length is unusual, but not necessarily impossible.

7. $z = (x - \mu)/\sigma$
    $z_{101} = (101 - 98.20)/0.62 = 4.52$
    Yes, that temperature is unusually high, since $4.52 > 2.00$. It suggests that either the person is healthy but has a very unusual temperature for a healthy person, or that person is sick (i.e., has an elevated temperature attributable to some cause).

8. $z = (x - \mu)/\sigma$
    $z_{259.0} = (259.1 - 178.1)/40.7 = 1.99$
    No, this level is not considered unusually high since $1.99 < 2.00$.
    NOTE: While this level is not unusually high compared to the general population, it still might be high enough to be of concern – i.e., it could be that the levels in the general population are unhealthily high.

9. In general $z = (x - \bar{x})/s$
    psychology: $z_{85} = (85 - 90)/10 = -5/10 = -0.50$
    economics: $z_{45} = (45 - 55)/5 = -10/5 = -2.00$
    The psychology score has the better relative position since $-0.50 > -2.00$.

10. In general, $z = (x - \mu)/\sigma$.
    a. $z_{144} = (144 - 128)/34 = .47$
    b. $z_{90} = (90 - 86)/18 = .22$
    c. $z_{18} = (18 - 15)/5 = .60$
    The score in part (c) has the highest z score and therefore is the highest relative score.

11. preliminary values: $n = 36$, $\Sigma x = 29.4056$, $\Sigma x^2 = 24.02111984$
    $\bar{x} = (\Sigma x)/n = 29.4056/36 = .81682$
    $s^2 = [n(\Sigma x^2) - (\Sigma x)^2]/[n(n-1)]$
    $= [36(24.02111984) - (29.4056)^2]/[36(35)] = .07100288/1260 = .000056351$
    $s = .007507$
    $z = (x - \bar{x})/s$
    $z_{.7901} = (.7901 - .81682)/.007507 = -3.56$
    Yes; since $-3.56 < -2.00$, .7901 is an unusual weight for regular Coke.

12. For the green M&M's, computer software gives the following relevant values.
    $n=7$ $\bar{x}=.9266$ $s=.0387$ max$=1.002$
    $z = (x - \bar{x})/s$
    $z_{1.002} = (1.002 - .9266)/.0387 = 1.95$
    No; since 1.95 2.00, 1.002 is not an unusual weight for a green M&M.

13. Let b = # of scores below x; n = total number of scores.
    In general, the percentile of score x is $(b/n) \cdot 100$.
    The percentile for a cotinine level of 149 is $(17/40) \cdot 100 = 42.5$.

14. Let b = # of scores below x; n = total number of scores.
    In general, the percentile of score x is $(b/n) \cdot 100$.
    The percentile for a cotinine level of 210 is $(24/40) \cdot 100 = 60$.

15. Let b = # of scores below x; n = total number of scores.
    In general, the percentile of score x is $(b/n) \cdot 100$.
    The percentile for a cotinine level of 35 is $(6/40) \cdot 100 = 15$.

16. Let b = # of scores below x; n = total number of scores.
In general, the percentile of score x is (b/n)·100.
The percentile for a cotinine level of 250 is (29/40)·100 = 72.5

17. To find P_{20}, L = (20/100)·40 = 8 – a whole number.
The mean of the 8th and 9th scores, P_{20} = (44+48)/2 = 46.

18. To find $Q_3 = P_{75}$, L = (75/100)·40 = 30 – a whole number.
The mean of the 30th and 31st scores, P_{75} = (250+253)/2 = 251.5.

19. To find P_{75}, L = (75/100)·40 = 30 – a whole number.
The mean of the 30th and 31st scores, P_{75} = (250+253)/2 = 251.5.

20. To find $Q_2 = P_{50}$, L = (50/100)·40 = 20 – a whole number.
The mean of the 20th and 21st scores, P_{50} = (167+173)/2 = 170.

21. To find P_{33}, L = (33/100)·40 = 13.2, rounded up to 14.
Since the 14th score is 121, P_{33} = 121.

22. To find P_{21}, L = (21/100)·40 = 8.4, rounded up to 9.
Since the 9th score is 48, P_{21} = 48.

23. To find P_{01}, L = (1/100)·40 = 0.4, rounded up to 1.
Since the 1st score is 0, P_{01} = 0.

24. To find P_{85}, L = (85/100)·40 = 34 – a whole number.
The mean of the 34th and 35th scores, P_{85} = (277+284)/2 = 280.5.

NOTE: For exercises 25-36, refer to the ordered list at the right.

#	level
1	2
2	8
3	44
4	62
5	62
6	89
7	94
8	98
9	98
10	112
11	123
12	125
13	126
14	130
15	146
16	149
17	149
18	173
19	175
20	181
21	207
22	223
23	237
24	254
25	264
26	267
27	271
28	280
29	293
30	301
31	309
32	318
33	325
34	384
35	447
36	462
37	531
38	596
39	600
40	920

25. Let b = # of scores below x; n = total number of scores
In general, the percentile of score x is (b/n)·100.
The percentile of score 123 is (10/40)·100 = 25.

26. Let b = # of scores below x; n = total number of scores
In general, the percentile of score x is (b/n)·100.
The percentile of score 309 is (30/40)·100 = 75.

27. Let b = # of scores below x; n = total number of scores
In general, the percentile of score x is (b/n)·100.
The percentile of score 271 is (26/40)·100 = 65.

28. Let b = # of scores below x; n = total number of scores
In general, the percentile of score x is (b/n)·100.
The percentile of score 126 is (12/40)·100 = 30.

29. To find P_{85}, L = (85/100)·40 = 34 – a whole number.
The mean of the 34th and 35th scores, P_{85} = (384+447)/2 = 415.5.

30. To find P_{35}, L = (35/100)·40 = 14 – a whole number.
The mean of the 14th and 15th scores, P_{35} = (130+146)/2 = 138.

31. To find $Q_1 = P_{25}$, L = (25/100)·40 = 10 – a whole number.
The mean of the 10th and 11th scores, P_{25} = (112+123)/2 = 117.5.

S-46 INSTRUCTOR'S SOLUTIONS Chapter 2

32. To find $Q_3 = P_{75}$, $L = (75/100) \cdot 40 = 30$ – a whole number.
The mean of the 30th and 31st scores, $P_{75} = (301+309)/2 = 305$.

33. To find P_{18}, $L = (18/100) \cdot 40 = 7.2$ rounded up to 8.
Since the 8th score is 98, $P_{18} = 98$.

34. To find P_{36}, $L = (36/100) \cdot 40 = 14.4$ rounded up to 15.
Since the 15th score is 146, $P_{36} = 146$.

35. To find P_{58}, $L = (58/100) \cdot 40 = 23.2$ rounded up to 24.
Since the 23rd score is 237, $P_{58} = 254$.

36. To find P_{96}, $L = (96/100) \cdot 40 = 38.4$ rounded up to 39.
Since the 39th score is 600, $P_{96} = 600$.

37. In general, z scores are not affected by the particular unit of measurement that is used. The relative position of a score (whether it is above or below the mean, its rank in an ordered list of the scores, etc.) is not affected by the unit of measurement, and relative position is what a z score communicates. Mathematically, the same units (feet, centimeters, dollars, etc.) appear in both the numerator and denominator of $(x-\mu)/\sigma$ and cancel out to leave the z score unit-free. In fact the z score is also called the *standard score* for that very reason – it is a standardized value that is independent of the unit of measure employed.

NOTE: In more technical language, changing from one unit of measure to another (feet to centimeters, °F to °C, dollars to pesos, etc.) is a linear transformation – i.e., if x is the score in one unit, then (for some appropriate values of a and b) $y = ax + b$ is the score in the other unit. In such cases it can be shown (see exercise 2.5 #36) that
$\mu_y = a\mu_x + b$ and
$\sigma_y = a\sigma_x$.
The new z score is the same the old one, since
$z_y = (y - \mu_y)/\sigma_y = [(ax + b) - (a\mu_x + b)]/a\sigma_x = [ax - a\mu_x]/a\sigma_x = (x - \mu_x)/\sigma_x = z_x$

38. Solve $z = (x - \mu)/\sigma$ to get
$x = \mu + z\sigma$
a. $x_{2.16} = 63.6 + (2.16)(2.5) = 63.6 + 5.4 = 69.0$ inches
b. $x_{-1.84} = 63.6 + (-1.84)(2.5) = 63.6 - 4.6 = 59.0$ inches

39.
a. uniform
b. bell-shaped
c. In general, the shape of a distribution of not affected if all values are converted to z scores. The shape of a distribution does not depend on the unit of measurement -- e.g., if the heights of a group of people form a bell-shaped distribution, they will do so whether measured in inches or centimeters. For a histogram, for example, using a different unit of measurement simply requires a re-labeling of the horizontal axis but does not change the shape of the figure. Since z scores are merely a re-labeling to standard units, converting all the scores to z scores will not change the shape of a distribution.

40. a. $n = 7$, $\Sigma x = 33$, $\Sigma x^2 = 273$
$\bar{x} = \Sigma x/n = 33/7 = 4.714$
$s^2 = [n(\Sigma x^2) - (\Sigma x)^2]/[n(n-1)]$
$= [7(273) - (33)^2]/[7(6)]$
$= (822)/42 = 19.571$
$s = 4.424$

x	$z = (x - \bar{x})/s$
1	-0.839584
1	-0.839584
2	-0.613542
3	-0.387500
5	0.064583
8	0.742709
13	1.872918
33	0

b. $n = 7$, $\Sigma z = 0$, $\Sigma z^2 = 6$
$\bar{z} = \Sigma z/n = 0/7 = 0$
$s^2 = [n(\Sigma z^2) - (\Sigma z)^2]/[n(n-1)] = [7(6) - (0)^2]/[7(6)] = (42)/42 = 1$
$s = 1$

c. Yes; as shown below, $\bar{z} = 0$ and $s_z = 1$ will be true for *any* set of z scores.
$\Sigma z = \Sigma(x-\bar{x})/s = (1/s) \cdot [\Sigma(x-\bar{x})] = (1/s) \cdot [\Sigma x - \Sigma \bar{x}] = (1/s) \cdot [n\bar{x} - n\bar{x}] = (1/s) \cdot 0 = 0$
$\Sigma z^2 = \Sigma[(x-\bar{x})/s]^2 = (1/s^2) \cdot [\Sigma(x-\bar{x})^2] = (1/s^2) \cdot [(n-1)s^2] = n-1$
$\bar{z} = (\Sigma z)/n = 0/n = 0$
$s_z^2 = [n \cdot \Sigma z^2 - (\Sigma z)^2]/[n(n-1)] = [n \cdot (n-1) - 0^2]/[n(n-1)] = 1$; $s_z = \sqrt{1} = 1$

41. a. The interquartile range is $Q_3 - Q_1$.
 For $Q_3 = P_{75}$, $L = (75/100) \cdot 40 = 30$ – a whole number.
 The mean of the 30th and 31st scores, $Q_3 = (250+253)/2 = 251.5$.
 For $Q_1 = P_{25}$, $L = (25/100) \cdot 40 = 10$ – a whole number.
 The mean of the 10th and 11th scores, $Q_1 = (86+87)/2 = 86.5$.
 The interquartile range is $251.5 - 86.5 = 165$.
 b. The midquartile is $(Q_1 + Q_3)/2 = (251.5+86.5)/2 = 169$.
 c. The 10-90 percentile range is $P_{90} - P_{10}$.
 For P_{90}, $L = (90/100) \cdot 40 = 36$ – a whole number.
 The mean of the 36th and 37th scores, $P_{90} = (289+290)/2 = 289.5$.
 For P_{10}, $L = (10/100) \cdot 40 = 4$ – a whole number.
 The mean of the 4th and 5th scores, $P_{10} = (3+17)/2 = 10$.
 The 10-90 percentile range is $289.5 - 10 = 279.5$
 d. Yes, $Q_2 = P_{50}$ by definition. Yes, they are always equal.
 e. For $Q_2 = P_{50}$, $L = (50/100) \cdot 40 = 20$ – a whole number.
 The mean of the 20th and 21st scores, $Q_2 = (167+173)/2 = 170$.
 No; in this case $170 = Q_2 \neq (Q_1 + Q_3)/2 = 169$, which demonstrates that the median does not necessarily equal the midquartile.

42. a. To find P_{35}, $L = (35/100) \cdot 54 = 18.9$.
 $P_{35} = x_{18.9} = x_{18} + .9 \cdot (x_{19} - x_{18}) = 114 + .9 \cdot (116 - 114) = 115.8$
 b. To find $Q_1 = P_{25}$, $L = (25/100) \cdot 54 = 13.5$.
 $Q_1 = x_{13.5} = x_{13} + .5 \cdot (x_{14} - x_{13}) = 80 + .5 \cdot (86 - 80) = 83.0$

43. a. $D_1 = P_{10}$, $D_5 = P_{50}$, $D_8 = P_{80}$
 b. In each case D_d, $L = (10d/100) \cdot 40 = 4d$ – a whole number.
 $D_d = (x_{4d}+x_{4d+1})/2$ for $d = 1,2,3,4,5,6,7,8,9$
 $D_1 = (x_4+x_5)/2 = (3+17)/2 = 10.0$
 $D_2 = (x_8+x_9)/2 = (44+48)/2 = 46.0$
 $D_3 = (x_{12}+x_{13})/2 = (103+112)/2 = 107.5$
 $D_4 = (x_{16}+x_{17})/2 = (130+131)/2 = 130.5$
 $D_5 = (x_{20}+x_{21})/2 = (167+173)/2 = 170.0$
 $D_6 = (x_{24}+x_{25})/2 = (208+210)/2 = 209.0$
 $D_7 = (x_{28}+x_{29})/2 = (234+245)/2 = 239.5$
 $D_8 = (x_{32}+x_{33})/2 = (265+266)/2 = 265.5$
 $D_9 = (x_{36}+x_{37})/2 = (289+290)/2 = 289.5$
 c. In each case Quintile_q, $L = (20q/100) \cdot 40 = 8q$ – a whole number.
 $\text{Quintile}_q = (x_{8d}+x_{8d+1})/2$ for $q = 1,2,3,4$
 $\text{Quintile}_1 = (x_8+x_9)/2 = (44+48)/2 = 46.0$
 $\text{Quintile}_2 = (x_{16}+x_{17})/2 = (130+131)/2 = 130.5$
 $\text{Quintile}_3 = (x_{24}+x_{25})/2 = (208+210)/2 = 209.0$
 $\text{Quintile}_4 = (x_{32}+x_{33})/2 = (265+266)/2 = 265.5$

S-48 INSTRUCTOR'S SOLUTIONS Chapter 2

2-7 Exploratory Data Analysis (EDA)

The exercises in this section may be done much more easily when ordered lists of the values are available. On the next page appear ordered lists for the data of exercises #1-8. The left-most column gives the ordered ID numbers 1-50. Data sets with more than 50 values (except for the 54 bear lengths in exercise #6) have more than one column – the second column being ordered values 51-100, etc.

NOTE: A boxplot can be misleading if an extreme value artificially extends the whisker beyond the reasonable data. This manual follows the convention of extending the whisker only to the highest and lowest values within $1.5(Q_3 - Q_1)$ of Q_2 – any point more than 1.5 times the width of the box away from the median will be considered an outlier and represented separately. This type of boxplot is discussed in exercise #13. The boxplots given in this manual can be converted to the ones presented in the text by extending the whisker to the most extreme outlier.

1. Consider the 40 digits.
   For $Q_1 = P_{25}$, $L = (25/100) \cdot 40 = 10$ – a whole number, use 10.5.
   For $\tilde{x} = Q_2 = P_{50}$, $L = (50/100) \cdot 40 = 20$ – a whole number, use 20.5.
   For $Q_3 = P_{75}$, $L = (75/100) \cdot 40 = 30$ – a whole number, use 30.5.
   $\min = x_1 = 0$
   $Q_1 = (x_{10} + x_{11})/2 = (2+2)/2 = 2$
   $Q_2 = (x_{20} + x_{21})/2 = (5+5)/2 = 5$
   $Q_3 = (x_{30} + x_{31})/2 = (7+7)/2 = 7$
   $\max = x_{40} = 9$

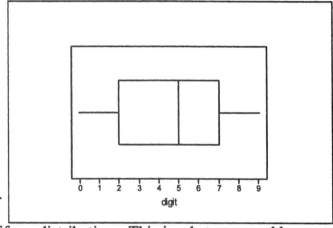

   Since the quartiles (indicated by the vertical bars) divide the figure into four approximately equal segments, the boxplot suggests the values follow a uniform distribution. This is what one would expect if the digits were selected with a random procedure.

2. Consider the 15 movie budgets, as given in millions of dollars.
   For $Q_1 = P_{25}$, $L = (25/100) \cdot 15 = 3.75$ rounded up to 4.
   For $\tilde{x} = Q_2 = P_{50}$, $L = (50/100) \cdot 15 = 7.5$ rounded up to 8.
   For $Q_3 = P_{75}$, $L = (75/100) \cdot 15 = 11.25$ rounded up to 12.
   $\min = x_1 = .25$
   $Q_1 = x_4 = 8$
   $Q_2 = x_8 = 18.5$
   $Q_3 = x_{12} = 70$
   $\max = x_{15} = 100$

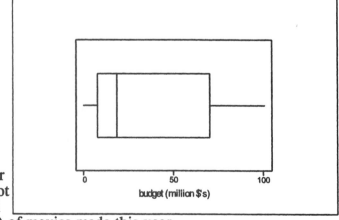

   Since some the films date from the 1970's, and the budgets presumably are the actual amounts spent in the year of production, the sample values are not likely to be representative of the actual dollar amounts (i.e., in current dollars) of movies made this year.

The following ordered lists are used for exercises #1-8.

	#1 dig	#2 bud	#3 cal	#4 nic	#5 wgt	#6 lgn	#7 alc	#8 temp	#8 temp	#8 temp
01	0	0.250	3.3	0.1	.870	36.0	0	96.5	98.3	99.1
02	0	0.325	3.5	0.2	.872	37.0	0	96.9	98.3	99.2
03	0	6.400	3.6	0.5	.874	40.0	0	97.0	98.4	99.4
04	0	8.000	3.6	0.6	.882	40.0	0	97.0	98.4	99.4
05	0	10.000	3.6	0.7	.888	41.0	0	97.0	98.4	99.5
06	0	15.000	3.6	0.7	.891	43.0	0	97.1	98.4	99.6
07	1	17.000	3.7	0.8	.897	43.5	0	97.1	98.4	
08	1	18.500	3.7	0.8	.898	46.0	0	97.1	98.4	
09	1	25.000	3.8	0.8	.908	46.0	0	97.2	98.4	
10	2	50.000	3.9	0.8	.908	47.0	0	97.3	98.4	
11	2	55.000	3.9	0.9	.908	48.0	0	97.3	98.4	
12	2	70.000	3.9	1.0	.911	48.0	0	97.3	98.4	
13	2	70.000	4.0	1.0	.912	49.0	0	97.4	98.4	
14	3	72.000	4.0	1.0	.913	50.0	0	97.4	98.4	
15	3	100.000	4.0	1.0	.920	52.0	0	97.4	98.5	
16	4		4.1	1.0	.924	52.5	0	97.5	98.5	
17	4			1.0	.924	53.0	0	97.5	98.5	
18	4			1.0	.933	53.0	0	97.6	98.5	
19	5			1.1	.936	54.0	0	97.6	98.6	
20	5			1.1	.952	57.3	0	97.6	98.6	
21	5			1.1	.983	57.5	0	97.6	98.6	
22	5			1.2		58.0	0	97.6	98.6	
23	5			1.2		59.0	0	97.6	98.6	
24	5			1.2		59.0	0	97.7	98.6	
25	5			1.2		59.0	0	97.8	98.6	
26	6			1.2		60.0	3	97.8	98.6	
27	6			1.3		60.5	4	97.8	98.6	
28	6			1.4		61.0	5	97.8	98.6	
29	7			1.4		61.0	7	97.8	98.6	
30	7					61.5	8	97.9	98.6	
31	7					62.5	13	97.9	98.6	
32	7					63.0	20	97.9	98.6	
33	7					63.0	28	98.0	98.6	
34	8					63.0	33	98.0	98.7	
35	8					63.5	34	98.0	98.7	
36	9					64.0	38	98.0	98.7	
37	9					64.0	39	98.0	98.7	
38	9					64.0	39	98.0	98.7	
39	9					65.0	46	98.0	98.7	
40	9					65.0	51	98.0	98.8	
41						66.5	72	98.0	98.8	
42						67.0	73	98.0	98.8	
43						67.5	74	98.0	98.8	
44						68.5	76	98.0	98.8	
45						70.0	80	98.0	98.8	
46						70.5	88	98.2	98.8	
47						72.0	113	98.2	98.9	
48						72.0	123	98.2	98.9	
49						72.0	142	98.2	99.0	
50						72.0	414	98.2	99.0	
						73.0				
						73.5				
						75.0				
						76.5				

3. Consider the 16 calorie amounts.
   For $Q_1 = P_{25}$, L = (25/100)·16 = 4 – a whole number, use 4.5.
   For $\tilde{x} = Q_2 = P_{50}$, L = (50/100)·16 = 8 – a whole number, use 8.5.
   For $Q_3 = P_{75}$, L = (75/100)·16 = 12 – a whole number, use 12.5.
   min = x_1 = 3.3
   $Q_1 = (x_4+x_5)/2 = (3.6+3.6)/2 = 3.6$
   $Q_2 = (x_8+x_9)/2 = (3.7+3.8)/2 = 3.75$
   $Q_3 = (x_{12}+x_{13})/2 = (3.9+4.0)/2 = 3.95$
   max = x_{16} = 4.1

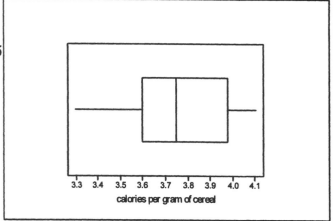

These values are not necessarily representative of the cereals consumed by the general population. Although cereals like "Corn Flakes" and "Bran Flakes" are not identified by manufacturer, it is likely that the values given are for cereals produced by the major companies (General Mills, Kellogg's, etc) and not the less expensive generic imitations used by many consumers. In addition, each cereals is given equal weight in the data set – regardless of how popular (or unpopular) it is with consumers.

4. Consider the 29 nicotine amounts.
   For $Q_1 = P_{25}$, L = (25/100)·29 = 7.25 rounded up to 8.
   For $\tilde{x} = Q_2 = P_{50}$, L = (50/100)·29 = 14.5 rounded up to 15.
   For $Q_3 = P_{75}$, L = (75/100)·29 = 21.75 rounded up to 22.
   min = x_1 = 0.1
   $Q_1 = x_8 = 0.8$
   $Q_2 = x_{15} = 1.0$
   $Q_3 = x_{22} = 1.2$
   max = x_{29} = 1.4

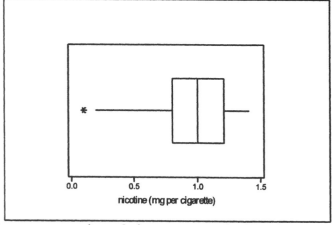

No; the sample values are not likely to be representative of cigarettes smoked by "an individual consumer" – no matter how that phrase is interpreted. The values would not be representative of cigarettes smoked by any one individual, since most consumers are faithful to a single brand and would not experience such a wide range of nicotine content. The values are not representative of those smoked by consumers in general because they include only cigarettes that are "100 mm long, filtered, and not menthol or light." In addition, each cigarette brand is given equal weight in the data set – regardless of how popular (or unpopular) it is with consumers.

5. Consider the 21 weights.
   For $Q_1 = P_{25}$, L = (25/100)·21 = 5.25 rounded up to 6.
   For $\tilde{x} = Q_2 = P_{50}$, L = (50/100)·21 = 10.5 rounded up to 11.
   For $Q_3 = P_{75}$, L = (75/100)·21 = 15.75 rounded up to 16.

min = x_1 = .870
Q_1 = x_6 = .891
Q_2 = x_{11} = .908
Q_3 = x_{16} = .924
max = x_{21} = .983

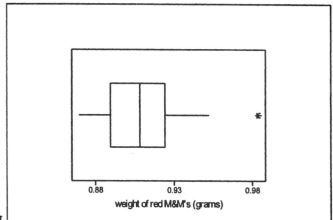

Yes; assuming there is no significant difference in the weights of the coloring agents, the weights of the red M&M's are likely to be representative of the weights of all the colors.

6. Consider the 54 lengths.
   For Q_1 = P_{25}, L = (25/100)·54 = 13.5 rounded up to 14.
   For $\tilde{x}$ = Q_2 = P_{50}, L = (50/100)·54 = 27 – a whole number, use 27.5
   For Q_3 = P_{75}, L = (75/100)·54 = 40.5 rounded up to 41.

min = x_1 = 36.0
Q_1 = x_{14} = 50.0
Q_2 = $(x_{27}+x_{28})2$ = (60.5+61.0)/2
     = 60.75
Q_3 = x_{41} = 66.5
max = x_{54} = 76.5

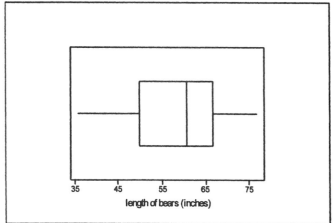

While the median is slightly toward the higher end (indicating a slight negative skew), the distribution is approximately symmetric.

7. Consider the 50 times.
   For Q_1 = P_{25}, L = (25/100)·50 = 12.5 rounded up to 13.
   For $\tilde{x}$ = Q_2 = P_{50}, L = (50/100)·50 = 25 -- an integer, use 25.5.
   For Q_3 = P_{75}, L = (75/100)·50 = 37.5 rounded up to 38.

min = x_1 = 0
Q_1 = x_{13} = 0
Q_2 = $x_{25.5}$ = (0+3)/2 = 1.5
Q_3 = x_{38} = 39
max = x_{50} = 414

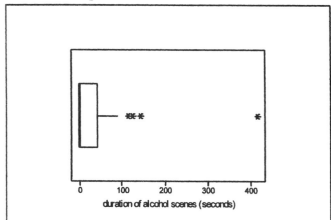

Based on the boxplot, the distribution appears to be extremely positively skewed.

8. Consider the 106 temperatures.
   For $Q_1 = P_{25}$, $L = (25/100) \cdot 106 = 26.5$ rounded up to 27.
   For $\tilde{x} = Q_2 = P_{50}$, $L = (50/100) \cdot 106 = 53$ -- an integer, use 53.5.
   For $Q_3 = P_{75}$, $L = (75/100) \cdot 106 = 79.5$ rounded up to 80.
   min = x_1 = 96.5
   $Q_1 = x_{27} = 97.8$
   $Q_2 = x_{53.5} = (98.4 + 98.4)/2 = 98.4$
   $Q_3 = x_{80} = 98.6$
   max = x_{106} = 99.6

   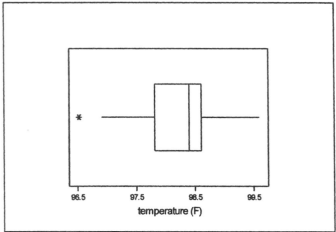

   The values are not centered around the 98.6, which is actually P_{75}.

On the next page appear ordered lists for the data of exercises #9-12. They follow the same format as the ordered lists given previously for exercises #1-8.

9. Consider the 39 actor and 39 actress values.
   For $Q_1 = P_{25}$, $L = (25/100) \cdot 39 = 9.75$ rounded up to 10.
   For $\tilde{x} = Q_2 = P_{50}$, $L = (50/100) \cdot 39 = 19.5$ rounded up to 20.
   For $Q_3 = P_{75}$, $L = (75/100) \cdot 39 = 29.25$ rounded up tp 30.
   For the actors,
   min = x_1 = 31
   $Q_1 = x_{10} = 37$
   $Q_2 = x_{20} = 43$
   $Q_3 = x_{30} = 51$
   max = x_{39} = 76
   For the actresses,
   min = x_1 = 21
   $Q_1 = x_{10} = 30$
   $Q_2 = x_{20} = 34$
   $Q_3 = x_{30} = 41$
   max = x_{39} = 80

   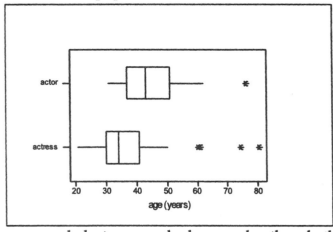

   The ages for the actresses cover a wider range and cluster around a lower value than do the ages of the actors.

10. Consider the 36 regular and 36 diet values.
    For $Q_1 = P_{25}$, $L = (25/100) \cdot 36 = 9$ – an integer, use 9.5.
    For $\tilde{x} = Q_2 = P_{50}$, $L = (50/100) \cdot 38 = 18$ – an integer, use 18.5.
    For $Q_3 = P_{75}$, $L = (75/100) \cdot 36 = 27$ – an integer, use 27.5.
    For the regular Coke,
    min = x_1 = .7901
    $Q_1 = x_{9.5} = (.8143 + .8150)/2 = .81465$
    $Q_2 = x_{18.5} = (.8170 + .8172)/2 = .8171$
    $Q_3 = x_{27.5} = (.8207 + .8211)/2 = .8209$
    max = x_{36} = .8295

The following ordered lists are used for exercises #9-12.

	#9 mal	#9 fem	#10 regu	#10 diet	#11 smo	#11 ets	#11 no	#12 clan	#12 rowl	#12 tols
01	31	21	.7901	.7758	0	0	0	43.9	70.9	51.9
02	32	24	.8044	.7760	1	0	0	58.2	74.0	52.8
03	32	25	.8062	.7771	1	0	0	64.4	78.6	58.4
04	32	26	.8073	.7773	3	0	0	69.4	79.2	64.2
05	33	26	.8079	.7802	17	0	0	72.7	79.5	65.4
06	35	26	.8110	.7806	32	0	0	72.9	80.2	68.5
07	36	26	.8126	.7811	35	0	0	73.1	82.5	69.4
08	36	27	.8128	.7813	44	0	0	73.4	83.7	71.4
09	37	28	.8143	.7822	48	0	0	76.3	84.3	71.6
10	37	30	.8150	.7822	86	1	0	76.4	84.6	72.2
11	38	30	.8150	.7822	87	1	0	79.8	85.3	73.6
12	39	31	.8152	.7826	103	1	0	89.2	86.2	74.4
13	39	31	.8152	.7830	112	1	0			
14	40	33	.8161	.7833	121	1	0			
15	40	33	.8161	.7837	123	1	0			
16	40	33	.8163	.7839	130	1	0			
17	41	34	.8165	.7844	131	1	0			
18	42	34	.8170	.7852	149	1	0			
19	42	34	.8172	.7852	164	1	0			
20	43	34	.8176	.7859	167	1	0			
21	43	35	.8181	.7861	173	2	0			
22	44	35	.8189	.7868	173	2	0			
23	45	35	.8192	.7870	198	3	0			
24	45	37	.8192	.7872	208	3	0			
25	46	37	.8194	.7874	210	3	0			
26	46	38	.8194	.7874	222	4	0			
27	47	39	.8207	.7879	227	13	0			
28	48	41	.8211	.7879	234	13	0			
29	48	41	.8299	.7881	245	17	0			
30	51	41	.8244	.7885	250	19	0			
31	53	42	.8244	.7892	253	45	0			
32	55	44	.8247	.7896	265	51	0			
33	56	49	.8251	.7907	266	69	0			
34	56	50	.8264	.7910	277	74	0			
35	60	60	.8284	.7923	284	178	1			
36	60	61	.8295	.7923	289	197	1			
37	61	61			290	241	9			
38	62	74			313	384	90			
39	76	80			477	543	244			
40					491	551	309			

For the diet Coke,
  min = x_1 = .7758
  Q_1 = $x_{9.5}$ = (.7822+.7822)/2
       = .7822
  Q_2 = $x_{18.5}$ = (.7852+.7852)/2
       = .7852
  Q_3 = $x_{27.5}$ = (.7879+.7879)/2
       = .7879
  max = x_{36} = .7923

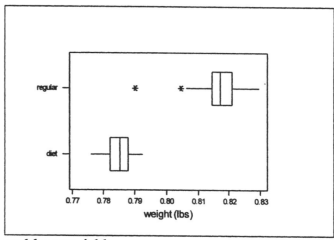

The weights for the diet sodas are lower and less variable.

11. Consider the 40 values for each group.
  For Q_1 = P_{25}, L = (25/100)·40 = 10 – an integer, use 10.5.
  For $\tilde{x}$ = Q_2 = P_{50}, L = (50/100)·40 = 20 – an integer, use 20.5.
  For Q_3 = P_{75}, L = (75/100)·40 = 30 – an integer, use 30.5.
For the smokers,
  min = x_1 = 0
  Q_1 = $x_{10.5}$ = (86+87)/2 = 86.5
  Q_2 = $x_{20.5}$ = (167+173)/2 = 170
  Q_3 = $x_{30.5}$ = (250+253)/2 = 251.5
  max = x_{40} = 491
For the ETS group,
  min = x_1 = 0
  Q_1 = $x_{10.5}$ = (1+1)/2 = 1
  Q_2 = $x_{20.5}$ = (1+2)/2 = 1.5
  Q_3 = $x_{30.5}$ = (19+45)/2 = 32
  max = x_{40} = 551
For the no ETS group,
  min = x_1 = 0
  Q_1 = $x_{10.5}$ = (0+0)/2 = 0
  Q_2 = $x_{20.5}$ = (0+0)/2 = 0
  Q_3 = $x_{30.5}$ = (0+0)/2 = 0
  max = x_{40} = 309

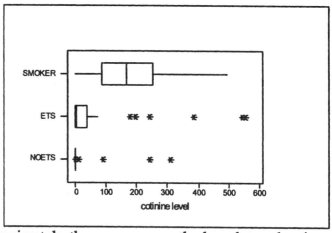

Even though all three groups cover approximately the same range, the boxplot makes it clear that the typical values are highest for the smokers and essentially zero (except for outliers) for the no ETS group.

12. Consider the 12 values for each group.
  For Q_1 = P_{25}, L = (25/100)·12 = 3 – an integer, use 3.5.
  For $\tilde{x}$ = Q_2 = P_{50}, L = (50/100)·12 = 6 – an integer, use 6.5.
  For Q_3 = P_{75}, L = (75/100)·12 = 9 – an integer, use 9.5.
For the Clancy book.
  min = x_1 = 43.9
  Q_1 = $x_{3.5}$ = (64.4+69.4)/2 = 66.9
  Q_2 = $x_{6.5}$ = (72.9+73.1)/2 = 73.0
  Q_3 = $x_{9.5}$ = (76.3+76.4)/2 = 76.35
  max = x_{12} = 89.2

For the Rowling book,
 min = x_1 = 70.9
 Q_1 = $x_{3.5}$ = (78.6+79.2)/2 = 78.9
 Q_2 = $x_{6.5}$ = (80.2+82.5)/2 = 81.35
 Q_3 = $x_{9.5}$ = (84.3+84.6)/2 = 84.45
 max = x_{12} = 86.2
For the Tolstoy book,
 min = x_1 = 51.9
 Q_1 = $x_{3.5}$ = (58.4+64.2)/2 = 61.3
 Q_2 = $x_{6.5}$ = (68.5+69.4)/2 = 68.95
 Q_3 = $x_{9.5}$ = (71.6+72.2)/2 = 71.9
 max = x_{12} = 74.4

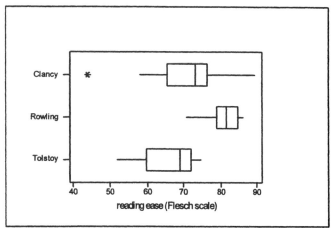

Yes; there does appear to be a difference in ease of reading. Yes; considering that high scores indicate easier reading and the intended audience of each author, the results are consistent with what most people would expect – Rowling writes for children, Clancy writes for as broad a range of adults as possible (trying to sell as many books as possible), Tolstoy writes for the more educated adults (caring more about producing a work of literature than selling it).

13. The following given values are also calculated in detail in exercise #11.
 Q_1 = 86.5 Q_2 = 170 Q_3 = 251.5
 a. IQR = Q_3 - Q_1 = 251.5 - 86.5 = 165
 b. Since 1.5(IQR) = 1.5(165) = 247.5, the modified boxplot line extends to
  the smallest value above Q_1 - 247.5 = 86.5 - 247.5 = -161, which is 0
  the largest value below Q_3+30 = 251.5+247.5 = 499, which is 491
 c. Since 3.0(IQR) = 3.0(165) = 495, mild outliers are x values for which
  Q_1 - 495 ≤ x < Q_1 - 247.5   or Q_3+247.5 < x ≤ Q_3+495
  -408.5 ≤ x < -161   499 < x ≤ 746.5
  There are no such values.
 d. Extreme outliers in this exercise are x values for which x < -408.5 or x > 746.5.
  There are no such values.

14. Choose battery #1 because it has the highest typical value (60) and the lowest variation (R = 29).

Review Exercises

1. The scores arranged in order are:
   42 43 46 46 47 48 49 49 50 51 51 51 51 51 52 52 54 54 54 54 54
   55 55 55 55 56 56 56 57 57 57 57 58 60 61 61 61 62 64 64 65 68 69
   preliminary values: n = 43, Σx = 2358, Σx^2 = 130,930
   a. $\bar{x}$ = (Σx)/n = (2358)/43 = 54.8
   b. $\tilde{x}$ = 55
   c. M = 51,54 [bi-modal]
   d. m.r. = (42 + 69)/2 = 55.5
   e. R = 69 - 42 = 27
   f. s = 6.2 (from part g)
   g. s^2 = [n(Σx^2) - ($\Sigma x)^2$]/[n(n-1)]
      = [43(130,930) - (2358)^2]/[43(42)]
      = (69,826)/1806 = 38.7
   h. For Q_1 = P_{25}, L = (25/100)·43 = 10.75 rounded up to 11. And so Q_1 = x_{11} = 51.

i. For $Q_3 = P_{75}$, $L = (75/100) \cdot 43 = 32.25$ rounded up to 33. And so $Q_3 = x_{33} = 58$.
j. For P_{10}, $L = (10/100) \cdot 43 = 4.3$ rounded up to 5. And so $P_{10} = x_5 = 47$.

2. a. $z = (x - \bar{x})/s$
   $z_{43} = (43 - 54.837)/6.218 = -1.90$
   b. No; Kennedy's inaugural age is not unusual since $-2.00 < -1.90 < -2.00$.
   c. According to the Range Rule of Thumb, the usual values are within 2s of $\bar{x}$
   usual minimum: $\bar{x} - 2s = 54.8 - 2(6.2) = 42.4$
   usual maximum: $\bar{x} + 2s = 54.8 + 2(6.2) = 67.2$
   The ages 42, 68 and 69 are unusual.
   d. Yes; an age of 35 would be unusual – even if $\bar{x}$ and s were recalculated with the value 35 included. It is likely that a presidential candidate of age 35 would find that age would be a campaign issue – especially during the primaries and other preliminary stages of the campaign, but probably not so much so if the candidate wins the nomination of a major party. Presumably a candidate who wins the nomination of a major party will have enough positive qualities to overcome an unusual age - after all, Ronald Reagan (the 69 on the list) was elected despite his "unusual" age.

3.
age	frequency
40 - 44	2
45 - 49	6
50 - 54	13
55 - 59	12
60 - 64	7
65 - 69	3
	43

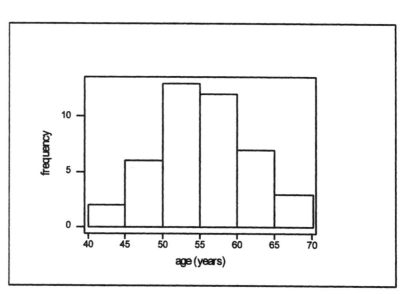

4. See the figure above at the right.
   NOTE: Unlike other data, age is not reported to the nearest unit. The bars in a histogram extend from class boundary to class boundary. Because of the way that ages are reported, the boundaries here are 40, 45, 50, etc -- i.e., someone 44.9 years old still reports an age of 44 and crosses into the next class only upon turning 45, not upon turning 44.5.

5. From results in exercise #1,
   min = x_1 = 42
   $Q_1 = x_{11} = 51$
   $Q_2 = x_{22} = 55$
   $Q_3 = x_{33} = 58$
   max = x_{43} = 69

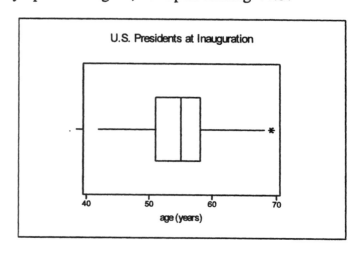

6. a. Since scores 48.6 to 61.0 are within 1·s of $\bar{x}$, the Empirical Rule for Data with a Bell-Shaped Distribution states that about 68% of such persons fall within those limits.
   b. Since scores 42.4 to 67.2 are within 2·s of $\bar{x}$, the Empirical Rule for Data with a Bell-Shaped Distribution states that about 95% of such persons fall within those limits.

7. In general, $z = (x - \bar{x})/s$.
   management, $z_{72} = (72 - 80)/12 = -0.67$
   production, $z_{19} = (19 - 20)/5 = -0.20$
   The score on the test for production employees has the better relative position since $-0.20 > -0.67$.

8. In general, the Range Rule of Thumb states that the range typically covers about 4 standard deviations -- with the lowest and highest scores being about 2 standard deviations below and above the mean respectively. While answers will vary, the following answers assume the cars driven by students have ages that range from $x_1 = 0$ years to $x_n = 12$ years.
   a. The estimated mean age is $(x_1 + x_n)/2 = (0 + 12)/2 = 6$ years.
   b. The estimated standard deviation of the ages is $R/4 = (x_n - x_1)/4 = (12 - 0)/4 = 3$ years.

9. Adding the same value to every time will increase all the times by the same amount. This changes the location of the times on the number line, but it does not affect the spread of the times or the shape of their distribution -- i.e., each measure of center will change by the amount added, but the measures of variation will not be affected.
   a. Since adding 5 minutes to every time moves all the times 5 units up on the number line, the mean will increase by 5 to $135 + 5 = 140$ minutes.
   b. Since adding 5 minutes to every time does not affect the spread of the times, the standard deviation will not change and remain 15 minutes.
   c. Since adding 5 minutes to every time does not affect the spread of the times, the variance will not change and will remain $15^2 = 225$.

10. Arranging the categories in decreasing order by frequency produces the following figure.

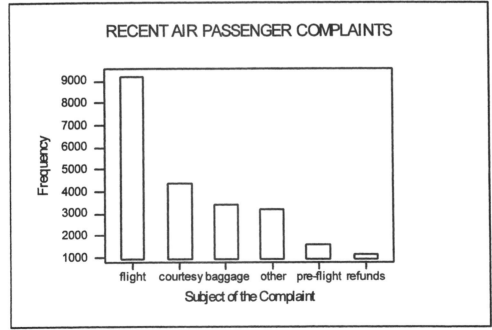

S-58   INSTRUCTOR'S SOLUTIONS Chapter 2

Cumulative Review Exercises

1. The scores arranged in order are:
   -241 -151 -125 -85 -65 20 20 27 30 41 80 105 140 186 325
   preliminary values: n = 15, Σx = 307, Σx^2 = 289313
   a. $\bar{x}$ = (Σx)/n = (307)/15 = 20.5
      $\tilde{x}$ = 27
      M = 20
      m.r. = (-241 + 325)/2 = 42
   b. R = 325 - (-241) = 566
      s^2 = [n(Σx^2) - (Σx)^2]/[n(n-1)]
         = [15(289313) - (307)^2]/[15(14)]
         = (4245446)/210 = 20216.4
      s = 142.2
   c. continuous. Even though the values were reported rounded to whole seconds, time can be any value on a continuum.
   d. ratio. Differences are consistent and there is a meaningful zero; 8 seconds is twice as much time as 4 seconds.

2. a. mode. The median requires at least ordinal level data, and the mean and the midrange require at least interval level data.
   b. convenience. The group was not selected by any process other than the fact that they happened to be the first names on the list.
   c. cluster. The population was divided into units (election precincts), some of which were selected at random to be examined in their entirety.
   d. The standard deviation, since it is the only measure of variation in the list. The standard deviation should be lowered to improve the consistency of the product.

3. No; using the mean of the state values counts each state equally, while states with more people will have a greater affect on the per capita consumption for the population in all 50 states combined. Use the state populations as weights to find the weighted mean.

Chapter 3

Probability

3-2 Fundamentals

1. A probability value is a number between 0 and 1 inclusive.
   a. 50-50 = 50% = .50
   b. 20% = .20
   c. no chance = 0% = 0

2. A probability value is a number between 0 and 1 inclusive.
   a. 90% = .90
   b. definitely = 100% = 1
   c. 1/10 = .10

3. Since $0 \leq P(A) \leq 1$ is always true, the following values less than 0 or greater than 1 cannot be probabilities.
   values less than 0: -1
   values greater than 1: 2 5/3 $\sqrt{2}$

4. a. 1.00   b. 0   c. 1/10   d. 1/2   e. 1/5

5. Assuming boys (B) and girls (G) are equally likely, the 8 equally likely possibilities for a family of 3 children are given at the right. Rule #2 can be used to find the desired probabilities as follows.
   a. P(1G) = 3/8
   b. P(2G's) = 3/8
   c. P(3G's) = 1/8

   BBB
   GBB
   BGB
   BBG
   GGB
   GBG
   BGG
   GGG

6. Let C = developing cancer of the brain or nervous system; use Rule #1
   P(C) = 135/420,000 = .000321
   No; this is not very different form the general population value of .000340. This suggests that cell phones are not the cause of such cancers.

7. Let H = getting a home run.
   P(H) = 73/476 = .1534
   Yes; this is very different from his lifetime probability of 567/7932 = .0715. It is about twice as high.

8. Let L = being struck by lightning; use Rule #1.
   P(L) = 389/281,421,906 = .00000138

9. Consider only the 85 women who were pregnant.
   a. Let W = the test wrongly concludes a woman is not pregnant.
      P(W) = 5/85 = .059
   b. No; since .059 > .05, it is not unusual for the test to be wrong for pregnant women.

10. Consider only the 14 women who were not pregnant.
    a. Let W = the test wrongly concludes a woman is pregnant.
       P(W) = 3/14 = .214
    b. No; since .214 > .05, it is not unusual for the test to be wrong for women who are not pregnant.

11. Let N = selecting someone who feels that secondhand smoke is not at all harmful.
    a. P(N) = 52/1038 = .0501
    b. By Rule #1, .0501 is an appropriate approximation for the proportion of people who believe that secondhand smoke is not at all harmful. No; since .0501 > .05, it is not unusual for someone to believe that secondhand smoke is not at all harmful.

12. Consider only the 19+844 = 863 patients who took the drug.
    Let F = a patient experiences flu symptoms.
    a. By Rule #1, an appropriate approximation is P(F) = 19/863 = .0220
    b. Yes; since .0220 < .05, it is considered unusual for a patient taking the drug to experience flu symptoms.

13. Consider only the 2624 + 168,262 = 170,886 American Airlines passenger who were bumped.
    Let I = a passenger is bumped involuntarily.
    a. By Rule #1, an appropriate approximation is P(I) = 2624/168,262 = .0154
    b. Yes; since .0164 < .05, such involuntary bumpings are considered unusual.

14. Let L = an American Airlines flight arrives late.
    a. By Rule #1, an appropriate estimate is P(L) = 42/150 = .280.
    b. No; since .280 > .05, it is not considered unusual for such a flight to arrive late.

15. a. Let C = selecting the correct birthdate.
       P(C) = 1/365
    b. Yes; since 1/365 = .003 < .05.
    c. Since P(C) = .003 is so small, C is unlikely to occur by chance. It appears that Mike had inside information and was not operating on chance.
    d. Depends. Most likely Mike was on the spot and gave an obviously incorrect response in an attempt to handle the situation humorously. If Kelly appreciates his spontaneity and sense of humor, the probability may be quite high.

16. NOTE: This problem can be approached two ways. Finding P(1 of your 2 numbers matches the winning number), the usual approach, requires using the Addition Rule of section 3-3. Finding P(the winning number matches one of your 2 numbers), which is an equivalent event, requires only the methods of this section.
    a. The winning number can be selected in 25,827,165 equally likely ways.
       Let W = you winning the grand prize
       W can occur 2 ways, if either of your numbers is selected; use Rule #2.
       P(W) = 2/25,827,165 = .0000000774
    b. Yes; since .0000000774 < .05, winning the grand prize is an unusual event.

17. a. Let B = a person's birthday is October 18.
       B is one of 365 (assumed) equally likely outcomes; use Rule #2.
       P(B) = 1/365 = .00274
    b. Let O = a person's birthday is in October.
       O includes 31 of 365 (assumed) equally likely outcomes; use Rule #2.
       P(N) = 31/365 = .0849
    c. Let D = a person's birthday was on a day that ends in the letter y.
       D includes all 7 of the (assumed) equally likely outcomes; use Rule #2.
       P(D) = 7/7 = 1
       NOTE: Since event D is a certainty, P(D) = 1 without reference to Rule #2 – i.e., even if all 7 outcomes are not equally likely

18. a. There were $831 + 18 = 849$ consumers surveyed.
   Let C = a consumer recognizes the Campbell's Soup brand name; use Rule #1.
   P(C) = 831/849 = .979
   b. Let B = a consumer recognizes the McDonald's brand name; use Rule #3.
   P(B) = .99 [This answer is subjective, and there is no single correct response.]
   c. Let V = a consumer recognizes the Veeco Instruments brand name; use Rule #3.
   P(V) = .01 [This answer is subjective, and there is no single correct response.]

19. There were $132 + 880 = 1012$ respondents.
   Let D = getting a respondent who gave the doorstop answer.
   D includes 132 of 1012 equally likely outcomes; use Rule #2.
   P(D) = 132/1012 = .130

20. Let A = selecting a driver in that age bracket has an accident.
   A occurred 136 times in a sample of 400; use Rule #1.
   P(A) ≈ 136/400 = .340
   Yes; since this indicates more than 1 such driver in every 3 is likely to have an accident, this should be a concern.
   NOTE: This is properly an estimate that such a person <u>had</u> an accident last year, and not that such a person <u>will have</u> an accident next year. To extend the estimate as desired requires the additional assumption that relevant factors (speed limits, weather, alcohol laws, the economy, etc.) remain the same.

21. The following table summarizes the relevant information from Data Set 27, where a win is defined as a positive result.

```
column        1   2   3   4   5   6   7   8   9  10  total
# of values  50  50  50  50  50  50  50  50  50  50   500
# of wins     6   6  10   8   7   6   8   9   6  11    77
# of 208's    0   1   1   3   1   1   3   2   0   1    13
```
   a. Let W = winning; use Rule #1.
   P(W) = 77/500 = .154
   b. Let R = running the deck; use Rule #1.
   P(R) = 13/500 = .026

22. There were $117 + 617 = 734$ patients in the clinical test.
   Let H = experiencing a headache; use Rule #1.
   P(H) = 117/734 = .159
   Perhaps; the .159 should be compared to the probability of experiencing a headache when not using Viagra – which cannot be assumed to be 0.00, especially if the clinical trial involved activity beyond merely taking the drug.

23. a. Assuming boys (B) and girls (G) are equally likely, the 4 equally likely possibilities for a family of 2 children are given at the right. Rule #2 can be used to find the desired probabilities as follows.
                 BB
                 BG
                 GB
                 GG
   b. P(2G's) = 1/4
   c. P(1 of each) = 2/4 = 1/2

24. a. Assuming boys (Br) and girls (Bl) are equally likely, the 4 equally likely possibilities are given at the right. Rule #2 can be used to find the desired probabilities as follows.
           BrBr
           BrBl
           BlBr
           BlBl
   b. P(BlBl) = 1/4
   c. P(brown eyes) = P(BrBr or BrBl or BlBr) = 3/4

25. a. net profit = (winnings received) − (amount bet) = $23 − $2 = $21
   b. payoff odds = (net profit):(amount bet) = 21:2
   c. actual odds = P(lose)/P(win) = (14/15)/(1/15) = 14/1 or 14:1

d. Payoff odds of 14:1 mean a profit of $14 for every $1 bet.
A winning bet of $2 means a profit of $28, which implies winnings received of $30.

26. Let O = an odd number occurs.
  a. P(O) = 18/38 = .474
  b. odds against winning = $P(\overline{O})/P(O)$ = (20/38)/(18/38) = 10/9 = 10:9
  c. Payoff odds of 1:1 mean a profit of $1 for every $1 bet.
  A winning bet of $18 means a profit of $18.
  d. Payoff odds of 10:9 mean a profit of $10 for every $9 bet.
  A winning bet of $18 means a profit of $20.

27. Chance fluctuations in testing situations typically prevent subjects from scoring exactly the same each time the test is given. When treated subjects show an improvement, there are two possibilities: (1) the treatment was effective and the improvement was due to the treatment or (2) the treatment was not effective and the improvement was due to chance. The accepted standard is that an event whose probability of occurrence is less than .05 is an unusual event. In this scenario, the probability that the treatment group shows an improvement even if the drug has no effect is calculated to be .04. If the treatment group does show improvement, then, either the treatment truly is effective or else an unusual event has occurred. Operating according to the accepted standard, one should conclude that the treatment was effective.

28. Probably, but not on the basis of those calculations. For a population that is 1/2 female, the probability that 20 randomly selected people are all women is $(1/2)^{20}$ = 1/(1,048,576). To determine bias by the court the relevant probability is not the probability that "20 randomly selected people are all women" but that "20 randomly selected people available for jury duty are all women" – i.e., those available for jury duty may not be evenly divided between the genders.

29. If the odds against A are a:b, then P(A) = b/(a+b).
  Let M = Millennium wins the next race.
  If the odds against M are 3:5, then a=3 and b=5 in the statement above.
  P(M) = 5/(3+5) = 5/8 = .625

30. p_t = P(member of the treatment group reporting a headache) = 117/734 = .1594
  p_c = P(member of the control group reporting a headache) = 29/725 = .0400
  NOTE: Do not work with rounded values. Store the above values (i.e., all the decimals as produced by the indicated quotients) in a calculator and recall them as needed.
  a. relative risk = p_t/p_c = .1594/.0400 = 3.99
  b. odds ratio = $\dfrac{p_t/(1-p_t)}{p_c/(1-p_c)} = \dfrac{.1594/.8406}{.0400/.9600} = \dfrac{.1896}{.0417} = 4.55$

31. Let O = a person is born on October 18.
  a. Every 4 years there are 1(366) + 3(365) = 1461 days.
   P(O) = 4/1461 [= .00273785, compare to 1/365 = .00273973]
  b. Every 400 years there are 97(366) + 303(365) =146097 days.
   P(O) = 400/146097 [= .00273791, compared to .00273785 and .00273973]

32. No matter where the two flies land, it is possible to cut the orange in half to include both flies on the same half. Since this is a certainty, the probability is 1. [Compare the orange to a globe. Suppose fly A lands on New York City, and consider all the circles of longitude. Wherever fly B lands, it is possible to slice the globe along some circle of longitude that places fly A and fly B on the same half.]
  NOTE: If the orange is marked into two predesignated halves before the flies land, the probability is different – one fly A lands, fly B has ½ a chance of landing on the same half. If both flies are to land on a specified one of the two predesignated halves, the probability is

different still – fly A has ½ a chance of landing on the specified half, and only ½ the time will fly B pick the same half: the final anser would be ½ of ½, or ¼.

33. This difficult problem will be broken into two events and a conclusion. Let L denote the length of the stick. Label the midpoint of the stick M, and label the first and second breaking points X and Y.
    - Event A: X and Y must be on opposite sides of M. If X and Y are on the same side of M, the side without X and Y will be longer than .5L and no triangle will be possible. Once X is set, P(Y is on same side as X) = P(Y is on opposite side from X) = .5.
    - Event B: The distance |XY| must be less than .5L. Assuming X and Y are on opposite sides of M, let Q denote the end closest to X and R denote the end closest to Y so that the following sketch applies.

    ```
    |-----|-----------|---|-------------|
    Q     X           M   Y             R
    ```

    In order to form a triangle, it must be true that |XY| = |XM| + |MY| < .5L. This happens only when |MY| < |QX|.
    With all choices random, there is no reason for |QX| to be larger than |MY|, or vice-versa. This means that P(|MY| < |QX|) = .5.
    - Conclusion: For a triangle, we need both events A and B. Since P(A) = .5 and P(B occurs assuming A has occurred) = .5, the probability of a triangle is (.5)(.5) = .25. In the notation of section 3-4, P(A and B) = P(A)·P(B|A) = (.5)(.5) = .25.

    NOTE: In situations where the laws of science and mathematics are difficult to apply (e.g., finding the likelihood that a particular thumbtack will land point up when it is dropped), probabilities are typically estimated using Rule #1. Suppose an instructor gives 100 sticks to a class of 100 students and says, "Break you stick at random 2 times so that you have 3 pieces." Does this mean that only about 25% of the students would be able to form a triangle with their 3 pieces? In theory, yes. In practice, no. It is likely the students would break the sticks "conveniently" and not "at random" – e.g., there would likely be fewer pieces less than one inch long than you would expect by chance, because such pieces would be difficult to break off by hand.

3-3 Addition Rule

1. a. No, it's possible for a cardiac surgeon to be a female.
   b. No, it's possible for a female college student to drive a motorcycle.
   c. Yes, a person treated with Lipitor would not be in the group that was given no treatment.

2. a. Yes, a person watching NBC would not be watching CBS at precisely the same time.
   b. Yes, a person opposing all government taxation would not approve of government taxation.
   c. No, there are currently United States senators who are females.

3. a. $P(\overline{A}) = 1 - P(A) = 1 - .05 = .95$
   b. Let B = a randomly selected woman over the age of 25 has a bachelor's degree.
   c. $P(\overline{B}) = 1 - P(B) = 1 - .218 = .782$

4. a. $P(\overline{A}) = 1 - P(A) = 1 - .0175 = .9825$
   b. Let B = selecting an American who believes that life exists elsewhere in the galaxy.
   c. $P(\overline{B}) = 1 - P(B) = 1 - .61 = .39$

5. Make a chart like the one on the right.
   Let G = getting a green pod.
   Let W = getting a white flower.

		POD		
		Green	Yellow	
FLOWER	Purple	5	4	9
	White	3	2	5
		8	6	14

There are two approaches.
* Use broad categories and allow for double-counting (i.e., the "formal addition rule").
P(G or W) = P(G) + P(W) - P(G and W)
= 8/14 + 5/14 - 3/14
= 10/14 = 5/7 or .714
* Use individual mutually exclusive categories that involve no double-counting (i.e., the "intuitive addition rule"). For simplicity in this problem, we use GP for "G and P" and GW for "G and W" and so on, as indicated in the table.
P(G or W) = P(GP or GW or YW)
= P(GP) + P(GW) + P(YW)
= 5/14 + 3/14 + 2/14
= 10/14 = 5/7 or .714

NOTE: In general, using broad categories and allowing for double-counting is a "more powerful" technique that "lets the formula do the work" and requires less analysis by the solver. Except when such detailed analysis is instructive, this manual uses the first approach.

NOTE: Throughout the manual we follow the pattern in exercise #5 of using the first letter [or other natural designation] of each category to represent that category. And so in exercises #9-13 P(S) = P(selecting a person who survived) and P(M) = P(selecting a man) and so on. If there is ever cause for ambiguity, the notation will be clearly defined. Since mathematics and statistics use considerable notation and formulas, it is important to clearly define what various letters stand for.

6. Refer to exercise #5.
P(Y or P) = P(Y) + P(P) - P(F and R)
= 6/14 + 9/14 - 4/14
= 11/14 or .786

7. Let O = a person's birthday falls on October 18; then P(O) = 1/365.
P($\bar{O}$) = 1 - P(O) = 1 - 1/365 = 364/365 or .997

8. Let O = a person's birthday falls in October; then P(O) = 31/365.
P($\bar{O}$) = 1 - P(O) = 1 - 31/365 = 334/365 or .915

9. Make a chart like the one on the right.
P(W or C) = P(W) + P(C) - P(W and C)
= 422/2223 + 109/2223 - 0/2223
= 531/2223 = .239
Note: Let C stand for child – i.e., boys and girls together. And so P(C) = 109/2223, P(C and S) = 56/2223, etc.

		FATE		
		Survived	Died	
GROUP	Men	332	1360	1692
	Women	318	104	422
	Boys	29	35	64
	Girls	27	18	45
		706	1517	2223

10. Refer to exercise #9.
P(M or S) = P(M) + P(S) - P(M and S)
= 1692/2223 + 706/2223 - 332/2223
= 2066/2223 = .929

11. Refer to exercise #9
P(C or S) = P(C) + P(S) - P(C and S)
= 109/2223 + 706/2223 - 56/2223
= 759/2223 = .341

12. Refer to exercise #9.
P(W or D) = P(W) + P(D) - P(W and D)
= 422/2223 + 1517/2223 - 104/2223
= 1835/2223 = .825

13. Make a chart like the one on the right.
    $P(\overline{A}) = 1 - P(A)$
    $= 1 - 40/100 = 60/100 = .60$

		Rh FACTOR		
		+	−	
	A	35	5	40
GROUP	B	8	2	10
	AB	4	1	5
	O	39	6	45
		86	14	100

14. Refer to exercise #13.
    $P(Rh-) = 14/100 = .14$

15. Refer to exercise #13.
    $P(A \text{ or } Rh-) = P(A) + P(Rh-) - P(A \text{ and } Rh-)$
    $= 40/100 + 14/100 - 5/100 = 49/100 = .49$

16. Refer to exercise #13.
    $P(A \text{ or } B) = P(A) + P(B)$
    $= 40/100 + 10/100 = 50/100 = .50$

17. Refer to exercise #13.
    $P(\overline{Rh+}) = 1 - P(Rh+)$
    $= 1 - 86/100$
    $= 14/100 = .14$

18. Refer to exercise #13.
    $P(B \text{ or } Rh+) = P(B) + P(Rh+) - P(B \text{ and } Rh+)$
    $= 10/100 + 86/100 - 8/100$
    $= 88/100 = .88$

19. Refer to exercise #13.
    $P(AB \text{ or } Rh+) = P(AB) + P(Rh+) - P(AB \text{ and } Rh+)$
    $= 5/100 + 86/100 - 4/100$
    $= 87/100 = .87$

20. Refer to exercise #13.
    $P(A \text{ or } O \text{ or } Rh+) = P(A) + P(O) + P(Rh+) - P(A \text{ and } Rh+) - P(O \text{ and } Rh+)$
    $= 40/100 + 45/100 + 86/100 - 35/100 - 39/100$
    $= 97/100 = .97$

21. Make a chart like the one on the right.
    $P(M \text{ or } 6) = P(M) + P(6) - P(M \text{ and } 6)$
    $= 6/20 + 6/20 - 2/20$
    $= 10/20 = .50$

		TRANSMISSION		
		A	M	
	4	6	3	9
CYLINDERS	6	4	2	6
	8	4	1	5
		14	6	20

22. Of the 107 subjects in Data Set 4, only four (#'s 19,46,66,83) are female non-smokers. Each of the other 103 subjects is either a male or a smoker. By Rule #2 of section 3.2, $P(M \text{ or } S) = 103/107 = .963$

23. Make a chart like the one on the right.
    Let A = a person is age 18-21
    N = a person does not respond
    $P(A \text{ or } N) = P(A) + P(N) - P(A \text{ and } N)$
    $= 84/359 + 31/359 - 11/359$
    $= 104/359 = .290$

		RESPOND?		
		Yes	No	
AGE	18-21	73	11	84
	22-29	255	20	275
		328	31	359

24. Refer to exercise #23.
    $P(A \text{ or } Y) = P(A) + P(Y) - P(A \text{ and } Y)$
    $= 84/359 + 328/359 - 73/359$
    $= 339/359 = .944$

25. The general formula is: P(A or B) = P(A) + P(B) - P(A and B).
Solving for P(A and B) yields: P(A and B) = P(A) + P(B) - P(A or B)
   a. For P(A) = 3/11 and P(B) = 4/11 and P(A or B) = 7/11, the Addition Rule produces
   P(A and B) = P(A) + P(B) - P(A or B)
   = 3/11 + 4/11 - 7/11
   = 0
   This means that A and B are mutually exclusive.
   b. For P(A) = 5/18 and P(B) = 11/18 and P(A or B) = 13/18, the Addition Rule produces
   P(A and B) = P(A) + P(B) - P(A or B)
   = 5/18 + 11/18 - 13/18
   = 3/18
   This means that A and B are not mutually exclusive.

26. No, as illustrated by the following example.
    Let A = has prostate cancer
    Let B = is a female
    Let C = has testicular cancer

27. If the *exclusive or* is used instead of the *inclusive or*, then the double-counted probability must be completely removed (i.e., must be subtracted twice) and the formula becomes
    P(A or B) = P(A) + P(B) - 2·P(A and B)

28. P(A or B or C)
    = P[(A or B) or C]
    = P(A or B) + P(C) - P[(A or B) and C]
    = [P(A) + P(B) - P(A and B)] + P(C) - P[(A and C) or (B and C)]
    = [P(A) + P(B) - P(A and B)] + P(C) - [P(A and C) + P(B and C) - P(A and B and C)]
    = P(A) + P(B) - P(A and B) + P(C) - P(A and C) - P(B and C) + P(A and B and C)
    = P(A) + P(B) + P(C) - P(A and B) - P(A and C) - P(A and B) + P(A and B and C)

3-4 Multiplication Rule: Basics

1. a. Independent, since getting a 5 when rolling a die does not affect the likelihood of getting heads when flipping a coin.
   b. Independent, since random selection means that the viewing preferences of the first person chosen in no way affect the viewing preferences of the second person.
   c. Dependent, since getting a positive response to a request for a date generally depends on the personal opinions of the one being asked – and dress is one of the factors that people consider when forming opinions of each other.
   NOTE: This manual views such subjective situations "in general" and without reading too much into the problem. The problem doesn't say, for example, how the invitation was extended – if it were extended over the phone, then dress would be irrelevant. And the problem doesn't say the nature of the relationship prior to the invitation – if the people already know each other fairly well, then the dress one happens to be wearing at the time of the invitation would probably have little if any affect on the response.

2. a. Independent, since the calculator (assumed to be battery operated) and the refrigerator (assumed to be electric) have difference power sources totally unrelated to each other.
   NOTE: One could argue that a refrigerator's failure may be due to improper maintenance – the same type of behavior that would put off replacing weak calculator batteries. If so, then a person's refrigerator not working suggests he may be one who doesn't properly maintain things – which suggests that his calculator might not be working either. This manual ignores such highly speculative hypothetical scenarios.

INSTRUCTOR'S SOLUTIONS Chapter 3   S-67

   b. Dependent, since the kitchen light and the refrigerator have the same power source – the electric service supplied to the house. Finding that one of them is not working makes it more likely that the other is not working – just as finding that one of them is working indicates there is electricity to the house and makes it likely that other electric appliances are working.
NOTE: If the light and the refrigerator are not working, they could have failed independently of each other – but that is irrelevant. The notion of statistical dependence rests upon whether the occurrence of one event <u>could</u> reasonably be related to the occurrence of another.
   c. Dependent, since impaired driving ability makes a car crash more likely.

3. Let T = getting tails when tossing a coin.
Let 3 = getting a three then rolling a die.
P(T and 3) = P(T)·P(3|T) = (1/2)·(1/6) = 1/12

4. Let K = randomly selecting K from the 26 letters in the alphabet.
Let 9 = randomly selecting 9 from the 10 base ten digits.
P(K and 9) = P(K)·P(9|K) = (1/26)·(1/10) = 1/260
This password would not be an effective deterrent against someone who knew that the password consisted of one letter followed by one digit – since there are only 260 possible combinations, which could all be tried in a reasonably short period of time. If the person trying to gain access knew only that the password could be any phrase of 8 or fewer characters, with at least 26+10 = 36 possibilities for each character, then K9 is one of millions of possibilities and would be as effective as any other password.

5. There are 1 blue + 3 green + 2 red + 1 yellow = 7 items.
Let A = the first item selected is colored green.
Let B = the second item selected is colored green.
In general, P(A and B) = P(A)·P(B|A)
   a. P(A and B) = P(A)·P(B|A) = (3/7)·(3/7) = 9/49 = .184
   b. P(A and B) = P(A)·P(B|A) = (3/7)·(2/6) = 6/42 = .143

6. Refer to exercise #5.
Let R = selecting an item colored red.
Let G = selecting an item colored green.
Let B = selecting an item colored blue.
   a. P(R_1 and G_2 and B_3) = P(R_1)·P(G_2|R_1)·P(B_3|R_1 and G_2)
                        = (2/7)·(3/7)·(1/7) = 6/343 = .0175
   b. P(R_1 and G_2 and B_3) = P(R_1)·P(G_2|R_1)·P(B_3|R_1 and G_2)
                        = (2/7)·(3/6)·(1/5) = 6/210 = .0286

7. Let D = selecting a defective gas mask
P(D) = 10322/19218 = .5371, for the first selection from the population
   a. P(D_1 and D_2) = P(D_1)·P(D_2|D_1) = (10322/19218)·(10322/19218) = .288477
   b. P(D_1 and D_2) = P(D_1)·P(D_2|D_1) = (10322/19218)·(10321/19217) = .288464
   c. The results differ only slightly and are identical to the usual accuracy with which probabilities are reported.
   d. Selecting without replacement. Replacement could lead to re-testing the same unit, which would not yield any new information.

8. Let O = selecting a a hunter who was wearing orange.
P(O) = 6/123 = .0488, for the first selection from the sample
   a. P(O_1 and O_2) = P(O_1)·P(O_2|O_1) = (6/123)·(6/123) = .00238
   b. P(O_1 and O_2) = P(O_1)·P(O_2|O_1) = (6/123)·(5/122) = .00200
   c. Selecting without replacement. Replacement could lead to re-interviewing the same hunter, which would lead to no new information.

S-68 INSTRUCTOR'S SOLUTIONS Chapter 3

9. Let C = a student guesses the correct response.
   $P(C) = \frac{1}{2}$ for each question.
   a. $P(C_1 \text{ and } C_2 \text{ and } C_3 \text{ and } C_4 \text{ and } C_5 \text{ and } C_6 \text{ and } C_7 \text{ and } \overline{C}_8 \text{ and } \overline{C}_9 \text{ and } \overline{C}_{10})$
   $= P(C_1) \cdot P(C_2) \cdot P(C_3) \cdot P(C_4) \cdot P(C_5) \cdot P(C_6) \cdot P(C_7) \cdot P(\overline{C}_8) \cdot P(\overline{C}_9) \cdot P(\overline{C}_{10})$
   $= (\frac{1}{2}) \cdot (\frac{1}{2}) \cdot (\frac{1}{2}) \cdot (\frac{1}{2}) \cdot (\frac{1}{2}) \cdot (\frac{1}{2}) \cdot (\frac{1}{2}) \cdot (\frac{1}{2}) \cdot (\frac{1}{2}) \cdot (\frac{1}{2}) = 1/1032$
   b. No; getting the first 7 correct and the last 3 incorrect is only one possible way of passing the test.

10. Let W = selecting a woman.
    $P(W) = 13/100$, for the first selection only
    $P(W_1 \text{ and } W_2 \text{ and } W_3) = P(W_1) \cdot P(W_2 | W_1) \cdot P(W_3 | W_1 \text{ and } W_2)$
    $= (13/100) \cdot (12/99) \cdot (11/98) = .00177$
    No; depending on the purpose for contacting the senators, the lobbyist would likely choose the ones that would best help him achieve that purpose.

11. a. Let N = a person is born on November 27.
       $P(N) = 1/365$, for each selection
       $P(N_1 \text{ and } N_2) = P(N_1) \cdot P(N_2)$
       $= (1/365) \cdot (1/365) = 1/133225 = .00000751$
    b. Let S = a person's birthday is favorable to being the same.
       $P(S) = 365/365$ for the first person, but then the second person must match
       P(the birthdays are the same) $= P(S_1 \text{ and } S_2) = P(S_1) \cdot P(S_2 | S_1)$
       $= (365/365) \cdot (1/365) = 1/365 = .00274$

12. a. Let J = being born on July 4.
       $P(J_1 \text{ and } J_2 \text{ and } J_3) = P(J_1) \cdot P(J_2) \cdot P(J_3)$
       $= (1/365) \cdot (1/365) \cdot (1/365) = .0000000206$
       Probably not. This probability is about 1 in 50 million, which means that the event can be expected to occur about once in every 50 million such situations.
    b. Let F = a birthday is favorable to producing the desired 3-way match.
       $P(F_1 \text{ and } F_2 \text{ and } F_3) = P(F_1) \cdot P(F_2 | F_1) \cdot P(F_3 | F_1 \text{ and } F_2)$
       $= (365/365) \cdot (1/365) \cdot (1/365) = .00000751$

13. Since the sample size is no more than 5% of the size of the population, treat the selections as being independent even though the selections are made without replacement.
    Let G = a selected CD is good.
    $P(G) = .97$, for each selection
    P(batch accepted) = P(all good)
    $= P(G_1 \text{ and } G_2 \text{ and}...\text{and } G_{12})$
    $= P(G_1) \cdot P(G_2) \cdot ... \cdot P(G_{12}) = (.97)^{12} = .694$

14. Let A = a poll is accurate as claimed.
    $P(A) = .95$, for each poll
    $P(A_1 \text{ and } A_2 \text{ and } A_3 \text{ and } A_4 \text{ and } A_5) = P(A_1) \cdot P(A_2) \cdot P(A_3) \cdot P(A_4) \cdot P(A_5)$
    $= (.95) \cdot (.95) \cdot (.95) \cdot (.95) \cdot (.95) = (.95)^5 = .774$
    No; the more polls one considers, the less likely it us that they all will be within the claimed margin of error.

15. Let G = a girl is born.
    $P(G) = .50$, for each couple
    $P(G_1 \text{ and } G_2 \text{ and}...\text{and } G_{10}) = P(G_1) \cdot P(G_2) \cdot ... \cdot P(G_{10}) = (.50)^{10} = .000977$
    Yes, the gender selection method appears to be effective. The choice is between two possibilities: (1) the gender selection method has no effect and a very unusual event occurred or (2) the gender selection method is effective.

INSTRUCTOR'S SOLUTIONS Chapter 3 S-69

16. Let O = the valve opens properly.
   P(O) = .9968, for each valve
   P(water gets through) = P(O_1 and O_2) = P(O_1)·P(O_2) = (.9968)·(.9968) = .9936
   Yes; since the systems is used only in emergencies, and not on a routine basis – but because the probability of a failure is .0064, the system can be expected to fail about 64 times in every 10,000 uses. That means if it is used 28 times a day [28x365 = 10,220 times a year], we expect about 64 failures annually – and after one such failure in a nuclear reactor, there might not be a next time for anything.

17. Let T = the tire named is consistent with all naming the same tire.
   P(T) = 4/4 for the first person, but then the others must match him
   P(all name the same tire) = P(T_1 and T_2 and T_3 and T_4)
   = P(T_1)·P(T_2|T_1)·P(T_3|T_1 and T_2)·P(T_4|T_1 and T_2 and T_3)
   = (4/4)(1/4)(1/4)(1/4) = 1/64 = .0156

18. Let F = a selection is favorable to producing the desired 9-way match
   P(F) = 5/5 for the first selection, and 1/5 thereafter
   P(F_1 and F_2 and...and F_9) = P(F_1)·P(F_2)·...·P(F_9)
   = (5/5)·(1/5)·...·(1/5) = 1·(1/5)^8 = .00000256
   There is no reasonable doubt if the voices are similar, but not otherwise. If the criminal had a high voice, for example, and only one of the five suspects had a high voice, all nine victims would probably select the suspect with the high voice even if he were not the guilty person.

19. Since the sample size is no more than 5% of the size of the population, treat the selections as being independent even though the selections are made without replacement.
   Let N = a CD is not defective
   P(N) = 1 – .02 = .98, for each selection
   P(N_1 and N_2 and...and N_{15}) = P(N_1)·P(N_2)·...·P(N_{15}) = (.98)^{15} = .739
   No, there is not strong evidence to conclude that the new process is better. Even if the new process made no difference, the observed result is not considered unusual and could be expected to occur about 74% of the time.

20. Let A = the alarm clock works properly.
   P(A) = .975, for each alarm clock
   a. P($\bar{A}$) = 1 – .975 = .025
   b. P($\bar{A}_1$ and $\bar{A}_2$) = P($\bar{A}_1$) P($\bar{A}_2$|$\bar{A}_1$) = (.025)·(.025) = .000625
   c. P(A_1 or A_2) = P(A_1) + P(A_2) – P(A_1 and A_2)
   = .975 + .975 – (.975)·(.975) = .999

21. Refer to the table at the right.
   P(Pos_1 and Pos_2) = P(Pos_1)·P(Pos_2)
   = (83/99)·(82/98)
   = .702

		TEST RESULT		
		Pos	Neg	
PREGNANT?	Yes	80	5	85
	No	3	11	14
		83	16	99

22. Refer to the table for exercise #21.
   P(Neg or No) = P(Neg) + P(No) – P(Neg and No)
   = 16/99 + 14/99 – 11/99 = 19/99 = .192

23. Refer to the table for exercise #21.
   P(Yes_1 and Yes_2) = P(Yes_1)·P(Yes_2) = (85/99)·(84/98) = .736

24. Refer to the table for exercise #21.
   P(Neg_1 and Neg_2 and Neg_3) = P(Neg_1)·P(Neg_2|Neg_1)·P(Neg_3|Neg_1 and Neg_2)
   = (16/99)·(15/98)·(14/97) = .00357

S-70 INSTRUCTOR'S SOLUTIONS Chapter 3

25. let D = a birthday is different from any yet selected
    $P(D_1) = 366/366$ NOTE: With nothing to match, it <u>must</u> be different.
    $P(D_2 | D_1) = 365/366$
    $P(D_3 | D_1 \text{ and } D_2) = 364/366$
    ...
    $P(D_5 | D_1 \text{ and } D_2 \text{ and}\ldots\text{and } D_4) = 362/366$
    ...
    $P(D_{25} | D_1 \text{ and } D_2 \text{ and}\ldots\text{and } D_{24}) = 342/366$
    a. P(all different) = $P(D_1 \text{ and } D_2 \text{ and } D_3)$
    $= P(D_1) \cdot P(D_2) \cdot P(D_3)$
    $= (366/366) \cdot (365/366) \cdot (364/366) = .992$
    b. P(all different) = $P(D_1 \text{ and } D_2 \text{ and}\ldots\text{and } D_5)$
    $= P(D_1) \cdot P(D_2) \cdot \ldots \cdot P(D_5)$
    $= (366/366) \cdot (365/366) \cdot \ldots \cdot (362/366) = .973$
    c. P(all different) = $P(D_1 \text{ and } D_2 \text{ and}\ldots\text{and } D_{25})$
    $= P(D_1) \cdot P(D_2) \cdot \ldots \cdot P(D_{25})$
    $= (366/366) \cdot (365/366) \cdot \ldots \cdot (342/366) = .432$
    NOTE: A program to perform this calculation can be constructed using a programming language, or using most spreadsheet or statistical software packages. In BASIC, for example, use
    ```
    10 LET P=1
    15 PRINT "How many birthdays?"
    20 INPUT D
    30   FOR K=1 TO D-1
    40   LET P=P*(366-K)/366
    50 NEXT K
    55 PRINT "The probability they all are different is"
    60 PRINT P
    70 END
    ```

26. let F = a birth is favorable to producing the desired 8-way match
    P(F) = 2/2 for the first birth, and 1/2 thereafter
    $P(F_1 \text{ and } F_2 \text{ and}\ldots\text{and } F_8) = P(F_1) \cdot P(F_2) \cdot \ldots \cdot P(F_8)$
    $= (2/2) \cdot (1/2) \cdot \ldots \cdot (1/2) = 1 \cdot (1/2)^7 = .00781$

27. This is problem can be done by two different methods. In either case,
    let A = getting an ace
    S = getting a spade
    * consider the sample space
    The first card could be any of 52 cards; for each first card, there are 51 possible second cards. This makes a total of 52·51 = 2652 equally likely outcomes in the sample space. How many of them are A_1S_2?
    The aces of hearts, diamonds and clubs can be paired with any of the 13 clubs for a total of 3·13 = 39 favorable possibilities. The ace of spades can only be paired with any of the remaining 12 members of that suit for a total of 12 favorable possibilities. Since there are 39 + 12 = 51 favorable possibilities among the equally likely outcomes,
    $P(A_1S_2) = 51/2652$
    $= .0192$
    * use the formulas
    Let As and Ao represent the ace of spades and the ace of any other suit respectively. Break A_1S_2 into mutually exclusive parts so the probability can be found by adding and without having to consider double-counting.
    $P(A_1S_2) = P[(As_1 \text{ and } S_2) \text{ or } (Ao_1 \text{ and } S_2)]$
    $= P(As_1 \text{ and } S_2) + P(Ao_1 \text{ and } S_2)$
    $= P(As_1) \cdot P(S_2 | As_1) + P(Ao_1) \cdot P(S_2 | Ao_1)$
    $= (1/52)(12/51) + (3/52)(13/51)$
    $= 12/2652 + 39/2652$
    $= 51/2652 = .0192$

INSTRUCTOR'S SOLUTIONS Chapter 3   S-71

28. The following English/logic facts are used in this exercise.
    * not (A or B) = not A and not B
       Is your sister either artistic or bright?
       No? Then she is not artistic and she is not bright.
    * not (A and B) = not A or not B
       Is your brother artistic and bright?
       No? Then either he is not artistic or he is not bright.
    a. P($\overline{\text{A or B}}$) = P($\overline{\text{A}}$ and $\overline{\text{B}}$)   [from the first fact above]
       or
       P($\overline{\text{A or B}}$) = 1 - P(A or B)   [rule of complimentary events]
    b. P($\overline{\text{A or B}}$) = 1 - P(A and B)   [from the second fact above]
    c. They are different: part (a) gives the complement of "A or B"
                while part (b) gives the complement of "A and B."

3-5 Multiplication Rule: Complements and Conditional Probability

1. If it is not true that "at least one of them has Group A blood," then "none of them has Group A blood."

2. If it is not true that "all of them are free of defects," then "at least one of them is defective."

3. If it is not true that "none of them is correct," then "at least one of them is correct."

4. If it is not true that "at least one of them accepts," then "none of them accepts."

5. Let L = selecting someone with long hair.
   Let W = selecting a woman.
   P(W|L) = .90 seems like a reasonable estimate. No; while someone with long hair is likely to be a woman, one cannot say such will "almost surely" be the case.

6. Let C = selecting someone who owns a motorcycle.
   Let M = selecting a male.
   P(M|C) = .95 seems like a reasonable estimate. Yes; it would be "reasonable to believe" that Pat is a male.

7. Let B = a child is a boy
   P(B) = .5, for each birth
   P(at least one girl) = 1 - P(all boys)
                       = 1 - P(B_1 and B_2 and B_3 and B_4 and B_5)
                       = 1 - P(B_1)·P(B_2)·P(B_3)·P(B_4)·P(B_5)
                       = 1 - (.5)·(.5)·(.5)·(.5)·(.5) = 1 - .03125 = .96875
   Yes; that probability is high enough to be "very confident" of getting at least one girl – since .03125 is less than .05, the accepted probability of an unusual event.

8. Refer to the notation of exercise #7.
   P(at least one girl) = 1 - P(all boys)
                       = 1 - P(B_1 and B_2 and...and B_{12})
                       = 1 - P(B_1)·P(B_2)·...·P(B_{12})
                       = 1 - (.5)·(.5)·...·(.5) = 1 - $(.5)^{12}$ = 1 - .000244 = .999756
   Either an extremely rare event has occurred or (more likely) there is some genetic factor present for this couple that makes P(B) > .5.

9. Let N = not getting cited for a traffic violation.
P(N) = .9 for each such infraction
P(at least violation citation) = 1 - P(no violation citations)
= 1 - P(N_1 and N_2 and N_3 and N_4 and N_5)
= 1 - P(N_1)·P(N_2)·P(N_3)·P(N_4)·P(N_5)
= 1 - (.9)·(.9)·(.9)·(.9)·(.9)
= 1 - .590 = .410

10. Let E = erroneously answering a question.
P(E) = 4/5 = .8, for each question
P(at least one correct) = 1 - P(all erroneous)
= 1 - P(E_1 and E_2 and E_3 and E_4)
= 1 - P(E_1)·P(E_2)·P(E_3)·P(E_4)
= 1 - (.8)·(.8)·(.8)·(.8) = 1 - .410 = .590
Since .590 > .5, you are more likely to pass than to fail – whether or not you can "reasonably expect" to pass depends on your perception of what is reasonable.

11. Let G = getting a girl.
P(G) = .5, for each birth
P(G_3 | $\overline{G}_1$ and $\overline{G}_2$) = P(G_3) = .5
No; this is not the same as P(G_1 and G_2 and G_3) = P(G_1)· P(G_2)· P(G_3)
= (.5)·(.5)·(.5) = .125

12. Make a chart like the one on the right.
Let Gr = getting a green pod.
Let Pu = getting a purple flower.
P(Pu | Gr) = 5/8 = .625

		POD		
		Green	Yellow	
FLOWER	Purple	5	4	9
	White	3	2	5
		8	6	14

13. Refer to the table at the right.
P(Neg | No) = 11/14 = .786
Since P(Neg | Yes) = 5/85 = .059 > .05,
it would not be considered unusual for
the test to read negative even if she were
pregnant. If a woman wanted to be more sure that she were not pregnant, she could take another test. Assuming independence of tests (i.e., that errors occur at random and there is no biological reason why the test should err in her particular case), P(2 false negatives) = $(.059)^2$ = .003.

		TEST RESULT		
		Pos	Neg	
PREGNANT?	Yes	80	5	85
	No	3	11	14
		83	16	99

14. Refer to the table for exercise #13.
P(No | Neg) = 11/16 = .6875
No; P(No | Neg) is not equal to P(Neg | No) calculated in exercise #13.

15. Let F = the alarm clock fails.
P(F) = .01, for each clock
P(at least one works) = 1 - P(all fail) = 1 - P(F_1 and F_2 and F_3)
= 1 - P(F_1)·P(F_2)·P(F_3)
= 1 - (.01)(.01)(.01) = 1 - .000001 = .999999
Yes; the student appears to gain, because now the probability of a timely alarm is virtually a certainty. There may be something else going on, however, if the student "misses many classes because of malfunctioning alarm clocks" when his clock works correctly 99% of the time.
NOTE: Rounded to 3 significant digits as usual, the answer is 1.00. In cases when rounding to 3 significant digits produces a probability of 1.00, this manual gives the answer with sufficient significant digits to distinguish the answer from a certainty.

16. Let G = a selected CD is good [i.e., the CD is good, not necessarily the music].
    P(G) = .97, for each CD
    P(reject the batch) = P(at least one defective)
    $= 1 - $ P(all good)
    $= 1 - $ P(G_1 and G_2 and...and G_{10})
    $= 1 - $ P(G_1)·P(G_2)·...·P(G_{10})
    $= 1 - (.97)(.97)...(.97) = 1 - (.97)^{10} = 1 - .737 = .263$

    NOTE: Since the sample is less than 5% of the population (i.e., 10/5000 = .002 < .05), treat the selections as independent and ignore that the sampling is without replacement.

17. Let N = a person is HIV negative
    P(N) = .9, for each person in the at-risk population
    P(HIV positive result) = P(at least person is HIV positive)
    $= 1 - $ P(all persons HIV negative)
    $= 1 - $ P(N_1 and N_2 and N_3)
    $= 1 - $ P(N_1)·P(N_2)·P(N_3)
    $= 1 - (.9)(.9)(.9) = 1 - .729 = .271$

    NOTE: This plan is very efficient. Suppose, for example, there were 3,000 people to be tested. Only in .271 = 27.1% of the groups would a retest need to be done for each of the 3 individuals. Those (.271)·(1,000) = 271 groups would generate 271·3 = 813 retests. The total number of tests required is then 1813 (1000 original tests + 813 retests), only 60% of the 3,000 tests that would have been required to test everyone individually.

18. Let S = a swimming pool passes the test.
    P(S) = .98, for each pool
    P(the combined sample fails) = P(at least one pool fails)
    $= 1 - $ P(all pools pass)
    $= 1 - $ P(S_1 and S_2 and...and S_6)
    $= 1 - $ P(S_1)·P(S_2)·...·P(S_6)
    $= 1 - (.98)(.98)...(.98) = 1 - (.98)^6 = 1 - .886 = .114$

    NOTE: This plan is very efficient. Suppose, for example, there were 60 pools to be tested. Only in .114 = 11.4% of the groups would a retest need to be done for each of the 6 pools. That (.114)·(10) = 1 case would generate 1·6 = 6 retests. The total number of tests required is then 16 (10 original tests + 6 retests), only 27% of the 60 tests that would have been required to test each pool individually.

19. Refer to the table at the right.
    This problem may be done two ways.
    * reading directly from the table
      P(M|D) = 1360/1517 = .897
    * using the formula
      P(M|D) = P(M and D)/P(D)
      $= (1360/2223)/(1517/2223)$
      $= .6118/.6824 = .897$

		FATE		
		Survived	Died	
	Men	332	1360	1692
GROUP	Women	318	104	422
	Boys	29	35	64
	Girls	27	18	45
		706	1517	2223

    NOTE: In general, the manual will use the most obvious approach -- which most often is the first one of applying the basic definition by reading directly from the table.

20. Refer to the table and comments for exercise #19.
    P(M|D) = 1360/1517 = .897

21. Refer to the table and comments for exercise #19.
    P([B or G]|S) = [29 + 27]/706 = 56/706 = .0793

22. Refer to the table and comments for exercise #19.
    P([M or W]|D) = [1360 + 104]/1517 = 1464/1517 = .965

S-74 INSTRUCTOR'S SOLUTIONS Chapter 3

23. Let F = getting a seat in the front row.
    P(F) = 2/24 = 1/12 = .083
    P(at least one F in n rides) = 1 - P(no F in n rides) = 1 - $(11/12)^n$.
    P(at least one F in n rides) > .95 requires that $(11/12)^n$ < .05.
    According to the table at the right, this first occurs for n = 35.

    $(11/12)^{31}$ = .0674
    $(11/12)^{32}$ = .0618
    $(11/12)^{33}$ = .0566
    $(11/12)^{34}$ = .0519
    $(11/12)^{35}$ = .0476

24. Make a chart like the one on the right.
    Let D = getting a defective one
    Let A = getting one made in Atlanta
    P(A|D) = 3/5 = .600

	RESULT		
	Good	Def.	
PLANT Atlanta	397	3	400
Balto.	798	2	800
	1195	5	1200

25. a. The table is given below: 300 = 0.3% of 100,000; 285 = 95% of 300; etc.

		TEST RESULT		
		Pos	Neg	
HIV?	Yes	285	15	300
	No	4985	94,715	99,700
		5270	94,730	100,000

    b. P(Yes|Pos) = 285/5270 = .0541

26. For 25 randomly selected people,
    a. P(no 2 share the same birthday) = .432 [see exercise #25c in section 3-4]
    b. P(at least 2 share the same birthday) = 1 - P(no 2 share the same birthday)
    = 1 - .432 = .568

27. When 2 coins are tossed, there are 4 equally likely possibilities: HH HT TH TT.
    There are two possible approaches.
    * counting the sample space directly
      If there is at least one H, that leaves only 3 equally likely possibilities: HH HT TH
      Since HH is one of those 3, P(HH|at least one H) = 1/3
    * applying an appropriate formula
      Use P(B|A) = P(A and B)/P(A), where B = HH
                                            A = at least one H
      P(HH|at least one H) = P(at least one H and HH)/P(at least one H)
                           = (1/4)/(3/4)
                           = 1/3

3-6 Probabilities Through Simulations

1. The digits 4,6,1,9,6 correspond to T,T,F,F,T.

2. The digits 4,6,1,9,6,9,9,4,3,8 correspond to 4,6,1,6,4,3.

3. Letting G and D represent good and defective fuses respectively, the digits 4,6,1,9,6 correspond to G,G,D,G,G.

4. The directions may be understood in either of two ways. Ignoring breaks between rows and taking the next usable digits in order yields the dates in the left column. Considering the digits a row at a time in groups of 5 yields the dates in the column at the right.
   211 = July 29   211 = July 29
   344 = December 9   044 = February 13
   044 = February 13   001 = January 1
   300 = October 26   191 = July 9
   151 = May 30   299 = October 25
   NOTE: Since grouping the digits in rows of 5 is an artificial convenience, this manual will use the first method in future exercises – unless specifically directed otherwise.

5. Using the 20 5-digit numbers to represent families of 5 children and letting odd digits represent girls produces the following list of 20 numbers of girls per family:
2 3 4 0 2 3 2 3 5 1 2 3 2 3 3 1 2 3 4 2
Letting x be the number of girls per simulated family gives the following distribution.

x	f	r.f.	true P(x)
0	1	.05	.03125
1	2	.10	.15625
2	7	.35	.31250
3	7	.35	.31250
4	2	.10	.15625
5	1	.05	.03125
	20	1.00	1.00000

From the simulation we estimate $P(x \geq 2) \approx 17/20 = .85$. This compares very favorably with the true value of .8125, and in this instance the simulation produced a good estimate.

6. Ignoring breaks between rows and taking the next usable digits in order in groups of 3 yields the following simulated rolls of 3 dice:
   461 643 211 344 446 631 516 431 155 664 462 125 525 251 452 544 652 625 323 335
The 20 corresponding totals are:
   11 13 4 11 14 10 12 8 11 16 12 8 12 8 11 13 13 13 8 11
Letting T be getting a total of 10, we estimate $P(T) \approx 1/20 = .05$. This is considerably lower than the true value of .125, and in this instance the simulation did not produce a good estimate.

7. Using the 20 5-digit rows to represent groups of 5 people and letting 0 represent a left-hander, produces the following list of 20 numbers of left-handers per group:
   0 0 0 1 0 2 1 1 0 1 0 1 0 0 0 1 0 0 0
Letting x be the number of left-handers per simulated group gives the following distribution.

x	f	r.f.	true P(x)
0	13	.65	.59049
1	6	.30	.32805
2	1	.05	.07290
3	0	.00	.00810
4	0	.00	.00045
5	0	.00	.00001
	20	1.00	1.00000

From the simulation we estimate $P(x \geq 1) \approx 7/20 = .35$. This compares very favorably to the true value of .40951, and in this instance the simulation produced a good estimate.

8. Ignoring breaks between rows and taking the next usable digits (1,2,3,4 – as suggested in the text) in order in groups of 2 yields the following pairs.
   41 43 21 13 44 44 31 14 31 14 42 12 22 14 24 42 22 32 33 34
The 20 corresponding numbers of yellow pods are:
   1 0 1 1 0 0 1 1 1 1 0 1 0 1 0 0 0 0 0 0
Letting x be the number of yellow pods per simulated pair gives the following distribution.

x	f	r.f.	true P(x)
0	11	.55	.5625
1	9	.40	.3750
2	0	.00	.0625
	20	1.00	1.0000

From the simulation we estimate $P(x \geq 1) \approx 9/20 = .45$. This compares very favorably to the true value of .4375, and in this instance the simulation produced a good estimate.

9. The following Minitab commands produce the desired simulation, where 0 represents a boy and 1 represents a girl.
```
MTB > RANDOM 100 C1-C5;
SUBC> INTEGER 0 1.
MTB > LET C6 = C1+C2+C3+C4+C5
MTB > TABLE C6
```
Letting x be the number of girls per simulated family of 5 gives the following distribution.

x	f	r.f.	true P(x)
0	2	.02	.03125
1	16	.16	.15625
2	27	.27	.31250
3	31	.31	.31250
4	19	.19	.15625
5	5	.05	.03125
	100	1.00	1.00000

From the simulation we estimate $P(x \geq 2) \approx 82/100 = .82$. This compares very favorably to the true value of .8125, and in this instance the simulation produced a good estimate.

10. The following Minitab commands produce the desired simulation.
```
MTB > RANDOM 100 C1-C3;
SUBC> INTEGER 0 6.
MTB > LET C4 = C1+C2+C3
MTB > TABLE C4
```
Letting x be the sum per simulated roll of 3 dice gives the distribution given below at the left, with the theoretical distribution for 216 such rolls given at the right.

x	f	r.f.	true P(x)
3	1	.01	1/216 = .005
4	0	.00	3/216 = .014
5	4	.04	6/216 = .028
6	5	.05	10/216 = .046
7	9	.09	15/216 = .069
8	9	.09	21/216 = .097
9	12	.12	25/216 = .116
10	10	.10	27/216 = .125
11	12	.12	27/216 = .125
12	12	.12	25/216 = .116
13	9	.09	21/216 = .097
14	10	.10	15/216 = .069
15	2	.02	10/216 = .046
16	4	.04	6/216 = .028
17	1	.01	3/216 = .014
18	0	.00	1/216 = .005
	100	1.00	216/216 = 1.000

Letting T be getting a sum of 10, we estimate $P(T) \approx 80/100 = .10$. This compares very favorably to the true value of .125, and in this instance the simulation produced a good estimate.

11. The following Minitab commands produce the desired simulation, where 0 represents a left-hander.
```
MTB > RANDOM 100 C1-C5;
SUBC> INTEGER 0 9.
MTB > CODE (1:9) TO 1 C1-C5 C1-C5
MTB > LET C6 = C1+C2+C3+C4+C5
MTB > LET C6 = 5-C6
MTB > TABLE C6
```

Letting x be the number of left-handers per simulated group gives the following distribution.

x	f	r.f.	true P(x)
0	60	.60	.59049
1	33	.33	.32805
2	5	.05	.07290
3	2	.02	.00810
4	0	.00	.00045
5	0	.00	.00001
	20	1.00	1.00000

From the simulation we estimate $P(x \geq 1) \approx .40$. This compares very favorably to the true value of .40951, and in this instance the simulation produced a good estimate.

12. The following Minitab commands produce the desired simulation, where 0 represents peas with a yellow pod.
```
MTB > RANDOM 100 C1-C2;
SUBC> INTEGER 0 3.
MTB > CODE (1:3) TO 1 C1-C2 C1-C2
MTB > LET C3 = C1+C2
MTB > LET C3 = 2-C3
MTB > TABLE C3
```
Letting x be the number of yellow pods per simulated pair gives the following distribution.

x	f	r.f.	true P(x)
0	56	.56	.5625
1	35	.35	.3750
2	9	.09	.0625
	100	1.00	1.0000

From the simulation we estimate $P(x \geq 1) \approx .44$. This compares very favorably to the true value of .4375, and in this instance the simulation produced a good estimate.

13. The probability of originally selecting the door with the car is 1/3. You can either not switch or switch.
    If you make a selection and do not switch, you expect to get the car 1/3 of the time.
    If you make a selection and switch under the conditions given, there are 2 possibilities.
    A = you originally had the winning door (and switched to a losing door).
    B = you originally had a losing door (and switched to a winning door). [You cannot switch from a losing door to another losing door. If you have a losing door and the door Monty Hall opens is a losing door, then the only door to switch to is the winning door.]
    If you switch, P(A) = 1/3 and P(B) = 2/3.
    Conclusion: the better strategy is to switch since
    P(winning without switching) = P(picking winning door originally) = 1/3
    P(winning with switching) = P(B above) = 2/3
    NOTE: While the above analysis makes the simulation unnecessary, it can be done as follows.
    1. For n trials, randomly select door 1,2 or 3 to hold the prize.
    2. For n trials, randomly select door 1,2 or 3 as your original selection.
    3. Determine m = the number of times the doors match.
    4. If you don't switch, you win m/n of the time. [m/n should be about 1/3]
    5. If you switch, you lose m/n of the time and you win (n-m)/n of the time. [(n-m)/n should be about 2/3]
    The following Minitab commands produce the desired simulation, where column 1 indicates the location of the winning door and column 2 indicates your choice.
```
MTB > RANDOM 100 C1-C2;
SUBC> INTEGER 1 3.
MTB > LET C3 = C1-C2
MTB > TABLE C3
```

S-78 INSTRUCTOR'S SOLUTIONS Chapter 3

A zero in C3 indicates that C1 and C2 match and that the contestant's original choice was correct. For this simulation, the n=100 trials produced 24 0's in C3. The simulated contestant originally picked the winning door 24/100 = 24% of the time and a losing door 76% of the time. If he doesn't switch, he wins 24% of the time – and loses 76% of the time. If he switches, he moves from a winner to a loser 24% of the time – and from a loser to a winner 76% of the time.

14. a. Let A = at least two of 50 random people have the same birth date. Generate 50 random numbers from 1 to 366. Order the numbers to see if there are any duplicates. Repeat this 100 times. P(A) ≈ (# of trials with duplicates)/100, which should be close to the true value of .970.
    b. Let B = at least three of 50 random people have the same birth date. Generate 50 random numbers from 1 to 366. Order the numbers to see if there are any triplicates. Repeat this 100 times. P(B) ≈ (# of trials with triplicates)/100, which should be close to the true value of .127.
    NOTE: The above simulations include February 29 with the same frequency as the other dates. A slightly more accurate simulation could be obtained by giving special consideration to the number 60 (which represents February 29 in the ordered list) as follows: If 60 occurs, flip a coin twice – if 2 heads occur [P(HH)=1/4] use the 60, otherwise ignore the 60 and choose another number.

15. No; his reasoning is not correct. No; the proportion of girls will not increase. Each family would consist of zero to several consecutive girls, followed by a boy. The types of possible families and their associated probabilities would be as follows:
    P(B) = ½ = 1/2   =16/32
    P(GB) = (½)·(½) = 1/4   =8/32
    P(GGB) = (½)·(½)·(½) = 1/8   =4/32
    P(GGGB) = (½)·(½)·(½)·(½) = 1/16   =2/32
    P(GGGGB) = (½)·(½)·(½)·(½)·(½)   =1/32
    etc.
Each collection of 32 families would be expected to look like this, where * represents one family with 5 or more girls and a boy:
    B   B   B   B   B   B   B   B
    B   B   B   B   B   B   B   B
    GB   GB   GB   GB   GB   GB   GB   GB
    GGB   GGB   GGB   GGB   GGGB   GGGB   GGGGB *
A gender count reveals 32 boys and 31 or more girls, an approximately equally distribution of genders. In practice, however, familes would likely stop after having a certain number of children – whether they had a boy or not. If that number was 5 for each family, then * = GGGGG and the expected gender count for the 32 families would be an exactly equal distribution of 31 boys and 31 girls.

3-7 Counting

1. $6! = 6·5·4·3·2·1 = 720$

2. $15! = 15·14·13·12·11·10·9·8·7·6·5·4·3·2·1 = 1,307,674,368,000$
   NOTE: It's staggering how such a small number like 15 can have such a large value for its factorial. In practice this means that if you have 15 books on a shelf you could arrange them in over 1.3 trillion different orders!

3. $_{25}P_2 = 25!/23! = (25·24·23!)/23! = 25·24 = 600$
   NOTE: This technique of "cancelling out" or "reducing" the problem by removing the factors $23! = 23·22·...·1$ from both the numerator and the denominator is preferred over

actually evaluating 25!, actually evaluating 23!, and then dividing those two very large numbers. In general, a smaller factorial in the denominator can be completely divided into a larger factorial in the numerator to leave only the "excess" factors not appearing the in the denominator. This is the technique employed in this manual -- e.g., see #5 below, where the 23! is cancelled from both the numerator and the denominator. In addition, $_nP_r$ and $_nC_r$ will always be integers; calculating a non-integer value for either expression indicates an error has been made. More generally, the answer to any <u>counting</u> problem (but not a <u>probability</u> problem) must always be a whole number; a fractional number indicates that an error has been made.

4. $_{100}P_3 = 100!/97! = (100 \cdot 99 \cdot 98 \cdot 97!)/97! = 100 \cdot 99 \cdot 98 = 970,200$

5. $_{25}C_2 = 25!/(2!23!) = (25 \cdot 24)/2! = 300$

6. $_{100}C_3 = 100!/(3!97!) = (100 \cdot 99 \cdot 98)/3! = 161,700$

7. $_{52}C_5 = 52!/(5!47!) = (52 \cdot 51 \cdot 50 \cdot 49 \cdot 48)/5! = 2,598,960$

8. $_{52}P_5 = 52!/47! = (52 \cdot 51 \cdot 50 \cdot 49 \cdot 48 \cdot 47!)/47! = 52 \cdot 51 \cdot 50 \cdot 49 \cdot 48 = 311,875,200$

9. Let W = winning the described lottery with a single selection.
    The total number of possible combinations is
    $_{49}C_6 = 49!/(43!6!) = (49 \cdot 48 \cdot 47 \cdot 46 \cdot 45 \cdot 44)/6! = 13,983,816$.
    Since only one combination wins, P(W) = 1/13,983,816.

10. Let W = winning the described lottery with a single selection.
    The total number of possible combinations is
    $_{69}C_6 = 69!/(63!6!) = (69 \cdot 68 \cdot 67 \cdot 66 \cdot 65 \cdot 64)/6! = 119,877,472$
    Since only one combination wins, P(W) = 1/119,877,472.

11. Let W = winning the described lottery with a single selection.
    The total number of possible combinations is
    $_{59}C_6 = 59!/(53!6!) = (59 \cdot 58 \cdot 57 \cdot 56 \cdot 55 \cdot 54)/6! = 45,057,474$
    Since only one combination wins, P(W) = 1/45,047,474.

12. Let W = winning the described lottery with a single selection.
    The total number of possible combinations is
    $_{39}C_5 = 39!/(34!5!) = (39 \cdot 38 \cdot 37 \cdot 36 \cdot 35)/5! = 575,757$.
    Since only one combination wins, P(W) = 1/575,757.

13. Let C = choosing the 4 oldest at random.
    The total number of possible groups of 4 is
    $_{32}C_4 = 32!/4!28! = 32 \cdot 31 \cdot 30 \cdot 29/4! = 35,960$.
    Since only one group of 4 is the 4 oldest, P(C) = 1/35,960 = .0000278.
    Yes; P(C) is low enough to suggest the event did not happen randomly.

14. There are $2 \cdot 2 \cdot 2 \cdot 2 \cdot 2 \cdot 2 \cdot 2 \cdot 2 = 2^8 = 256$ possible sequences. Yes; the keyboard being used to prepare this manual has 47 character keys – considering both lower and upper cases, it appears that $2 \cdot 47 = 94$ sequences would be needed.

15. Let W = winning the described lottery with a single selection.
    The total number of possible permutations is
    $_{42}C_6 = 42!/36! = 42 \cdot 41 \cdot 40 \cdot 39 \cdot 38 \cdot 37 = 3,776,965,920$.
    Since only one permutation wins, P(W) = 1/3,776,965,920.

16. Let S = rolling a 6.
    P(S) = 1/6, for each trial.
    P(S_1 S_2 S_3 S_4 S_5) = P(S_1)·P(S_2)·P(S_3)·P(S_4)·P(S_5) = $(1/6)^5$ = 1/7776 = .000128
    Yes; that probability is low enough to suggest the event did not occur by chance alone.

17. Let Y = the 6 youngest are chosen.
    The total number of possible groups of 6 is $_{15}C_6$ = 15!/(6!9!) = 5005.
    Since only of the possibilities consists of the 6 youngest, P(Y) = 1/5005 = .000200.
    Yes; that probability is low enough to suggest the event did not occur by chance alone.

18. $_{10}P_3$ = 10!/7! = 10·9·8 = 720

19. For 3 cities there are 3! = 6 possible sequences; and so there are 4 more possible routes.
    For 8 cities there are 8! = 40,320 possible sequences.

20. The total number of possibilities is 10^9 = 1,000,000,000.
    P(randomly generating any particular number) = 1/1,000,000,000

21. $_5C_2$ = 5!/(2!3!) = (5·4)/2! = 10

22. a. Since a different arrangement of selected names is a different slate, use
       $_{12}P_4$ = 12!/8! = 12·11·10·9 = 11,880
    b. Since a different arrangement of selected names is the same committee, use
       $_{12}C_4$ = 12!/(8!4!) = (12·11·10·9)/4! = 495

23. For 6 letters there are 6! = 720 possible sequences.
    The only English word possible from those letters is "satire."
    Since there is only one correct sequence, P(correct sequence by chance) = 1/720.

24. The first note is given by *. There are 3 possibilities (R,U,D) for each of the next 15 notes. There are 3·3·3·3·3·3·3·3·3·3·3·3·3·3·3 = 3^{15} = 14,348,907 possible sequences.
    NOTE: This assumes each song has at least 16 notes, and it does not guarantee that two different melodies will not have the same representation – if one goes up steps every time the other goes up one step, for example, they both will show a U in that position.

25. Let O = opening the lock on the first try.
    The total number of possible "combinations" is 50·50·50 = 125,000.
    Since only of the possibilities is correct, P(O) = 1/125,000.

26. This problem may be approached in two different ways.
    * general probability formulas of earlier sections
       Let F = selecting a card favorable for getting a flush.
       P(F_1) = 52/52, since the first card could be anything
       P(F_2) = 12/51, since the second card must be from the same suit as the first
       etc.
       P(getting a flush) = P(F_1 and F_2 and F_3 and F_4 and F_5)
                         = P(F_1)·P(F_2)·P(F_3)·P(F_4)·P(F_5)
                         = (52/52)·(12/51)·(11/50)·(10/49)·(9/48) = .00198
    * counting techniques of this section
       The total number of possible 5 card selections is
          $_{52}C_5$ = 52!/(47!5!) = 2,598,960
       The total number of possible 5 card selections from one particular suit is
          $_{13}C_5$ = 13!/(8!5!) = 1287
       The total number of possible 5 card selections from any of the 4 suits is
          4·1287 = 5148
       P(getting a flush) = 5148/2598960 = .00198

27. a. There are 2 possibilities (B,G) for each baby. The number of possible sequences is
    $2 \cdot 2 \cdot 2 \cdot 2 \cdot 2 \cdot 2 \cdot 2 \cdot 2 = 2^8 = 256$
    b. The number of ways to arrange 8 items consisting of 4 B's and 4 G's is $8!/(4!4!) = 70$.
    c. Let E = getting equal numbers of boys and girls in 8 births.
    Based on parts (a) and (b), P(E) 70/256 = .273

28. a. There are 2 possibilities (B,G) for each baby. The number of possible sequences is
    $2 \cdot 2 \cdot 2 \cdot 2 \cdot 2 \cdot 2 \cdot 2 \cdot 2 \cdot 2 \cdot 2 \cdot 2 \cdot 2 \cdot 2 \cdot 2 \cdot 2 \cdot 2 \cdot 2 \cdot 2 \cdot 2 \cdot 2 = 2^{20} = 1,048,576$
    b. The number of ways to arrange 20 items consisting of 10 B's and 10 G's is
    $20!/(10!10!) = 20 \cdot 19 \cdot 18 \cdot 17 \cdot 16 \cdot 15 \cdot 14 \cdot 13 \cdot 12 \cdot 11/10! = 184,756$
    c. Let E = getting equal numbers of boys and girls in 20 births.
    Based on parts (a) and (b), P(E) = 184756/1048576 = .176.
    d. The preceding results indicate that getting 10 boys and 10 girls when 20 babies are randomly selected is not an unusual event – for one such random selection. But the probability of getting that result twice in a row would be $(.176) \cdot (.176) = .0310$, which would be considered unusual since .0310 < .05. Getting 10 boys and 10 girls "consistently" should arouse suspicion.

29. The 1st digit may be any of 8 (any digit except 0 or 1).
    The 2nd digit may be any of 2 (0 or 1).
    The 3rd digit may be any of 9 (any digit except whatever the 2nd digit is).
    There are $8 \cdot 2 \cdot 9 = 144$ possibilities.

30. The total number of ways of selecting 5 eggs from among 12 is
    $_{12}C_5 = 12!/(7!5!) = (12 \cdot 11 \cdot 10 \cdot 9 \cdot 8)/5! = 792$.
    a. Let A = selecting all 3 of the cracked eggs. The total number of ways of selecting 3 eggs from among the 3 cracked ones and 2 eggs from among the 9 good ones is
    $_3C_3 \cdot _9C_2 = [3!/(0!3!)] \cdot [9!/(7!2!)] = [1] \cdot [36] = 36$
    P(A) = 36/792 = .0455
    b. Let N = selecting none of the cracked eggs. The total number of ways of selecting 0 eggs from among the 3 cracked ones and 5 eggs from among the 9 good ones is
    $_3C_0 \cdot _9C_5 = [3!/(3!0!)] \cdot [9!/(4!5!)] = [1] \cdot [126] = 126$
    P(N) = 126/792 = .1591
    c. Let T = selecting 2 of the cracked eggs. The total number of ways of selecting 2 eggs from among the 3 cracked ones and 3 eggs from among the 9 good ones is
    $_3C_2 \cdot _9C_3 = [3!/(1!2!)] \cdot [9!/(6!3!)] = [3] \cdot [84] = 252$
    P(T) = 252/792 = .3182

NOTE 1: Since the answer in part (a) required 4 decimal places to provide 3 significant digit accuracy, the answers in parts (b) and (c) are also given to 4 decimal places for consistency and ease of comparison. In general, this pattern will apply throughout the manual.

NOTE 2: There is only one possibility not covered in parts (a)-(c) above -- viz., one of the cracked eggs is selected. As a check, it is recommended that this probability be calculated. If these 4 mutually exclusive and exhaustive probabilities do not sum to 1.000, then an error has been made.
    Let O = selecting 1 of the cracked eggs.
    The total number of ways of selecting 1 egg from among the 3 cracked ones and 4 eggs from among the 9 good ones is $_3C_1 \cdot _9C_4 = [3!/(2!1!)] \cdot [9!/(5!4!)] = [3] \cdot [126] = 378$
    P(O) = 378/792 = .4773
P(N or O or T or A) = P(N) + P(O) + P(T) + P(A)
    = .1591 + .4773 + .3182 + .0455
    = 1.000 [the 4 values actually sum to 1.0001 due to round-off]
And so everything seems in order. Performing an extra calculation that can provide a check to both the methodology and the mathematics is a good habit that is well worth the effort.

31. Winning the game as described requires two events A and B as follows.
    A = selecting the correct 5 numbers between 1 and 47
        There are $_{47}C_5$ = 47!/(42!5!) = 1,533,939 possible selections.
        Since there is only 1 winning selection, P(A) = 1/1,533,939
    B = selecting the correct number between 1 and 27 in a separate drawing
        There are $_{27}C_1$ = 27!/(26!1!) = 27 possible selections.
        Since there is only 1 winning selection, P(B) = 1/27.
    P(winning the game) = P(A and B)
                       = P(A)·P(B|A)
                       = (1/1,533,939)·(1/27) = 1/41,416,353

32. a. Since there are 64 teams at the start and only 1 team left at the end, 63 teams must be eliminated. Since each game eliminates 1 team, it takes 63 games to eliminate 63 teams.
    b. Let W = picking all 63 winners by random guessing.
       Since there are 2 outcomes for each of the 63 games, there are 2^{63} = 9.223·10^{18} possible sets of results. Since only one such result gives all the correct winners,
       P(W) = $1/2^{63}$ = 1/9.223·10^{18} = 1.084·10^{-19}.
    c. Let E = picking all 63 winners by expert guessing, where each guess has a 70% chance.
       P(E) = $(.7)^{63}$ = 1.743·10^{-10}. This is about 10^9 (one billion) times more likely than winning by random guessing, but it still represents a chance of only 1 in $1/(.7)^{63}$ = 1/(1.743·10^{-10}) = 5,738,831,575 of correctly picking all 63 winners.
       NOTE: The logic employed above is as follows: If E is to have 1 chance in x of occurring, then P(E) = 1/x and solving for x yields x = 1/P(E).

33. There are 26 possible first characters, and 36 possible characters for the other positions. Find the number of possible names using 1,2,3,...,8 characters and then add to get the total.

    ```
    characters    possible names
        1         26                          =                 26
        2         26·36                       =                936
        3         26·36·36                    =             33,696
        4         26·36·36·36                 =          1,213,056
        5         26·36·36·36·36              =         43,670,016
        6         26·36·36·36·36·36           =      1,572,120,576
        7         26·36·36·36·36·36·36        =     56,596,340,736
        8         26·36·36·36·36·36·36·36     =  2,037,468,266,496
                                     total    =  2,095,681,645,538
    ```

34. a. The number of handshakes is the number of ways 2 people can be chosen from 5,
       $_5C_2$ = 5!/(3!2!) = (5·4)/2! = 10.
    b. The number of handshakes is the number of ways 2 people can be chosen from n,
       $_nC_2$ = n!/[(n-2)!2!] = n·(n-1)/2! = n·(n-1)/2.
    c. Visualize the people entering one at a time, each person sitting to the right of the previous person who entered. Where the first person sits is irrelevant – i.e., it merely establishes a reference point but does not affect the number of arrangements. Once the 1st person sits, there are 4 possibilities for the person at his right. Once the 2nd person sits, there are 3 possibilities for the person at his right. Once the 3rd person sits, there are 2 possibilities for the person at his right. Once the 4th person sits, there is only 1 remaining person to sit as his right. And so the number of possible arrangements is 4·3·2·1 = 24.
    d. Use the same reasoning as in part (c). Once the 1st person sits, there are n-1 possibilities for the person at his right. Once the 2nd person sits, there are n-2 possibilities for the person at his right. And so on... Once the next to the last person sits, there is only 1 possibility for the person at his right. And so the number of possible arrangements is (n-1)·(n-2)·...·1 = (n-1)!.

35. a. The calculator factorial key gives $50! = 3.04140932 \times 10^{64}$.
    Using the approximation, $K = (50.5) \cdot \log(50) + .39908993 - .43429448(50)$
    $= 85.79798522 + .39908933 - 21.71472400$
    $= 64.48235115$
    and then $50! = 10^k$
    $= 10^{64.48235115}$
    $= 3.036345215 \times 10^{64}$
    NOTE: The two answers differ by 5.1×10^{61} (i.e., by 51 followed by 60 zeros – which is "zillions and zillions"). While the error may seem large, the numbers being dealt with are so large that the error is only $(5.1 \times 10^{61})/(3.04 \times 10^{64}) = 1.7\%$
    b. The number of possible routes is $300!$.
    Using the approximation, $K = (300.5) \cdot \log(300) + .39908993 - .43429448(300) = 614.4$
    and then $300! = 10^{614.5}$
    Since the number of digits in 10^x is the next whole number above x, $300!$ has 615 digits.

36. a. Assuming the judge knows there are to be 4 computers and 4 humans and frames his guesses accordingly, the possible ways he cold guess is the number of ways 4 of the 8 could be labeled as "computer" and is given by $_8C_4 = 8!/(4!4!) = (8 \cdot 7 \cdot 6 \cdot 5)/4! = 70$. The probability of guessing the right combination by chance alone is therefore 1/70.
    NOTE: If the judge guesses on each one individually, either not knowing or not considering that there should be 4 computers and 4 humans, then the probability of guessing all 8 correctly is $(½)^8 = 1/256$.
    b. Under the assumptions of part (a), the probability that all 10 judges make all correct guesses is $(1/70)^{10} = 3.54 \times 10^{-19}$.

Review Exercises

1. Refer to the table at the right.
    Let PT = the polygraph indicates truth
    PL = the polygraph indicates lie
    AT = the subject is actually telling the truth
    AL = the subject is actually telling a lie
    $P(AL) = 20/100 = .200$

		POLYGRAPH		
		Truth	Lie	
SUBJECT	Truth	65	15	80
	Lie	3	17	20
		68	32	100

2. Refer to the table and notation of exercise #1.
    $P(PL) = 32/100 = .320$

3. Refer to the table and notation of exercise #1.
    $P(AL \text{ or } PL) = P(AL) + P(PL) - P(AL \text{ and } PL)$
    $= 20/100 + 32/100 - 17/100 = 35/100 = .350$

4. Refer to the table and notation of exercise #1.
    $P(AT \text{ or } PT) = P(AT) + P(PT) - P(AT \text{ and } PT)$
    $= 80/100 + 68/100 - 65/100 = 83/100 = .830$

5. Refer to the table and notation of exercise #1.
    $P(AT_1 \text{ and } AT_2) = P(AT_1) \cdot P(AT_2 | AT_1)$
    $= (80/100) \cdot (79/99) = .638$

6. Refer to the table and notation of exercise #1.
    $P(PL_1 \text{ and } PL_2) = P(PL_1) \cdot P(PL_2 | PL_1)$
    $= (32/100) \cdot (31/99) = .100$

7. Refer to the table and notation of exercise #1.
   $P(AT|PL) = 15/32 = .469$

8. Refer to the table and notation of exercise #1.
   $P(PL|AT) = 15/80 = .1875$

9. Let B = a computer breaks down during the first two years
   a. $P(B) = 992/4000 = .248$
   b. $P(B_1 \text{ and } B_2) = P(B_1) \cdot P(B_2) = (.248) \cdot (.248) = .0615$
   c. P(at least 1 breaks down) = 1 - P(all good)
   $= 1 - P(\bar{B}_1 \text{ and } \bar{B}_2 \text{ and } \bar{B}_3)$
   $= 1 - P(\bar{B}_1) \cdot P(\bar{B}_2) \cdot P(\bar{B}_3) = 1 - (.752)^3 = .575$

10. Since the sample size is no more than 5% of the size of the population, treat the selections as being independent even though the selections are made without replacement.
    Let G = a CD is good.
    $P(G) = .98$, for each selection
    P(reject batch) = P(at least one is defective)
    $= 1 - P(\text{all are good})$
    $= 1 - P(G_1 \text{ and } G_2 \text{ and } G_3 \text{ and } G_4)$
    $= 1 - P(G_1) \cdot P(G_2) \cdot P(G_3) \cdot P(G_4) = 1 - (.98)^4 = .0776$
    NOTE: P(reject batch) = $1 - (2450/2500) \cdot (2449/2499) \cdot (2448/2498) \cdot (2447/2497) = .0777$
    if the problem is done considering the effect of the non-replacement.

11. Let G = getting a girl
    $P(G) = ½$, for each child
    P(all girls) = $P(G_1 \text{ and } G_2 \text{ and } G_3 \text{ and}...\text{and } G_{12})$
    $= P(G_1) \cdot P(G_2) \cdot P(G_3) \cdot ... \cdot P(G_{12}) = (½)^{12} = 1/4096$
    Yes; since the probability of the getting all girls by chance alone is so small, it appears the company's claim is valid.

12. a. Let W = the committee of 3 is the 3 wealthiest members.
        The number of possible committees of 3 is $_{10}C_3 = 10!/(7!3!) = 120$.
        Since only one of those committees is the 3 wealthiest members,
        $P(W) = 1/120 = .00833$
    b. Since order is important, the number of possible slates of officers is
        $_{10}P_3 = 10!/7! = 720$

13. Let E = a winning even number occurs.
    a. $P(E) = 18/38 [=9/19]$
    b. odds against E = $P(\bar{E})/P(E)$
        $= (20/38)/(18/38) = 20/18 = 10/9$, usually expressed as 10:9
    c. payoff odds = (net profit):(amount bet)
        If the payoff odds are 1:1, the net profit is $5 for a winning $5 bet.

14. Since the sample size is no more than 5% of the size of the population, treat the selections as being independent even though the selections are made without replacement.
    Let R = getting a Republican.
    $P(R) = .30$, for each selection
    $P(R_1 \text{ and } R_2 \text{ and}...\text{and } R_{12}) = P(R_1) \cdot P(R_2) \cdot ... \cdot P(R_{12}) = (.30)^{12} = .000000531$
    No, the pollster's claim is not correct.

15. Let L = a selected 27 year old male lives for one more year.
    $P(L) = .9982$, for each such person
    $P(L_1 \text{ and } L_2 \text{ and}...\text{and } L_{12}) = P(L_1) \cdot P(L_2) \cdot ... \cdot P(L_{12}) = (.9982)^{12} = .979$

INSTRUCTOR'S SOLUTIONS Chapter 3 S-85

16. a. Let W = winning the described lottery with a single selection.
  The total number of possible combinations is
  $_{52}C_6 = 52!/(46!6!) = (52 \cdot 51 \cdot 50 \cdot 49 \cdot 48 \cdot 47)/6! = 20{,}358{,}520$.
  Since only one combination wins, P(W) = 1/20,358,520.
  b. Let W = winning the described lottery with a single selection.
  The total number of possible combinations is
  $_{30}C_5 = 30!/(25!5!) = (30 \cdot 29 \cdot 28 \cdot 27 \cdot 26)/5! = 142{,}506$.
  Since only one combination wins, P(W) = 1/142,506.
  c. Two events need to occur to win the Big Game.
  *Let A = selecting the winning 5 from among the 50.
    The total number of possible combinations is
    $_{50}C_5 = 50!/(45!5!) = (50 \cdot 49 \cdot 48 \cdot 47 \cdot 46)/5! = 2{,}118{,}760$.
    Since only one combination wins, P(A) = 1/2,118,760.
  *Let B = selecting the winning 1 from among the 36, and P(B) = 1/36.
  P(winning the Big Game) = P(A and B)
  $= P(A) \cdot P(B|A)$
  $= (1/2{,}118{,}760) \cdot (1/36)$
  $= 1/76{,}275{,}360$

Cumulative Review Exercises

1. scores arranged in order: 0 0 0 2 3 3 3 3 4 4 4 5 5 5 5 6 6 6 6 7 7
  summary statistics: n = 21, $\Sigma x = 84$, $\Sigma x^2 = 430$
  a. $\bar{x} = (\Sigma x)/n = 84/21 = 4.0$
  b. $\tilde{x} = 4$, the 11th score in the ordered list
  c. $s^2 = [n \cdot (\Sigma x^2) - (\Sigma x)^2]/[n(n-1)] = [21(430) - (84)^2]/[(21)(20)] = 4.7$
  $s = 2.2$
  d. $s^2 = 4.7$ [from part (c)]
  e. Yes; all of the scores are either 0 or positive.
  f. Let G = randomly selecting a positive value
  P(G) = 18/21 = 6/7.
  g. Let G = randomly selecting a positive value
  $P(G_1 \text{ and } G_2) = P(G_1) \cdot P(G_2|G_1)$
  $= (18/21) \cdot (17/20) = .729$
  h. Let E = the treatment is effective.
  P(E) = ½, under the given assumptions, for each person
  $P(E_1 \text{ and } E_2 \text{ and} \ldots \text{and } E_{18}) = P(E_1) \cdot P(E_2) \cdot \ldots \cdot P(E_{18}) = (½)^{18} = 1/262{,}144$
  Yes; this probability is low enough to reject the idea that the treatment is ineffective.
  Yes; the treatment appears to be effective.

2. Let x = the height selected.
  The values identified on the boxplot are: x_1 [minimum] = 56.1
  $P_{25} = 62.2$
  $P_{50} = 63.6$
  $P_{75} = 65.0$
  x_n [maximum] = 71.1
  NOTE: Assume that the number of heights is so large that (1) repeated sampling does not affect the probabilities for subsequent selections and (2) it can be said that P_a has a% of the heights below it and (100-a)% of the heights above it.
  a. The exact values of $\bar{x}$ and s cannot be determined from the information given. The Range Rule of Thumb [the highest and lowest values are approximately 2 standard deviations above and below the mean], however, indicates that $s \approx (x_n - x_1)/4$ and $\bar{x} \approx (x_1 + x_n)/2$. In this case we estimate $\bar{x} \approx (56.1 + 71.1)/2 = 63.6$

NOTE: The boxplot, which gives $\tilde{x} = P_{50} = 63.6$, is symmetric. This suggests that the original distribution is approximately so. In a symmetric distribution the mean and median are equal. This also suggests $\bar{x} \approx 63.6$.

b. $P(56.1 < x < 62.2) = P(x_1 < x < P_{25})$
$\phantom{P(56.1 < x < 62.2)} = .25 - 0 = .25$

c. $P(x < 62.2 \text{ or } x > 63.6) = P(x < 62.2) + P(x > 63.6)$
$\phantom{P(x < 62.2 \text{ or } x > 63.6)} = P(x < P_{25}) + P(x > P_{50})$
$\phantom{P(x < 62.2 \text{ or } x > 63.6)} = .25 + .50 = .75$

NOTE: No height is less than 62.2 <u>and</u> greater than 63.6 – i.e., the addition rule for mutually exclusive events can be used.

d. Let B = selecting a height between 62.2 and 63.6.
$P(B) = P(P_{25} < x < P_{50})$
$ = .50 - .25 = .25$
$P(B_1 \text{ and } B_2) = P(B_1) \cdot P(B_2 | B_1)$
$\phantom{P(B_1 \text{ and } B_2)} = (.25) \cdot (.25) = .0625$

e. Let S = a selected woman is shorter than the mean.
   T = a selected woman is taller than the mean.
   E = the event that a group of 5 women consists of 3 T's and 2 S's.
There are $2 \cdot 2 \cdot 2 \cdot 2 \cdot 2 = 32$ equally likely [see the NOTE below] sequences of T's and S's. The number of arrangements of 3 T's and 2 S's is $5!/(3!2!) = 10$.
$P(E) = 10/32 = .3125$

NOTE: The exact value of the mean cannot be determined from the information given. From part (a) we estimate $\bar{x} \approx 63.6$, which is also the median. If this is true then *in this particular problem* $P(x < \bar{x}) = P(S) = .50$ and $P(x > \bar{x}) = P(T) = .50$. This means that all arrangements of T's and S's are equally likely and
$P(E) = (\# \text{ of ways E can occur})/(\text{total number of possible outcomes})$

Chapter 4

Probability Distributions

4-2 Random Variables

1. a. Continuous, since height can be any value on a continuum.
   b. Discrete, since the number of points must be an integer.
   c. Continuous, since time can be any value on a continuum.
   d. Discrete, since the number of players must be an integer.
   e. Discrete, since salary is typically stated in whole dollars – and it cannot be stated in units smaller than whole number of cents.

2. a. Discrete, since monetary cost must be stated in units of .01 dollars.
   b. Discrete, since the number of movies must be an integer.
   c. Continuous, since time can be any value on a continuum.
   d. Discrete, since the number of actors must be an integer.
   e. Continuous, since weight can be any value on a continuum.

NOTE: When working with probability distributions and formulas in the exercises that follow, always keep these important facts in mind.
* If one of the conditions for a probability distribution does not hold, the formulas do not apply and produce numbers that have no meaning.
* $\Sigma x \cdot P(x)$ gives the mean of the x values and must be a number between the highest and lowest x values.
* $\Sigma x^2 \cdot P(x)$ gives the mean of the x^2 values and must be a number between the highest and lowest x^2 values.
* $\Sigma P(x)$ must always equal 1.000.
* Σx and Σx^2 have no meaning and should not be calculated.
* The quantity $[\Sigma x^2 \cdot P(x) - \mu^2]$ cannot possibly be negative; if it is, then there is a mistake.
* Always be careful to use the <u>unrounded</u> mean in the calculation of the variance and to take the square root of the <u>unrounded</u> variance to find the standard deviation.

3. This is a probability distribution since $\Sigma P(x) = 1$ is true and $0 \le P(x) \le 1$ is true for each x.

x	P(x)	x·P(x)	x^2	x^2P(x)	
0	.125	0	0	0	$\mu = \Sigma x \cdot P(x)$ = 1.500, rounded to 1.5
1	.375	.375	1	.375	$\sigma^2 = \Sigma x^2 \cdot P(x) - \mu^2$
2	.375	.750	4	1.500	= 3.000 − (1.500)²
3	.125	.375	9	1.125	= .750
	1.000	1.500		3.000	σ = .866, rounded to 0.9

4. This is not a probability distribution since $\Sigma P(x) = .977 \ne 1$.

5. This is not a probability distribution since $\Sigma P(x) = .94 \ne 1$.

6. This is a probability distribution since $\Sigma P(x) = 1$ is true and $0 \le P(x) \le 1$ is true for each x.

x	P(x)	x·P(x)	x^2	x^2P(x)	
0	.0000	0	0	0	$\mu = \Sigma x \cdot P(x)$ = 3.9598, rounded to 4.0
1	.0001	.0001	1	.0001	$\sigma^2 = \Sigma x^2 \cdot P(x) - \mu^2$
2	.0006	.0012	4	.0024	= 15.7204 − (3.9598)²
3	.0387	.1161	9	.3483	= .0404
4	.9606	3.8424	16	15.3696	σ = .2010, rounded to 0.2
	1.0000	3.9598		15.7204	

S-88 INSTRUCTOR'S SOLUTIONS Chapter 4

7. This is a probability distribution since $\Sigma P(x)=1$ is true and $0 \leq P(x) \leq 1$ is true for each x.

x	P(x)	x·P(x)	x^2	x^2P(x)
0	.512	0	0	0
1	.301	.301	1	.301
2	.132	.264	4	.528
3	.055	.165	9	.495
	1.000	.730		1.324

$\mu = \Sigma x \cdot P(x)$
$ = .730$, rounded to 0.7
$\sigma^2 = \Sigma x^2 \cdot P(x) - \mu^2$
$ = 1.324 - (.73)^2$
$ = .791$
$\sigma = .889$, rounded to 0.9

8. This is not a probability distribution since $\Sigma P(x) = .986 \neq 1$.

9. This is a probability distribution since $\Sigma P(x)=1$ is true and $0 \leq P(x) \leq 1$ is true for each x.

x	P(x)	x·P(x)	x^2	x^2P(x)
4	.1809	.7236	16	2.8944
5	.2234	1.1170	25	5.5850
6	.2234	1.3404	36	8.0424
7	.3723	2.6061	49	18.2427
	1.0000	5.7871		34.7645

$\mu = \Sigma x \cdot P(x)$
$ = 5.7871$, rounded to 5.8
$\sigma^2 = \Sigma x^2 \cdot P(x) - \mu^2$
$ = 34.7645 - (5.7871)^2$
$ = 1.2740$
$\sigma = 1.129$, rounded to 1.1

No; since $.1809 > .05$, a four game sweep is not considered an unusual event.

10. This is not a probability distribution since $\Sigma P(x) = .9686 \neq 1$.

11.
x	P(x)	x·P(x)
5	244/495	2.4646
-5	251/495	-2.5353
	495/495	-0.0707

$E = \Sigma x \cdot P(x)$
$ = \$-0.0707$ [i.e., a loss of 7.07¢]

Since the expected loss is 7.07¢ for a $5 bet, the expected (or long run) loss is $7.07/5 = 1.41$¢ for each $1 a person bets.

12. a.
x	P(x)	x·P(x)
175	1/38	175/38
-5	37/38	-185/38
	38/38	-10/38

$E = \Sigma x \cdot P(x)$
$ = -10/38$
$ = \$-0.263$ [i.e., a loss of 26.3¢]

The expected loss is 26.3¢ for a $5 bet.

b.
x	P(x)	x·P(x)
5	18/38	90/38
-5	20/38	-100/38
	38/38	-10/38

$E = \Sigma x \cdot P(x)$
$ = -10/38$
$ = \$-0.263$ [i.e., a loss of 26.3¢]

The expected loss is 26.3¢ for a $5 bet.

c. Since the bets in parts (a) and (b) each have (exactly the same) negative expectation, the best option is not to bet at all.

NOTE: Problems whose probabilities do not "come out even" as decimals can be very sensitive to rounding errors. Since the P(x) values are used in several subsequent calculations, express them as exact fractions instead of rounded decimals. If it is not convenient to continue with fractions throughout the entire problem, then use sufficient decimal places (typically one more than "usual") in the x·P(x) and x^2·P(x) columns to guard against cumulative rounding errors.

13. a. Mike "wins" $100,000 - $250 = $99,750 if he dies.
    Mike "loses" $250 if he lives.

b.
x	P(x)	x·P(x)
99750	.0015	149.625
-250	.9985	-249.625
	1.0000	-100.000

$E = \Sigma x \cdot P(x)$
$ = -100.000$ [i.e., a loss of $100]

c. Since the company is making a $100 profit at this price, it would break even if it sold the policy for $100 less – i.e. for $150. NOTE: This oversimplified analysis ignores the cost of doing business. If the costs (printing, salaries, etc.) associated with offering the policy is $25, for example, then the company's profit is only $75 and selling the

policy for $75 less (i.e., for $175) would represent the break even point for the company.
d. Buying life insurance is similar to purchasing other services. You let someone make a profit off you by shoveling the snow in your driveway because the cost is worth the effort and frustration you save. You let someone make a profit off you by selling you life insurance because the cost is worth the peace you have from providing for the financial security of your heirs.

14.

x	P(x)	x·P(x)
1,000,000	.000000011	.011111111
100,000	.000000009	.000909091
25,000	.000000009	.000227273
5,000	.000000027	.000136362
2,500	.000000036	.000090909
0	.999999908	0
	1.000000000	.012474746

a. $E = \Sigma x \cdot P(x)$
  $= .012$ [i.e., 1.2 cents]
b. mean loss
  $=$ price of stamp $- 1.2¢$
  $= 37¢ - 1.2¢$
  $= 35.8¢$ [as of spring 2003]

No; from a purely financial point of view, it is not worth entering this contest.

15. a. $10 \cdot 10 \cdot 10 \cdot 10 = 10,000$
    b. $1/10,000 = .0001$
    c.

x	P(x)	x·P(x)
2787.50	.0001	.27875
−.50	.9999	−.49995
	1.0000	−.22120

$E = \Sigma x \cdot P(x)$
$= -.2212$ [i.e., a loss of 22.12¢]

d. The Pick 4 game is the better of the two. Even though they both are losing propositions, the expected value is higher for the Pick 4 game – i.e., -22.12 > -22.5.

16. a. $10 \cdot 10 \cdot 10 = 1,000$
    b. $1/1,000 = .001$
    c.

x	P(x)	x·P(x)
249.50	.001	.2495
−.50	.999	−.4995
	1.000	−.2500

$E = \Sigma x \cdot P(x)$
$= -.25$ [i.e., a loss of 25¢]

d. The New Jersey Pick 3 is the better of the two. Even though they both are losing propositions, the expected value is higher for the New Jersey version – i.e., -22.5 > -25.

17. a. $P(x=9) = .122$
    b. $P(x \geq 9) = P(x=9 \text{ or } x=10 \text{ or } x=11 \text{ or } x=12 \text{ or } x=13 \text{ or } x=14)$
       $= P(x=9) + P(x=10) + P(x=11) + P(x=12) + P(x=13) + P(x=14)$
       $= .122 + .061 + .022 + .006 + .001 + .000 = .212$
    c. Part (b). In determining whether a particular event in some hierarchy of events is unusual, one should consider both that event and the events that are more extreme. Even though the probability of a particular event may be very low, that event might not be unusual in a practical sense. P(35 girls in 60 births) = .045 < .05, for example, but getting 35 girls in 60 births is one of the more common outcomes – it's just that there are so many possible outcomes that the probability of any one particular outcome is small.
    d. No; this is not evidence that the technique is effective. Getting 9 or more girls by chance alone is not unusual, since $P(x \geq 9) = .212 > .05$.

18. a. $P(x=12) = .006$
    b. $P(x \geq 12) = P(x=12 \text{ or } x=13 \text{ or } x=14)$
       $= P(x=12) + P(x=13) + P(x=14)$
       $= .006 + .001 + .000 = .007$
    c. Part (b). In determining whether a particular event in some hierarchy of events is unusual, one should consider both that event and the events that are more extreme. Even though the probability of a particular event may be very low, that event might not be unusual in a practical sense. P(35 girls in 60 births) = .045 < .05, for example, but

getting 35 girls in 60 births is one of the more common outcomes – it's just that there are so many possible outcomes that the probability of any one particular outcome is small.

d. Yes; this is evidence that the technique is effective. Getting 12 or more girls by chance alone would be unusual, since $P(x \geq 12) = .007 < .05$.

19. a. $P(x \geq 11) = P(x=11 \text{ or } x=12 \text{ or } x=13 \text{ or } x=14)$
$= P(x=11) + P(x=12) + P(x=13) + P(x=14)$
$= .022 + .006 + .001 + .000 = .029$

b. Yes; this is evidence that the technique is effective. Getting 11 or more girls by chance alone would be unusual, since $P(x \geq 11) = .029 < .05$.

20. a. $P(x \geq 10) = P(x=10 \text{ or } x=11 \text{ or } x=12 \text{ or } x=13 \text{ or } x=14)$
$= P(x=10) + P(x=11) + P(x=12) + P(x=13) + P(x=14)$
$= .061 + .022 + .006 + .001 + .000 = .090$

b. No; this is not evidence that the technique is effective. Getting 10 or more girls by chance alone is not unusual, since $P(x \geq 10) = .090 > .05$.

21. $P(x \geq 8) = P(x=8 \text{ or } x=9 \text{ or } x=10 \text{ or } x=11 \text{ or } x=12 \text{ or } x=13 \text{ or } x=14)$
$= P(x=8) + P(x=9) + P(x=10) + P(x=11) + P(x=12) + P(x=13) + P(x=14)$
$= .183 + .122 + .061 + .022 + .006 + .001 + .000 = .395$

No; Bob's claim is not valid. Getting 8 or more answers correct by chance alone is not unusual, since $P(x \geq 8) = .395 > .05$.

22. $P(x \leq 2) = P(x=0 \text{ or } x=1 \text{ or } x=2)$
$= P(x=0) + P(x=1) + P(x=2)$
$= .000 + .001 + .006 = .007$

Yes; the number of Bob's correct answers is unusual. Getting 2 or fewer answers correct by chance alone would be unusual, since $P(x \leq 2) = .007 < .05$.

No; Bob's claim that he can pass true/false tests by guessing is not valid. Since his number of correct answers was unusually low rather than unusually high, he would not pass.

23. For every $1000, Bond A gives a profit of $(.06)(\$1000) = \$60$ with probability .99.

```
    x        P(x)      x·P(x)          E = Σx·P(x)
   60        .99       59.40             = 49.40   [i.e., a profit of $49.40]
-1000        .01      -10.00
            1.00       49.40
```

For every $1000, Bond B gives a profit of $(.08)(\$1000) = \$80$ with probability .95.

```
    x        P(x)      x·P(x)          E = Σx·P(x)
   80        .95       76.00             = 26.00   [i.e., a profit of $26.00]
-1000        .05      -50.00
            1.00       26.00
```

Bond A is the better bond since it was the higher expected value – i.e., 49.40 > 26.00. Since both bonds have positive expectations, either one would be a reasonable selection. Although bond A has a higher expectation, a person willing to assume more risk in hope of a higher payoff might opt for Bond B.

24. The 16 equally like outcomes in the sample space are given at the right. If x represents the number of girls, counting the numbers of favorable outcomes indicates

$P(x = 0) = 1/16 = .0625$
$P(x = 1) = 4/16 = .2500$
$P(x = 2) = 6/16 = .3750$
$P(x = 3) = 4/16 = .2500$
$P(x = 4) = 1/16 = .0625$

outcome	x	outcome	x
BBBB	0	GBBB	1
BBBG	1	GBBG	2
BBGB	1	GBGB	2
BGBB	1	GGBB	2
BGGB	2	GGGB	3
BGBG	2	GGBG	3
BBGG	2	GBGG	3
BGGG	3	GGGG	4

x	P(x)	x·P(x)	x^2	x^2P(x)
0	.0625	0	0	0
1	.2500	.2500	1	.2500
2	.3750	.7500	4	1.5000
3	.2500	.7500	9	2.2500
4	.0625	.2500	16	1.0000
	1.0000	2.0000		5.0000

$\mu = \Sigma x \cdot P(x)$
   $= 2.0000$, rounded to 2.0
$\sigma^2 = \Sigma x^2 \cdot P(x) - \mu^2$
   $= 5.0000 - (2.0000)^2$
   $= 1.0000$
$\sigma = 1.0000$, rounded to 1.0

25. Let C and N represent correctly and not correctly calibrated altimeters respectively. For 8 C's and 2 N's, there are 7 possible samples of size n=3 – given with their probabilities at the right. Letting x be the number of N's produced the distribution and calculations given below.

P(GGG) = (8/10)·(7/9)·(6/8) = 336/720
P(GGN) = (8/10)·(7/9)·(2/8) = 112/720
P(GNG) = (8/10)·(2/9)·(7/8) = 112/720
P(NGG) = (2/10)·(8/9)·(7/8) = 112/720
P(GNN) = (8/10)·(2/9)·(1/8) = 16/720
P(NGN) = (2/10)·(8/9)·(1/8) = 16/720
P(NNG) = (2/10)·(1/9)·(8/8) = 16/720
                                720/720

x	P(x)	x·P(x)	x^2	x^2P(x)
0	336/720	0	0	0
1	336/720	336/720	1	336/720
2	48/720	96/720	4	192/720
	720/720	432/720		528/720

$\mu = \Sigma x \cdot P(x)$
   $= 432/720 = .600$, rounded to 0.6
$\sigma^2 = \Sigma x^2 \cdot P(x) - \mu^2$
   $= (528/720) - (432/720)^2 = .3733$
$\sigma = .611$, rounded to 0.6

26. a. For random selections with replacement, P(x) = .1 for each x = 0,1,2,3,4,5,6,7,8,9.

x	P(x)	x·P(x)	x^2	x^2P(x)
0	.1	0	0	0
1	.1	.1	1	.1
2	.1	.2	4	.4
3	.1	.3	9	.9
4	.1	.4	16	1.6
5	.1	.5	25	2.5
6	.1	.6	36	3.6
7	.1	.7	49	4.9
8	.1	.8	64	6.4
9	.1	.9	81	8.1
	1.0	4.5		28.5

$\mu = \Sigma x \cdot P(x)$
   $= 4.5$
$\sigma^2 = \Sigma x^2 \cdot P(x) - \mu^2$
   $= 28.5 - (4.5)^2$
   $= .0404$
$\sigma = 2.8723$, rounded to 2.9

b. The z score for each x is $z = (x-\mu)/\sigma = (x-4.5)/2.8723$.

z	P(z)	z·P(z)	z^2	z^2P(z)
-1.5667	.1	-.15667	2.45455	.245455
-1.2185	.1	-.12185	1.48485	.148485
-0.8704	.1	-.08704	.75758	.075758
-0.5222	.1	-.05222	.27272	.027272
-0.1741	.1	-.01741	.03030	.003030
0.1741	.1	.01741	.03030	.003030
0.5222	.1	.05222	.27272	.027272
0.8704	.1	.08704	.75758	.075758
1.2185	.1	.12185	1.48485	.148485
1.5667	.1	.15667	2.45455	.245455
	1.0	0		1.000000

$\mu = \Sigma z \cdot P(z)$
   $= 0$
$\sigma^2 = \Sigma z^2 \cdot P(z) - \mu^2$
   $= 1 - (0)^2$
   $= 1$
$\sigma = 1$

c. Yes; for the z scores of any population of size n,
$\Sigma z = \Sigma(x-\mu)/\sigma = (1/\sigma) \cdot \Sigma(x-\mu) = (1/\sigma) \cdot [\Sigma x - \Sigma \mu] = (1/\sigma) \cdot [n\mu - n\mu] = 0$
$\Sigma z^2 = \Sigma[(x-\mu)/\sigma]^2 = (1/\sigma^2) \cdot \Sigma(x-\mu)^2 = (1/\sigma^2) \cdot n\sigma^2 = n$
$\mu = \Sigma z/n = 0/n = 0$ and $\sigma^2 = \Sigma(z-\mu)^2/n = \Sigma z^2/n = n/n = 1$

27. The distribution and calculations are given below.

x	P(x)	x·P(x)	x^2	x^2P(x)
1	.2	.2	1	.2
2	.2	.4	2	.8
3	.2	.6	9	1.8
4	.2	.8	16	3.2
5	.2	1.0	25	5.0
	1.0	3.0		11.0

$\mu = \Sigma x \cdot P(x)$
   $= 3.0$
$\sigma^2 = \Sigma x^2 \cdot P(x) - \mu^2$
   $= 11.0 - (3.0)^2$
   $= 2.0$
$\sigma = 1.414$

a. For n=5, $\mu = (n+1)/2 = 6/2 = 3.0$. This agrees with the calculations above.
b. For n=5, $\sigma^2 = (n^2-1)/12 = 24/12 = 2$.
$\sigma = \sqrt{2} = 1.414$. This agrees with the calculations above.
c. For n=20,
$\mu = (n+1)/2 = 21/2 = 10.5$.
$\sigma^2 = (n^2-1)/12 = 399/12 = 33.25$
$\sigma = \sqrt{33.25} = 5.766$, rounded to 5.8

28. Since each die has 6 faces, there will be 6·6 = 36 possible outcomes. If each of the 12 sums 1,2,3,...12 is to be equally likely, each sum must occur exactly 3 times. As a starting point, suppose that one of the dice is normal. If one die contains the usual digits 1,2,3,4,5,6, the other die
    (1) must have three 0's to pair with the 1 to get three sums of 1.
    (2) must have three 6'3 to pair with the 6 to get three sums of 12.
The 36 possible outcomes generated by such dice would be

1-0	2-0	3-0	4-0	5-0	6-0
1-0	2-0	3-0	4-0	5-0	6-0
1-0	2-0	3-0	4-0	5-0	6-0
1-6	2-6	3-6	4-6	5-6	6-6
1-6	2-6	3-6	4-6	5-6	6-6
1-6	2-6	3-6	4-6	5-6	6-6

Inspection indicates that each of the sums 1,2,3,...,12 appears exactly 3 times so that $P(x) = 3/36 = 1/12$ for x = 1,2,3,...,12. The solution, therefore, is to mark one die normally 1,2,3,4,5,6 and mark the other die 0,0,0,6,6,6.

4-3 Binomial Experiments

NOTE: The four requirements for a binomial experiment are
#1 There are a fixed number of trials.
#2 The trials are independent.
#3 Each trial has two possible named outcomes.
#4 The probabilities remain constant for each trial.

1. No; requirement #3 is not met. There are more than 2 possible outcomes.

2. Yes; all four requirements are met.

3. No; requirement #3 is not met. There are more than 2 possible outcomes.

4. Yes; all four requirements are met.

5. Yes; all four requirements are met.

6. Yes; all four requirements are met.

7. No; requirement #3 is not met. There are more than 2 possible outcomes.

8. Yes; all four requirements are met.

9. let W = guessing the wrong answer
       C = guessing the correct answer
   P(W) = 4/5, for each question
   P(C) = 1/5, for each question

a. P(WWC) = P(W_1 and W_2 and C_3)
   = P(W_1)·P(W_2)·P(C_3)
   = (4/5)·(4/5)·(1/5) = 16/125 = .128
b. There are 3 possible arrangements: WWC, WCW, CWW
   Following the pattern in part (a)
   P(WWC) = P(W_1 and W_2 and C_3) = P(W_1)·P(W_2)·P(C_3) = (4/5)·(4/5)·(1/5) = 16/125
   P(WCW) = P(W_1 and C_2 and W_3) = P(W_1)·P(C_2)·P(W_3) = (4/5)·(1/5)·(4/5) = 16/125
   P(CWW) = P(C_1 and W_2 and W_3) = P(C_1)·P(W_2)·P(W_3) = (1/5)·(4/5)·(4/5) = 16/125
c. P(exactly one correct answer) = P(WWC or WCW or CWW)
   = P(WWC) + P(WCW) + P(CWW)
   = 16/125 + 16/125 + 16/125 = 48/125 = .384

10. let W = guessing the wrong answer
    C = guessing the correct answer
    P(W) = 3/4, for each question
    P(C) = 1/4, for each question
    a. P(WWCCCC) = P(W_1 and W_2 and C_3 and C_4 and C_5 and C_6)
       = P(W_1)·P(W_2)·P(C_3)·P(C_4)·P(C_5)·P(C_6)
       = (3/4)·(3/4)·(1/4)·(1/4)·(1/4)·(1/4) = 9/4096 = .00220
    b. There are 6!/(2!4!) = 15 possible arrangements as follows. The probability for each arrangement will include the factor (3/4) two times and the factor (1/4) four times to give the result 9/4096 = .00220 of part (a)

    | WWCCCC | CWWCCC | CCWWCC | CCCWWC | CCCCWW |
    | WCWCCC | CWCWCC | CCWCWC | CCCWCW | |
    | WCCWCC | CWCCWC | CCWCCW | | |
    | WCCCWC | CWCCCW | | | |
    | WCCCCW | | | | |

    c. P(exactly 4 correct answers) = 15·(9/4096) = .0330

11. From Table A-1 in the .01 column and the 2-0 row, .980.

12. From Table A-1 in the .01 column and the 7-2 row, .002.

13. From Table A-1 in the .95 column and the 4-3 row, .171.

14. From Table A-1 in the .99 column and the 6-5 row, .057.

15. From Table A-1 in the .95 column and the 10-4 row, $.000^+$.

16. From Table A-1 in the .05 column and the 11-7 row, $.000^+$.

NOTE: To use the binomial formula, one must identify 3 quantities: n,x,p. Table A-1, for example, requires only these 3 values to supply a probability. Since what the text calls "q" always equals 1-p, it can be so designated without introducing unnecessary notation [just as no special notation is utilized for the quantity n-x, even though it appears twice in the binomial formula]. This has the additional advantage of ensuring that the probabilities p and 1-p sum to 1.00 and protecting against an error in the separate calculation and/or identification of "q." In addition, reversing the order of (n-x)! and x! in the denominator of the $_nC_x$ coefficient term seems appropriate. That agrees with the n!/(n_1!n_2!) logic of the "permutation rule when some objects are alike" for n_1 objects of one type and n_2 objects of another type, and that places the "x" and "n-x" in the same order in both the denominator of the coefficient term and the exponents. Such a natural ordering also leads to fewer errors. Accordingly, this manual expresses the binomial formula as

P(x) = [n!/x!(n-x)!]·p^x·$(1-p)^{n-x}$

17. $P(x) = [n!/x!(n-x)!] \cdot p^x \cdot (1-p)^{n-x}$
    $P(x=4) = [6!/4!2!] \cdot (.55)^4 \cdot (.45)^2$
    $\quad\quad = [15] \cdot (.55)^4 (.45)^2 = [15] \cdot (.0915) \cdot (.2025) = .2779$

IMPORTANT NOTE: The intermediate values of 15, .0915 and .2025 are given to help those with incorrect answers to identify the portion of the problem in which the mistake was made. This practice will be followed in most problems (i.e., not just binomial problems) throughout the manual. In practice, all calculations can be done in one step on the calculator. You may choose to (or be asked to) write down such intermediate values for your own (or the instructor's) benefit, but never round off in the middle of a problem. Do not write the values down on paper and then re-enter them in the calculator – use the memory to let the calculator remember with complete accuracy any intermediate values that will be used in subsequent calculations. In addition, always make certain that the quantity [n!/x!(n-x)!] is a whole number and that the final answer is between 0 and 1.

18. $P(x) = [n!/x!(n-x)!] \cdot p^x \cdot (1-p)^{n-x}$
    $P(x=2) = [6!/2!4!] \cdot (.45)^2 \cdot (.55)^4$
    $\quad\quad = [15] \cdot (.45)^2 \cdot (.55)^4 = .2780$

19. $P(x) = [n!/x!(n-x)!] \cdot p^x \cdot (1-p)^{n-x}$
    $P(x=3) = [8!/3!5!] \cdot (1/4)^3 \cdot (3/4)^5$
    $\quad\quad = [56] \cdot (.25)^3 \cdot (.75)^5 = .208$

20. $P(x) = [n!/x!(n-x)!] \cdot p^x \cdot (1-p)^{n-x}$
    $P(x=8) = [10!/8!2!] \cdot (1/3)^8 \cdot (2/3)^2$
    $\quad\quad = [45] \cdot (1/6561) \cdot (4/9) = .00305$

21. $P(x \geq 5) = P(x=5 \text{ or } x=6) = P(x=5) + P(x=6)$
    $\quad\quad\quad\quad = .3283 + .1428 = .4711$
    No; since .4711 > .05, this is not an unusual occurrence.

22. $P(x \leq 2) = P(x=0 \text{ or } x=1 \text{ or } x=2) = P(x=0) + P(x=1) + P(x=2)$
    $\quad\quad\quad\quad = .0005 + .0071 + .0462 = .0538.$
    No; since .0538 > .05, this is not an unusual occurrence.

23. $P(x > 1) = 1 - P(x \leq 1)$
    $\quad\quad\quad = 1 - P(x=0 \text{ or } x=1)$
    $\quad\quad\quad = 1 - [P(x=0) + P(x=1)]$
    $\quad\quad\quad = 1 - [.0005 + .0071]$
    $\quad\quad\quad = 1 - [.0076] = .9924$
    NOTE: This also could be found as $P(x > 1) = P(x=2 \text{ or } x=3 \text{ or } x=4 \text{ or } x=5 \text{ or } x=6)$. In general the manual will choose the most efficient technique for solving problems that may be approached in more than one way.
    Yes; since P(not having more than one on time) = 1 - .9924 = .0076 < .05, not having more than one on time would be an unusual occurrence.

24. $P(x \geq 1) = 1 - P(x=0)$
    $\quad\quad\quad = 1 - .0005 = .9995$
    Yes; since P(not having at least one on time) = 1 - .9995 = .0005 < .05, not having at least one on time would be an unusual occurrence.

25. Let x = the number of men that are color blind.
    binomial problem: n = 6 and p = .09, use the binomial formula
    $P(x) = [n!/x!(n-x)!] \cdot p^x \cdot (1-p)^{n-x}$
    $P(x=2) = [6!/2!4!] \cdot (.09)^2 \cdot (.91)^4$
    $\quad\quad = [15] \cdot (.0081) \cdot (.6857) = .0833$

26. Let x = the number of defective bulbs.
    binomial problem: n = 24 and p = .04, use the binomial formula
    $P(x) = [n!/x!(n-x)!] \cdot p^x \cdot (1-p)^{n-x}$
    P(accept shipment) = P(x≤1)
                             = P(x=0) + P(x=1)
                             = $[24!/0!24!] \cdot (.04)^0 \cdot (.96)^{24} + [24!/1!23!] \cdot (.04)^1 \cdot (.96)^{23}$
                             = [1]·(1)·(.3754) + [24]·(.04)·(.3911)
                             = .3754 + .3754 = .7508

27. Let x = the number of taxpayers that are audited.
    binomial problem: n = 5 and p = .01, use Table A-1
    a. P(x=3) = .000⁺
    b. P(x≥3) = P(x=3) + P(x=4) + P(x=5)
                 = .000⁺ + .000⁺ + .000⁺ = .000⁺
    c. Since .000⁺ < .05, "three or more are audited" is an unusual result and we conclude that factors other than chance are at work. It appears that either customers of the Hemingway Company are being targeted by the IRS for audits or that customers with the types of returns that are often audited are selecting the Hemingway Company to do their returns.
    NOTE: A classic scenario of this type is the hospital that has higher than average patient mortality. This can be due either to the care being below average or to the accepting of a higher than average proportion of more serious cases.

28. Let x = the number of wrong answers.
    binomial problem: n = 10 and p = .15, use the binomial formula
    a. $P(x) = [n!/x!(n-x)!] \cdot p^x \cdot (1-p)^{n-x}$
        $P(x=1) = [10!/1!9!] \cdot (.15)^1 \cdot (.85)^9$
                    = [10]·(.1500)·(.2316) = .347
    b. $P(x=0) = [10!/0!10!] \cdot (.15)^0 \cdot (.85)^{10}$
                    = [1]·(1)·(.1969) = .197
        P(x≤1) = P(x=0) + P(x=1)
                 = .197 + .347 = .544
    c. No; since .544 > .05, "at most one wrong number" is not an unusual occurrence when the error rate is 15%.

29. Let x = the number of booked passengers that actually arrive.
    binomial problem: n = 15 and p = .85, use the binomial formula
    $P(x) = [n!/x!(n-x)!] \cdot p^x \cdot (1-p)^{n-x}$
    $P(x=15) = [15!/15!0!] \cdot (.85)^{15} \cdot (.15)^0$
                    = [1]·(.0874)·(1) = .0874
    No; the probability is not low enough to be of no concern – since .0874 > .05, not having enough seats is not an unusual event.

30. Let x = the number of Viagra users that experienced a headache.
    binomial problem: n = 8 and p = .04, use the binomial formula
    a. $P(x) = [n!/x!(n-x)!] \cdot p^x \cdot (1-p)^{n-x}$
        $P(x=3) = [8!/3!5!] \cdot (.04)^3 \cdot (.96)^5$
                    = [56]·(.000064)·(.8154) = .00292
    b. $P(x) = [n!/x!(n-x)!] \cdot p^x \cdot (1-p)^{n-x}$
        $P(x=8) = [8!/8!0!] \cdot (.04)^8 \cdot (.96)^0$
                    = $[1] \cdot (.04)^8 (1) = 6.55 \times 10^{-12}$ = .00000000000655
    c. Yes; if the 4% rate applied to Viagra users, there is almost no chance that all 8 would experience headaches.

31. Let x = the number of households tuned to *60 Minutes*.
    binomial problem: n = 10 and p = .10, use Table A-1
    a. P(x=0) = .107
    b. P(x≥1) = 1 - P(x=0)
            = 1 - .107 = .893
    c. P(x≤1) = P(x=0) + P(x=1)
            = .107 + .268 = .375
    d. No; since .375 > .05, it would not be unusual to find at most one household watching *60 Minutes* when that show had a 20% share of the market.

32. Let x = the number of special program students who graduated.
    binomial problem: n = 10 and p = .94, use the binomial formula
    P(x) = [n!/x!(n-x)!]·p^x·$(1-p)^{n-x}$
    a. P(x≥9) = P(x=9) + P(x=10)
             = [10!/9!1!]·$(.94)^9$·$(.06)^1$ + [10!/10!0!]·$(.94)^{10}$·$(.06)^0$
             = [10]·(.5730)·(.06) + [1]·(.5386)·(1)
             = .344 + .539 = .883
    b. P(x=8) = [10!/8!2!]·$(.94)^8$·$(.06)^2$
             = [36]·(.6096)·(.0036) = .079
       P(x≤7) = 1 - P(x>7)
             = 1 - [P(x=8) + P(x=9) + P(x=10)]
             = 1 - [.079 + .344 + .539]
             = 1 - [.962] = .038
    Yes; since P(x≤7) = .038 < .05, getting only 7 that graduated would be an unusual result.

33. Let x = the number of women hired.
    binomial problem: n = 20 and p = .50, use the binomial formula
    P(x) = [n!/x!(n-x)!]·p^x·$(1-p)^{n-x}$
    P(x≤2) = P(x=0) + P(x=1) + P(x=2)
          = [20!/0!20!]·$(.5)^0$·$(.5)^{20}$ + [20!/1!19!]·$(.5)^1$·$(.5)^{19}$ + [20!/2!18!]·$(.5)^2$·$(.5)^{18}$
          = [1]·(1)·(.00000095) + [20]·(.5)·(.00000191) + [190]·(.25)·(.00000381)
          = .00000095 + .00001907 + .00018120 = .000201
    Yes, the small probability of this result occurring by chance when all applicants are otherwise equally qualified does support a charge of gender discrimination.

34. Let x = the number of jackpots hit.
    binomial problem: n = 5 and p = 1/2000 = .0005, use the binomial formula
    a. P(x) = [n!/x!(n-x)!]·p^x·$(1-p)^{n-x}$
       P(x=2) = [5!/2!3!]·$(.0005)^2$·$(.9995)^3$
             = [10]·(.00000025)·(.9985) = .000002496
    b. P(x≥2) = 1 - [P(x=0) + P(x=1)]
             = 1 - {[5!/0!5!]·$(.0005)^0$·$(.9995)^5$ + [5!/1!4!]·$(.0005)^1$·$(.9995)^4$}
             = 1 - {[1]·(1)·(.997502499) + [5]·(.0005)·(.9980015)}
             = 1 - {.997502499 + .002495004}
             = 1 - {.999997503} = .000002497
    c. No; if the slot machine is functioning properly with P(jackpot) = 1/2000, then either the guest is not telling the truth or an extremely rare event as occurred.

35. The requested table is given in the .50 column of the second page of Table A-1 in the back of the book.
    Let x = the number of girls in 12 births.
    P(x≥9) = P(x=9) + P(x=10) + P(x=11) + P(x=12)
          = .054 + .016 + .003 + $.000^+$ = .073
    No; this is not evidence that the gender selection technique is effective. Since .073 > .05, getting 9 or more girls is not an unusual event and could occur by chance alone.

36. Let x = the number of selected persons that have so taken courses.
    binomial: n = 5 and p = .57, use the binomial formula
    $P(x) = [n!/x!(n-x)!] \cdot p^x \cdot (1-p)^{n-x}$
    $P(x \le 1) = P(x=0) + P(x=1)$
    $ = [5!/0!0!] \cdot (.57)^0 \cdot (.43)^5 + [5!/1!0!] \cdot (.57)^1 \cdot (.43)^4$
    $ = [1] \cdot (1) \cdot (.0147) + [5] \cdot (.57) \cdot (.0342)$
    $ = .0147 + .0974 = .1121$
    No; this is not evidence that the 57% rate is wrong. Since .1121 > .05, the described event is not unusual and could reasonably occur by chance alone when the rate is 57%.

37. Let x = the number of components tested to find 1st defect
    geometric problem: p = .2, use the geometric formula
    $P(x) = p \cdot (1-p)^{x-1}$
    $P(x=7) = (.2) \cdot (.8)^6 = (.2) \cdot (.2621) = .0524$

38. Let x = the number of winning selections
    A = 6 (winning numbers), B = 48 (losing numbers), n = 6 (selections)
    Use the hypergeometric formula.
    $P(x) = [A!/(A-x)!x!] \cdot [B!/(B-n+x)!(n-x)!] \div [(A+B)!/(A+B-n)!n!]$
    a. $P(x=6) = [6!/0!6!] \cdot [48!/48!0!] \div [54!/48!/6!]$
       $ = [1] \cdot [1] \div [25,827,165] = .0000000387$
    b. $P(x=5) = [6!/1!5!] \cdot [48!/47!1!] \div [54!/48!/6!]$
       $ = [6] \cdot [48] \div [25,827,165] = .0000112$
    c. $P(x=3) = [6!/3!3!] \cdot [48!/45!3!] \div [54!/48!/6!]$
       $ = [20] \cdot [17296] \div [25,827,165] = .0134$
    d. $P(x=0) = [6!/6!0!] \cdot [48!/42!6!] \div [54!/48!/6!]$
       $ = [1] \cdot [12,271,512] \div [25,827,165] = .475$
    NOTE: The formula is really $[_AC_x] \cdot [_BC_{n-x}] \div [_{A+B}C_n]$ and comes from the methods of Chapter 3.

39. Extending the pattern to cover 6 types of outcomes, where $\Sigma x = n$ and $\Sigma p = 1$,
    $P(x_1,x_2,x_3,x_4,x_5,x_6) = [n!/(x_1!x_2!x_3!x_4!x_5!x_6!)] \cdot p_1^{x_1} \cdot p_2^{x_2} \cdot p_3^{x_3} \cdot p_4^{x_4} \cdot p_5^{x_5} \cdot p_6^{x_6}$
    n = 20
    $p_1 = p_2 = p_3 = p_4 = p_5 = p_6 = 1/6$
    $x_1 = 5, x_2 = 4, x_3 = 3, x_4 = 2, x_5 = 3, x_6 = 3$
    Use the multinomial formula.
    $P(x_1,x_2,x_3,x_4,x_5,x_6) = [n!/(x_1!x_2!x_3!x_4!x_5!x_6!)] \cdot p_1^{x_1} \cdot p_2^{x_2} \cdot p_3^{x_3} \cdot p_4^{x_4} \cdot p_5^{x_5} \cdot p_6^{x_6}$
    $ = [20!/(5!4!3!2!3!3!)] \cdot (1/6)^5 \cdot (1/6)^4 \cdot (1/6)^3 \cdot (1/6)^2 \cdot (1/6)^3 \cdot (1/6)^3$
    $ = [20!/(5!4!3!2!3!3!)] \cdot (1/6)^{20}$
    $ = [1.955 \cdot 10^{12}] \cdot (2.735 \cdot 10^{-16}) = .000535$

4.4 Mean, Variance and Standard Deviation for the Binomial Distribution

1. $\mu = n \cdot p = (400) \cdot (.2) = 80.0$
   $\sigma^2 = n \cdot p \cdot (1-p) = (400) \cdot (.2) \cdot (.8) = 64.0; \sigma = 8.0$
   minimum usual value = $\mu - 2\sigma = 80.0 - 2 \cdot (8.0) = 64.0$
   maximum usual value = $\mu + 2\sigma = 80.0 + 2 \cdot (8.0) = 96.0$

2. $\mu = n \cdot p = (250) \cdot (.45) = 112.5$
   $\sigma^2 = n \cdot p \cdot (1-p) = (250) \cdot (.45) \cdot (.55) = 61.875; \sigma = 7.86$
   minimum usual value = $\mu - 2\sigma = 112.5 - 2 \cdot (7.86) = 96.8$
   maximum usual value = $\mu + 2\sigma = 112.5 + 2 \cdot (7.86) = 128.2$

3. $\mu = n \cdot p = (1984) \cdot (3/4) = 1488.0$
   $\sigma^2 = n \cdot p \cdot (1-p) = (1984) \cdot (3/4) \cdot (1/4) = 372.0; \sigma = 19.3$
   minimum usual value = $\mu - 2\sigma = 1488.0 - 2 \cdot (19.3) = 1449.4$
   maximum usual value = $\mu + 2\sigma = 1488.0 + 2 \cdot (19.3) = 1526.6$

4. $\mu = n \cdot p = (767) \cdot (1/6) = 127.83$
   $\sigma^2 = n \cdot p \cdot (1-p) = (767) \cdot (1/6) \cdot (5/6) = 106.53; \sigma = 10.32$
   minimum usual value = $\mu - 2\sigma = 127.83 - 2 \cdot (10.32) = 107.2$
   maximum usual value = $\mu + 2\sigma = 127.83 + 2 \cdot (10.32) = 148.5$

5. Let x = the number of correct answers.
   binomial problem: n = 10, p = .5
   a. $\mu = n \cdot p = (10) \cdot (.5) = 5.0$
      $\sigma^2 = n \cdot p \cdot (1-p) = (10) \cdot (.5) \cdot (.5) = 2.50; \sigma = 1.6$
   b. maximum usual value = $\mu + 2\sigma = 5.0 + 2 \cdot (1.6) = 8.2$
      No; since 7 is less than or equal to the maximum usual value, it would not be unusual for a student to pass by getting at least 7 correct answers.

6. let x = the number of correct answers
   binomial problem: n = 10, p = .2
   a. $\mu = n \cdot p = (10) \cdot (.2) = 2.0$
      $\sigma^2 = n \cdot p \cdot (1-p) = (10) \cdot (.2) \cdot (.8) = 1.60; \sigma = 1.3$
   b. maximum usual value = $\mu + 2\sigma = 2.0 + 2 \cdot (1.3) = 4.6$
      Yes; since 7 is greater than the maximum usual value, it would be unusual for a student to pass by getting at least 7 correct answers.

7. Let x = the number of wins.
   binomial problem: n = 100, p = 1/38
   a. $\mu = n \cdot p = (100) \cdot (1/38) = 2.6$
      $\sigma^2 = n \cdot p \cdot (1-p) = (100) \cdot (1/38) \cdot (37/38) = 2.56; \sigma = 1.6$
   b. minimum usual value = $\mu - 2\sigma = 2.6 - 2 \cdot (1.6) = -0.6$ [use 0, by practical constraints]
      No; since 0 is greater than or equal to the minimum usual value, it would not be unusual not to win once in the 100 trials.

8. Let x = the number of left-handers.
   binomial problem: n = 25, p = .1
   a. $\mu = n \cdot p = (25) \cdot (.1) = 2.5$
      $\sigma^2 = n \cdot p \cdot (1-p) = (25) \cdot (.1) \cdot (.9) = 2.25; \sigma = 1.5$
   b. maximum usual value = $\mu + 2\sigma = 2.5 + 2 \cdot (1.5) = 5.5$
      No; since 5 is less than or equal to the maximum usual value, it would not be unusual for a class of 25 students to include 5 left-handers.

9. Let x = the number of girls among the 15 births.
   binomial problem: n=15, p = .82
   a. The table at the right was constructed using the .50 column and the final group of values (for n=15) in Table A-1.
   b. $\mu = n \cdot p = (15) \cdot (.50) = 7.5$
      $\sigma^2 = n \cdot p \cdot (1-p) = (15) \cdot (.50) \cdot (.50) = 3.75; \sigma = 1.9$
   c. minimum usual value = $\mu - 2\sigma = 7.5 - 2 \cdot (1.9) = 3.7$
      maximum usual value = $\mu + 2\sigma = 7.5 + 2 \cdot (1.9) = 11.3$
      No; since 10 is between the minimum and maximum usual values, it would not be unusual to get that many girls. NOTE: Since the problem doesn't specify whether the couples were trying to conceive boys or girls, both extremes need to be considered.

x	P(x)
0	.000+
1	.000+
2	.003
3	.014
4	.042
5	.092
6	.153
7	.196
8	.196
9	.153
10	.092
11	.042
12	.014
13	.003
14	.000+
15	.000+
	1.000

10. Let x = the number of letter r's per page.
    binomial problem: n = 2600, p = .077
    a. μ = n·p = (2600)·(.077) = 200.2
       σ^2 = n·p·(1-p) = (2600)·(.077)·(.923) = 184.78; σ = 13.6
    b. Unusual values are those outside $\mu \pm 2\cdot\sigma$
       200.2 $\pm$ 2·(13.6)
       200.2 $\pm$ 27.2
       173.0 to 227.4
    No; since 175 is within the above limits it would not be considered an unusual result.

11. Let x = the number of customers filing complaints.
    binomial problem: n = 850, p = .032
    a. μ = n·p = (850)·(.032) = 27.2
       σ^2 = n·p·(1-p) = (850)·(.032)·(.968) = 26.33; σ = 5.1
    b. minimum usual value = $\mu - 2\sigma$ = 27.2 - 2·(5.1) = 17.0
    Yes; since 7 is less than the minimum usual value, it would be unusual to get that many complaints if the program had no effect and the rate were still 3.2%. This is sufficient evidence to conclude that the program is effective in lowering the rate of complaints.

12. Let x = the number of M&M's that are blue.
    binomial problem: n = 100, p = .10
    a. μ = n·p = (100)·(.10) = 10.0
       σ^2 = n·p·(1-p) = (100)·(.10)·(.90) = 9.0; σ = 3.0
    b. Unusual values are those outside $\mu \pm 2\cdot\sigma$
       10.0 $\pm$ 2·(3.0)
       10.0 $\pm$ 6.0
       4.0 to 16.0
    No; since 5 is within the above limits it would not be considered an unusual result. No; it does not seem that the claimed rate of 10% is wrong.

13. Let x = the number that develop cancer.
    binomial problem: n = 420,000, p = .000340
    a. μ = n·p = (420,000)·(.000340) = 142.8
       σ^2 = n·p·(1-p) = (420,000)·(.000340)·(.999660) = 142.75; σ = 11.9
    b. Unusual values are those outside $\mu \pm 2\cdot\sigma$
       142.8 $\pm$ 2·(11.9)
       142.8 $\pm$ 23.8
       119.0 to 166.6
    No; since 135 is within the above limits, it would not be considered an unusual result.
    c. These results do not support the publicized concern that cell phones are such a risk.

14. Let x = the number receiving the drug who experience flu symptoms
    binomial problem: n = 863, p = .019
    a. μ = n·p = (863)·(.019) = 16.4
       σ^2 = n·p·(1-p) = (863)·(.019)·(.981) = 16.09; σ = 4.0
    b. Unusual values are those outside $\mu \pm 2\cdot\sigma$
       16.4 $\pm$ 2·(4.0)
       16.4 $\pm$ 8.0
       8.4 to 24.4
    No; since 19 is within these limits, it would not be considered an unusual result.
    c. No; these results do not suggest that persons receiving the drug experience flu symptoms any differently from persons not receiving the drug.

15. Let x = the number of adults who believe human cloning should not be allowed.
    a. (1012)·(.89) = 901
       NOTE: Actually, x/1012 rounds to 89% for 896$\leq$x$\leq$905.

b. binomial problem: n = 1012, p = .50
   $\mu = n \cdot p = (1012) \cdot (.50) = 506.0$
   $\sigma^2 = n \cdot p \cdot (1-p) = (1012) \cdot (.50) \cdot (.50) = 253.0$; $\sigma = 15.9$
c. maximum usual value = $\mu + 2\sigma$ = 506.0 + 2·(15.9) = 537.8
   Yes; the results (at least 896 believe that human cloning should not be allowed) of the Gallup poll are unusually high if the assumed rate of 50% is true. Yes; this is evidence that the majority of adults believe that human cloning should not be allowed.

16. Let x = the number of 500 such NYC drivers that had an accident in the last year.
    a. (500)·(.42) = 210
       NOTE: Actually, x/210 rounds to 42% for 208≤x≤212.
    b. binomial problem: n = 500, p = .34
       $\mu = n \cdot p = (500) \cdot (.34) = 170.0$
       $\sigma^2 = n \cdot p \cdot (1-p) = (500) \cdot (.34) \cdot (.66) = 112.2$; $\sigma = 10.6$
    c. maximum usual value = $\mu + 2\sigma$ = 170.0 + 2·(10.6) = 191.2
       Yes; the results (at least 208 with accidents) for the NYC drivers are unusually high if the 34% rate for the general population applies. Yes; it appears that the higher rates for NYC drivers are justified.

17. Let x = the number of baby girls.
    binomial problem: n = 100, p = .5
    $\mu = n \cdot p = (100) \cdot (.5) = 50.0$
    $\sigma^2 = n \cdot p \cdot (1-p) = (100) \cdot (.5) \cdot (.5) = 25.0$; $\sigma = 5.0$
    a. Yes: a probability histogram would reveal an approximately bell-shaped distribution. The most likely value is x=50, with the likelihoods decreasing symmetrically for x values smaller or larger than 50.
    b. Since 40 and 60 correspond to $\mu \pm 2 \cdot \sigma$, the empirical rule indicates that approximately 95% of the time x will fall between 40 and 60.
    c. Since 35 and 65 correspond to $\mu \pm 3 \cdot \sigma$, the empirical rule indicates that approximately 99.7% of the time x will fall between 35 and 65.
    d. Since 40 and 60 correspond to $\mu \pm 2 \cdot \sigma$, use Chebyshev's theorem with k=2. The portion of cases within k·σ of μ is at least $1 - 1/k^2$ = 1- 1/4 = 3/4 (i.e., at least 75%).
    NOTE: Parts (b) and (d) agree, since 95% is "at ;east 75%." Chebyshev's theorem applies to all distributions and must be more general; the empirical rule applies only to bell-shaped distributions and can be more specific.

18. NOTE: Because mathematical formulation of this problem involves complicated algebra that does not promote better understanding of the concepts involved, use a trial-and-error approach using Table A-1. The procedure given below may not be the most efficient, but it is easy to follow and promotes better understanding of the concepts involved.
    Let x = the number of edible pizzas.
    binomial problem: n is unknown, p = .8, we want P(x≥5) ≥ .99
    for n=5, P(x≥5) = P(x=5)
                   = .328
    for n=6, P(x≥5) = P(x=5) + P(x=6)
                   = .393 + .262 = .655
    for n=7, P(x≥5) = P(x=5) + P(x=6) + P(x=7)
                   = .275 + .367 + .210 = .852
    for n=8, P(x≥5) = P(x=5) + P(x=6) + P(x=7) + P(x=8)
                   = .147 + .294 + .336 + .168 = .945
    for n=9, P(x≥5) = P(x=5) + P(x=6) + P(x=7) + P(x=8) + P(x=9)
                   = .066 + .176 + .302 + .302 + .134 = .980
    for n=10, P(x≥5)= P(x=5) + P(x=6) + P(x=7) + P(x=8) + P(x=9) + P(x=10)
                   = .026 + .088 + .201 + .302 + .268 + .107 = .992
    The minimum number of pizzas necessary to be at least 99% sure that there will be 5 edible pizzas available is n = 10.

4-5 The Poisson Distribution

1. $P(x) = \mu^x \cdot e^{-\mu}/x!$ with $\mu = 2$ and $x = 3$
   $P(x=3) = (2)^3 \cdot e^{-2}/3!$
   $= (8) \cdot (.1353)/6 = .180$

2. $P(x) = \mu^x \cdot e^{-\mu}/x!$ with $\mu = .5$ and $x = 2$
   $P(x=2) = (.5)^2 \cdot e^{-.5}/2!$
   $= (.25) \cdot (.6065)/2 = .0758$

3. $P(x) = \mu^x \cdot e^{-\mu}/x!$ with $\mu = 100$ and $x = 99$
   $P(x=99) = (100)^{99} \cdot e^{-100}/99!$
   This cannot be evaluated directly on most calculators because the numbers are too large. Even many statistical software packages cannot evaluate this.
   Using Excel, however, yields Poisson(100,99,0) = .039661.
   An approximation is possible using logarithms and exercise #35 of section 3-7 as follows.
   $n! \approx 10^k$, where $k = (n+.5) \cdot (\log n) + .39908993 - .43429448 \cdot n$
   for 99, $k = (99.5) \cdot (\log 99) + .39908993 - .43429446 \cdot (99) = 155.9696383$
   $P(x=99) \approx (100)^{99} \cdot e^{-100}/10^{155.9696383}$
   $\log [P(x=99)] \approx 99 \cdot (\log 100) - 100 \cdot (\log e) - 155.9696383 \cdot (\log 10)$
   $\approx 99 \cdot (2) - 100 \cdot (.434294482) - 155.9696383 \cdot (1)$
   ≈ -1.399086463
   $P(x=99) \approx 10^{-1.399086463}$
   $\approx .03989$
   NOTE: Statdisk gives $P(x=99) = .03986$.

4. $P(x) = \mu^x \cdot e^{-\mu}/x!$ with $\mu = 500$ and $x = 512$
   $P(x=512) = (500)^{512} \cdot e^{-500}/512!$
   This cannot be evaluated directly on most calculators because the numbers are too large. Even many statistical software packages cannot evaluate this.
   Using Statdisk, however, yields $P(x=512) = .01528$.
   An approximation is possible using logarithms and exercise #35 of section 3-7 as follows.
   $n! \approx 10^k$, where $k = (n+.5) \cdot (\log n) + .39908993 - .43429448 \cdot n$
   for 512, $k = (512.5) \cdot (\log 512) + .39908993 - .43429446 \cdot (512) = 1166.541171$
   $P(x=512) \approx (500)^{512} \cdot e^{-500}/10^{1166.541171}$
   $\log [P(x=512)] \approx 512 \cdot (\log 500) - 500 \cdot (\log e) - 1166.541171 \cdot (\log 10)$
   $\approx 512 \cdot (2.698970004) - 500 \cdot (.434294482) - 1166.541171 \cdot (1)$
   ≈ -1.81576995
   $P(x=512) \approx 10^{-1.81576995}$
   $\approx .01528$

NOTE: In the problems that follow, remember to store the unrounded value for μ for use in the Poisson calculations.

5. a. The number of atoms lost is 1,000,000 - 977,287 = 22,713.
      $\mu = 22,713/365 = 62.2$
   b. $P(x) = \mu^x \cdot e^{-\mu}/x!$ with $\mu = 62.2$ and $x = 50$
      $P(x=50) = (62.2)^{50} \cdot e^{-62.2}/50! = .0155$

6. a. $\mu = 11/365 = .0301$
   b. $P(x) = \mu^x \cdot e^{-\mu}/x!$ with $\mu = .0301$ and $x = 0$
      $P(x=0) = (.0301)^0 \cdot e^{-.0301}/0! = .9703$
   c. $P(x \geq 1) = 1 - P(x=0)$
      $= 1 - .9703 = .0297$
   d. The personnel should be called as needed. Yes; this does mean that women giving birth might not get the immediate attention available in more populated areas.

7. Let x = the number of horse-kick deaths per corps per year
   $P(x) = \mu^x \cdot e^{-\mu}/x!$ with $\mu = 196/280 = .70$
   a. $P(x=0) = (.70)^0 \cdot e^{-.70}/0! = (1) \cdot (.4966)/1 = .4966$
   b. $P(x=1) = (.70)^1 \cdot e^{-.70}/1! = (.70) \cdot (.4966)/1 = .3476$
   c. $P(x=2) = (.70)^2 \cdot e^{-.70}/2! = (.4900) \cdot (.4966)/2 = .1217$
   d. $P(x=3) = (.70)^3 \cdot e^{-.70}/3! = (.3430) \cdot (.4966)/6 = .0284$
   e. $P(x=4) = (.70)^4 \cdot e^{-.70}/4! = (.2401) \cdot (.4966)/24 = .0050$
   The following table compares the actual relative frequencies to the Poisson probabilities.

x	f	r.f.	P(x)	
0	144	.5143	.4966	
1	91	.3250	.3476	note: r.f. = f/Σf
2	32	.1143	.1217	
3	11	.0393	.0284	
4	2	.0071	.0050	
5 or more	0	.0000	.0007	(by subtraction)
	280	1.0000	1.0000	

   The agreement between the observed relative frequencies and the probabilities predicted by the Poisson formula is very good.

NOTE: The observed/predicted comparison in the above exercise and the ones that follow is made using relative frequencies. It could have been made using frequencies. The choice is arbitrary.

8. Let x = the number of homicides per day
   $P(x) = \mu^x \cdot e^{-\mu}/x!$ with $\mu = 116/365 = .3178$
   a. $P(x=0) = (.3178)^0 \cdot e^{-.3178}/0! = (1) \cdot (.7277)/1 = .7277$
   b. $P(x=1) = (.3178)^1 \cdot e^{-.3178}/1! = (.3178) \cdot (.7277)/1 = .2313$
   c. $P(x=2) = (.3178)^2 \cdot e^{-.3178}/2! = (.1010) \cdot (.7277)/2 = .0368$
   d. $P(x=3) = (.3178)^3 \cdot e^{-.3178}/3! = (.0321) \cdot (.7277)/6 = .0038$
   e. $P(x=4) = (.3178)^4 \cdot e^{-.3178}/4! = (.0102) \cdot (.7277)/24 = .0003$
   The following table compares the actual relative frequencies to the Poisson probabilities.

x	f	r.f.	P(x)	
0	268	.7342	.7277	
1	79	.2164	.2313	note: r.f. = f/Σf
2	17	.0466	.0368	
3	1	.0027	.0038	
4 or more	0	.0000	.0004	(by subtraction)
	365	1.0000	1.0000	

   The agreement between the observed relative frequencies and the probabilities predicted by the Poisson formula is very good.

9. Let x = the number of wins in 200 trials.
   binomial problem: n = 200, p = 1/38
   Poisson approximation appropriate since n = 200 ≥ 100 and np = 200·(1/38) = 5.26 ≤ 10.
   $P(x) = \mu^x \cdot e^{-\mu}/x!$ with $\mu = np = 200 \cdot (1/38) = 5.263$
   a. $P(x=0) = (5.263)^0 \cdot e^{-5.263}/0! = (1) \cdot (.00518)/1 = .00518$
   b. $P(x \geq 1) = 1 - P(x=0) = 1 - .00518 = .99482$
   c. $P(x=1) = (5.263)^1 \cdot e^{-5.263}/1! = (5.263) \cdot (.00518)/1 = .02726$
      $P(x=2) = (5.263)^2 \cdot e^{-5.263}/2! = (27.701) \cdot (.00518)/2 = .07173$
      $P(x=3) = (5.263)^3 \cdot e^{-5.263}/3! = (145.794) \cdot (.00518)/6 = .12584$
      $P(x=4) = (5.263)^4 \cdot e^{-5.263}/4! = (767.336) \cdot (.00518)/24 = .16558$
      $P(x=5) = (5.263)^5 \cdot e^{-5.263}/5! = (4038.611) \cdot (.00518)/120 = .17430$
      P(lose money) = P(x=0) + P(x=1) + P(x=2) + P(x=3) + P(x=4) + P(x=5)
         = .00518 + .02726 + .07173 + .12584 + .16558 + .17430 = .5699
   d. P(win money) = 1 - .5699 = .4301

10. Let x = the number of major earthquakes per year.
$P(x) = \mu^x \cdot e^{-\mu}/x!$ with $\mu = 93/100 = .93$
  a. $P(x=0) = (.93)^0 \cdot e^{-.93}/0! = (1) \cdot (.3946)/1 = .3946$
  b. $P(x=1) = (.93)^1 \cdot e^{-.93}/1! = (.93) \cdot (.3946)/1 = .3669$
  c. $P(x=2) = (.93)^2 \cdot e^{-.93}/2! = (.8649) \cdot (.3946)/2 = .1706$
  d. $P(x=3) = (.93)^3 \cdot e^{-.93}/3! = (.8044) \cdot (.3946)/6 = .05289$
  e. $P(x=4) = (.93)^4 \cdot e^{-.93}/4! = (.7481) \cdot (.3946)/24 = .01230$
  f. $P(x=5) = (.93)^5 \cdot e^{-.93}/5! = (.6957) \cdot (.3946)/120 = .002287$
  g. $P(x=6) = (.93)^6 \cdot e^{-.93}/6! = (.6470) \cdot (.3946)/720 = .0003545$
  h. $P(x=7) = (.93)^7 \cdot e^{-.93}/7! = (.6017) \cdot (.3946)/5040 = .00004710$

The following table compares the actual relative frequencies to the Poisson probabilities.

x	f	r.f.	P(x)	
0	47	.4700	.3946	
1	31	.3100	.3669	note: r.f. = f/Σf
2	13	.1300	.1706	
3	5	.0500	.0529	
4	2	.0200	.0123	
5	0	.0000	.0023	
6	1	.0100	.0004	
7	1	.0100	.0000+	
8 or more	0	.0000	.0000+	(by subtraction)
	100	1.0000	1.0000	

The agreement between the observed relative frequencies and the probabilities predicted by the Poisson formula is relatively good.

11. Let x = the number of successes in 100 trials.
binomial problem: n = 100, p = .1
Poisson approximation appropriate since n = 100 ≥ 100 and np = 100·(.1) = 10 ≤ 10.
$P(x) = \mu^x \cdot e^{-\mu}/x!$ with $\mu = np = 100 \cdot (.1) = 10$
$P(x=101) = (10)^{101} \cdot e^{-10}/101!$

This cannot be evaluated directly on most calculators because the numbers are too large. Even many statistical software packages cannot evaluate this.
Using Excel, however, yields Poisson(10,101,0) = 4.17x10⁻³¹
An approximation is possible using logarithms and exercise #35 of section 3-7 as follows.
$n! \approx 10^k$, where $k = (n+.5) \cdot (\log n) + .39908993 - .43429448 \cdot n$
for 101, k = (101.5)·(log 101) + .39908993 - .43429446·(101) = 159.9739689
$P(x=101) \approx (10)^{101} \cdot e^{-10}/10^{159.9739689}$
log [P(x=101)] ≈ 101·(log 10) - 10·(log e) - 159.97396789·(log 10)
           ≈ 101·(1) - 10·(.434294482) - 159.9739689·(1)
           ≈ -63.31691373
$P(x=101) \approx 10^{-63.31691373}$
           ≈ 4.82x10⁻⁶⁴

It appears that the Poisson value is so small that it can be considered 0, agreeing with the fact that x=101 is impossible in such a binomial distribution.

12. As illustrated by the table at the right, the Poisson approximations are not acceptable. The binomial probabilities [taken from Table A-1 using n=10 and p=.5] are symmetric around x=5. The Poisson probabilities [calculated using $P(x) = \mu^x \cdot e^{-\mu}/x!$ with $\mu=np=10 \cdot (.5)=5$] are positively skewed around x=4.5. For x=4, for example, the error is approximately 14%. In addition, the Poisson shows that the impossible event x≥11 has probability 1 - .9863 = .0137.

x	binomial P(x)	Poisson P(x)
0	.001	.0067
1	.010	.0337
2	.044	.0842
3	.117	.1404
4	.205	.1755
5	.246	.1755
6	.205	.1462
7	.117	.1044
8	.044	.0653
9	.010	.0363
10	.001	.0181
	1.000	.9863

Review Exercises

1. a. A random variable is a characteristic that assumes a single value (usually a numerical value), determined by chance, for each outcome of an experiment.
    b. A probability distribution is a statement of the possible values a random variable can assume and the probability associated with each of those values. To be valid it must be true for each value x in the distribution that $0 \le P(x) \le 1$ and that $\Sigma P(x) = 1$.
    c. Yes, the given table is a valid probability distribution because it meets the definition and conditions in part (b) above.
    NOTE: The following table summarizes the calculations for parts (d) and (e).

x	P(x)	x·P(x)	x^2	x^2P(x)
0	.08	0	0	0
1	.05	.05	1	.05
2	.10	.20	4	.40
3	.13	.39	9	1.17
4	.15	.60	16	2.40
5	.21	1.05	25	5.25
6	.09	.54	26	3.24
7	.19	1.33	49	9.31
	1.00	4.16		21.82

    d. $\mu = \Sigma x \cdot P(x) = 4.16$, rounded to 4.2
    e. $\sigma^2 = \Sigma x^2 \cdot P(x) - \mu^2 = 21.82 - (4.16)^2 = 4.5144$; $\sigma = 2.1247$, rounded to 2.1
    f. No; since P(x=0) = .08 > .05.

2. Let x = the number of TV's tuned to *West Wing*.
    binomial problem: n = 20 and p = .15, use the binomial formula
    a. $E(x) = \mu = n \cdot p = (20) \cdot (.15) = 3.0$
    b. $\mu = n \cdot p = (20) \cdot (.15) = 3.0$
    c. $\sigma^2 = n \cdot p \cdot (1-p) = (20) \cdot (.15) \cdot (.85) = 2.55$; $\sigma = 1.6$
    d. $P(x) = [n!/x!(n-x)!] \cdot p^x \cdot (1-p)^{n-x}$
    $P(x=5) = [20!/5!15!] \cdot (.15)^5 \cdot (.85)^{15}$
    $= [15504] \cdot (.0000759) \cdot (.0874) = .103$
    e. approach #1: minimum usual value = $\mu - 2\sigma = 3.0 - 2 \cdot (1.6) = -0.2$
    No; since 0 is not less than the minimum usual value, it would not be unusual to find that 0 sets are tuned to *West Wing*.
    approach #2: $P(x=5) = [20!/0!20!] \cdot (.15)^0 \cdot (.85)^{20} = (1) \cdot (1) \cdot (.0388) = .0388$
    Yes; since .0388 < .05, it would be unusual to find that 0 sets are tuned to *West Wing*.
    resolution: Both approaches are guidelines, and not hard and fast rules. Approach #1 best applies when the probability distribution is approximately bell-shaped and the underlying question is whether it is considered unusual to get a result as extreme as or more extreme than the one under consideration. Approach #2 best applies when the underlying question is about one particular result only, often because that result is the lower or upper limit of the probability distribution. In this instance, follow approach #2.

3. Let x = the number of companies that test for drug abuse.
    binomial problem: n = 10 and p = .80, use Table A-1
    a. P(x=5) = .026
    b. $P(x \ge 5) = P(x=5) + P(x=6) + P(x=7) + P(x=8) + P(x=9) + P(x=10)$
    $= .026 + .088 + .201 + .302 + .268 + .107 = .992$
    c. $\mu = n \cdot p = (10) \cdot (.80) = 8.0$
    $\sigma^2 = n \cdot p \cdot (1-p) = (10) \cdot (.80) \cdot (.20) = 1.60$; $\sigma = 1.3$

d. Unusual values are those outside $\mu \pm 2\cdot\sigma$
$$8.0 \pm 2\cdot(1.3)$$
$$8.0 \pm 2.6$$
$$5.4 \text{ to } 10.6$$
No; since 6 is within these limits, it would not be unusual to find that 6 of the 10 companies test for substance abuse.

4. Let x = the number of workers fired for inability to get along with others.
binomial problem: n = 5 and p = .17, use the binomial formula
$P(x) = [n!/x!(n-x)!]\cdot p^x \cdot (1-p)^{n-x}$
   a. $P(x=4) = [5!/4!1!]\cdot(.17)^4\cdot(.83)^1 = (5)\cdot(.000835)\cdot(.83) = .00347$
   $P(x=5) = [5!/5!0!]\cdot(.17)^5\cdot(.83)^0 = (1)\cdot(.000142)\cdot(1) = .000142$
   $P(x\geq 4) = P(x=4) + P(x=5)$
   $= .00347 + .00014 = .00361$
   b. Yes; since .00361 < .05, x≥4 would be an unusual result for a company with the standard rate of 17%.

5. Let x = the number of deaths per year.
Poisson problem: $P(x) = \mu^x \cdot e^{-\mu}/x!$
   a. $\mu = 7/365 = .019$
   b. $P(x=0) = (.019)^0\cdot e^{-.019}/0! = (1)\cdot(.9810)/1 = .9810$
   c. $P(x=1) = (.019)^1\cdot e^{-.019}/1! = (.109)\cdot(.9810)/1 = .0188$
   d. $P(x\geq 1) = 1 - [P(x=0) + P(x=1)]$
   $= 1 - [.9810 + .0188]$
   $= 1 - .9998 = .0002$
   e. No; since .0002 < .05 by such a wide margin, having more than one death per day is such a rare event that no such contingency plans are necessary.
   NOTE: This decision was based on the deaths of village residents, not on deaths occurring within the village limits. If there are other circumstances to consider (e.g., a nearby highway prone to fatal accidents), then having such contingency plans would be advisable.

Cumulative Review Exercises

1. a. The table below at the left was used to calculate the mean and standard deviation.

x	f	f·x	f·x²		x	r.f.
0	47	0	0		0	.644
1	3	3	3		1	.041
2	1	2	4		2	.014
3	0	0	0		3	.000
4	3	12	48		4	.041
5	11	55	275		5	.151
6	3	18	108		6	.041
7	3	21	147		7	.041
8	1	8	64		8	.014
9	1	9	81		9	.014
	73	128	730			1.000

$\bar{x} = [\Sigma(f\cdot x)]/[\Sigma f] = 128/73 = 1.8$
$s^2 = \{[\Sigma f]\cdot[\Sigma(f\cdot x^2)] - [\Sigma(f\cdot x)]^2\}/\{[(\Sigma f)]\cdot[(\Sigma f) - 1]\}$
$= [(73)\cdot(730) - (128)^2]/[(73)\cdot(72)]$
$= 36906/5256 = 7.022$
$s = 2.6$

b. The table is given above at the right and was constructed using r.f. = f/(Σf) = f/70.

c. The first two columns of the table below constitute the requested probability distribution.
The other columns were added to calculate the mean and the standard deviation.

x	P(x)	x·P(x)	x^2	x^2·P(x)
0	.1	0	0	0
1	.1	.1	1	.1
2	.1	.2	4	.4
3	.1	.3	9	.9
4	.1	.4	16	1.6
5	.1	.5	25	2.5
6	.1	.6	36	3.6
7	.1	.7	49	4.9
8	.1	.8	64	6.4
9	.1	.9	81	8.1
	1	4.5		28.5

$\mu = \Sigma x \cdot P(x)$
$= 4.5$
$\sigma^2 = \Sigma x^2 \cdot P(x) - \mu^2$
$= 28.5 - (4.5)^2$
$= 8.25$
$\sigma = 2.8723$, rounded to 2.9

d. The presence of so many 0's make it clear that the last digits do not represent a random sample. Factors other than accurate measurement (which, like social security numbers and other such data, would represent random final digits) determined the recorded distances.

2. Let x = the number of such cars that fail the test.
binomial problem: n = 20 and p = .01, use the binomial formula
a. $E(x) = \mu = n \cdot p = (20) \cdot (.01) = 0.2$
b. $\mu = n \cdot p = (20) \cdot (.01) = 0.2$
$\sigma^2 = n \cdot p \cdot (1-p) = (20) \cdot (.01) \cdot (.99) = .198; \sigma = 0.4$
c. $P(x) = [n!/x!(n-x)!] \cdot p^x \cdot (1-p)^{n-x}$
$P(x=0) = [20!/0!20!] \cdot (.01)^0 \cdot (.99)^{20} = (1) \cdot (1) \cdot (.818) = .818$
$P(x \geq 1) = 1 - P(x=0)$
$= 1 - .818 = .182$
d. Unusual values are those outside $\mu \pm 2 \cdot \sigma$
$0.2 \pm 2 \cdot (0.4)$
0.2 ± 0.8
-0.6 to 1.0
Yes; since 3 is outside these limits, it would be considered an unusual result.
e. Let F = a randomly selected car fails the test.
P(F) = .01, for each selection
$P(F_1 \text{ and } F_2) = P(F_1) \cdot P(F_2)$, for independent events
$= (.01) \cdot (.01)$
$= .0001$

Chapter 5

Normal Probability Distributions

5-2 The Standard Normal Distribution

1. The height of the rectangle is .5. Probability corresponds to area, and the area of a rectangle is (width)·(height).
    $P(x < 50.3)$ = (width)·(height)
    $= (50.3 - 50.0) \cdot (.5)$
    $= (.3) \cdot (.5)$
    $= .15$

    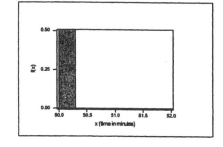

2. The height of the rectangle is .5. Probability corresponds to area, and the area of a rectangle is (width)·(height).
    $P(x > 51.0)$ = (width)·(height)
    $= (52.0 - 51.0) \cdot (.5)$
    $= (1.0) \cdot (.5)$
    $= .50$

    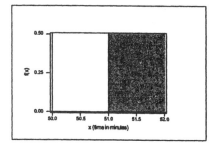

3. The height of the rectangle is .5. Probability corresponds to area, and the area of a rectangle is (width)·(height).
    $P(50.5 < x < 50.8)$ = (width)·(height)
    $= (50.8 - 50.5) \cdot (.5)$
    $= (.3) \cdot (.5)$
    $= .15$

    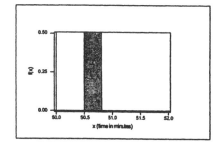

4. The height of the rectangle is .5. Probability corresponds to area, and the area of a rectangle is (width)·(height).
    $P(50.5 < x < 51.8)$ = (width)·(height)
    $= (51.8 - 50.5) \cdot (.5)$
    $= (1.3) \cdot (.5)$
    $= .65$

5. The height of the rectangle is 1/6. Probability corresponds to area, and the area of a rectangle is (width)·(height).
    $P(x > 10)$ = (width)·(height)
    $= (12 - 10) \cdot (1/6)$
    $= (2) \cdot (1/6)$
    $= .333$

    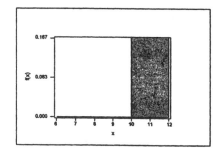

S-108 INSTRUCTOR'S SOLUTIONS Chapter 5

6. The height of the rectangle is 1/6. Probability corresponds to area, and the area of a rectangle is (width)·(height).
P(x < 11) = (width)·(height)
= (11-6)·(1/6)
= (5)·(1/6)
= .833

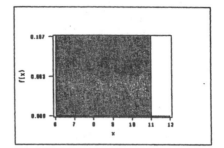

7. The height of the rectangle is 1/6. Probability corresponds to area, and the area of a rectangle is (width)·(height).
P(7 < x < 10) = (width)·(height)
= (10-7)·(1/6)
= (3)·(1/6)
= .5

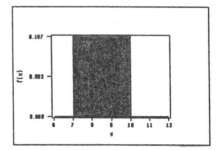

8. The height of the rectangle is 1/6. Probability corresponds to area, and the area of a rectangle is (width)·(height).
P(6.5 < x < 8) = (width)·(height)
= (8-6.5)·(1/6)
= (1.5)·(1/6)
= .250

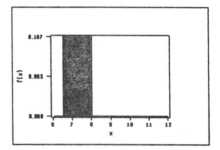

NOTE: The sketch is the key to exercises 9-28. It tells whether to subtract two Table A-2 probabilities, to subtract a Table A-2 probability from 1.0, etc. For the remainder of chapter 5, THE ACCOMPANYING SKETCHES ARE NOT TO SCALE and are intended only as aides to help the reader understand how to use the tabled values to answers the questions. In addition, the probability of any single point in a continuous distribution is zero – i.e., P(x=a) = 0 for any single point a. For normal distributions, therefore, this manual ignores P(x=a) and uses P(x<a) = 1 - P(x>a).

9. P(z<-.25) = .4013

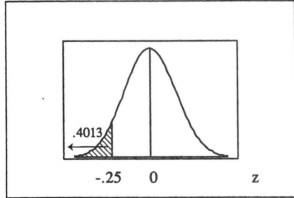

10. P(z<-2.75) = .0030

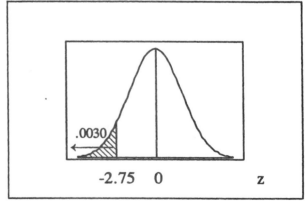

11. P(z < .25) = .5987

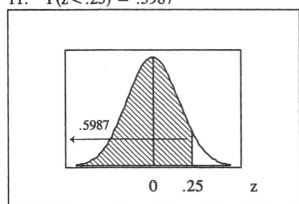

12. P(z < 2.75) = .9970

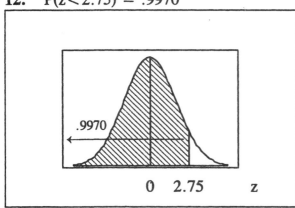

13. P(z > 2.33)
    = 1 - P(z < 2.33)
    = 1 - .9901
    = .0099

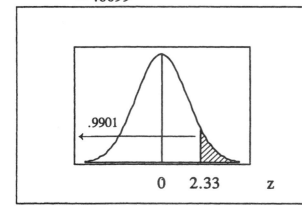

14. P(z > 1.96)
    = 1 - P(z < 1.96)
    = 1 - .9750
    = .0250

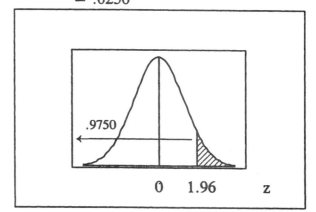

15. P(z > -2.33)
    = 1 - P(z < -2.33)
    = 1 - .0099
    = .9901

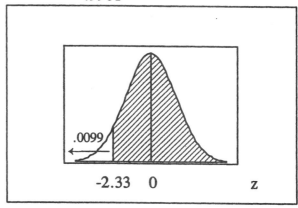

16. P(z > -1.96)
    = 1 - P(z < -1.96)
    = 1 - .0250
    = .9750

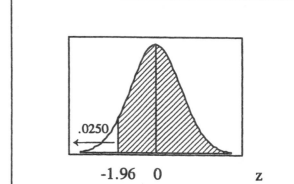

17. P(.50 < z < 1.50)
    = P(z < 1.50) − P(z < .50)
    = .9332 − .6915
    = .2417

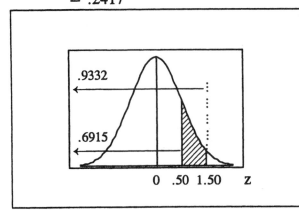

18. P(1.50 < z < 2.50)
    = P(z < 2.50) − P(z < 1.50)
    = .9938 − .9332
    = .0606

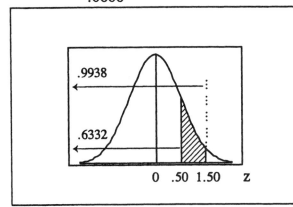

19. P(−2.00 < z < −1.00)
    = P(z < −1.00) − P(z < −2.00)
    = .1587 − .0228
    = .1359

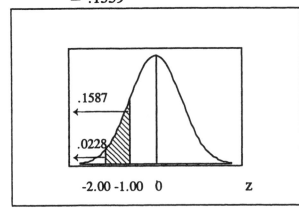

20. P(2.00 < z < 2.34)
    = P(z < 2.34) − P(z < 2.00)
    = .9904 − .9772
    = .0132

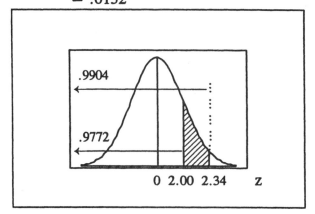

21. P(−2.67 < z < 1.28)
    = P(z < 1.28) − P(z < −2.67)
    = .8997 − .0038
    = .8959

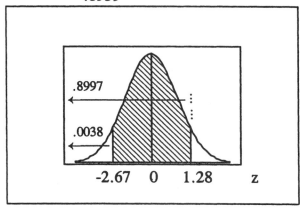

22. P(−1.18 < z < 2.15)
    = P(z < 2.15) − P(z < −1.18)
    = .9842 − .1190
    = .8652

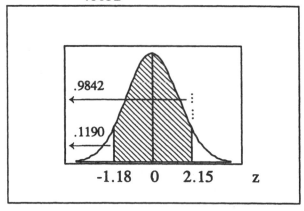

23. P(−.52 < z < 3.75)
    = P(z < 3.75) − P(z < −.52)
    = .9999 − .3015
    = .6984

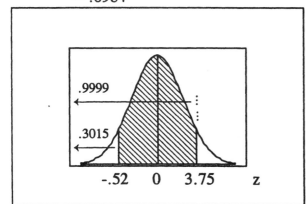

24. P(−3.88 < z < 1.07)
    = P(z < 1.07) − P(z < −3.88)
    = .8577 − .0001
    = .8576

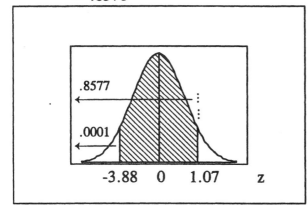

25. P(z > 3.57)
    = 1 − P(z < 3.57)
    = 1 − .9999
    = .0001

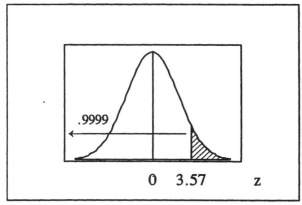

26. P(z < −3.61) = .0001

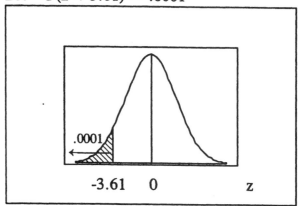

27. P(z > 0)
    = 1 − P(z < 0)
    = 1 − .5000
    = .5000

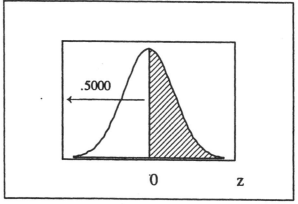

28. P(z < 0) = .5000

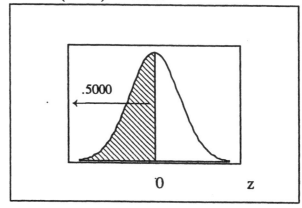

29. $P(-1.00 < z < 1.00)$
    $= P(z < 1.00) - P(z < -1.00)$
    $= .8413 - .1587$
    $= .6826$ or 68.26%

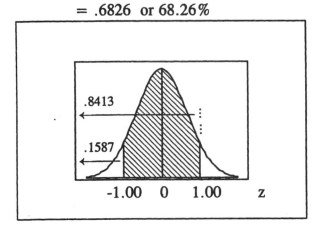

30. $P(-2.00 < z < 2.00)$
    $= P(z < 2.00) - P(z < -2.00)$
    $= .9772 - .0228$
    $= .9544$ or 95.44%

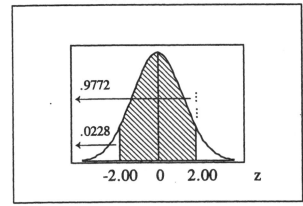

31. $P(-3.00 < z < 3.00)$
    $= P(z < 3.00) - P(z < -3.00)$
    $= .9987 - .0013$
    $= .9974$ or 99.74%

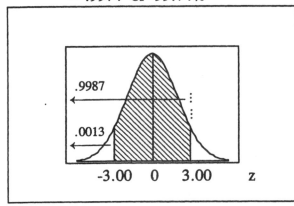

32. $P(-3.50 < z < 3.50)$
    $= P(z < 3.50) - P(z < -3.50)$
    $= .9999 - .0001$
    $= .9998$ or 99.98%

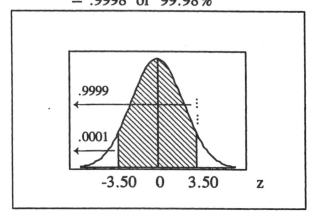

33. $P(-1.96 < z < 1.96)$
    $= P(z < 1.96) - P(z < -1.96)$
    $= .9750 - .0250$
    $= .9500$

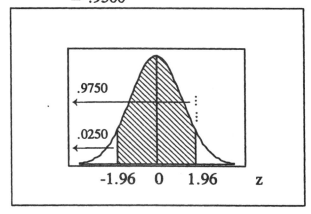

34. $P(z < 1.645) = .9500$

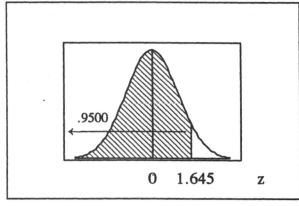

35. $P(z > -2.575)$
    $= 1 - P(z < -2.575)$
    $= 1 - .0050$
    $= .9950$

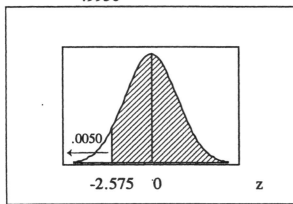

36. $P(1.96 < z < 2.33)$
    $= P(z < 2.33) - P(z < 1.96)$
    $= .9901 - .9750$
    $= .0151$

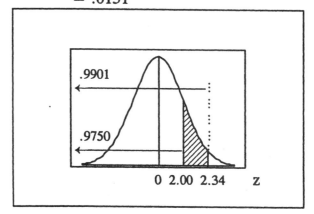

NOTE: The sketch is the key to exercises 37-40. It tells what probability (i.e., cumulative area) to look up when reading Table A-2 "backwards." It also provides a check against gross errors by indicating at a glance whether a z score is above or below 0.

37. For P_{90}, the cumulative area is .9000. The closest entry is .8997, for which $z = 1.28$.

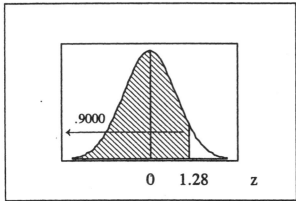

39. For the lowest 5%, the cumulative area is .0500 – indicated by an asterisk, for which $z = -1.645$.

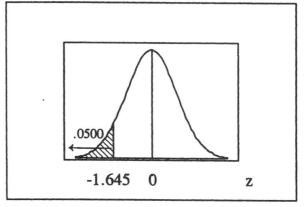

38. For P_{20}, the cumulative area is .2000. The closest entry is .2005, for which $z = -.84$.

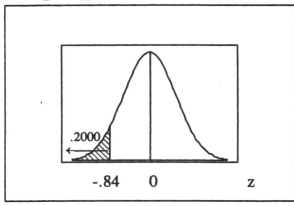

40. For the highest and lowest 3%, the cumulative areas are .9700 and .0003. The closest entries are .9699 and .0301, for which $z = 1.88$ and -1.88.

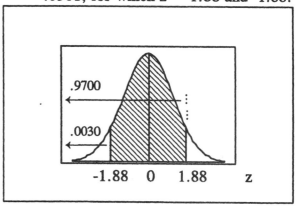

41. Rewrite each of the given statements in terms of z, recalling that z is the number of standard deviations a score is from the mean.

   a. P(-1.00 < z < 1.00)
      = P(z < 1.00) - P(z < -1.00)
      = .8413 - .1587
      = .6826 or 68.26%

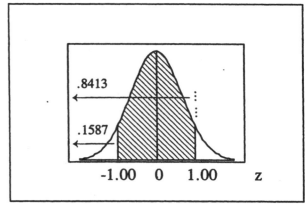

   b. P(-1.96 < z < 1.96)
      = P(z < 1.96) - P(z < -1.96)
      = .9750 - .0250
      = .9500 or 95.00%

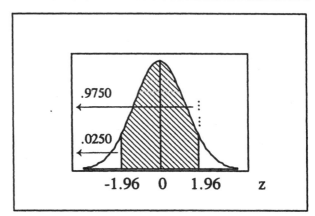

   c. P(-3.00 < z < 3.00)
      = P(z < 3.00) - P(z < -3.00)
      = .9987 - .0013
      = .9974 or 99.74%

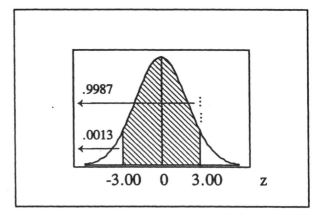

   d. P(-1.00 < z < 2.00)
      = P(z < 2.00) - P(z < -1.00)
      = .9772 - .1587
      = .8185 or 81.85%

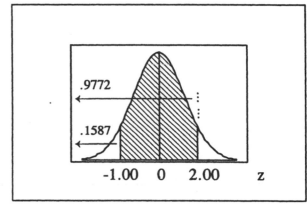

e. $P(z < -2.00$ or $z > 2.00)$
    $= P(z < -2.00) + P(z > 2.00)$
    $= P(z < -2.00) + [1 - P(z < 2.00)]$
    $= .0228 + [1 - .9772]$
    $= .0228 + .0228$
    $= .0456$

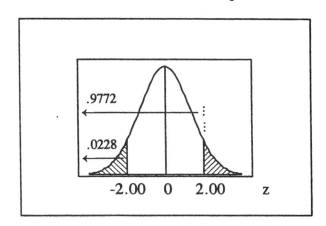

42. Since area = 1 for the entire distribution,
    (width)·(height) = 1
    $(2\sqrt{3})$·(height) = 1
         height = $1/2\sqrt{3}$ = .2887
    a. $P(-1 < x < 1)$ = (width)·(height)
             = (2)·(.2887)
             = .5774

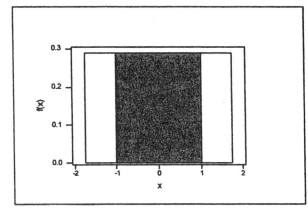

b. $P(-1.00 < z < 1.00)$
    $= P(z < 1.00) - P(z < -1.00)$
    $= .8413 - .1587$
    $= .6826$

c. Yes; the error is more than .10 [i.e., 10%].

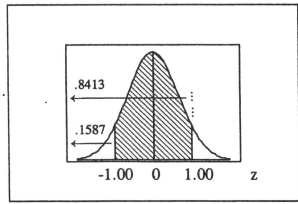

43. The sketches are the key. They tell what probability (i.e., cumulative area) to look up when reading Table A-2 "backwards." They also provides a check against gross errors by indicating whether a score is above or below zero.
    a. $P(0 < z < a) = P(z < a) - P(z < 0)$
       $.3907 = P(z < a) - .5000$
       $.8907 = P(z < a)$
            $a = 1.23$

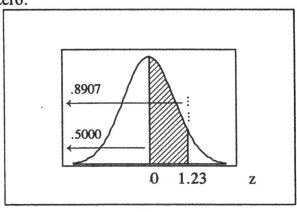

b. Since $P(-b<z<b) = .8664$ and $\Sigma P(z) = 1$,
   $P(z<-b) + .8664 + P(z>b) = 1$.
   By symmetry, $P(z<-b) = P(z>b)$.
   Together these imply,
   $$2 \cdot P(z<-b) + .8664 = 1$$
   $$2 \cdot P(z<-b) = .1336$$
   $$P(z<-b) = .0668$$
   $$-b = -1.50$$
   $$b = 1.50$$

   Observe that $.9332 - .0668 = .8664$, as given in the problem.

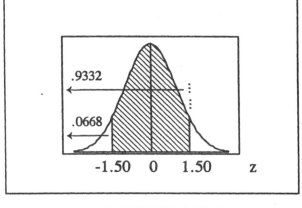

c. $P(z>c) = .0643$
   $P(z<c) = 1 - .0643$
   $\phantom{P(z<c)} = .9357$
   $c = 1.52$

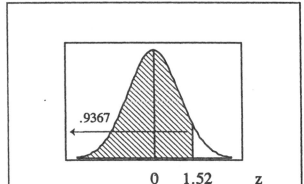

d. $P(z>d) = .9922$
   $P(z<d) = 1 - .9922$
   $\phantom{P(z<d)} = .0078$
   $d = -2.42$

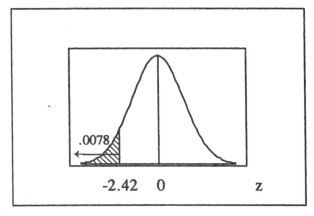

e. $P(z<e) = .4500$ [closest entry is .4483]
   $e = -.13$

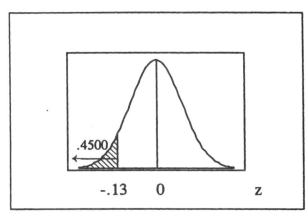

44. In figure 5-2, the minimum is 50.0 and the maximum is 52.0.
$\mu = (50.0 + 52.0)/2 = 102.0/2 = 51.0$
$\sigma = (52.0 - 50.0)/\sqrt{12} = 2/\sqrt{12} = .577$

45. a. Moving across the x values from the minimum to the maximum will accumulate probability at a constant rate from 0 to 1. The result will be a straight line (i.e., with constant slope) from (minimum x, 0) to (maximum x, 1). The sketch is given below on the left.

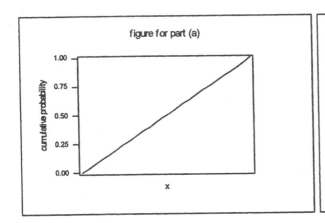
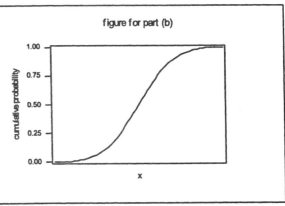

b. Moving across the x values from the minimum to the maximum will accumulate probability slowly at first, and a larger rate near the middle, and more slowly again at the upper end. The result will be a curve that has a variable slope – starting near 0, gradually increasing until reaching a maximum at μ, and then gradually decreasing again toward 0. The sketch is given above on the right.

46. NOTE: Since the vertical axis gives the height of the figure above an x value and not the probability of an x value, avoid confusion by using the label f(x) instead of P(x). It is correct from the figure to say f(5) = c, for example, but not P(x=5) = c.
a. Since the area of a triangle is (½)·(base)·(height) and $\Sigma P(x) = 1$,
$(½)·(5-0)·c + (½)·(10-5)· = 1$
$2.5c + 2.5c = 1$
$5c = 1$
$c = .20$
b. Since 3 is 3/5 of the way from the left end to the peak, f(3) = (3/5)·(.20) = .12.
$P(0 < x < 3) = (½)·(base)·(height)$
$= (½)·(3)·(.12)$
$= .18$
c. Since 2 is 2/5 of the way from the left end to the peak, f(2) = (2/5)·(.20) = .08.
Since 9 is 1/5 of the way from the left end to the peak, f(9) = (1/5)·(.20) = .04.
$P(2 < x < 9) = 1 - P(x < 2) - P(x > 9)$
$= 1 - (½)·(base)·(height) - (½)·(base)·(height)$
$= 1 - (½)·(2)·(.08) - (½)·(1)·(.04)$
$= 1 - .08 - .02$
$= 1 - .10$
$= .90$

5-3 Applications of Normal Distributions

NOTE: In each nonstandard normal distribution, x scores are converted to z scores using the formula $z = (x-\mu)/\sigma$ and rounded to two decimal places. For "backwards" problems, solving the preceding formula for x yields $x = \mu + z\sigma$. The area (cumulative probability) given in Table A-2 will be designated by A. As in the previous section, drawing and labeling the sketch is the key to successful completion of the exercises.

1. $\mu = 100$
   $\sigma = 15$
   $P(x < 115)$
   $\quad = P(z < 1.00)$
   $\quad = .8413$

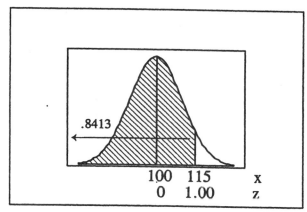

2. $\mu = 100$
   $\sigma = 15$
   $P(x > 131.5)$
   $\quad = P(z > 2.10)$
   $\quad = 1 - P(z < 2.10)$
   $\quad = 1 - .9821$
   $\quad = .0179$

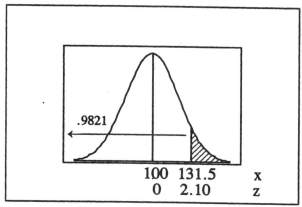

3. $\mu = 100$
   $\sigma = 25$
   $P(90 < x < 110)$
   $\quad = P(-.67 < z < .67)$
   $\quad = P(z < .67) - P(z < -.67)$
   $\quad = .7486 - .2514$
   $\quad = .4972$

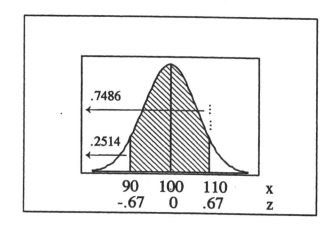

4. $\mu = 100$
$\sigma = 15$
$P(110 < x < 120)$
$\quad = P(.67 < z < 1.33)$
$\quad = P(z < 1.33) - P(z < .67)$
$\quad = .9082 - .7486$
$\quad = .1596$

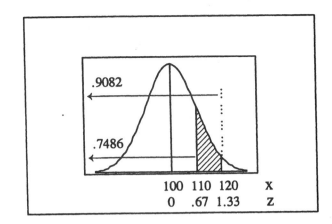

5. $\mu = 100$
$\sigma = 15$
For P_{20}, $A = .2000$ [.2005] and $z = -.84$.
$x = \mu + z\sigma$
$\quad = 100 + (-.84)(15)$
$\quad = 100 - 12.6$
$\quad = 87.4$

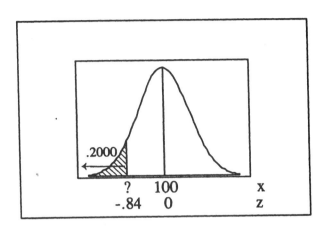

6. $\mu = 100$
$\sigma = 15$
For P_{80}, $A = .8000$ [.7995] and $z = .84$.
$x = \mu + z\sigma$
$\quad = 100 + (.84)(15)$
$\quad = 100 + 12.6$
$\quad = 112.6$

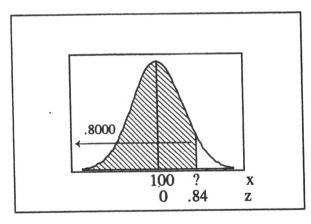

7. $\mu = 100$
$\sigma = 15$
For the top 15%,
$\quad A = .8500$ [.8508] and $z = 1.04$.
$x = \mu + z\sigma$
$\quad = 100 + (1.04)(15)$
$\quad = 100 + 15.6$
$\quad = 115.6$

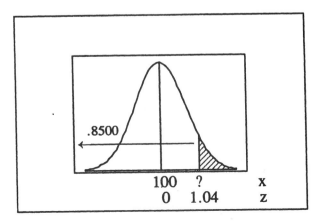

8. $\mu = 100$
   $\sigma = 15$
   For the top 55%,
   A=.4500 [.4483] and z=-.13.
   $x = \mu + z\sigma$
   $= 100 + (-.13)(15)$
   $= 100 - 1.95$
   $= 98.05$

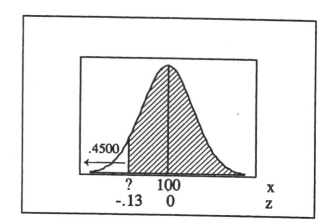

9. $\mu = 98.20$
   $\sigma = .62$
   a. P(x > 100.6)
   $= P(z > 3.87)$
   $= 1 - P(<3.87)$
   $= 1 - .9999$
   $= .0001$
   Yes; the cutoff is appropriate in that there is a small probability of saying that a healthy person has a fever, but many people with low grade fevers may erroneously be labeled healthy.

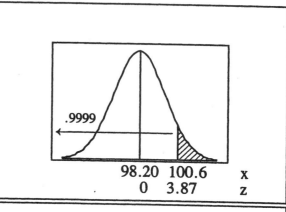

   b. For the top 5%, A=.9500 and z=1.645.
   $x = \mu + z\sigma$
   $= 98.20 + (1.645)(.62)$
   $= 98.20 + 1.02$
   $= 99.22$

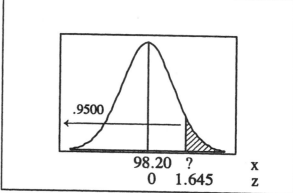

10. $\mu = 268$
    $\sigma = 15$
    a. P(x > 308)
    $= P(z > 2.67)$
    $= 1 - P(z < 2.67)$
    $= 1 - .9962$
    $= .0038$
    The result suggests an unusual event has occurred – but certainly not an impossible one, as about 38 of every 10,000 pregnancies can be expected to last as long.

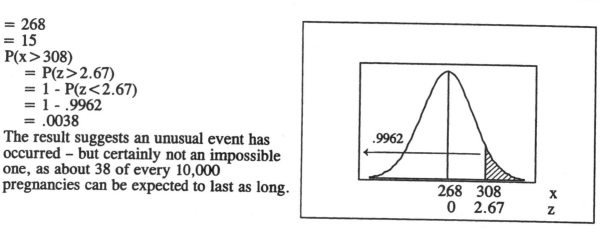

b. For the lowest 4%,
 A = .0400 [.0401] and z = -1.75.
 $x = \mu + z\sigma$
 $= 268 + (-1.75)(15)$
 $= 268 - 26$
 $= 242$

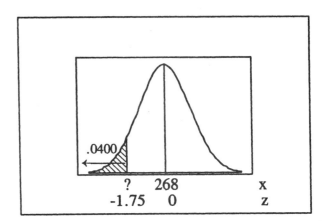

11. $\mu = 998$
 $\sigma = 202$
 a. $P(x < 1100)$
 $= P(z < .50)$
 $= .6915$ or 69.15%

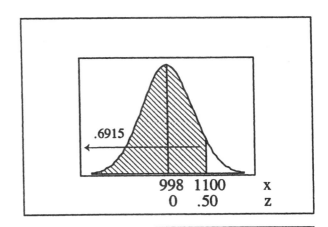

b. For the top 40%,
 A = .6000 [.5987] and z = .25.
 $x = \mu + z\sigma$
 $= 998 + (.25)(202)$
 $= 998 + 50.5$
 $= 1048.5$

The minimum required score would change each year, and it could not be determined until all the test results were in. In addition, there could be problems because the applicants might feel they were competing against each other rather than a standard.

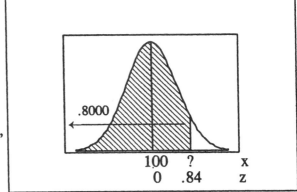

12. $\mu = 6.0$
 $\sigma = 1.0$
 For the smallest 2.5%, A = .0250 and z = -1.96.
 For the largest 2.5%, A = .9750 and z = 1.96.
 $x_S = \mu + z\sigma$
 $= 6.0 + (-1.96)(1.0)$
 $= 6.0 - 1.96$
 $= 4.04$, rounded to 4.0
 $x_L = \mu + z\sigma$
 $= 6.0 + (1.96)(1.0)$
 $= 6.0 + 1.96$
 $= 7.96$, rounded to 8.0

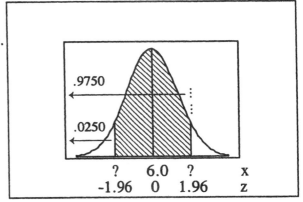

13. $\mu = 8.2$
    $\sigma = 1.1$
   a. $P(x<5.0)$
      $= P(z<-2.91)$
      $= .0018$

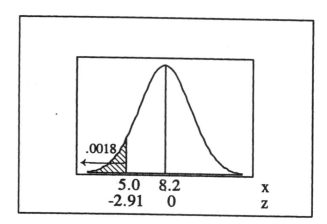

   b. For the shortest 1%,
      $A = .0100$ [.0099] and $z = -2.33$.
      $x = \mu + z\sigma$
      $= 8.2 + (-2.33)(1.1)$
      $= 8.2 - 2.6$
      $= 5.6$ years
      For a "nicer" figure, consider a warranty of 5.5 years = 66 months [for which $P(x<5.5) < P(x<5.6) = .01$].

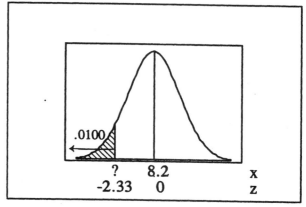

14. $\mu = 7.1$
    $\sigma = 1.4$
   a. $P(x<8.0)$
      $= P(z<.64)$
      $= .7389$

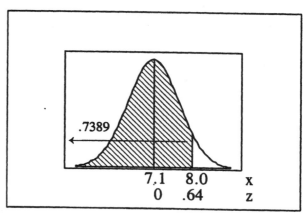

   b. For the lowest 2%,
      $A = .0200$ [.0202] and $z = -2.05$
      $x = \mu + z\sigma$
      $= 7.1 + (-2.05)(1.4)$
      $= 7.1 - 2.87$
      $= 4.23$, rounded to 4.2
      For a "nicer" figure, consider a warranty of 50 months = 4.17 years [for which $P(x<4.17) < P(x<4.2) = .02$].

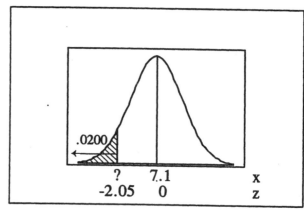

15. $\mu = .91470$
    $\sigma = .03691$
    a. $P(x < .88925)$
       $= P(z < -.69)$
       $= .2451$ or 24.51%, close to 25%

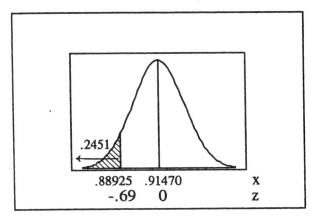

b. For the lowest 25%,
   $A = .2500$ [.2514] and $z = -.67$.
   $x = \mu + z\sigma$
   $= .91470 + (-.67)(.03691)$
   $= .91470 - .02473$
   $= .88997$, close to $.88925$

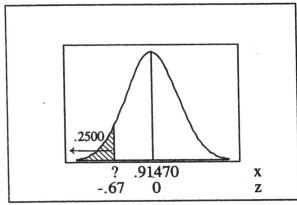

NOTE: Comparing the results using the normal distribution in parts (a) and (b) above to the actual values is not quite fair for two reasons.
1. There are differences in precision. While the weights in the problem were measured to the nearest .00001 gram, percentiles are stated to the whole percent (i.e., the nearest .01) and using the closest entry in Table A-2 limits the accuracy to 3 digits (since the z scores are accurate to the nearest 0.01). A more appropriate comparison uses 2 digits to see that $25\% = 25\%$ in part (a) and 3 digits to see that $.890 \approx .889$ in part (b).
2. The normal distribution is continuous, but the actual data is discrete. Since the 25th percentile for $n = 100$ scores is $(x_{25} + x_{26})/2$, it might be more appropriate to use $A = .2550$ instead of $A = .2500$ in any comparisons. Adjustments to handle discrepancies between continuous and discrete variables are covered in section 5-6.

16. $\mu = .816822$
    $\sigma = .007507$
    a. $P(x > .8250)$
       $= P(z > 1.09)$
       $= 1 - P(z < 1.09)$
       $= 1 - .8621$
       $= .1379$

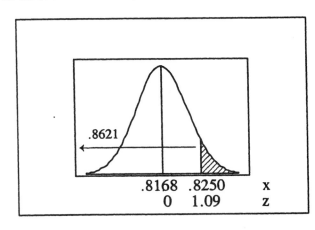

b. For the bottom 2.5%, A=.0250 & z=-1.96.
For the top 2.5%, A=.9750 & z=1.96.
$x_B = \mu + z\sigma$
   $= .816822 + (-1.96)(.007507)$
   $= .816822 - .014714$
   $= .80211$
$x_T = \mu + z\sigma$
   $= .816822 + (1.96)(.007507)$
   $= .816822 + .014714$
   $= .82154$

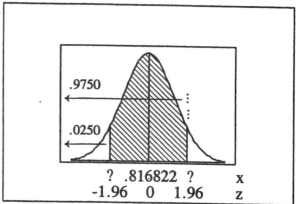

17. $\mu = 63.6$
    $\sigma = 2.5$
    $P(x>70)$
    $= P(z>2.56)$
    $= 1 - P(z<2.56)$
    $= 1 - .9948$
    $= .0052$ or .52%

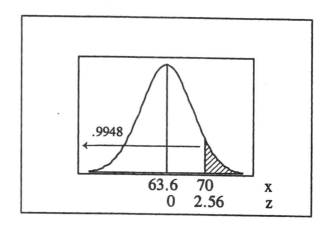

18. $\mu = 63.6$
    $\sigma = 2.5$
    $P(58<x<80)$
    $= P(-2.24<z<6.56)$
    $= P(z<6.56) - P(z<-2.24)$
    $= .9999 - .0125$
    $= .9874$ or 98.74%
    No; only 1.26% are too short or too tall.

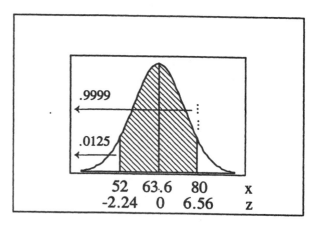

19. $\mu = 63.6$
    $\sigma = 2.5$
    $P(66.5<x<71.5)$
    $= P(1.16<z<3.16)$
    $= P(z<3.16) - P(z<1.16)$
    $= .9992 - .8770$
    $= .1222$
    The percentage meeting the requirement is 12.22%. Yes; the Rockettes are taller than the general population of women.

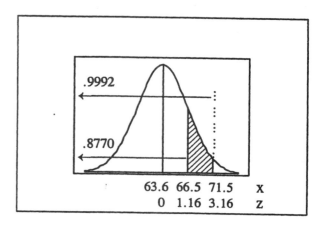

20. μ = 63.6
    σ = 2.5
    For the shortest 20%,
        A = .2000 [.2005] and z = −.84.
    For the tallest 20%,
        A = ..8000 [.7995] and z = .84.
    $x_S = \mu + z\sigma$
        = 63.6 + (−.84)(2.5)
        = 63.6 − 2.1
        = 61.5
    $x_T = \mu + z\sigma$
        = 63.6 + (.84)(2.5)
        = 63.6 + 2.1
        = 65.7

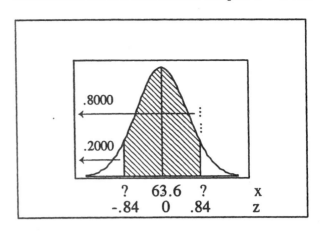

21. normal distribution with μ = 143 lbs and σ = 29 lbs
    a. The z scores are always unit-free. Because the numerator and the denominator of the fraction $z = (x-\mu)/\sigma$ have the same units, the units will divide out.
    b. For a population of size N, $\mu = \Sigma x/N$ and $\sigma^2 = \Sigma(x-\mu)^2/N$.
       As shown below, $\mu_z = 0$ and $\sigma_z = 1$ will be true for *any* set of z scores.
       $\Sigma z = \Sigma[(x-\mu)/\sigma] = (1/\sigma)\cdot[\Sigma(x-\mu)]$
                $= (1/\sigma)\cdot[\Sigma x - \Sigma\mu] = (1/\sigma)\cdot[N\mu - N\mu] = (1/\sigma)\cdot 0 = 0$
       $\Sigma z^2 = \Sigma[(x-\mu)/\sigma]^2 = (1/\sigma^2)\cdot[\Sigma(x-\mu)^2] = (1/\sigma^2)\cdot[N\sigma^2] = N$
       $\mu_z = (\Sigma z)/N = 0/N = 0$
       $\sigma_z^2 = \Sigma(z-\mu_z)^2/N = \Sigma(z-0)^2/N = \Sigma z^2/N = N/N = 1$; $\sigma_z = \sqrt{1} = 1$
       Since rescaling the scores will not change the basic shape of the distribution, the z scores will still have a normal distribution.
    c. For a population of size N, $\mu = \Sigma x/N$ and $\sigma^2 = \Sigma(x-\mu)^2/N$.
       As shown below, multiplying each value by a constant multiplies the mean and the standard deviation by that constant.
       Let $y = k \cdot x$
       $\mu_y = \Sigma y/N = [\Sigma(k\cdot x)]/N = [k\cdot(\Sigma x)]/N = k\cdot[(\Sigma x)/N] = k\cdot\mu_x$
       $\sigma_y^2 = \Sigma[y - \mu_y]^2/N = \Sigma[k\cdot x - k\cdot\mu_x]^2/N = k^2\cdot\Sigma[x-\mu_x]^2/N = k^2\cdot\sigma_x^2$; $\sigma_y = k\cdot\sigma_x$
       Since changing from pounds to kilograms multiplies each weight by .4536,
       μ = (.4536)(143) = 64.9 kg
       σ = (.4536)(29) = 13.2 kg

22. normal distribution with μ = 100 and σ = 15
    a. P(x > 105) = P(z > .33) = 1 − P(z < .33) = 1 − .6293 = .3707
    b. P(x > 105) = P_c(x > 105.5) = P(z > .36) = 1 − P(z < .37) = 1 − .6443 = .3557
    c. There is a difference of .3707 − .3557 = .015, an error of .015/.3557 = 4.2% when the correction is not used.

23. normal distribution with μ = 25 and σ = 5
    a. For a population of size N, $\mu = \Sigma x/N$ and $\sigma^2 = \Sigma(x-\mu)^2/N$.
       As shown below, adding a constant to each score will increase the mean by that amount but not affect the standard deviation. In non-statistical terms, shifting everything by k units does not affect the spread of the scores
       Let $y = x + k$
       $\mu_y = [\Sigma(x+k)]/N = [\Sigma x + Nk]/N = (\Sigma x)/N + (Nk)/N = \mu_x + k$
       $\sigma_y^2 = \Sigma[y - \mu_y]^2/N = \Sigma[(x+k) - (\mu_x+k)]^2/N = \Sigma[x - \mu_x]^2/N = \sigma_x^2$; $\sigma_y = \sigma_x$
       If the teacher adds 50 to each grade
       new mean = 25 + 50 = 75
       new standard deviation = 5 (same as before)

b. No; curving the scores should consider the variation. Had the test been more appropriately constructed, it is not likely that every student would score exactly 50 points higher. If the typical student score increased by 50, we would expect the better students to increase by more than 50 and the poorer students to increase by less than 50. This would make the scores more spread out and increase the standard deviation.

c. For the top 10%, A=.9000 [.8997] and z=1.28.
$x = \mu + z\sigma = 25 + (1.28)(5) = 25 + 6.4 = 31.4$
For the bottom 70%, A=.7000 [.6985] and z=.52.
$x = \mu + z\sigma = 25 + (.52)(5) = 25 + 2.6 = 27.6$
For the bottom 30%, A=.3000 [.3015] and z=-.52.
$x = \mu + z\sigma = 25 + (-.52)(5) = 25 - 2.6 = 22.4$
For the bottom 10%, A=.1000 [.1003] and z=-1.28.
$x = \mu + z\sigma = 25 + (-1.28)(5) = 25 - 6.4 = 18.6$
This produces the grading scheme given at the right.

A: higher than 31.4
B: 27.6 to 31.4
C: 22.4 to 27.6
D: 18.6 to 22.4
E: less than 18.6

d. The curving scheme in part (c) is fairer because it takes into account the variation as discussed in part (b). Assuming the usual 90-80-70-60 letter grade cutoffs, the percentage of A's under the scheme in part (a) with $\mu=75$ and $\sigma=5$ is
$P(x>90) = P(z>3.00)$
$= 1 - P(z<3.00)$
$= 1 - .9987$
$= .0013$ or .13%.
This is considerably less than the 10% A's under the scheme in part (c) and reflects the fact that the variation in part (a) is unrealistically small.

24. normal distribution with $\mu=1017$ and $Q_1=880$.
a. For Q_1, A=.2500 [.2514] and z=-.67.
$z = (x - \mu)/\sigma$
$\sigma = (x - \mu)/z$
$= (880 - 1017)/(-.67)$
$= 204.48$

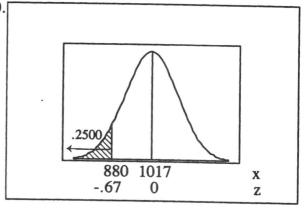

b. For P_{99}, A=.9900 [.9901] and z=2.33
$x = \mu + z\sigma$
$= 1017 + (2.33)(204.48)$
$= 1017 + 476$
$= 1493$

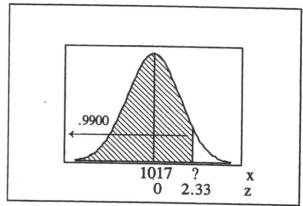

25. a. For P_{67}, $A = .6700$ and $z = .44$
$$x_{SAT} = \mu + z\sigma$$
$$= 998 + (.44)(202)$$
$$= 998 + 89$$
$$= 1087$$
$$x_{ACT} = \mu + z\sigma$$
$$= 20.9 + (.44)(4.6)$$
$$= 20.9 + 2.0$$
$$= 22.9$$

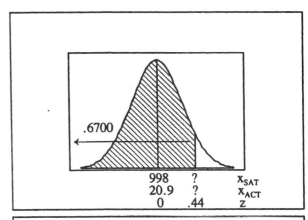

b. $z_{SAT} = (x - \mu)/\sigma$
$$= (1220 - 998)/202 = 1.10$$
$$x_{ACT} = \mu + z\sigma$$
$$= 20.9 + (1.10)(4.6) = 26.0$$

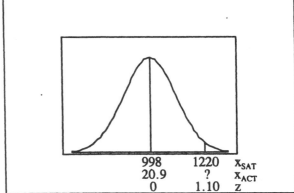

5-4 Sampling Distributions and Estimators

1. No; while the long run average of the values of the sample proportion equals the population proportion, sampling variability produces a distribution of sample proportions. The value of any one sample proportion is a member of that sampling distribution, but not necessarily equal to its long run average.

2. It is the distribution of the sample means from all possible samples of size 12. If the book has 300 pages, for example, there would be $_{300}C_{12} = 300!/(288!12!) = 8.88 \times 10^{20}$ possible samples of size 12. The value 5.08 is one of those 8.88×10^{20} values, whose long rung average is equal to the true value for all 300 pages.

3. No; the histogram from one sample reflects the shape of the population distribution of individual scores. The shape of the sampling distribution of sample means is not necessarily the same as the shape of the population distribution of individual scores.

4. a. The 10% is a statistic, because it was calculated from a sample and not from the entire population.
   b. The sampling distribution of which this value is a part consists of the sample proportions from all possible samples of size 1015 from the population.
   c. Yes; as the sample size increases, the sampling variability decreases – so that any one value is more likely to be closer to the true population value. For a sample of size $n=1$, the sampling distribution is the original population and the sampling variance is the same as the population variance. If the sample is the entire population, there is only one possible sample and the sampling "distribution" is a single value with variance zero. As the sample size increases from 1 to the population size, the sampling variance decreases from the original population variance to zero.

5. For the original population,
   $\mu = \Sigma x/N = (10+6+5)/3 = 21/3 = 7$.
   a. The 9 equally likely samples and their means are given at the immediate right.
   b. Each of the 9 means has probability 1/9. The sampling distribution, a list of the means and the total probability of each, is given at the far right
   c. $\mu_{\bar{x}} = \Sigma \bar{x} \cdot P(\bar{x}) = 63/9 = 7$.
   d. Yes. Yes, the sample mean is an unbiased estimator of the population mean.

sample	$\bar{x}$
5,5	5.0
5,6	5.5
5,10	7.5
6,5	5.5
6,6	6.0
6,10	8.0
10,5	7.5
10,6	8.0
10,10	10.0
	63.0

$\bar{x}$	$P(\bar{x})$	$\bar{x} \cdot P(\bar{x})$
5.0	1/9	5/9
5.5	2/9	11/9
6.0	1/9	6/9
7.5	2/9	15/9
8.0	2/9	16/9
10.0	1/9	10/9
	9/9	63/9

6. For the original population,
   $\mu = \Sigma x/N = (1+11+9+3)/4 = 24/4 = 6$.
   a. The 16 equally likely samples and their means are given at the immediate right.
   b. Each of the 16 means has probability 1/16. The sampling distribution, a list of the means and the total probability of each, is given at the far right
   c. $\mu_{\bar{x}} = \Sigma \bar{x} \cdot P(\bar{x}) = 96/16 = 6$
   d. Yes. Yes, the sample mean is an unbiased estimator of the population mean.

sample	$\bar{x}$
1,1	1
1,3	2
1,9	5
1,11	6
3,1	2
3,3	3
3,9	6
3,11	7
9,1	5
9,3	6
9,9	9
9,11	10
11,1	6
11,3	7
11,9	10
11,11	11
	96

$\bar{x}$	$P(\bar{x})$	$\bar{x} \cdot P(\bar{x})$
1	1/16	1/16
2	2/16	4/16
3	1/16	3/16
5	2/16	10/16
6	4/16	24/16
7	2/16	14/16
9	1/16	9/16
10	2/16	20/16
11	1/16	11/16
	16/16	96/16

7. For the original population,
   $\mu = \Sigma x/N$
   $= (85+79+82+73+78)/5$
   $= 397/5 = 79.4$
   a. The 25 equally likely samples and their means are given at the immediate right.
   b. Each of the 25 means has probability 1/25. The sampling distribution, a list of the means and the total probability of each, is given at the far right
   c. $\mu_{\bar{x}} = \Sigma \bar{x} \cdot P(\bar{x}) = 1985/25 = 79.4$
   d. Yes. Yes, the sample mean is an unbiased estimator of the population mean.

sample	$\bar{x}$
73,73	73.0
73,78	75.5
73,79	76.0
73,82	77.5
73,85	79.0
78,73	75.5
78,78	78.0
78,79	78.5
78,82	80.0
78,85	81.5
79,73	76.0
79,78	78.5
79,79	79.0
79,82	80.5
79,85	82.0
82,73	77.5
82,78	80.0
82,79	80.5
82,82	82.0
82,85	83.5
85,73	79.0
85,78	81.5
85,79	82.0
85,82	83.5
85,85	85.0
	1985.0

$\bar{x}$	$P(\bar{x})$	$\bar{x} \cdot P(\bar{x})$
73.0	1/25	73/25
75.5	2/25	151/25
76.0	2/25	152/25
77.5	2/25	155/25
78.0	1/25	78/25
78.5	2/25	157/25
79.0	3/25	237/25
80.0	2/25	160/25
80.5	2/25	161/25
81.5	2/25	163/25
82.0	3/25	246/25
83.5	2/25	167/25
85.0	1/25	85/25
	25/25	1985/25

INSTRUCTOR'S SOLUTIONS Chapter 5   S-129

8. For the original population,
$\mu = \Sigma x/N$
   $= (62+46+68+64+57)/5$
   $= 297/5 = 59.4$
   a. The 25 equally likely samples and their means are given at the immediate right.
   b. Each of the 25 means has probability 1/25. The sampling distribution, a list of the means and the total probability of each, is given at the far right
   c. $\mu_{\bar{x}} = \Sigma \bar{x} \cdot P(\bar{x}) = 1485/25 = 59.4$
   d. Yes. Yes, the sample mean is an unbiased estimator of the population mean.

sample	$\bar{x}$	$\bar{x}$	$P(\bar{x})$	$\bar{x} \cdot P(\bar{x})$
46,46	46.0	46.0	1/25	46/25
46,57	51.5	51.5	2/25	103/25
46,62	54.0	54.0	2/25	108/25
46,64	55.0	55.0	2/25	110/25
46,68	57.0	57.0	3/25	171/25
57,46	51.5	59.5	2/25	119/25
57,57	57.0	60.5	2/25	121/25
57,62	59.5	62.0	1/25	62/25
57,64	60.5	62.5	2/25	125/25
57,68	62.5	63.0	2/25	126/25
62,46	54.0	64.0	1/25	64/25
62,57	59.5	65.0	2/25	130/25
62,62	62.0	66.0	2/25	132/25
62,64	63.0	68.0	1/25	68/25
62,68	65.0		25/25	1485/25
64,46	55.0			
64,57	60.5			
64,62	63.0			
64,64	64.0			
64,68	66.0			
68,46	57.0			
68,57	62.5			
68,62	65.0			
68,64	66.0			
68,68	68.0			
	1485.0			

9. For the original population, the proportion of females is p = 3/4.
   a. Selecting with replacement, there are $4^2 = 16$ equally likely samples.
      Let x = the number of females per sample.
      binomial problem: n = 2 and p = 3/4, use the binomial formula
      $P(x) = [n!/x!(n-x)!] \cdot p^x \cdot (1-p)^{n-x}$
      $P(x=0) = [2!/0!2!] \cdot (3/4)^0 \cdot (1/4)^2 = [1] \cdot (1) \cdot (1/16) = 1/16$
      $P(x=1) = [2!/1!1!] \cdot (3/4)^1 \cdot (1/4)^1 = [2] \cdot (3/4) \cdot (1/4) = 6/16$
      $P(x=2) = [2!/2!0!] \cdot (3/4)^2 \cdot (1/4)^0 = [1] \cdot (9/16) \cdot (1) = 9/16$
      This means that of the 16 samples, there are
         1 with $\hat{p} = 0/2 = 0$ [namely, MM]
         6 with $\hat{p} = 1/2 = .50$ [namely, MA MB MC AM BM CM]
         9 with $\hat{p} = 2/2 = 1.00$ [namely, AA AB AC BA BB BC CA CB CC]
   b. The sampling distribution, a list of the sample proportions and the probability of each, is as follows.

$\hat{p}$	$P(\hat{p})$	$\hat{p} \cdot P(\hat{p})$
0.00	1/16	0/16
0.50	6/16	3/16
1.00	9/16	9/16
	16/16	12/16

   c. $\mu_{\hat{p}} = \Sigma \hat{p} \cdot P(\hat{p}) = 12/16 = 3/4$
   d. Yes. Yes, the sample proportion is an unbiased estimator of the population proportion.

10. For the original population, the proportion of defects is p = 2/5.
    a. Selecting with replacement, there are $5^2 = 25$ equally likely samples.
       Let x = the number of defects per sample.
       binomial problem: n = 2 and p = 2/5, use the binomial formula
       $P(x) = [n!/x!(n-x)!] \cdot p^x \cdot (1-p)^{n-x}$
       $P(x=0) = [2!/0!2!] \cdot (2/5)^0 \cdot (3/5)^2 = [1] \cdot (1) \cdot (9/25) = 9/25$
       $P(x=1) = [2!/1!1!] \cdot (2/5)^1 \cdot (3/5)^1 = [2] \cdot (2/5) \cdot (3/5) = 12/25$
       $P(x=2) = [2!/2!0!] \cdot (2/5)^2 \cdot (3/5)^0 = [1] \cdot (4/25) \cdot (1) = 4/25$
       This means that of the 25 samples, there are
          9 with $\hat{p} = 0/2 = 0$
         12 with $\hat{p} = 1/2 = .50$
          4 with $\hat{p} = 2/2 = 1.00$

b. The sampling distribution, a list of the sample proportions and the probability
of each, is as follows.

$\hat{p}$	$P(\hat{p})$	$\hat{p} \cdot P(\hat{p})$
0.00	9/25	0/25
0.50	12/25	6/25
1.00	4/25	4/25
	25/25	10/25

c. $\mu_{\hat{p}} = \Sigma \hat{p} \cdot P(\hat{p}) = 10/25 = 2/5$

d. Yes. Yes, the sample proportion is an unbiased estimator of the population proportion.

11. For the original population, the proportion of Democrats is $10/13 = .769$.
   a. One procedure is as follows. Generate a list of random numbers, find their remainders when divided by 13, let the 10 remainders 0-9 correspond to the 10 Democrats and let the 3 remainders 10-12 correspond to the 3 Republicans. So that each remainder has the same chance to occur, limit the usable random numbers to a whole multiple of 13. Using the 7x13 = 91 two-digit numbers from 00 to 90 and discarding all others, for example, allows each remainder 7 chances to appear among the usable random numbers. Applying this procedure to the first two digits of each 5-digit sequence in the list of random numbers for the exercises in section 3-6 yields the following.

   46 gives remainder 7, a Democrat
   99 is not usable
   72 gives remainder 7, a Democrat
   44 gives remainder 5, a Democrat
   86 gives remainder 8, a Democrat
   00 gives remainder 0, a Democrat

   b. The proportion of Democrats in the sample in part (a) is $5/5 = 1.00$.
   c. The proportion in part (b) is a statistic, because it was calculated from a sample.
   d. No; the above sample proportion does not equal the true population proportion of .769. No; samples of size n=5 only produce sample proportions of 0,.20,.40,.60,.80 or 1.00.
   e. The sample proportion is an unbiased estimator of the population proportion. If all possible $13^5 = 371,293$ samples were listed, the mean of those sample proportions would equal the true population proportion.

12. For the original population, the proportion of senators from Maine is 2/3.
    Label the senators ME1, ME2 and TX. Selecting without replacement there are $3 \cdot 2 = 6$ equally likely possible samples given at the left below.

sample	$\hat{p}$
ME1,ME2	2/2 = 1.00
ME1,TX	1/2 = .50
ME2,ME1	2/2 = 1.00
ME2,TX	1/2 = .50
TX,ME1	1/2 = .50
TX,ME2	1/2 = .50

$\hat{p}$	$P(\hat{p})$	$\hat{p} \cdot P(\hat{p})$
0.50	4/6	2/6
1.00	2/6	2/6
	6/6	4/6

The sampling distribution given at the right above verifies that $\mu_{\hat{p}} = \Sigma \hat{p} \cdot P(\hat{p}) = 4/6 = 2/3$, which equal to the population proportion.

13. The population mean is $\mu = \Sigma x/N = (62+46+68+64+57) = 297/5 = 59.4$.
   a. All possible samples of size n=2 form a population of $N = {}_5C_2 = 10$ members as follows.

sample	$\overline{x}$	$\overline{x}-\mu_{\overline{x}}$	$(\overline{x}-\mu_{\overline{x}})^2$
46-57	51.5	-7.9	62.41
46-62	54.0	-5.4	29.16
46-64	55.0	-4.4	19.36
46-68	57.0	-2.4	5.76
57-62	59.5	.1	.01
57-64	60.5	1.1	1.21
57-68	62.5	3.1	9.61
62-64	63.0	3.6	12.96
62-68	65.0	5.6	31.36
64-68	66.0	6.6	43.56
	594.0	0	215.40

$\mu_{\overline{x}} = \Sigma \overline{x}/N = 594.0/10 = 59.4$

$\sigma_{\overline{x}}^2 = \Sigma(\overline{x}-\mu_{\overline{x}})^2/N = 215.40/10 = 21.540$

$\sigma_{\overline{x}} = 4.64$

b. All possible samples of size n=3 form a population of $N={}_5C_3=10$ members as follows.

```
sample     x̄      x̄-μx̄    (x̄-μx̄)²
46-57-62   55.00  -4.40    19.36
46-57-64   55.67  -3.73    13.94
46-57-68   57.00  -2.40     5.76
46-62-64   57.33  -2.07     4.27
46-62-68   58.67   -.73      .54
46-64-68   59.33   -.07      .00
57-62-64   61.00   1.60     2.56
57-62-68   62.33   2.93     8.60
57-64-68   63.00   3.60    12.96
62-64-68   64.67   5.27    27.74
           594.00     0    95.73
```

$\mu_{\bar{x}} = \Sigma\bar{x}/N = 594.00/10 = 59.4$

$\sigma_{\bar{x}}^2 = \Sigma(\bar{x}-\mu_{\bar{x}})^2/N = 95.73/10 = 9/573$

$\sigma_{\bar{x}} = 3.09$

c. All possible samples of size n=4 form a population of $N={}_5C_4=5$ members as follows.

```
sample        x̄      x̄-μx̄    (x̄-μx̄)²
46-57-62-64   57.25  -2.15    4.6225
46-57-62-68   58.25  -1.15    1.3225
46-57-64-68   58.75   -.65     .4225
46-62-64-68   60.00    .60     .3600
57-62-64-68   62.75   3.35   11.2225
              297.00     0   17.9500
```

$\mu_{\bar{x}} = \Sigma\bar{x}/N = 297.00/5 = 59.4$

$\sigma_{\bar{x}}^2 = \Sigma(\bar{x}-\mu_{\bar{x}})^2/N = 17.9500/5 = 3.5900$

$\sigma_{\bar{x}} = 1.89$

d. Yes; each of the above sampling distributions has a mean equal to the population mean.

e. As the sample size increases, the variation of the sampling distribution decreases.

14. The MAD of the population is 1.56, as illustrated by the following table.

```
x    |x-x̄|
1    5/3       x̄ = Σx/N = 8/3
2    2/3       MAD = Σ|x-x̄|/N
5    7/3           = (14/3)/3 = 14/9 = 1.56
8    14/3
```

The sampling distribution of the 9 possible equally likely sample MAD's has a mean of .89, as illustrated by the following tables.

```
sample   x̄     |x-x̄|'s    MAD
1,1      1.0    0,0         0
1,2      1.5    .5,.5       .5
1,5      3.0    2,2        2.0
2,1      1.5    .5,.5       .5
2,2      2.0    0,0         0
2,5      3.5    1.5,1.5    1.5
5,1      3.0    2,2        2.0
5,2      3.5    1.5,1.5    1.5
5,5      5.0    0,0         0
```

```
MAD   P(MAD)   MAD·P(MAD)
 0     3/9       0/9          μ_MAD = ΣMAD·P(MAD)
 .5    2/9       1/9                = 8/9 = .89
1.5    2/9       3/9
2.0    2/9       4/9
       9/9       8/9
```

The sample MAD is not a good estimator of the population MAD.

15. The 27 equally likely samples of size n=3 are listed below, followed by their medians. The sampling distribution of the medians is given at the right.

```
111-1  121-1  151-1  211-1  221-2  251-2  511-1  521-2  551-5
112-1  122-2  152-2  212-2  222-2  252-2  512-2  522-2  552-5
115-1  125-2  155-5  215-2  225-2  255-5  515-5  525-5  555-5
```

```
m   P(m)    m·P(m)
1   7/27     7/27
2   13/27   26/27
5   7/27    35/27
    27/27   68/27
```

The means for the above samples are as follows. The sampling distribution for the means is given at the right.

```
3/3   4/3   7/3   4/3   5/3   8/3   7/3   8/3   11/3
4/3   5/3   8/3   5/3   6/3   9/3   8/3   9/3   12/3
7/3   8/3  11/3   8/3   9/3  12/3  11/3  12/3   15/3
```

$\mu_m = \Sigma m \cdot P(m) = 68/27 = 2.52$

$\mu_{\bar{x}} = \Sigma\bar{x} \cdot P(\bar{x}) = 216/81 = 2.67$

Since $\mu = 8/3 = 2.67$, the sample mean is a good estimator for the population mean, but the sample median is not. The sampling distributions indicate that

```
x̄       P(x̄)    x̄·P(x̄)
3/3     1/27     3/81
4/3     3/27    12/81
5/3     3/27    15/81
6/3     1/27     6/81
7/3     3/27    21/81
8/3     6/27    48/81
9/3     3/27    27/81
11/3    3/27    33/81
12/3    3/27    36/81
15/3    1/27    15/81
        27/27  216/81
```

5-5 The Central Limit Theorem

NOTE: When using individual scores (i.e., making a statement about one x score from the original distribution), convert x to z using the mean and standard deviation of the x's and $z = (x-\mu)/\sigma$. When using a sample of n scores (i.e., making a statement about), convert $\bar{x}$ to z using the mean and standard deviation of the $\bar{x}$'s and $z = (\bar{x}-\mu_{\bar{x}})/\sigma_{\bar{x}}$.

IMPORTANT NOTE: After calculating $\sigma_{\bar{x}}$, <u>STORE IT</u> in the calculator to recall it with total accuracy whenever it is needed in subsequent calculations. <u>DO NOT</u> write it down on paper rounded off (even to several decimal places) and then re-enter it in the calculator whenever it is needed. This avoids both round-off errors and recopying errors

1. a. normal distribution
   $\mu = 172$
   $\sigma = 29$
   $P(x<167)$
   $= P(z<-.17)$
   $= .4325$

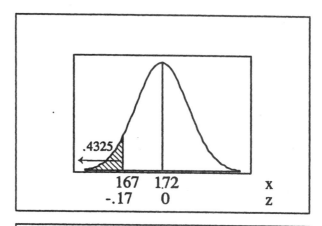

   b. normal distribution,
      since the original distribution is so
   $\mu_{\bar{x}} = \mu = 172$
   $\sigma_{\bar{x}} = \sigma/\sqrt{n} = 29/\sqrt{36} = 4.833$
   $P(\bar{x}<167)$
   $= P(z<-1.03)$
   $= .1515$

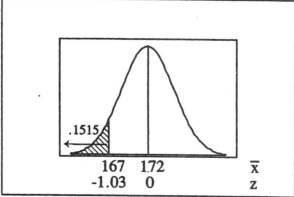

NOTE: Remember that the figures are for illustrative purposes, as an aid to solving the problem, and are not to scale. In scaled drawings, the shaded area would be very large (almost ½ the figure) in part 1(a) and very small in part 1(b).

2. a. normal distribution
   $\mu = 172$
   $\sigma = 29$
   $P(x>180)$
   $= P(z>.28)$
   $= 1 - P(z<.28)$
   $= 1 - .6103$
   $= .3897$

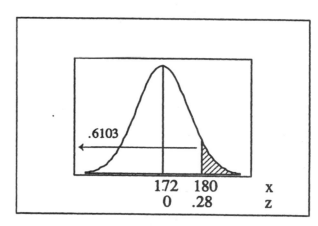

b. normal distribution,
   since the original distribution is so
   $\mu_{\bar{x}} = \mu = 172$
   $\sigma_{\bar{x}} = \sigma/\sqrt{n} = 29/\sqrt{100} = 2.9$
   $P(\bar{x} > 180)$
   $= P(z > 2.76)$
   $= 1 - P(z < 2.76)$
   $= 1 - .9971$
   $= .0029$

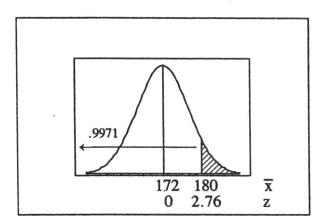

3. a. normal distribution
   $\mu = 172$
   $\sigma = 29$
   $P(170 < x < 175)$
   $= P(-.07 < z < .10)$
   $= P(z < .10) - P(z < -.07)$
   $= .5398 - .4721$
   $= .0677$

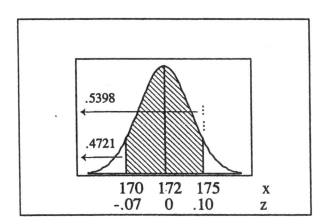

b. normal distribution,
   since the original distribution is so
   $\mu_{\bar{x}} = \mu = 172$
   $\sigma_{\bar{x}} = \sigma/\sqrt{n} = 29/\sqrt{64} = 3.625$
   $P(170 < \bar{x} < 175)$
   $= P(-.55 < z < .83)$
   $= P(z < .83) - P(z < -.55)$
   $= .7967 - .2912$
   $= .5055$

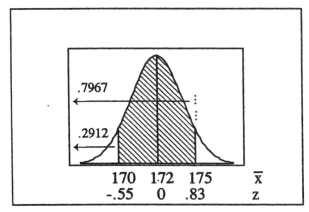

4. a. normal distribution
   $\mu = 172$
   $\sigma = 29$
   $P(100 < x < 165)$
   $= P(-2.48 < z < -.24)$
   $= P(z < -.24) - P(z < -2.48)$
   $= .4052 - .0066$
   $= .3986$

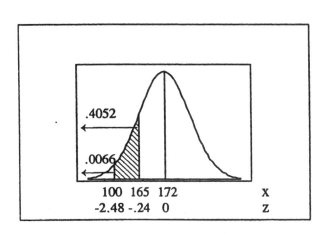

b. normal distribution,
   since the original distribution is so
   $\mu_{\bar{x}} = \mu = 172$
   $\sigma_{\bar{x}} = \sigma/\sqrt{n} = 29/\sqrt{81} = 3.222$
   $P(100 < \bar{x} < 165)$
   $= P(-22.34 < z < -2.17)$
   $= P(z < -2.17) - P(z < -22.34)$
   $= .1050 - .0001$
   $= .0149$

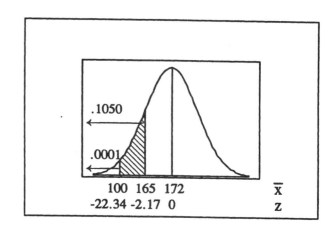

5. a. normal distribution,
      since the original distribution is so
      $\mu_{\bar{x}} = \mu = 172$
      $\sigma_{\bar{x}} = \sigma/\sqrt{n} = 29/\sqrt{25} = 5.8$
      $P(\bar{x} > 160)$
      $= P(z > -2.07)$
      $= 1 - P(z < -2.07)$
      $= 1 - .0192$
      $= .9808$
   b. When the original distribution is normal, the sampling distribution of the $\bar{x}$'s is normal regardless of the sample size.

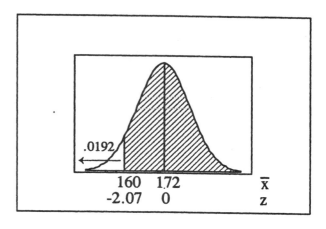

6. a. normal distribution,
      since the original distribution is so
      $\mu_{\bar{x}} = \mu = 172$
      $\sigma_{\bar{x}} = \sigma/\sqrt{n} = 29/\sqrt{4} = 14.5$
      $P(160 < \bar{x} < 180)$
      $= P(-.83 < z < .55)$
      $= P(z < .55) - P(z < -.83)$
      $= .7088 - .2033$
      $= .5055$
   b. When the original distribution is normal, the sampling distribution of the $\bar{x}$'s is normal regardless of the sample size.

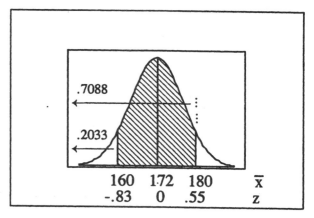

7. a. normal distribution
      $\mu = 143$
      $\sigma = 29$
      $P(140 < x < 211)$
      $= P(-.10 < z < 2.34)$
      $= P(z < 2.34) - P(z < -.10)$
      $= .9904 - .4602$
      $= .5302$

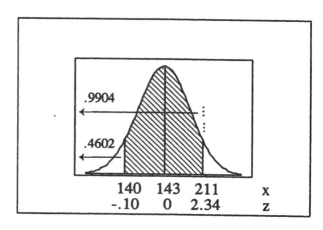

b. normal distribution,
   since the original distribution is so
   $\mu_{\bar{x}} = \mu = 143$
   $\sigma_{\bar{x}} = \sigma/\sqrt{n} = 29/\sqrt{36} = 4.833$
   $P(140 < \bar{x} < 211)$
   $= P(-.62 < z < 14.07)$
   $= P(z < 14.07) - P(z < -.62)$
   $= .9999 - .2676$
   $= .7323$

c. The information from part (a) is more relevant, since the seat will be occupied by one woman at a time.

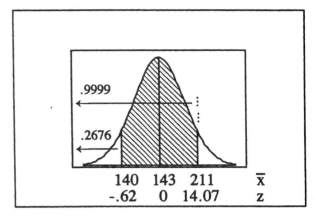

8. a. normal distribution
   $\mu = 6.0$
   $\sigma = 1.0$
   $P(x < 6.2)$
   $= P(z < .20)$
   $= .5793$

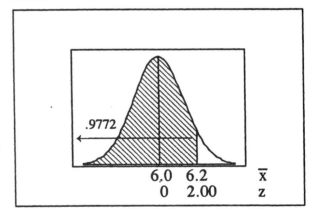

b. normal distribution,
   since the original distribution is so
   $\mu_{\bar{x}} = \mu = 6.0$
   $\sigma_{\bar{x}} = \sigma/\sqrt{n} = 1.0/\sqrt{100} = .10$
   $P(\bar{x} < 6.2)$
   $= P(z < 2.00)$
   $= .9772$

c. The information from part (a) is more relevant, since the helmets will be worn by one man at a time.

9. a. normal distribution,
   since the original distribution is so
   $\mu_{\bar{x}} = \mu = 14.4$
   $\sigma_{\bar{x}} = \sigma/\sqrt{n} = 1.0/\sqrt{2} = .707$
   $P(\bar{x} > 16.0)$
   $= P(z > 2.26)$
   $= 1 - P(z < 2.26)$
   $= 1 - .9881$
   $= .0119$

b. No. Yes; even if all the seats were occupied by 2 men, there would be a problem only about 1% 0f the time – in practice, most men probably sit with a female or a child.

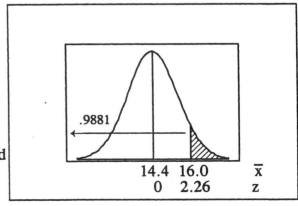

10. normal distribution,
  by the Central Limit Theorem
  $\mu_{\bar{x}} = \mu = .500$
  $\sigma_{\bar{x}} = \sigma/\sqrt{n} = .289/\sqrt{100} = .0289$
  $P(\bar{x} > .57)$
  $= P(z > 2.42)$
  $= 1 - P(z < 2.42)$
  $= 1 - .9922$
  $= .0078$

b. Yes; since .0078 < .05, generating such random numbers would be considered an unusual event for a properly working random number generator.

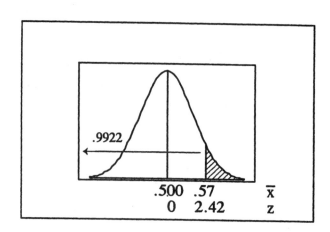

11. a. normal distribution,
  by the Central Limit Theorem
  $\mu_{\bar{x}} = \mu = 12.00$
  $\sigma_{\bar{x}} = \sigma/\sqrt{n} = .11/\sqrt{36} = .0183$
  $P(\bar{x} \geq 12.19)$
  $= P(z \geq 10.36)$
  $= 1 - P(z < 10.36)$
  $= 1 - .9999$
  $= .0001$

b. No; if $\mu = 12$, then an extremely rare event has occurred. No; there is no cheating, since the actual amount is more than the advertised amount.

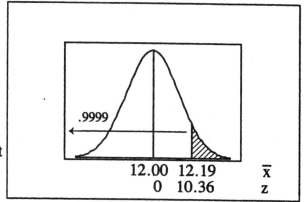

12. a. normal distribution
  $\mu = 100$
  $\sigma = 15$
  $P(x > 133)$
  $= P(z > 2.20)$
  $= .1 - P(z < 2.20)$
  $= 1 - .9861$
  $= .0139$

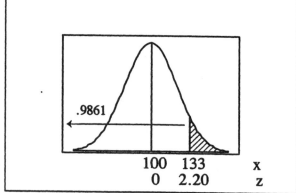

b. normal distribution,
  since the original distribution is so
  $\mu_{\bar{x}} = \mu = 100$
  $\sigma_{\bar{x}} = \sigma/\sqrt{n} = 15/\sqrt{9} = 5.0$
  $P(\bar{x} > 133)$
  $= P(z > 6.60)$
  $= 1 - P(z < 6.60)$
  $= 1 - .9999$
  $= .0001$

c. No; if the mean is 133, all we know is that the scores were clustered around 133. Some scores could have been considerably higher, and others considerably lower – in particular, some scores could have been lower than 131.5.

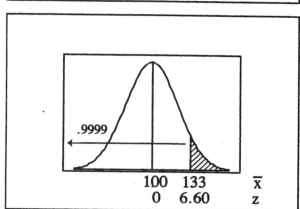

13. a. normal distribution,
       by the Central Limit Theorem
    $\mu_{\bar{x}} = \mu = 8.2$
    $\sigma_{\bar{x}} = \sigma/\sqrt{n} = 1.1/\sqrt{50} = .156$
    $P(\bar{x} \leq 7.8)$
       $= P(z \leq -2.57)$
       $= .0051$
    b. Yes; it appears that the TV sets sold by the Portland Electronics store are of less than average quality.

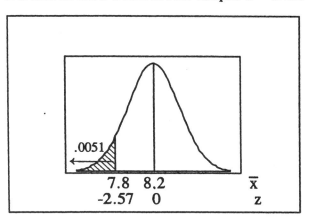

14. a. normal distribution
    $\mu = 114.8$
    $\sigma = 13.1$
    $P(x > 140)$
       $= P(z > 1.92)$
       $= 1 - P(z < 1.92)$
       $= 1 - .9726$
       $= .0274$

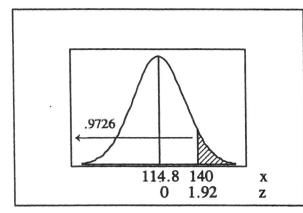

   b. normal distribution,
      since the original distribution is so
    $\mu_{\bar{x}} = \mu = 114.8$
    $\sigma_{\bar{x}} = \sigma/\sqrt{n} = 13.1/\sqrt{4} = 6.55$
    $P(\bar{x} > 140)$
       $= P(z > 3.85)$
       $= 1 - P(z < 3.85)$
       $= 1 - .9999$
       $= .0001$
   c. When the original distribution is normal, the sampling distribution of the $\bar{x}$'s is normal regardless of the sample size.
   d. No; the mean can be less than 140 even though one or more of the values is greater than 140.

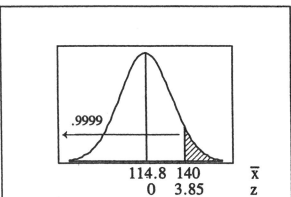

15. a. normal distribution,
       by the Central Limit Theorem
    $\mu_{\bar{x}} = \mu = .941$
    $\sigma_{\bar{x}} = \sigma/\sqrt{n} = .313/\sqrt{40} = .0495$
    $P(\bar{x} \leq .882)$
       $= P(z < -1.19)$
       $= .1170$
    b. No; since $.1170 > .05$, getting such a mean under the previous conditions would not be considered unusual.

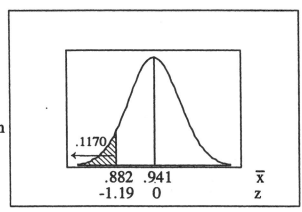

16. a. normal distribution
   $\mu = 509$
   $\sigma = 112$
   $P(x \geq 590)$
   $= P(z \geq .72)$
   $= 1 - P(z < .72)$
   $= 1 - .7642$
   $= .2358$

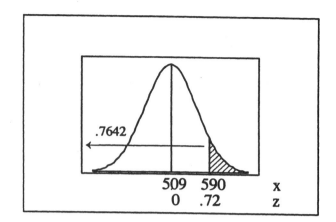

   b. normal distribution,
      since the original distribution is so
   $\mu_{\bar{x}} = \mu = 509$
   $\sigma_{\bar{x}} = \sigma/\sqrt{n} = 112/\sqrt{16} = 28$
   $P(\bar{x} \geq 590)$
   $= P(z \geq 2.89)$
   $= 1 - P(z < 2.89)$
   $= 1 - .9981$
   $= .0019$

   c. When the original distribution is normal, the sampling distribution of the $\bar{x}$'s is normal regardless of the sample size.

   d. Yes; since .0019 < .05 by quite a bit, there is strong evidence to support the claim that the course is effective -- assuming, as stated, that the 16 participants were truly a random sample (and not self-selected or otherwise chosen in some biased manner).

17. normal distribution,
   since the original distribution is so
   $\mu_{\bar{x}} = \mu = 27.44$
   $\sigma_{\bar{x}} = \sigma/\sqrt{n} = 12.46/\sqrt{4872} = .179$
   $P(\bar{x} > 27.88)$
   $= P(z > 2.46)$
   $= 1 - P(z < 2.46)$
   $= 1 - .9931$
   $= .0069$
   The system is currently acceptable and can expect to be overloaded only $(.0069)(52) = .36$ weeks a year -- or about once every three years.

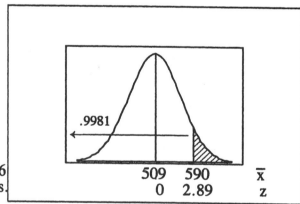

18. $\mu = .9147$
   $\sigma = .0369$
   a. This cannot be answered without knowing the shape of the distribution.
      NOTE: If the weights of the M&M's are normally distributed, then Table A-2 can be used and $P(x > .9085) = P(z > -.17)$
      $= 1 - P(z < -.17)$
      $= 1 - .4325$
      $= .5675$.

b. normal distribution,
   by the Central Limit Theorem
   $\mu_{\bar{x}} = \mu = .9147$
   $\sigma_{\bar{x}} = \sigma/\sqrt{n} = .0369/\sqrt{1498} = .000953$
   $P(\bar{x} \geq .9085)$
   $= P(z \geq -6.50)$
   $= 1 - P(z < -6.50)$
   $= 1 - .0001$
   $= .9999$
c. Yes; the company is providing at least the amount claimed on the label with almost absolute certainty.

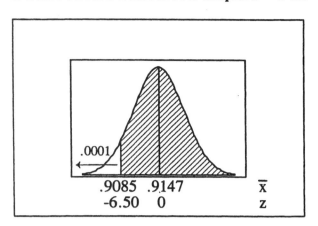

19. normal distribution,
    since the original distribution is so
    $\mu_{\bar{x}} = \mu = 172$
    $\sigma_{\bar{x}} = \sigma/\sqrt{n} = 29/\sqrt{16} = 7.25$
    This is a "backwards" normal problem, with A = .9760 and z = 1.96.
    $\bar{x} = \mu_{\bar{x}} + z\sigma_{\bar{x}}$
    $= 172 + (1.96)(7.25)$
    $= 172 + 14.21$
    $= 186.21$
    For 16 men, the total weight is
    $(16)(186.21) = 2979$ lbs.

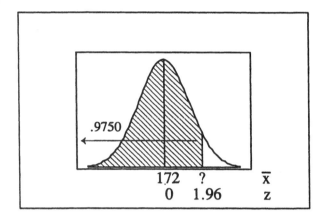

20. a. normal distribution,
    since the original distribution is so
    $\mu_{\bar{x}} = \mu = 14.4$
    $\sigma_{\bar{x}} = \sigma/\sqrt{n} = 1.0/\sqrt{18} = .236$
    This is a "backwards" normal problem, with A = .9760 and z = 1.96.
    $\bar{x} = \mu_{\bar{x}} + z\sigma_{\bar{x}}$
    $= 14.4 + (1.96)(.236)$
    $= 14.4 + .46$
    $= 14.86$
    For 18 men, the total width is
    $(18)(14.86) = 267.5$ in.

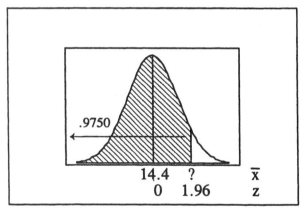

b. The 14.4 inch figure applies to the general male population, not to football players (who would tend to be larger than normal anyway) wearing pads. Furthermore, the calculations in part (a) do not consider any spacing between the players.

21. normal distribution, since the original distribution is so
    NOTE: Since $8/120 = .0667 > .05$, use the finite population correction factor
    $\mu_{\bar{x}} = \mu = 143$
    $\sigma_{\bar{x}} = [\sigma/\sqrt{n}] \cdot \sqrt{(N-n)/(N-1)}$
    $= [29/\sqrt{8}] \cdot \sqrt{(120-8)/(120-1)}$
    $= [29/\sqrt{8}] \cdot \sqrt{(112)/(119)}$
    $= 9.947$

a. A total weight of 1300 for 8 women corresponds to $\bar{x} = 1300/8 = 162.5$.
$P(\bar{x} \leq 162.5)$
$= P(z \leq 1.96)$
$= .9750$

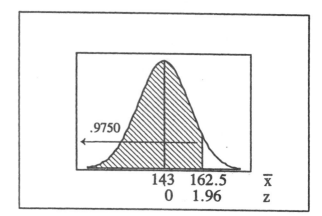

b. normal distribution,
since the original distribution is so
$\mu_{\bar{x}} = \mu = 143$
$\sigma_{\bar{x}} = 9.947$ (as calculated above)
This is a "backwards" normal problem,
with A = .9900 [.9901] and z = 2.33.
$\bar{x} = \mu_{\bar{x}} + z\sigma_{\bar{x}}$
$= 143 + (2.33)(9.947)$
$= 143 + 23.18$
$= 166.18$
For 8 women, the total weight is
$(8)(166.18) = 1329$ lbs.

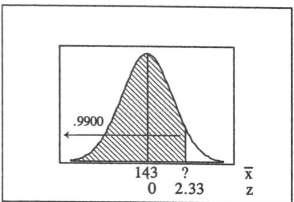

22. a.

x	x-μ	(x-μ)²
2	-6	36
3	-5	25
6	-2	4
8	0	0
11	3	9
18	10	100
48	0	174

$\mu = (\Sigma x)/N = 48/6 = 8$

$\sigma^2 = \Sigma(x-\mu)^2/N = 174/6 = 29$

$\sigma = 5.385$

b. and c.

sample	$\bar{x}$	$\bar{x} - \mu_{\bar{x}}$	$(\bar{x} - \mu_{\bar{x}})^2$
2,3	2.5	-5.5	30.25
2,6	4.0	-4.0	16.00
2,8	5.0	-3.0	9.00
2,11	6.5	-1.5	2.25
2,18	10.0	2.0	4.00
3,6	4.5	-3.5	12.25
3,8	5.5	-2.5	6.25
3,11	7.0	-1.0	1.00
3,18	10.5	2.5	6.25
6,8	7.0	-1.0	1.00
6,11	8.5	.5	.25
6,18	12.0	4.0	16.00
8,11	9.5	1.5	2.25
8,18	13.0	5.0	25.00
11,18	14.5	6.5	42.25
	120.0	0.0	174.00

d. Refer to the 15 $\bar{x}$ values and the two "extra" columns in the table for parts (b) and (c).

$\mu_{\bar{x}} = \Sigma\bar{x}/15 = 120/15 = 8$

$\sigma^2_{\bar{x}} = \Sigma(\bar{x}-\mu_{\bar{x}})^2/15 = 174/15 = 11.6$

$\sigma_{\bar{x}} = \sqrt{11.6} = 3.406$

e. $\mu = 8 = \mu_{\bar{x}}$
$[\sigma/\sqrt{n}] \cdot \sqrt{(N-n)/(N-1)} = [5.385/\sqrt{2}] \cdot \sqrt{(6-2)/(6-1)} = 3.406 = \sigma_{\bar{x}}$

23. normal distribution,
    by the Central Limit Theorem
    $\mu_{\bar{x}} = \mu = .500$
    $\sigma_{\bar{x}} = \sigma/\sqrt{n} = .289/\sqrt{100} = .0289$
    $P(.499 < \bar{x} < .501)$
    $= P(-.03 < z < .03)$
    $= P(z < .03) - P(z < -.03)$
    $= .5120 - .4880$
    $= .0240$

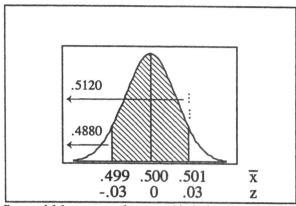

Since $.0240 < .05$, the event is "unusual" in that the sample mean will not fall within those narrow limits very often. Since the probability of getting a result "as extreme or more extreme" would be very close to ½ (whether the sample mean was less than or greater than $\mu = .5000$), the event is not of the "extreme" sort that leads one to question the assumptions associated with the procedure. The key idea is the difference between "unusual in the sense of rare" and "unusual in the sense of extreme." Since there are so many possible results, the probability of any one them (even the most "normal" ones) occurring is small.

5-6 Normal Distribution as Approximation to Binomial Distribution

NOTE: As in the previous sections, P(E) represents the probability of an event E; this manual uses $P_c(E)$ to represent the probability of an event E with the continuity correction applied.

1. the area to the right of 15.5; in symbols, $P(x > 15) = P_c(x > 15.5)$

2. the area to the right of 23.5; in symbols, $P(x \geq 24) = P_c(x > 23.5)$

3. the area to the left of 99.5; in symbols, $P(x < 100) = P_c(x < 99.5)$

4. the area from 26.5 to 27.5; in symbols, $P(x = 57) = P_c(26.5 < x < 27.5)$

5. the area to the left of 4.5; in symbols, $P(x \leq 4) = P_c(x < 4.5)$

6. the area from 14.5 to 20.5; in symbols, $P(15 \leq x \leq 20) = P_c(14.5 < x < 20.5)$

7. the area from 7.5 to 10.5; in symbols, $P(8 \leq x \leq 10) = P_c(7.5 < x < 10.5)$

8. the area from 2.5 to 3.5; in symbols, $P(x = 3) = P_c(2.5 < x < 3.5)$

IMPORTANT NOTE: As in the previous sections, store σ in the calculator so that it may be recalled with complete accuracy whenever it is needed in subsequent calculations.

9. binomial: $n = 14$ and $p = .50$
    a. from Table A-1, $P(x = 9) = .122$
    b. normal approximation appropriate since
       $np = 14(.50) = 7 \geq 5$
       $n(1-p) = 14(.50) = 7 \geq 5$
       $\mu = np = 14(.50) = 7$
       $\sigma = \sqrt{np(1-p)} = \sqrt{14(.50)(.50)} = 1.871$
       $P(x = 9)$
       $= P_c(8.5 < x < 9.5)$
       $= P(.80 < z < 1.34)$
       $= P(z < 1.34) - P(z < .80)$
       $= .9099 - .7881$
       $= .1218$

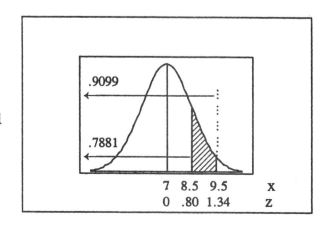

10. binomial: n = 12 and p = .80
   a. from Table A-1, P(x=7) = .053
   b. normal approximation <u>not</u> appropriate since
      n(1-p) = 12(.20) = 2.4 < 5

11. binomial: n = 15 and p = .90
   a. from Table A-1, P(x≥14) = P(x=14) + P(x=15)
                              = .343 + .206
                              = .549
   b. normal approximation <u>not</u> appropriate since n(1-p) = 15(.10) = 1.5 < 5

12. binomial: n = 13 and p = .40
   a. from Table A-1,
      P(x<3) = P(x=0) + P(x=1) + P(x=2)
             = .001 + .011 + .045
             = .057
   b. normal approximation appropriate since
      np = 13(.40) = 5.2 ≥ 5
      n(1-p) = 13(.60) = 7.8 ≥ 5
      $\mu = np = 13(.40) = 5.2$
      $\sigma = \sqrt{np(1-p)} = \sqrt{13(.40)(.60)} = 1.766$
      P(x<3)
         = $P_c(x<2.5)$
         = P(z<-1.53)
         = .0630

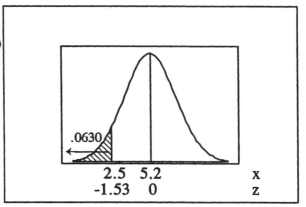

13. let x = the number of girls born
   binomial: n = 100 and p = .50
   normal approximation appropriate since
      np = 100(.50) = 50 ≥ 5
      n(1-p) = 100(.50) = 50 ≥ 5
   $\mu = np = 100(.50) = 50$
   $\sigma = \sqrt{np(1-p)} = \sqrt{100(.50)(.50)} = 5.000$
   P(x>55)
      = $P_c(x>55.5)$
      = P(z>1.10)
      = 1 - P(z<1.10)
      = 1 - .8643
      = .1357

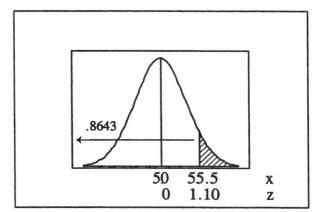

No; since .1357 > .05, it is not unusual to get more than 55 girls in 100 births.

14. let x = the number of girls born
   binomial: n = 100 and p = .50
   normal approximation appropriate since
      np = 100(.50) = 50 ≥ 5
      n(1-p) = 100(.50) = 50 ≥ 5
   $\mu = np = 100(.50) = 50$
   $\sigma = \sqrt{np(1-p)} = \sqrt{100(.50)(.50)} = 5.000$
   P(x≥65)
      = $P_c(x>64.5)$
      = P(z>2.90)
      = 1 - P(z<2.90)
      = 1 - .9981
      = .0019

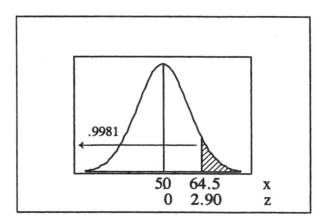

Yes: since .0019 < .05, it is unusual to get at least 65 girls in 100 births.

INSTRUCTOR'S SOLUTIONS Chapter 5   S-143

15. let x = the number of correct responses
   binomial: n = 100 and p = .50
   normal approximation appropriate since
      np = 100(.50) = 50 ≥ 5
      n(1-p) = 100(.50) = 50 ≥ 5
   μ = np = 100(.50) = 50
   σ = $\sqrt{np(1-p)}$ = $\sqrt{100(.50)(.50)}$ = 5.000
   P(x ≥ 60)
      = P_c(x > 59.5)
      = P(z > 1.90)
      = 1 - P(z < 1.90)
      = 1 - .9713
      = .0287

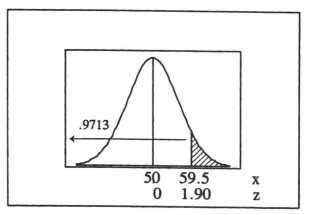

No; the probability of passing by chance is small – since .0287 < .05, it would be unusual.

16. let x = the number of correct responses
   binomial: n = 25 and p = .2
   normal approximation appropriate since
      np = 25(.2) = 5 ≥ 5
      n(1-p) = 25(.8) = 20 ≥ 5
   μ = np = 25(.2) = 5
   σ = $\sqrt{np(1-p)}$ = $\sqrt{25(.2)(.8)}$ = 2.000
   P(3 ≤ x ≤ 10)
      = P_c(2.5 < x < 10.5)
      = P(-1.25 < z < 2.75)
      = P(z < 2.75) - P(z < -1.25)
      = .9970 - .1056
      = .8914

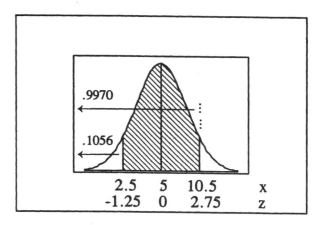

17. let x = the number with yellow pods
   binomial: n = 580 and p = .25
   normal approximation appropriate since
      np = 580(.25) = 145 ≥ 5
      n(1-p) = 580(.75) = 435 ≥ 5
   μ = np = 580(.25) = 145
   σ = $\sqrt{np(1-p)}$ = $\sqrt{580(.25)(.75)}$ = 10.428
   P(x ≥ 152)
      = P_c(x > 151.5)
      = P(z > .62)
      = 1 - P(z < .62)
      = 1 - .7324
      = .2676

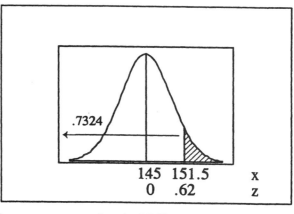

No; such as result would not be unusual when the true proportion is 25%.

18. let x = the number experiencing flu symptoms
   binomial: n = 863 and p = .019
   normal approximation appropriate since
      np = 863(.019) = 16.397 ≥ 5
      n(1-p) = 863(.981) = 846.603 ≥ 5
   μ = np = 863(.019) = 16.397
   σ = $\sqrt{np(1-p)}$ = $\sqrt{683(.019)(.981)}$ = 4.011
   P(x ≥ 19)
      = P_c(x > 18.5)
      = P(z > .52)
      = 1 - P(z < .52)
      = 1 - .6985 = .3015

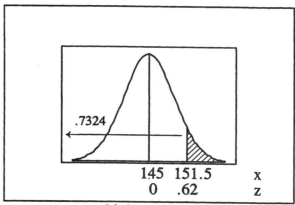

Those taking the drug do not appear to have flu symptoms at a higher than normal rate.

19. let x = the number of color blind men
    binomial: n = 600 and p = .09
    normal approximation appropriate since
       np = 600(.09) = 54 ≥ 5
       n(1-p) = 600(.91) = 546 ≥ 5
    μ = np = 600(.09) = 54
    σ = √np(1-p) = √600(.09)(.91) = 7.010
    P(x ≥ 50)
       = P_c(x > 49.5)
       = P(z > -.64)
       = 1 - P(z < -.64)
       = 1 - .2611
       = .7389

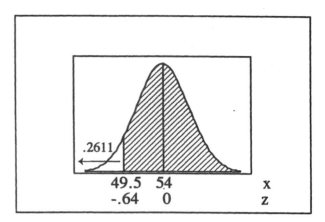

No; they will get the desired number about 3/4 of the time, which cannot be described as "very confident."

20. let x = the number developing cancer
    binomial: n = 420,000 and p = .000340
    normal approximation appropriate since
       np = 420,000(.000340) = 142.8 ≥ 5
       n(1-p) = 420,000(.999660) = 419,857.2 ≥ 5
    μ = np = 420,000(.000340) = 142.8
    σ = √np(1-p) = √420,000(.000340)(.999660)
              = 11.949
    P(x ≤ 135)
       = P_c(x < 135.5)
       = P(z < -.61)
       = .2709

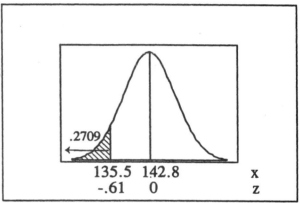

The results suggest that finding "135 or fewer cancer cases" is not an unusual event, and so the cancer rate among cell phone uses appears not to differ significantly from that of the general population.
NOTE: To test the media claim that cell phone usage is associated with a higher rate, one should calculate the probability of finding "135 or more cancer cases."

21. let x = the number of booked arrivals
    binomial: n = 400 and p = .85
    normal approximation appropriate since
       np = 400(.85) = 340 ≥ 5
       n(1-p) = 400(.15) = 60 ≥ 5
    μ = np = 400(.85) = 340
    σ = √np(1-p) = √400(.85)(.15) = 7.141
    P(x > 350)
       = P_c(x > 350.5)
       = P(z > 1.47)
       = 1 - P(z < 1.47)
       = 1 - .9292
       = .0708

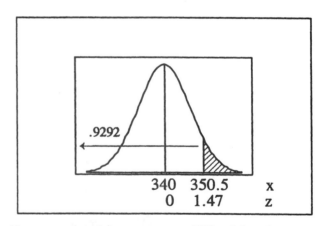

Yes; the airline would probably be willing to allow overbooking to occur 7% of the time (i.e., about 1 in every 14 flights) in order to fly with fewer empty seats.

22. let x = the number that arrive on time
binomial: n = 40 and p = .723
normal approximation appropriate since
  np = 40(.723) = 28.92 ≥ 5
  n(1-p) = 40(.277) = 11.08 ≥ 5
μ = np = 40(.723) = 28.92
σ = $\sqrt{np(1-p)}$ = $\sqrt{40(.723)(.277)}$ = 2.830
P(x ≤ 19)
  = P_c(x < 19.5)
  = P(z < -3.33)
  = .0004
Yes; since .0004 < .05, getting 19 or fewer on-time flights would be an unusual event.

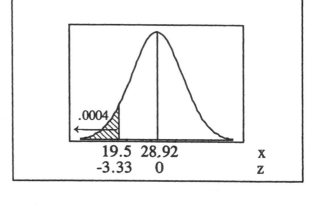

23. let x = the number of women hired
binomial: n = 62 and p = .50
normal approximation appropriate since
  np = 62(.50) = 31 ≥ 5
  n(1-p) = 62(.50) = 31 ≥ 5
μ = np = 62(.50) = 31
σ = $\sqrt{np(1-p)}$ = $\sqrt{62(.50)(.50)}$ = 3.937
P(x ≤ 21)
  = P_c(x < 21.5)
  = P(z < -2.41)
  = .0080
Yes; since the resulting probability is so small, it does support a charge of such discrimination.

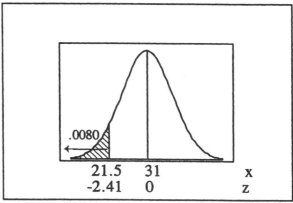

24. let x = the number of blue M&M's
binomial: n = 100 and p = .10
normal approximation appropriate since
  np = 100(.10) = 10 ≥ 5
  n(1-p) = 100(.90) = 90 ≥ 5
μ = np = 100(.10) = 10
σ = $\sqrt{np(1-p)}$ = $\sqrt{100(.10)(.90)}$ = 3.000
P(x ≤ 5)
  = P_c(x < 5.5)
  = P(z < -1.50)
  = .0668
No; since .0668 > .05, it is not very unusual to get 5 or fewer blue M&M's among 100.

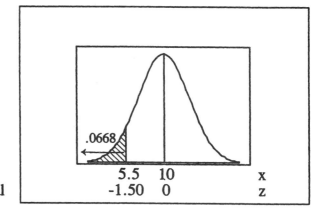

25. let x = the number with group O blood
binomial: n = 400 and p = .45
normal approximation appropriate since
  np = 400(.45) = 180 ≥ 5
  n(1-p) = 400(.55) = 220 ≥ 5
μ = np = 400(.45) = 180
σ = $\sqrt{np(1-p)}$ = $\sqrt{400(.45)(.55)}$ = 9.950
P(x ≥ 177)
  = P_c(x > 176.5)
  = P(z > -.35)
  = 1 - P(z < -.35)
  = 1 - .3632 = .6368
Yes; the pool is "likely" to be sufficient, but it wouldn't be unusual for it not to be sufficient either.

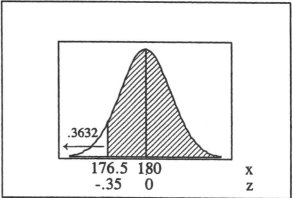

26. let x = the number of defects found
binomial: n = 50 and p = .10
normal approximation appropriate since
np = 50(.10) = 5 ≥ 5
n(1-p) = 50(.90) = 45 ≥ 5
μ = np = 50(.10) = 5
σ = $\sqrt{np(1-p)}$ = $\sqrt{50(.10)(.90)}$ = 2.12
P(x≥2)
 = P_c(x>1.5)
 = P(z>-1.65)
 = 1 - P(z<-1.65)
 = 1 - .0495
 = .9505
Yes; it will do so about 95% of the time.

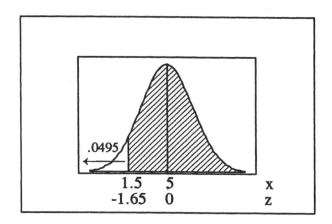

27. let x = the number of accidents in the group
binomial: n = 500 and p = .34
normal approximation appropriate since
np = 500(.34) = 170 ≥ 5
n(1-p) = 500(.66) = 330 ≥ 5
μ = np = 500(.34) = 170
σ = $\sqrt{np(1-p)}$ = $\sqrt{500(.34)(.66)}$ = 10.592
A 40% accident rate implies
x = 500(.40) = 200.
P(x≥200)
 = P_c(x>199.5)
 = P(z>2.79)
 = 1 - P(z<2.79)
 = 1 - .9974
 = .0026
Yes; either a very unusual sample occurred, or the true NYC rate us higher than 34%.
NOTE: The number of in the survey that had accidents last year was not given. Any 198≤x≤202 rounds to 40%. The conclusion will be the same for any value within those limits.

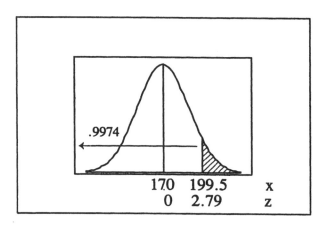

28. let x = the number opposed to human cloning
binomial: n = 1012 and p = .50
normal approximation appropriate since
np = 1012(.50) = 506 ≥ 5
n(1-p) = 1012(.50) = 506 ≥ 5
μ = np = 1012(.50) = 506
σ = $\sqrt{np(1-p)}$ = $\sqrt{1012(.50)(.50)}$ = 15.906
An 89% opposition rate implies
x = 1012(.89) = 901
P(x≥901)
 = P_c(x>900.5)
 = P(z>24.80)
 = 1 - P(z<24.80)
 = 1 - .9999
 = .0001
Yes; this is strong evidence that the majority of the population opposes human cloning.
NOTE: The number of in the survey that opposed human cloning was not given. Any 896≤x≤905 rounds to 89%. The conclusion will be the same for any value within those limits.

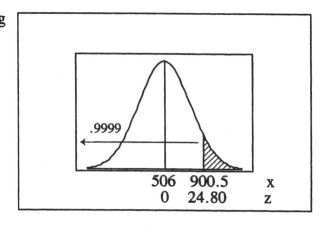

29. let x = the number of times Marc wins $35
    binomial: n = 200 and p = 1/38
    normal approximation appropriate since
        np = 200(1/38) = 5.26 ≥ 5
        n(1-p) = 200(37/38) = 194.76 ≥ 5
    μ = np = 200(1/38) = 5.26
    σ = $\sqrt{np(1-p)}$ = $\sqrt{200(1/38)(37/38)}$ = 2.264
    Marc needs at least 6 $35 wins for a profit.
    P(x≥6)
        = P_c(x>5.5)
        = P(z>.10)
        = 1 - P(z<.10)
        = 1 - .5398
        = .4602

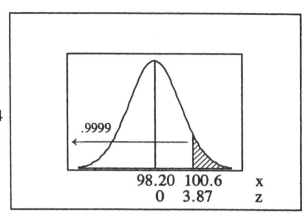

30. This is a two-step problem. Use the normal distribution to find the probability a set lasts more than 10 years, and then use that probability as p in a binomial problem with n=250.
    let t = time (in years) of TV life before replacement
    normal: μ = 8.2 and σ = 1.1
    P(t>10.0)
        = P(z>1.64)
        = 1 - P(z<1.64)
        = 1 - .9495
        = .0505

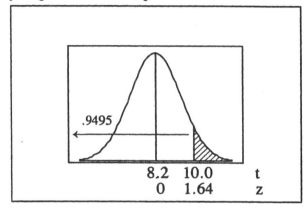

    let x = the number of TV's with t>10
    binomial: n = 250 and p = .0505
    normal approximation appropriate since
        np = 250(.0505) = 12.625 ≥ 5
        n(1-p) = 250(.9495) = 237.375 ≥ 5
    μ = np = 250(.0505) = 12.625
    σ = $\sqrt{np(1-p)}$ = $\sqrt{250(.0505)(.9495)}$ = 3.462
    P(x≥15)
        = P_c(x>14.5)
        = P(z>.54)
        = 1 - P(z<.54)
        = 1 - .7054
        = .2946

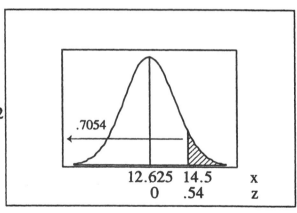

31. a. binomial: n = 4 and p = .350
    P(x≥1) = 1 - P(x=0)
        = 1 - [4!/0!4!]·$(.350)^0(.650)^4$
        = 1 - .1785
        = .8215

b. binomial: n = 56·4 = 224 and p = .350
   normal approximation appropriate since
   np = 224(.350) = 78.4 ≥ 5
   n(1-p) = 224(.650) = 145.6 ≥ 5
   μ = np = 224(.350) = 78.4
   σ = $\sqrt{np(1-p)}$ = $\sqrt{224(.350)(.650)}$ = 7.139
   P(x≥56)
   = P_c(x>55.5)
   = P(z>-3.20)
   = 1 - P(z<-3.20)
   = 1 - .0007
   = .9993

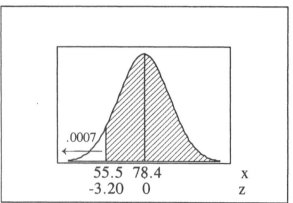

c. let H = getting at least one hit in 4 times at bat
   P(H) = .8215 [from part (a) above]
   for 56 consecutive games, $[P(H)]^{56}$ = $[.8125]^{56}$ = .0000165

d. The solution below employs the methods and notation of parts (a) and (c) above.
   for $[P(H)]^{56}$ > .10, it is required that
   P(H) > $(.10)^{1/56}$
   P(H) > .9597
   for P(H) = P(x≥1) > .9597, it is required that
   1 - P(x=0) > .9597
   .0403 > P(x=0)
   .0403 > [4!/0!4!]·$p^0(1-p)^4$
   .0403 > $(1-p)^4$
   $(.0403)^{1/4}$ > 1 - p
   p > 1 - $(.0403)^{1/4}$
   p > 1 - .448
   p > .552

32. Letting x represent the number of persons with advanced reservation who <u>do</u> show up for a flight, the problem is a binomial one with n unknown and p=.93. The airline wants to find the largest n for which P(x≤250) = .95. The problem may be solved two ways
   (a) increasing n and calculating P(x≤250) until P(x≤250) drops below .95
   (b) solving directly for the value n at which P(x≤250) = .95.

a. The following table summarized the procedure for finding P(x≤250) = P_c(x<250.5), where p=.93.

n	μ=np	σ=$\sqrt{np(1-p)}$	z=(250.5-μ)/σ	P(x<250.5)
260	241.8	4.11	2.11	.9826
261	242.7	4.12	1.89	.9706
262	243.7	4.13	1.66	.9515
263	244.6	4.14	1.43	.9236
264	245.5	4.15	1.20	.8849
265	246.5	4.15	.98	.8365
266	247.4	4.16	.75	.7734
267	248.3	4.17	.53	.7019
268	249.2	4.18	.30	.6179
269	250.2	4.18	.08	.5319
270	251.1	4.19	-.14	.4443
271	252.0	4.20	-.36	.3594
272	253.0	4.21	-.58	.2810
273	253.9	4.22	-.80	.2119
274	254.8	4.22	-1.02	.1539
275	255.8	4.23	-1.24	.1975
276	256.7	4.24	-1.46	.0721
277	257.6	4.25	-1.67	.0475

And so for n=262 the is a 95.15% probability that x≤250 and that everyone who shows up with advanced reservations will have a seat. For n>262, the probability falls below that acceptable level.

b. From Table A-2, the z score below which 95% of the probability occurs is 1.645. Solving to find the n for which 250.5 persons with advanced reservations show up 95% of the time produces the following sequence of steps.
$$(x-\mu)/\sigma = z$$
$$(250.5 - .93n)/\sqrt{n(.93)(.07)} = 1.645$$
$$(250.5 - .93n) = 1.645 \cdot \sqrt{n(.93)(.07)}$$
$$62750 - 465.93n + .8649n^2 = .1762n$$
$$.8649n^2 - 466.1062n + 62750.25 = 0$$
Solving for n using the quadratic formula,
$$n = [466.1062 \pm \sqrt{(-466.1062)^2 - 4(.8649)(62750.25)}]/2(.8649)$$
$$= [466.1062 \pm 12.8136]/1.7298$$
$$= 453.2925/1.7298 \text{ or } 478.9198/1.7298$$
$$= 262.05 \text{ or } 276.86$$
And so z=1.645 when n=262.05. If n>262.05, then z<1.645 and the probability is less than 95%– and so we round down to n=262. The extraneous root n=276.86 [which was introduced when both sides of the equation were squared] corresponds to z=-1.645. This corresponds to the n for P(x≤250) = .05 and agrees with the table in part (a).

5-7 Determining Normality

1. Not normal. The points exhibit a systematic "snaking" pattern around a straight lin passing through the data.

2. Not normal. It would require a "broken" line with four segments of four different slopes to pass through the data.

3. Not normal. It would require a "broken" line with four segments of four different slopes to pass through the data.

4. Normal. The data form a pattern that can be approximated by a single straight line.

NOTE: The tables on the next page give the information for working exercises #5-8 and #9-12 by hand. While these exercises can be answered using computer software, seeing the actual procedure and calculations involved promotes a better understanding of the concepts and processes of this section. For each problem,
  x = the n original scores arranged in order
  cp = the cumulative probability values 1/2m, 3/2m, 5/2n,..., (2n-1)/2n
  z = the z scores have the cp value to their left.

5. No. The frequency distribution and histogram for the Wednesday rain data are given below. The distribution is not normal because it is not symmetric – being positively skewed and having no values below the modal class.

rainfall	frequency
-.05 - .05	41
.05 - .15	7
.15 - .25	1
.25 - .35	2
.35 - .45	0
.45 - .55	0
.55 - .65	2
	53

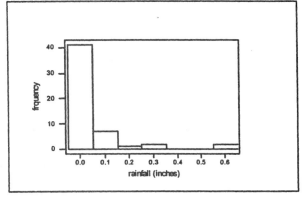

#	5. rainfall x	cp	z	6. heads x	cp	z	7. M&M's x	cp	z	8. conductivity x	cp	z
1	.00	.009	-2.35	35.5	.01	-2.33	.856	.015	-2.17	28.1	.008	-2.40
2	.00	.028	-1.91	35.7	.03	-1.88	.858	.045	-1.69	29.3	.025	-1.97
3	.00	.047	-1.67	39.2	.05	-1.64	.860	.076	-1.43	30.2	.041	-1.74
4	.00	.066	-1.51	39.6	.07	-1.48	.866	.106	-1.25	30.3	.057	-1.58
5	.00	.085	-1.37	39.7	.09	-1.34	.867	.136	-1.10	30.5	.074	-1.45
6	.00	.104	-1.26	39.8	.11	-1.23	.871	.167	-0.97	32.1	.090	-1.34
7	.00	.123	-1.16	39.9	.13	-1.13	.875	.197	-0.85	32.7	.107	-1.25
8	.00	.142	-1.07	40.1	.15	-1.04	.876	.227	-0.75	32.8	.123	-1.16
9	.00	.160	-0.99	40.2	.17	-0.95	.889	.258	-0.65	33.5	.139	-1.08
10	.00	.179	-0.92	40.2	.19	-0.88	.897	.288	-0.56	33.8	.156	-1.01
11	.00	.198	-0.85	40.2	.21	-0.81	.898	.318	-0.47	35.9	.172	-0.95
12	.00	.217	-0.78	40.4	.23	-0.74	.900	.348	-0.39	40.5	.189	-0.88
13	.00	.236	-0.72	40.4	.25	-0.67	.902	.379	-0.31	40.6	.205	-0.82
14	.00	.255	-0.66	40.7	.27	-0.61	.902	.409	-0.23	41.5	.221	-0.77
15	.00	.274	-0.60	40.9	.29	-0.55	.904	.439	-0.15	42.4	.238	-0.71
16	.00	.292	-0.55	40.9	.31	-0.50	.909	.470	-0.08	43.2	.254	-0.66
17	.00	.311	-0.49	40.9	.33	-0.44	.909	.500	0.00	44.3	.270	-0.61
18	.00	.330	-0.44	40.9	.35	-0.39	.914	.530	0.08	45.6	.287	-0.56
19	.00	.349	-0.39	40.9	.37	-0.33	.914	.561	0.15	46.5	.303	-0.51
20	.00	.368	-0.34	41.0	.39	-0.28	.919	.591	0.23	46.7	.320	-0.47
21	.00	.387	-0.29	41.0	.41	-0.23	.920	.621	0.31	46.7	.336	-0.42
22	.00	.406	-0.24	41.0	.43	-0.18	.921	.652	0.39	47.1	.352	-0.38
23	.00	.425	-0.19	41.0	.45	-0.13	.923	.682	0.47	48.1	.369	-0.33
24	.00	.443	-0.14	41.0	.47	-0.08	.928	.712	0.56	48.3	.385	-0.29
25	.00	.462	-0.09	41.1	.49	-0.03	.930	.742	0.65	48.5	.402	-0.25
26	.00	.481	-0.05	41.1	.51	0.03	.930	.773	0.75	48.5	.418	-0.21
27	.00	.500	0.00	41.2	.53	0.08	.932	.803	0.85	48.5	.434	-0.17
28	.00	.519	0.05	41.3	.55	0.13	.936	.833	0.97	48.6	.451	-0.12
29	.00	.538	0.09	41.4	.57	0.18	.955	.864	1.10	49.0	.467	-0.08
30	.00	.557	0.14	41.7	.59	0.23	.965	.894	1.25	49.0	.484	-0.04
31	.00	.575	0.19	41.7	.61	0.28	.976	.924	1.43	49.7	.500	0.00
32	.00	.594	0.24	41.7	.63	0.33	.988	.955	1.69	49.8	.516	0.04
33	.00	.613	0.29	41.7	.65	0.39	1.033	.985	2.17	49.8	.533	0.08
34	.00	.632	0.34	41.8	.67	0.44				49.9	.549	0.12
35	.00	.651	0.39	41.9	.69	0.50				49.9	.566	0.17
36	.00	.670	0.44	41.9	.71	0.55				49.9	.582	0.21
37	.00	.689	0.49	42.0	.73	0.61				50.3	.598	0.25
38	.01	.708	0.55	42.0	.75	0.67				50.4	.615	0.29
39	.01	.726	0.60	42.2	.77	0.74				50.5	.631	0.33
40	.01	.745	0.66	42.2	.79	0.81				51.0	.648	0.38
41	.02	.764	0.72	42.3	.81	0.88				51.0	.664	0.42
42	.06	.783	0.78	42.3	.83	0.95				51.2	.680	0.47
43	.06	.802	0.85	42.4	.85	1.04				51.3	.697	0.51
44	.06	.821	0.92	42.5	.87	1.13				51.7	.713	0.56
45	.08	.840	0.99	42.6	.89	1.23				51.9	.730	0.61
46	.08	.858	1.07	42.8	.91	1.34				52.0	.746	0.66
47	.12	.877	1.16	42.8	.93	1.48				52.1	.762	0.71
48	.14	.896	1.26	42.8	.95	1.64				52.2	.779	0.77
49	.18	.915	1.37	43.2	.97	1.88				52.4	.795	0.82
50	.27	.934	1.51	43.2	.99	2.33				53.6	.811	0.88
51	.31	.953	1.67							55.2	.828	0.95
52	.64	.972	1.91							56.8	.844	1.01
53	.64	.991	2.35							56.8	.861	1.08
54										57.0	.877	1.16
55										57.1	.893	1.25
56										57.3	.910	1.34
57										57.7	.926	1.45
58										57.8	.943	1.58
59										57.8	.959	1.74
60										58.4	.975	1.97
61										59.2	.992	2.40

6. Yes. The frequency distribution and histogram for the male head circumferences are given below. The circumferences appear approximately normally distributed, with frequencies tapering off in both directions about the modal class.

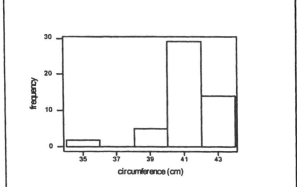

```
circumference   frequency
34.0 - 35.9         2
36.0 - 37.9         0
38.0 - 39.9         5
40.0 - 41.9        29
42.0 - 43.9        14
                   50
```

7. Yes. The frequency distribution and histogram for the M&M weights are given below. The weights appear approximately normally distributed, with frequencies tapering off in both directions about the modal class.

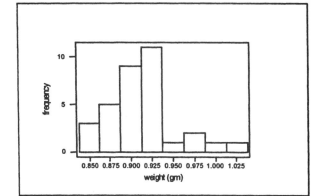

```
weight (mg)    frequency
 838 -  862        3
 863 -  887        5
 888 -  912        9
 913 -  937       11
 938 -  962        1
 963 -  987        2
 988 - 1012        1
1013 - 1037        1
                  33
```

8. No. The frequency distribution and histogram for the conductivity values are given below. The distribution is not normal because it is not bell-shaped – being more like a uniform distribution with a single spike.

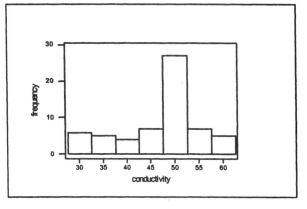

```
conductivity   frequency
27.5 - 32.5        6
32.5 - 37.5        5
37.5 - 42.5        4
42.5 - 47.5        7
47.5 - 52.5       27
52.5 - 57.5        7
57.5 - 62.5        5
                  61
```

NOTE: The normal quantile plots for exercises #9-12 may be constructed using the appropriate columns from the table on the previous page. Exercise #10 most clearly illustrates the process: *each of the 50 scores is 2% of the data set and the first score covers .00-.02, the midpoint of which is .01.* Plot the x values on the horizontal axis and the corresponding z values (calculated as indicated from the cp values) on the vertical axis. Making a "vertically consistent-scaled" plot with the z scores determined from the normal probability distribution is equivalent to making a "vertically stretch-scaled" plot using the cumulative probabilities. As with most exercises in this section, the judgment as to whether the points can be approximated by a straight line is subjective.

9. No. Since the points do not lie close to a straight line, conclude that the population distribution is not approximately normal.

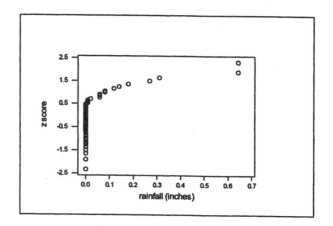

10. Yes. Since the points lie close to a straight line, conclude that the population distribution is approximately normal. This is especially obvious if the two lowest scores are ignored. In practice, one should check to see if there are legitimate reasons for eliminating such scores – for example, they may be measurements from babies that are exceptions to any normal patterns because they were born very prematurely.

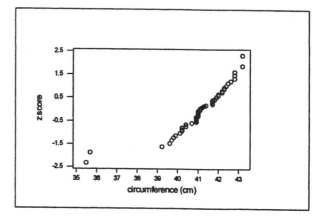

11. Yes. Since the points lie close to a straight line, conclude that the population distribution is approximately normal.

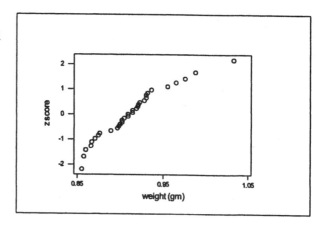

12. No. Since the points do not lie close to a straight line, conclude that the population distribution is not approximately normal. While it may seem possible to draw a reasonable straight line through the points, consider the pattern of the points about such a line. The fact that the points would "snake" around such a line in an obvious pattern (as opposed to being randomly distributed around the line) indicates some underlying non-normal structure.

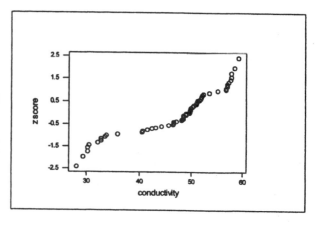

13. The two histograms are given below. The heights (on the left) appear to be approximately normally distributed, while the cholesterol levels (at the right) appear to be positively skewed. Many natural phenomena are normally distributed. Height is a natural phenomenon unaffected by human input; cholesterol levels are humanly influenced (by diet, exercise, medication, etc.) in ways that might alter any naturally occurring distribution.

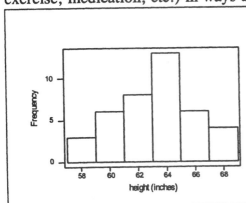 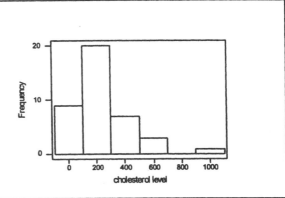

14. The two histograms are given below. The systolic blood pressure levels (on the left) appear to be positively skewed, while the elbow breadths (at the right) appear to be more normally distributed. Many natural phenomena are normally distributed. Elbow breadth is a natural phenomenon unaffected by human input (especially for women, who may be assumed not to engage in body-building programs); blood pressure is humanly influenced (by diet, exercise, medication, etc.) in ways that might alter any naturally occurring distribution.

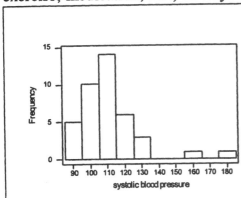 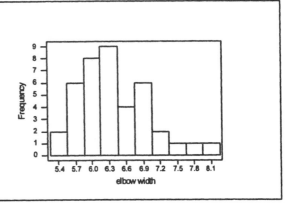

15. The corresponding z scores in the table below at the left were determined as follows.
    (1) Arrange the n scores in order and place them in the x column.
    (2) For each x_i, calculate the cumulative probability using $cp_i = (2i-1)/2n$ for $i = 1,2,\ldots,n$.
    (3) For each cp_i, find the z_i for which $P(z<z_i) = cp_i$ for $i = 1,2,\ldots,n$.

    The resulting normal quantile plot at the right indicates the data appear to come from a population with a normal distribution.

i	x	cp	z
1	73	.10	-1.28
2	78	.30	-.52
3	79	.50	.00
4	82	.70	.52
5	85	.90	1.28

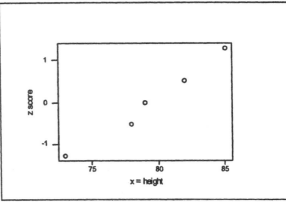

16. The corresponding z scores in the table below at the left were determined as follows.
(1) Arrange the n scores in order and place them in the x column.
(2) For each x_i, calculate the cumulative probability using $cp_i = (2i-1)/2n$ for $i = 1,2,\ldots,n$.
(3) For each cp_i, find the z_i for which $P(z<z_i) = cp_i$ for $i = 1,2,\ldots,n$.
The resulting normal quantile plot at the right indicates the data do not appear to come from a population with a normal distribution.

i	x	cp	z
1	0.42	.083	-1.38
2	0.48	.250	-.67
3	0.73	.417	-.21
4	1.10	.583	.21
5	1.10	.750	.67
6	5.40	.917	1.38

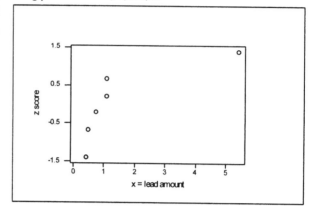

17. No. The z scores from the cumulative probability must be used, and not the z scores for the raw data. The z formula is a linear transformation.
$$z = (x - \mu)/\sigma = (1/\sigma) \cdot x - (\mu/\sigma) = a \cdot x + b$$
The (x,z) pairs will all lie on the straight line $z = (1/\sigma) \cdot x - (\mu/\sigma)$ regardless of the shape of the data.

18. The corresponding z scores in the table below at the left were determined as follows.
(1) Arrange the n scores in order and place them in the x column.
(3) For each x_i, calculate $\ln(x_i)$
(3) For each x_i, calculate the cumulative probability using $cp_i = (2i-1)/2n$ for $i = 1,2,\ldots,n$.
(4) For each cp_i, find the z_i for which $P(z<z_i) = cp_i$ for $i = 1,2,\ldots,n$.
Since the cp and z columns are based only on i, they apply to both the x and ln(x) values.

The resulting normal quantile plots at the right indicate the x data do not appear to come from a population with a normal distribution, but the ln(x) data do appear to come from a population with a normal distribution.

i	x	ln(x)	cp	z
1	21.3	3.06	.025	-1.96
2	31.5	3.45	.075	-1.44
3	31.8	3.46	.125	-1.15
4	47.9	3.87	.175	-.93
5	51.9	3.95	.225	-.76
6	54.1	3.99	.275	-.60
7	55.1	4.01	.325	-.45
8	57.4	4.05	.375	-.32
8	72.2	4.28	.425	-.19
10	73.7	4.30	.475	-.06
11	75.9	4.33	.525	.06
12	75.9	4.33	.575	.19
13	87.4	4.47	.625	.32
14	127.7	4.85	.675	.45
15	130.3	4.87	.725	.60
16	138.1	4.93	.775	.76
17	160.8	5.08	.825	.93
18	210.6	5.35	.875	1.15
19	403.4	6.00	.925	1.44
20	454.9	6.12	.975	1.96

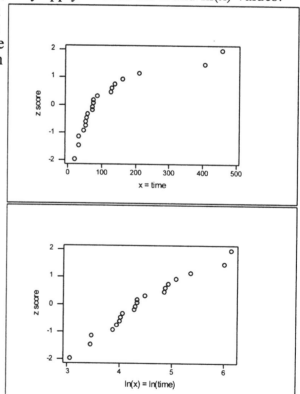

Review Exercises

1. a. normal distribution
   $\mu = 178.1$
   $\sigma = 40.7$
   $P(x > 260)$
   $= P(z > 2.01)$
   $= 1 - P(z < 2.01)$
   $= 1 - .9778$
   $= .0222$

   b. normal distribution
   $\mu = 178.1$
   $\sigma = 40.7$
   $P(170 < x < 200)$
   $= P(-.20 < z < .54)$
   $= P(z < .54) - P(z < -.20)$
   $= .7054 - .4207$
   $= .2847$

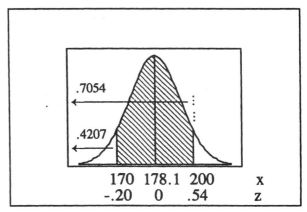

   c. normal distribution,
   since the original distribution is so
   $\mu_{\bar{x}} = \mu = 178.1$
   $\sigma_{\bar{x}} = \sigma/\sqrt{n} = 40.7/\sqrt{9} = 13.567$
   $P(170 < \bar{x} < 200)$
   $= P(-.60 < z < 1.61)$
   $= P(z < 1.61) - P(z < -.60)$
   $= .9463 - .2743$
   $= .6720$

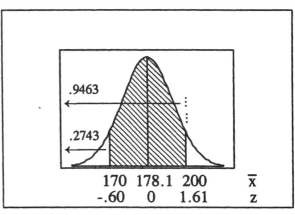

   d. For the top 3%,
   $A = .9700 [.9699]$ and $z = 1.88$
   $x = \mu + z\sigma$
   $= 178.1 + (1.88)(40.7)$
   $= 178.1 + 76.5$
   $= 254.6$

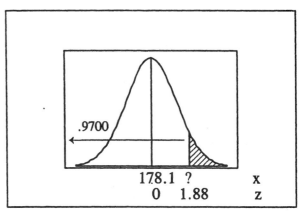

2. a. normal distribution
   $\mu = 3420$
   $\sigma = 495$
   $P(x < 2200)$
   $= P(z < -2.46)$
   $= .0069$ or $.69\%$

   For 900 births, we expect
   $(.0069)(900) \approx 6$ to be at risk.

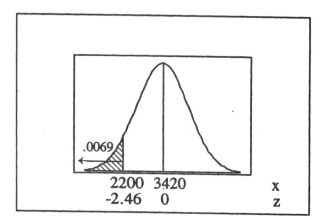

b. For the lowest 2%,
   $A = .0200$ [.0202] and $z = -2.05$
   $x = \mu + z\sigma$
   $= 3420 + (-2.05)(495)$
   $= 3420 - 1015$
   $= 2405$

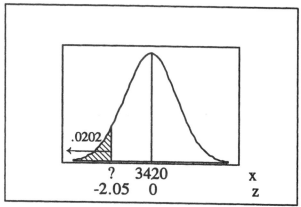

c. normal distribution,
   since the original distribution is so
   $\mu_{\bar{x}} = \mu = 3420$
   $\sigma_{\bar{x}} = \sigma/\sqrt{n} = 495/\sqrt{16} = 123.75$
   $P(\bar{x} > 3700)$
   $= P(z > 2.26)$
   $= 1 - P(z < 2.26)$
   $= 1 - .9881$
   $= .0019$

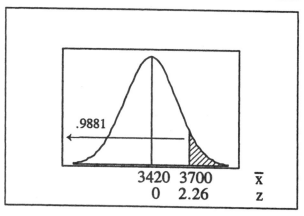

d. normal distribution,
   since the original distribution is so
   $\mu_{\bar{x}} = \mu = 3420$
   $\sigma_{\bar{x}} = \sigma/\sqrt{n} = 495/\sqrt{49} = 70.714$
   $P(3300 < \bar{x} < 3700)$
   $= P(-1.70 < z < 3.96)$
   $= P(z < 3.96) - P(z < -1.70)$
   $= .9999 - .0446$
   $= .9553$

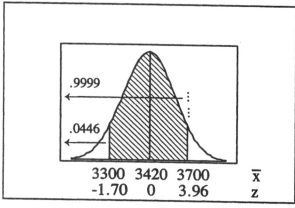

3. let x = the number of offspring with blue eyes
   binomial: n = 100 and p = .25
   normal approximation appropriate since
   np = 100(.25) = 25 ≥ 5
   n(1-p) = 100(.75) = 75 ≥ 5
   μ = np = 100(.25) = 25
   σ = $\sqrt{np(1-p)}$ = $\sqrt{100(.25)(.75)}$ = 4.330
   P(x ≤ 19)
   = P_c(x < 19.5)
   = P(z < -1.27)
   = .1020
   No; since .1020 > .05, it is not very unusual to get 19 or fewer such offspring among 100.

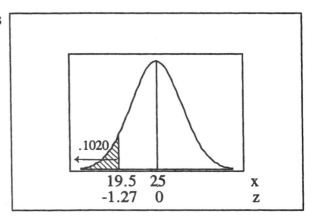

4. a. normal distribution
   μ = 69.0
   σ = 2.8
   P(64 < x < 78)
   = P(-1.79 < z < 3.21)
   = P(z < 3.21) - P(z < -1.79)
   = .9993 - .0367
   = .9626

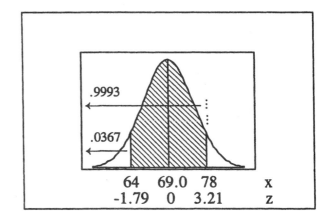

   b. For the shortest 2%,
   A = .0200 [.0202] and z = -2.05.
   x = μ + zσ
   = 69.0 + (-2.05)(2.8)
   = 69.0 - 5.7
   = 63.3
   For the tallest 2%,
   A = .9800 [.9798] and z = 2.05.
   x = μ + zσ
   = 69.0 + (2.05)(2.8)
   = 6.90 + 5.7
   = 74.7
   The new minimum and maximum requirements would be 63.3 inches and 74.7 inches respectively.

   c. normal distribution,
   since the original distribution is so
   $\mu_{\bar{x}}$ = μ = 69.0
   $\sigma_{\bar{x}}$ = $\sigma/\sqrt{n}$ = 2.8/$\sqrt{64}$ = .350
   P($\bar{x}$ > 68.0)
   = P(z > -2.86)
   = 1 - P(z < -2.86)
   = 1 - .0021
   = .9979

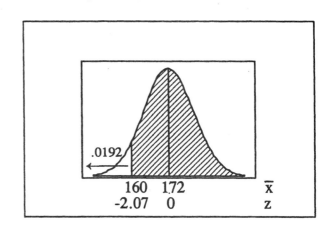

5. Probability corresponds to area. The calculations at the right verify that the height is 2.5 from 11.8 to 12.2. Since there is no probability outside those limits, the height outside those limits is 0.

total area = 1
(width)(height) = 1
(12.2 - 11.8)(height) = 1
.4(height) = 1
height = 1/.4 = 2.5

a. P(x < 12.0) = (width)(height)
= (12.0-11.8)(2.5)
= (.2)(2.5)
= .50

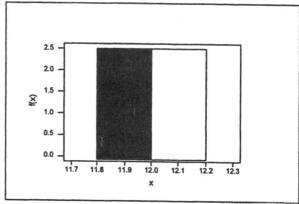

b. P(11.2 < x < 12.7) = P(11.2 < x < 11.8) + P(11.8 < x < 12.2) + P(12.2 < x < 12.7)
= (width)(height) + (width)(height) + (width)(height)
= (11.8 - 11.2)(0) + (12.2 - 11.8)(2.5) + (12.7 - 12.2)(0)
= (.6)(0) + (.4)(2.5) + (.5)(0)
= 0 + 1 + 0
= 1, a certainty

c. P(x > 12.2) = (width)(height)
= (width)·0
= 0, an impossibility

d. P(11.9 < x < 12.0) = (width)(height)
= 12.0 - 11.9)(2.5)
= (.1)(2.5)
= .25

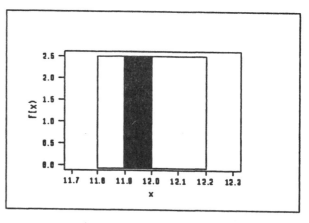

6. a. Since n=100, the Central Limit Theorem guarantees that the sampling distribution of the means of different samples will be normal – regardless of the shape of the original distribution.
b. $\sigma_{\bar{x}} = \sigma/\sqrt{n} = 512/\sqrt{100} = 51.2$ lb
c. Let x = the number who voted in the last election.
   binomial: n = 1200 and p = ?
For each per person define the variable y as follows:
   y = 1 if the person voted
   y = 0 if the person did not vote.
The sample proportion x/n is really a sample mean, since x/n = Σy/n. Since n=1200, the Central Limit Theorem guarantees that the sampling distribution will be normal – regardless of the shape of the original distribution.

7. let x = the number of women selected
   binomial: n = 20 and p = .30
   normal approximation appropriate since
   $\quad$ np = 20(.30) = 6 ≥ 5
   $\quad$ n(1-p) = 20(.70) = 14 ≥ 5
   μ = np = 20(.25) = 6
   σ = $\sqrt{np(1-p)}$ = $\sqrt{20(.30)(.70)}$ = 2.049
   P(x≤2)
   $\quad$ = P_c(x<2.5)
   $\quad$ = P(z<-1.71)
   $\quad$ = .0436

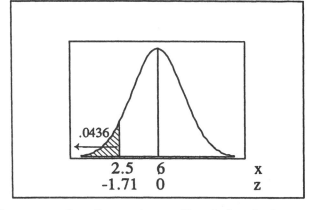

   Yes; since .0436 < .05, it would be unusual
   to get 2 or fewer women by chance alone.
   Either an unusual event has occurred or there is some other factor (e.g., discrimination)
   affecting the hiring process.
   NOTE: Since the above probability calculated using the normal approximation is close to
   .05, it would be good to calculate the more accurate exact binomial probability.
   P(x≤2) = P(x=0) + P(x=1) + P(x=2)
   $\quad\quad\quad$ = $[20!/(0!20!)](.30)^0(.70)^{20}$ + $[20!/(1!19!)](.30)^1(.70)^{19}$ + $[20!/(2!18!)](.30)^2(.70)^{18}$
   $\quad\quad\quad$ = [1](1)(.00080) + [20](.30)(.00114) + [190](.09)(.00162)
   $\quad\quad\quad$ = .00080 + .00684 + .02785
   $\quad\quad\quad$ = .0355
   This probability is even smaller, confirming the previous conclusion.

8. Arrange the n=70 scores in order, calculate the cumulative probability for each score, and
   determine the z score for each cumulative probability. The frequency histogram (bell-
   shaped) and normal quantile plot (points approximately in a straight line) suggest that the
   data come from a population with a normal distribution.

i	x	cp	z	i	x	cp	z
01	3.407	.007	-2.45	36	3.588	.507	.02
02	3.450	.021	-2.03	37	3.590	.521	.05
03	3.464	.036	-1.80	38	3.598	.536	.09
04	3.468	.050	-1.64	39	3.600	.550	.13
05	3.475	.064	-1.52	40	3.601	.564	.16
06	3.482	.079	-1.41	41	3.604	.579	.20
07	3.491	.093	-1.32	42	3.604	.593	.23
08	3.492	.107	-1.24	43	3.611	.607	.27
09	3.494	.121	-1.17	44	3.617	.621	.31
10	3.494	.136	-1.10	45	3.621	.636	.35
11	3.506	.150	-1.04	46	3.622	.650	.39
12	3.507	.164	-.98	47	3.625	.664	.42
13	3.508	.179	-.92	48	3.632	.679	.46
14	3.511	.193	-.87	49	3.635	.693	.50
15	3.516	.207	-.82	50	3.635	.707	.55
16	3.521	.221	-.77	51	3.638	.721	.59
17	3.522	.236	-.72	52	3.639	.736	.63
18	3.522	.250	-.67	53	3.643	.750	.67
19	3.526	.264	-.63	54	3.643	.764	.72
20	3.531	.279	-.59	55	3.645	.779	.77
21	3.532	.293	-.55	56	3.647	.793	.82
22	3.535	.307	-.50	57	3.654	.807	.87
23	3.542	.321	-.46	58	3.660	.821	.92
24	3.545	.336	-.42	59	3.665	.836	.98
25	3.548	.350	-.39	60	3.666	.850	1.04
26	3.569	.364	-.35	61	3.667	.864	1.10
27	3.573	.379	-.31	62	3.671	.879	1.17
28	3.576	.393	-.27	63	3.673	.893	1.24
29	3.577	.407	-.23	64	3.678	.907	1.32
30	3.580	.421	-.20	65	3.687	.921	1.41
31	3.582	.436	-.16	66	3.688	.936	1.52
32	3.582	.450	-.13	67	3.718	.950	1.64
33	3.583	.464	-.09	68	3.723	.964	1.80
34	3.585	.479	-.05	69	3.725	.979	2.03
35	3.588	.493	-.02	70	3.726	.993	2.45

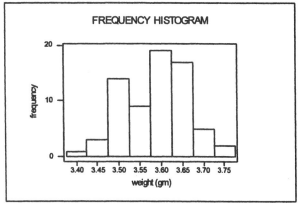

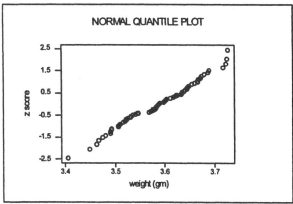

Cumulative Review Exercises

1. The n=8 ordered scores are: 55 59 62 63 66 66 66 67
   summary values: $\Sigma x = 504, \Sigma x^2 = 31{,}876$
   a. $\bar{x} = (\Sigma x)/n = (504)/8 = 63.0$ mm
   b. $\tilde{x} = (63 + 66)/2 = 64.5$ mm
   c. M = 66 mm
   d. $s^2 = [n(\Sigma x^2) - (\Sigma x)^2]/[n(n-1)]$
   $= [8(31{,}876) - (504)^2]/[8(7)]$
   $= 992/56$
   $= 17.714$
   $s = 4.209$ rounded to 4.2 mm
   e. $z = (x-\bar{x})/s$
   $z_{59} = (59 - 63.0)/4.209 = -.95$
   f. r.f. of scores greater than 59 is 6/8 = .75 = 75%
   g. normal: $\mu = 63.0$ and $\sigma = 4.299$
   $P(x > 59) = P(z > -.95)$
   $= 1 - P(z < .95)$
   $= 1 - .1711$
   $= .8289$ or 82.89%
   h. Ratio, since distances between scores are meaningful and there is a natural zero.
   i. Continuous, since length can be any value on a continuum.

2. a. Let L = a person is left-handed.
   P(L) = .10, for each random selection
   $P(L_1 \text{ and } L_2 \text{ and } L_3) = P(L_1) \cdot P(L_2) \cdot P(L_3) = (.10) \cdot (.10) \cdot (.10) = .001$
   b. Let N = a person is not left-handed
   P(N) = .90, for each random selection
   P(at least one left-hander) = 1 - P(no left-handers)
   $= 1 - P(N_1 \text{ and } N_2 \text{ and } N_3)$
   $= 1 - P(N_1) \cdot P(N_2) \cdot P(N_3)$
   $= 1 - (.90) \cdot (.90) \cdot (.90)$
   $= 1 - .729$
   $= .271$
   c. binomial: n = 3 and p = P(left-hander) = .10
   normal approximation <u>not</u> appropriate since
   np = 3(.10) = 0.3 < 5
   d. binomial: n = 50 and p = .10
   $\mu = np = 50(.10) = 5$
   e. binomial: n = 50 and p = .10
   $\sigma = \sqrt{np(1-p)} = \sqrt{50(.10)(.90)} = 2.121$
   f. An unusual score is one that is more than 2 standard deviations from the mean.
   $z = (x-\mu)/\sigma$
   $z_8 = (8-5)/2.212$
   $= 1.41$.
   Since 8 is 1.41 standard deviations from the mean, it would <u>not</u> be an unusual score.

Chapter 6

Estimates and Sample Sizes

6-1 Estimating a Population Proportion

IMPORTANT NOTE: When calculating confidence intervals using the formula
$$\hat{p} \pm E$$
$$\hat{p} \pm z_{\alpha/2}\sqrt{\hat{p}\hat{q}/n}$$
do not round off in the middle of the problem. This may be accomplished conveniently on most calculators having a memory as follows.
(1) Calculate $\hat{p} = x/n$ and STORE the value
(2) Calculate E as 1 - RECALL = * RECALL = ÷ n = √ * $z_{\alpha/2}$ =
(3) With the value of E showing on the display, the upper confidence limit is calculated by + RECALL.
(4) With the value of the upper confidence limit showing on the display, the lower confidence limit is calculated by - RECALL ± + RECALL

You must become familiar with your own calculator. [Do your homework using the same type of calculator you will be using for the exams.] The above procedure works on most calculators; make certain you understand why it works and verify whether it works on your calculator. If it does not seem to work on your calculator, or if your calculator has more than one memory so that you can STORE both $\hat{p}$ and E at the same time, ask your instructor for assistance.

NOTE: It should be true that $0 \leq \hat{p} \leq 1$ and that $E \leq .5$ [usually, <u>much</u> less than .5]. If such is not the case, an error has been made.

1. For .99 confidence, $\alpha = 1-.99 = .01$ and $\alpha/2 = .01/2 = .005$.
   For the upper .005, A = .9950 and z = 2.575.
   $z_{\alpha/2} = z_{.005} = 2.575$

2. For .90 confidence, $\alpha = 1-.90 = .10$ and $\alpha/2 = .10/2 = .05$.
   For the upper .05, A = .9500 and z = 1.645.
   $z_{\alpha/2} = z_{.05} = 1.645$

3. For .98 confidence, $\alpha = 1-.98 = .02$ and $\alpha/2 = .02/2 = .01$.
   For the upper .01, A = .9900 [.9901] and z = 2.33.
   $z_{\alpha/2} = z_{.01} = 2.33$

4. For .92 confidence, $\alpha = 1-.92 = .08$ and $\alpha/2 = .08/2 = .04$.
   For the upper .04, A = .9600 [.9599] and z = 1.75.
   $z_{\alpha/2} = z_{.04} = 1.75$

5. Let L = the lower confidence limit; U = the upper confidence limit.
   $\hat{p} = (L + U)/2 = (.220 + .280)/2 = .500/2 = .250$
   $E = (U - L)/2 = (.280 - .220)/2 = .060/2 = .030$
   The interval can be expressed as $.250 \pm .030$.

6. Let L = the lower confidence limit; U = the upper confidence limit.
   $\hat{p} = (L + U)/2 = (.456 + .496)/2 = .952/2 = .476$
   $E = (U - L)/2 = (.496 - .456)/2 = .040/2 = .020$
   The interval can be expressed as $.476 \pm .020$.

7. Let L = the lower confidence limit; U = the upper confidence limit.
   $\hat{p} = (L + U)/2 = (.604 + .704)/2 = 1.308/2 = .654$
   $E = (U - L)/2 = (.704 - .604)/2 = .100/2 = .050$
   The interval can be expressed as $.654 \pm .050$.

8. Let L = the lower confidence limit; U = the upper confidence limit.
   $L = \hat{p} - E = .742 - .030 = .712$
   $U = \hat{p} + E = .742 + .030 = .772$
   The interval can be expressed as $.712 < p < .772$.

9. Let L = the lower confidence limit; U = the upper confidence limit.
   $\hat{p} = (L + U)/2 = (.444 + .484)/2 = .928/2 = .464$
   $E = (U - L)/2 = (.484 - .444)/2 = .040/2 = .020$

10. Let L = the lower confidence limit; U = the upper confidence limit.
    $\hat{p} = (L + U)/2 = (.278 + .338)/2 = .616/2 = .308$
    $E = (U - L)/2 = (.338 - .278)/2 = .060/2 = .030$

11. Let L = the lower confidence limit; U = the upper confidence limit.
    $\hat{p} = (L + U)/2 = (.632 + .678)/2 = 1.310/2 = .655$
    $E = (U - L)/2 = (.678 - .632)/2 = .046/2 = .023$

12. Let L = the lower confidence limit; U = the upper confidence limit.
    $\hat{p} = (L + U)/2 = (.887 + .927)/2 = 1.814/2 = .907$
    $E = (U - L)/2 = (.927 - .887)/2 = .040/2 = .020$

13. $\alpha = .05$ and $\hat{p} = x/n = 200/800 = .25$
    $E = z_{.025}\sqrt{\hat{p}\hat{q}/n} = 1.960\sqrt{(.25)(.75)/800} = .0300$

14. $\alpha = .01$ and $\hat{p} = x/n = 400/1200 = .33$
    $E = z_{.005}\sqrt{\hat{p}\hat{q}/n} = 2.575\sqrt{(.33)(.67)/1200} = .0350$

15. $\alpha = .01$ and $\hat{p} = x/n = [450/1000] = .45$
    $E = z_{.005}\sqrt{\hat{p}\hat{q}/n} = 2.575\sqrt{(.45)(.55)/1000} = .0405$

16. $\alpha = .05$ and $\hat{p} = x/n = [400/500] = .80$
    $E = z_{.025}\sqrt{\hat{p}\hat{q}/n} = 1.96\sqrt{(.80)(.20)/500} = .0351$

17. $\alpha = .05$ and $\hat{p} = x/n = 300/400 = .750$
    $\hat{p} \pm z_{.025}\sqrt{\hat{p}\hat{q}/n}$
    $.750 \pm 1.96\sqrt{(.750)(.250)/400}$
    $.750 \pm .042$
    $.708 < p < .792$

18. $\alpha = .01$ and $\hat{p} = x/n = 200/1200 = .1667$
    $\hat{p} \pm z_{.005}\sqrt{\hat{p}\hat{q}/n}$
    $.1667 \pm 2.575\sqrt{(.1667)(.8333)/1200}$
    $.1667 \pm .0277$
    $.139 < p < .194$

19. $\alpha = .02$ and $\hat{p} = x/n = 176/1655 = .1063$
    $\hat{p} \pm z_{.01}\sqrt{\hat{p}\hat{q}/n}$
    $.1063 \pm 2.33\sqrt{(.1063)(.8937)/1655}$
    $.1063 \pm .0177$
    $.089 < p < .124$

INSTRUCTOR'S SOLUTIONS Chapter 6   S-163

20. $\alpha = .10$ and $\hat{p} = x/n = 1776/2001 = .8876$
    $\hat{p} \pm z_{.05}\sqrt{\hat{p}\hat{q}/n}$
    $.8876 \pm 1.645\sqrt{(.8876)(.1124)/2001}$
    $.8876 \pm .0116$
    $.176 < p < .899$

21. $\hat{p}$ unknown, use $\hat{p} = .5$
    $n = [(z_{.005})^2\hat{p}\hat{q}]/E^2 = [(2.575)^2(.5)(.5)]/(.060)^2 = 460.46$, rounded up to 461

22. $\hat{p}$ unknown, use $\hat{p} = .5$
    $n = [(z_{.025})^2\hat{p}\hat{q}]/E^2 = [(1.96)^2(.5)(.5)]/(.038)^2 = 665.10$, rounded up to 666

23. $\hat{p} = .15$
    $n = [(z_{.025})^2\hat{p}\hat{q}]/E^2 = [(1.96)^2(.185)(.815)]/(.05)^2 = 231.69$, rounded up to 232

24. $\hat{p} = .33$
    $n = [(z_{.05})^2\hat{p}\hat{q}]/E^2 = [(1.645)^2(.08)(.92)]/(.03)^2 = 221.29$, rounded up to 222

25. a. We are 95% certain that the interval from 4.891% to 5.308% contains the true percentage of 18-20 year old males who drove within the last month while impaired from alcohol.
    b. Yes, since any rate greater than 0% presents a risk to society and is preventable.
    c. The upper confidence interval limit of 5.308% would be a responsible conservative estimate.

26. a. We are 99% certain that the interval from 93.053% to 94.926% contains the true percentage of U.S. households having telephones.
    b. Based on the preceding result, pollsters should not be too concerned about results from surveys conducted by telephone if (1) proper techniques are used and (2) issues that would elicit differing responses from phone and non-phone households are not at stake. Proper techniques involve follow up of non-responses (due to work schedules, etc.), making certain the appropriate person in the household (and not just whoever answers the phone, etc.) is responding, etc. Issues involving agricultural policies, for example, would miss the opinions of the Amish (who are primarily involved in farming and do not have phones for religious reasons). And issues involving welfare policies might under-survey those using such services (since multiple poverty-level families tend to share housing units and have only one phone among them). In general, however, one can probably assume that those households with phones would be the households most likely to buy, use or be familiar with the subject of the survey. afety, for example, while 98% of the drivers driving while chewing gum may not be a threat to safety.

27. NOTE: This problem is limited to whole percent accuracy. Any x value from 293 (293/1025 = .2859) to 302 (302/1025 = .2946) gives a $\hat{p}$ that rounds to the given 29%. Since the exact value of x was not given, three decimal accuracy is not possible.
    a. $\hat{p} = x/n = [297.25/1025] = .29 = 29\%$
    b. $\hat{p} \pm z_{.005}\sqrt{\hat{p}\hat{q}/n}$
       $.29 \pm 2.575\sqrt{(.29)(.71)/1025}$
       $.29 \pm .04$
       $.25 < p < .33$
       $25\% < p < 33\%$
    c. The upper confidence interval limit of 33% would be a responsible estimate for determining the maximum impact.

28. NOTE: This problem is limited to whole percent accuracy. Any x value from 317 (317/491 = .6456) to 321 (321/491 = .6538) gives a $\hat{p}$ that rounds to the given 65%. Since the exact value of x was not given, three decimal accuracy is not possible.

a. $\hat{p} = x/n = [319.15/491] = .65 = 65\%$
b. $\hat{p} \pm z_{.025}\sqrt{\hat{p}\hat{q}/n}$
   $.65 \pm 1.96\sqrt{(.65)(.35)/491}$
   $.65 \pm .04$
   $.61 < p < .69$
   $61\% < p < 69\%$
c. Yes; since the interval in which we have 95% confidence is entirely above 50%.

29. a. $\alpha = .05$ and $\hat{p} = x/n = 152/(152+428) = 152/580 = .262$
   $\hat{p} \pm z_{.025}\sqrt{\hat{p}\hat{q}/n}$
   $.262 \pm 1.96\sqrt{(.262)(.738)/580}$
   $.262 \pm .036$
   $.226 < p < .298$
   $22.6\% < p < 29.8\%$
   b. No; since the confidence interval includes 25%, there is no contradiction.

30. a. $\alpha = .01$ and $\hat{p} = x/n = 701/1002 = .700$
   $\hat{p} \pm z_{.005}\sqrt{\hat{p}\hat{q}/n}$
   $.700 \pm 2.575\sqrt{(.700)(.300)/1002}$
   $.700 \pm .037$
   $.662 < p < .737$
   b. No; since the confidence interval does not include 61%, the survey results are not consistent with the actual voter turnout.

31. a. $\alpha = .01$ and $\hat{p} = x/n = 7/221 = .032$
   NOTE: Since $\hat{p} = x/n = 3.2\%$ was given, we may assume $x = \hat{p}n = .032 \cdot 221 = 7$. No other integer value x/221 rounds to .032.
   $\hat{p} \pm z_{.005}\sqrt{\hat{p}\hat{q}/n}$
   $.032 \pm 2.575\sqrt{(.032)(.968)/221}$
   $.032 \pm .030$
   $.00134 < p < .0620$
   $0.134\% < p < 6.20\%$
   b. Since 1.8% is within the interval in part (a), it cannot be ruled out as the correct value for the true proportion of Ziac users that experience dizziness. The test does not provide evidence that Ziac users experience any more dizziness than non-users – i.e., the test does not provide evidence that dizziness is an adverse reaction to Ziac.

32. a. $\alpha = .02$ and $\hat{p} = x/n = 144/785 = .183$
   NOTE: Since $\hat{p} = x/n = 18.3\%$ was given, we may assume $x = \hat{p}n = .183 \cdot 785 = 144$. No other integer value x/785 rounds to .183.
   $\hat{p} \pm z_{.01}\sqrt{\hat{p}\hat{q}/n}$
   $.183 \pm 2.33\sqrt{(.183)(.817)/785}$
   $.183 \pm .032$
   $.151 < p < .216$
   $15.1\% < p < 21.6\%$
   b. Yes; since 27% is not within the interval in part (a), it is not a reasonable possibility for the true smoking rate of those with four years of college.

33. $\hat{p}$ unknown, use $\hat{p} = .50$
   $n = [(z_{.005})^2\hat{p}\hat{q}]/E^2 = [(2.575)^2(.50)(.50)]/(.02)^2 = 4144.14$, rounded up to 4145

34. a. $\hat{p}$ unknown, use $\hat{p} = .5$
   $n = [(z_{.005})^2\hat{p}\hat{q}]/E^2 = [(2.575)^2(.5)(.5)]/(.025)^2 = 2652.25$, rounded up to 2653
   b. $\hat{p} = .15$
   $n = [(z_{.005})^2\hat{p}\hat{q}]/E^2 = [(2.575)^2(.15)(.85)]/(.025)^2 = 1352.65$, rounded up to 1353

c. In general, results from a self-selected sample are not valid. It is not appropriate to assume that those who choose to respond will be representative of the general population. In this instance it would not be unreasonable to think that right-handed golfers would tend to ignore the survey as unimportant, while left-handed golfers would be more likely to call in to "stand up" for themselves – in which case the results would over-estimate the proportion of golfers that are left-handed.

35. a. $\hat{p} = .86$
  $n = [(z_{.03})^2\hat{p}\hat{q}]/E^2 = [(1.88)^2(.86)(.14)]/(.03)^2 = 472.82$, rounded up to 473
  b. $\hat{p}$ unknown, use $\hat{p} = .5$
  $n = [(z_{.03})^2\hat{p}\hat{q}]/E^2 = [(1.88)^2(.5)(.5)]/(.03)^2 = 981.78$, rounded up to 982
  c. In general, results from a self-selected sample are not valid. It is not appropriate to assume that those who choose to respond will be representative of the general population.

36. a. $\hat{p} = .07$
  $n = [(z_{.025})^2\hat{p}\hat{q}]/E^2 = [(1.96)^2(.07)(.93)]/(.03)^2 = 277.88$, rounded up to 278
  b. $\hat{p}$ unknown, use $\hat{p} = .5$
  $n = [(z_{.025})^2\hat{p}\hat{q}]/E^2 = [(1.96)^2(.5)(.5)]/(.03)^2 = 1067.11$, rounded up to 1068

37. a. $\alpha = .10$ and $\hat{p} = x/n = 7/80 = .0875$
  $\hat{p} \pm z_{.01}\sqrt{\hat{p}\hat{q}/n}$
  $.0875 \pm 1.645\sqrt{(.0875)(.9125)/80}$
  $.0875 \pm .0520$
  $.036 < p < .139$
  b. $\hat{p} = .0875$
  $n = [(z_{.02})^2\hat{p}\hat{q}]/E^2 = [(2.05)^2(.0875)(.9125)]/(.03)^2 = 372.83$, rounded up to 373
  c. Yes; since $0.25\% = .0025$ is not in within the confidence interval in part (a), we may safely conclude that women have a lower rate of red/green color blindness than men do.

38. **NOTE:** This problem is limited to whole percent accuracy. Any x value between 780 (780/4000 = .1950) and 820 (820/4000 = .2050) gives a $\hat{p}$ that rounds to the given 20%. Since the exact value of x was not given, three decimal accuracy is not possible.
  a. $\alpha = .03$ and $\hat{p} = x/n = [800/4000] = .20$
  $\hat{p} \pm z_{.015}\sqrt{\hat{p}\hat{q}/n}$
  $.20 \pm 2.17\sqrt{(.20)(.80)/4000}$
  $.20 \pm .01$
  $.19 < p < .21$
  b. $\hat{p}$ unknown, use $\hat{p} = .5$
  $n = [(z_{.005})^2\hat{p}\hat{q}]/E^2 = [(2.575)^2(.5)(.5)]/(.005)^2 = 66306.25$, rounded up to 66307
  c. Yes; since $11\% = .11$ is not in within the confidence interval in part (a), we may safely conclude that 60 Minutes had a greater share of the audience than the wrestling special did. No; exposing professional wrestling was probably not necessary.
  d. In general, results from a self-selected sample are not valid. It is not appropriate to assume that those who choose to respond will be representative of the general population.

39. a. $\alpha = .05$ and $\hat{p} = x/n = 135/420000 = .000321$
  $\hat{p} \pm z_{.025}\sqrt{\hat{p}\hat{q}/n}$
  $.000321 \pm 1.96\sqrt{(.000321)(.999679)/420000}$
  $.000321 \pm .000054$
  $.000267 < p < .000376$
  $.0267\% < p < .0376\%$
  b. No; since the confidence interval includes .0340%, the results are consistent with those for the general population.

40. NOTE: This problem is limited to tenth of a percent accuracy. Any x value from 434 (434/8411 = .05160) to 441 (441/8411 = .05243) gives a $\hat{p}$ that rounds to the given 5.2%. Since the exact value of x was not given, four decimal accuracy is not possible.

a. $\alpha = .05$ and $\hat{p} = x/n = [437.372/8411] = .052$
$\hat{p} \pm z_{.025}\sqrt{\hat{p}\hat{q}/n}$
$.052 \pm 1.96\sqrt{(.052)(.948)/8411}$
$.052 \pm .005$
$.047 < p < .057$
$4.7\% < p < 5.7\%$

b. Yes; since the confidence intervals do not overlap, the death rates appear to differ substantially. The occurrence of an explosion or fire on the ground increases the death rate significantly.

c. Use 5.7%, the reasonable upper limit for the true percentage of pilot deaths in all crash landings.

41. a. $\alpha = .05$ and $\hat{p} = x/n = 6/123 = .0488$
$\hat{p} \pm z_{.025}\sqrt{\hat{p}\hat{q}/n}$
$.0488 \pm 1.96\sqrt{(.0488)(.9512)/123}$
$.0488 \pm .0381$
$.0107 < p < .0868$
$1.07\% < p < 8.68\%$

b. $\alpha = .05$ and $\hat{p} = x/n = 811/1115 = .727$
$\hat{p} \pm z_{.025}\sqrt{\hat{p}\hat{q}/n}$
$.727 \pm 1.96\sqrt{(.727)(.273)/1115}$
$.727 \pm .026$
$.701 < p < .753$
$70.1\% < p < 75.3\%$

c. Yes; the confidence intervals do not overlap. About 72.7% of the hunters in general wear orange. If wearing orange is unrelated to some other event involving hunters (e.g., eating eggs for breakfast), we expect about 72.7% of the hunters involved in that event to wear orange. If being mistaken for game is unrelated to wearing orange, we expect about 72.7% of those mistaken for game to be wearing orange. Since the proportion of hunters mistaken for game while wearing orange is significantly less, wearing orange appears to lessen the chances of being mistaken for game.
NOTE: While the true percentage who actually wear orange is likely less than the percentage who say they routinely wear orange [see exercise #30 for a comparable situation], the differences here are too great to be explained away in that manner.

42. NOTE: All parts of this problem are limited to whole percent accuracy. For n=651, more than one value for x rounds to the given $\hat{p}$ values of 94% and 75%. Since the exact value of x was not given, three decimal accuracy is not possible.

a. $\alpha = .10$ and $\hat{p} = x/n = [611.94/651] = .94 = 94\%$
$\hat{p} \pm z_{.05}\sqrt{\hat{p}\hat{q}/n}$
$.94 \pm 1.645\sqrt{(.94)(.06)/651}$
$.94 \pm .02$
$.92 < p < .96$
$92\% < p < 96\%$

b. $\alpha = .10$ and $\hat{p} = x/n = [488.25/651] = .75 = 75\%$
$\hat{p} \pm z_{.05}\sqrt{\hat{p}\hat{q}/n}$
$.75 \pm 1.645\sqrt{(.75)(.25)/651}$
$.75 \pm .03$
$.72 < p < .78$
$72\% < p < 78\%$

c. Yes; since the confidence intervals do not overlap, we may infer that the true percentage is higher for sloppy dressing than for unstylish dressing.

43. $\alpha = .05$ and $\hat{p} = x/n = 21/100 = .210$
$\hat{p} \pm z_{.025}\sqrt{\hat{p}\hat{q}/n}$
$.210 \pm 1.960\sqrt{(.210)(.790)/100}$
$.210 \pm .080$
$.130 < p < .290$
$13.0\% < p < 29.0\%$
Yes; this result is consistent with the 20% rate reported by the candy maker.

44. a. $\alpha = .05$ and $\hat{p} = x/n = 28/50 = .560$
$\hat{p} \pm z_{.025}\sqrt{\hat{p}\hat{q}/n}$
$.560 \pm 1.96\sqrt{(.560)(.440)/50}$
$.560 \pm .138$
$.422 < p < .698$
$42.2\% < p < 69.8\%$
b. $\alpha = .05$ and $\hat{p} = x/n = 25/50 = .500$
$\hat{p} \pm z_{.025}\sqrt{\hat{p}\hat{q}/n}$
$.500 \pm 1.96\sqrt{(.500)(.500)/50}$
$.500 \pm .139$
$.361 < p < .639$
$36.1\% < p < 63.9\%$
c. No; since the confidence intervals have considerable overlap, there is no significant difference between the two population percentages. The two percentages could be equal.
d. The analysis presented in parts (a) and (b) considers only the presence of tobacco or alcohol usage, and ignores the length of the usage. [Even an analysis that considers the length of the usage fails to consider the nature of the usage – i.e., whether the product is presented as a positive or negative factor.]

45. For n=829, any x value from 419 [419/829 = .5054] to 426 [426/829 = .5138] rounds to 51%. The two confidence intervals associated with those extreme possibilities are as follows.
$\alpha = .05$ and $\hat{p} = x/n = 419/829 = .505$
$\hat{p} \pm z_{.025}\sqrt{\hat{p}\hat{q}/n}$
$.505 \pm 1.96\sqrt{(.505)(.495)/829}$
$.505 \pm .034$
$.471 < p < .539$
$\alpha = .05$ and $\hat{p} = x/n = 426/829 = .514$
$\hat{p} \pm z_{.025}\sqrt{\hat{p}\hat{q}/n}$
$.514 \pm 1.96\sqrt{(.514)(.486)/829}$
$.514 \pm .034$
$.480 < p < .548$
While the results do not differ substantially from the $.476 < p < .544$ given in the text using $\hat{p} = .51$, they do illustrate the folly of giving an interval with 3 decimal "accuracy" when the original problem was stated to the nearest percent. In truth, even the second decimal is not certain – as the two intervals above round to $.47 < p < .54$ and $.48 < p < .55$.

46. $\hat{p}$ unknown, use $\hat{p} = .5$
$n = [N\hat{p}\hat{q}(z_{.005})^2]/[\hat{p}\hat{q}(z_{.005})^2 + (N-1)E^2]$
$= [(10000)(.5)(.5)(2.575)^2]/[(.5)(.5)(2.575)^2 + (9999)(.005)^2]$
$= [16576.56]/[1.907631] = 8689.61$, rounded up to 8690

47. $\alpha = .05$ and $\hat{p} = x/n = 630/750 = .840$
$\hat{p} - z_{.05}\sqrt{\hat{p}\hat{q}/n}$
$.840 - 1.645\sqrt{(.840)(.160)/750}$
$.840 - .022$
$.818 < p$
The interval is expressed as $p > .818$. The desired figure is 81.8%

48. $\alpha = .05$ and $\hat{p} = x/n = 3/8 = .375$
$\hat{p} \pm z_{.025}\sqrt{\hat{p}\hat{q}/n}$
$.375 \pm 1.96\sqrt{(.375)(.625)/8}$
$.375 \pm .335$
$.040 < p < .710$
Yes; the results are "reasonably close" – being shifted down 4.5% from the true interval $.085 < p < .755$ – but depending on the context such an error could be serious.

49. $\alpha = .01$ and $\hat{p} = x/n = 95/100 = .950$
$\hat{p} \pm z_{.005}\sqrt{\hat{p}\hat{q}/n}$
$.950 \pm 2.575\sqrt{(.950)(.050)/100}$
$.950 \pm .056$
$.894 < p < 1.006$
This interval is noteworthy because the upper limit is greater than 1, the maximum possible value for p in any problem. This occurs because of using the normal approximation to the binomial – which is just barely appropriate since $n(1-p) \approx 100(.05) = 5$, the minimum acceptable value to approximate the binomial distribution with the normal. In such cases the interval should be reported as $.894 < p < 1$ [and not as $.894 < p \le 1$, because the presence of 5 tails indicates that $p=1$ is not true].

50. a. If $\hat{p} = x/n = 0/n$, then
  (1) $np \approx 0 < 5$, and the normal distribution does not apply.
  (2) $E = z_{\alpha/2}\sqrt{\hat{p}\hat{q}/n} = 0$, and there is no meaningful interval.
  b. Since $\hat{p} = x/n = 0/20 = 0$, use the upper limit $3/n = 3/20 = .15$ to produce the interval $0 \le p < .15$ [and not $0 < p < .15$, because the failure to observe any successes in the sample does not rule out $p=0$ as the true population proportion].

51. Use the normal distribution with $\mu = 63.6$ and $\sigma = 2.5$ to estimate p.
$\hat{p} = P(x > 60)$
$= P(z > -1.44)$
$= 1 - P(z < -1.44)$
$= 1 - .0749$
$= .9251$
$n = [(z_{.01})^2\hat{p}\hat{q}]/E^2 = [(2.33)^2(.9251)(.0749)]/(.025)^2 = 601.87$, rounded up to 602

52. Since 19 of 20 implies $19/20 = .95 = 95\%$ confidence, use $\alpha = .05$.
$\hat{p}$ unknown, use $\hat{p} = .5$
$n = [(z_{.025})^2\hat{p}\hat{q}]/E^2 = [(1.96)^2(.5)(.5)]/(.01)^2 = 9604$

6-3 Estimating a Population Mean: σ Known

1. For .98 confidence, $\alpha = 1-.98 = .02$ and $\alpha/2 = .02/2 = .01$.
For the upper .01, $A = .9900$ [.9901] and $z = 2.33$.
$z_{\alpha/2} = z_{.01} = 2.33$

2. For .95 confidence, $\alpha = 1-.95 = .05$ and $\alpha/2 = .05/2 = .025$.
For the upper .025, $A = .9750$ and $z = 1.96$.
$z_{\alpha/2} = z_{.025} = 1.96$

3. For .96 confidence, $\alpha = 1-.96 = .04$ and $\alpha/2 = .04/2 = .02$.
For the upper .02, $A = .9800$ [.9798] and $z = 2.05$.
$z_{\alpha/2} = z_{.02} = 2.05$

INSTRUCTOR'S SOLUTIONS Chapter 6   S-169

4. For .995 confidence, α = 1-.995 = .005 and α/2 = .005/2 = .0025.
   For the upper .0025, A = .9975 and z = 2.81.
   $z_{\alpha/2} = z_{.0025} = 2.81$

5. Yes, if the sample is a simple random sample.

6. No, because σ is not known.

7. Yes, if the sample is a simple random sample.

8. No, because σ is not known.

NOTE: When the sample mean is given (i.e., not calculated from raw data), the accuracy with which $\bar{x}$ is reported determines the accuracy for the endpoints of the confidence interval.

9. α = .05, α/2 = .025, $z_{.025}$ = 1.96
   a. E = $z_{.025} \cdot \sigma/\sqrt{n}$
      = 1.96·(12,345)/√100
      = 2419.62 dollars
   b. $\bar{x} \pm E$
      95,000 ± 2,420
      92,580 < μ < 97,420 (dollars)

10. α = .01, α/2 = .005, $z_{.005}$ = 2.575
    a. E = $z_{.005} \cdot \sigma/\sqrt{n}$
       = 2.575·(4.6)/√50
       = 1.68 years
    b. $\bar{x} \pm E$
       80.5 ± 1.7
       78.8 < μ < 82.2 (years)

11. α = .10, α/2 = .05, $z_{.05}$ = 1.645
    a. E = $z_{.05} \cdot \sigma/\sqrt{n}$
       = 1.645·(2.50)/√25
       = .8225 seconds
    b. $\bar{x} \pm E$
       5.24 ± .82
       4.42 < μ < 6.06 (seconds)

12. α = .05, α/2 = .025, $z_{.025}$ = 1.96
    a. E = $z_{.025} \cdot \sigma/\sqrt{n}$
       = 1.96·(9900)/√28
       = 3667.01 dollars
    b. $\bar{x} \pm E$
       45,678 ± 3,667
       42,011 < μ < 49,345 (dollars)

13. n = $[z_{.025} \cdot \sigma/E]^2$ = [1.96·(500)/125]² = 61.47, rounded up to 62

14. n = $[z_{.005} \cdot \sigma/E]^2$ = [2.575·(15)/3]² = 165.77, rounded up to 166

15. n = $[z_{.05} \cdot \sigma/E]^2$ = [1.645·(48)/5]² = 249.38, rounded up to 250

16. n = $[z_{.03} \cdot \sigma/E]^2$ = [1.88·(9877)/500]² = 1379.20, rounded up to 1380

17. $\bar{x}$ = 318.1

18. 262.09 < μ < 374.11

19. First find the value of E as follows.
    width = 2·E
    374.11 - 262.09 = 2·E
    112.02 = 2·E
    56.01 = E
    Then use $\bar{x}$ and E to give the interval as: 318.1 ± 56.01

20. We have 95% confidence that the interval from 262.09 to 374.11 contains μ, the true value of the population mean.

21. $\bar{x} \pm z_{.025} \cdot \sigma/\sqrt{n}$
    $30.4 \pm 1.96 \cdot (1.7)/\sqrt{61}$
    30.4 ± 0.4
    $30.0 < \mu < 30.8$
    It is unrealistic to know the value of σ.

22. $\bar{x} \pm z_{.005} \cdot \sigma/\sqrt{n}$
    $182.9 \pm 2.575 \cdot (121.8)/\sqrt{54}$
    182.9 ± 42.7
    $140.2 < \mu < 225.6$
    It is unrealistic to know the value of σ.

23. $\bar{x} \pm z_{.05} \cdot \sigma/\sqrt{n}$
    $172.5 \pm 1.645 \cdot (119.5)/\sqrt{40}$
    172.5 ± 31.1
    $141.4 < \mu < 203.6$
    It is unrealistic to know the value of σ.

24. $\bar{x} \pm z_{.005} \cdot \sigma/\sqrt{n}$
    $40.6 \pm 2.575 \cdot (1.6)/\sqrt{100}$
    40.6 ± 0.4
    $40.2 < \mu < 41.0$
    It is unrealistic to know the value of σ.

25. $n = [z_{.025} \cdot \sigma/E]^2 = [1.96 \cdot (15)/2]^2 = 216.09$, rounded up to 217

26. $n = [z_{.005} \cdot \sigma/E]^2 = [2.575 \cdot (.068)/.025]^2 = 49.06$, rounded up to 50

27. $n = [z_{.025} \cdot \sigma/E]^2 = [1.96 \cdot (6250)/500]^2 = 600.25$, rounded up to 601

28. $n = [z_{.02} \cdot \sigma/E]^2 = [2.05 \cdot (112.2)/15]^2 = 235.13$, rounded up to 236

29. Using the range rule of thumb, $\sigma \approx$ (range)/4 = (70,000-12,000)/4 = 58,000/4 = 14,500.
    $n = [z_{.025} \cdot \sigma/E]^2 = [1.96 \cdot (14,500)/100]^2 = 80,769.64$, rounded up to 80,770
    No, this sample size is not practical. Of the three terms in the formula for n, $z_{\alpha/2}$ and E are set by the expectations of the researcher while σ is determined by the variable under consideration and not under control of the researcher. Either lower the level confidence (which will make $z_{\alpha/2}$ less than 1.96, thus decreasing n) or increase the margin of error (which will make E larger than 100, thus decreasing n).

30. Answers will vary, but a reasonable estimate for the minimum and maximum ages of college textbooks in use might be 0 years and 10 years respectively.
    Using the range rule of thumb, $\sigma \approx$ (range)/4 = (10-0)/4 = 10/4 = 2.5.
    $n = [z_{.05} \cdot \sigma/E]^2 = [1.645 \cdot (2.5)/.25]^2 = 270.60$, rounded up to 271

31. The minimum and maximum values in the data set are 56 and 96 respectively.
    Using the range rule of thumb, $\sigma \approx$ (range)/4 = (96-56)/4 = 40/4 = 10 and
    $n = [z_{.025} \cdot \sigma/E]^2 = [1.96 \cdot (10)/2]^2 = 96.04$, rounded up to 97.
    Using the sample standard deviation for the male pulse values [11.297] to estimate σ,
    $n = [z_{.025} \cdot \sigma/E]^2 = [1.96 \cdot (11.297)/2]^2 = 122.56$, rounded up to 123.
    The two values are relatively close. Since s (which considers all the data) is a better estimator for σ than R/4 (which is based entirely on the extreme values), the sample size of 123 should be preferred.

32. The minimum and maximum values in the data set are 41 and 102 respectively.
    Using the range rule of thumb, $\sigma \approx$ (range)/4 = (102-41)/4 = 61/4 = 15.25 and
    $n = [z_{.025} \cdot \sigma/E]^2 = [1.96 \cdot (15.25)/3]^2 = 99.26$, rounded up to 100.
    Using the sample standard deviation of the female diastolic values [11.626] to estimate σ,
    $n = [z_{.025} \cdot \sigma/E]^2 = [1.96 \cdot (11.626)/3]^2 = 57.69$, rounded up to 58.
    The two values seem significantly different. Since s (which considers all the data) is a better estimator for σ than R/4 (which is based entirely on the extreme values), the sample size of 58 should be preferred.

33. $\bar{x} \pm [z_{.025} \cdot \sigma/\sqrt{n}] \cdot \sqrt{(N-n)/(N-1)}$
    $110 \pm [1.96 \cdot (15)/\sqrt{35}] \cdot \sqrt{(250-35)/(250-1)}$
    $110 \pm [4.9695] \cdot [.9292]$
    110 ± 4.617
    $105 < \mu < 115$

34. From exercise #25, $\alpha = .05$ and $E = 2$ and $\sigma = 15$. Use $N = 200$ to find n as follows.
$$n = N\sigma^2(z_{.025})^2/[(N-1)E^2 + \sigma^2(z_{.025})^2]$$
$$= 200 \cdot (15)^2(1.96)^2/[(200-1)(2)^2 + (15)^2(1.96)^2]$$
$$= 172872/[1660.36]$$
$$= 104.12, \text{ rounded up to } 105$$

6-4 Estimating a Population Mean: σ Not Known

IMPORTANT NOTE: This manual uses the following conventions.
(1) The designation "df" stands for "degrees of freedom."
(2) Since the t value depends on both the degrees of freedom and the probability lying beyond it, double subscripts are used to identify points on t distributions. The t distribution with 15 degrees of freedom and .025 beyond it, for example, is designated $t_{15,.025} = 2.132$.
(3) Always use the closest entry in Table A-3. When the desired df is exactly halfway between the two nearest tabled values, be conservative and choose the one with the lower df.
(4) As the degrees of freedom increase, the t distribution approaches the standard normal distribution – and the "large" row of the t table actually gives z values. Consequently, the z scores for certain "popular" α and $\alpha/2$ values may be found by reading Table A-3 "frontwards" instead of reading Table A-2 "backwards." This is not only easier but also more accurate, since Table A-3 includes one more decimal place. Note, for example, that $t_{large,.05} = 1.645 = z_{.05}$ (as found from the z table) and $t_{large,.01} = 2.326 = z_{.01}$ (more accurate than the 2.33 as found from the z table). The manual uses this technique from this point on. [That $t_{large,.005} = 2.576 \neq 2.575 = z_{.005}$ is discrepancy caused by using different mathematical approximation techniques to construct the tables, and not a true difference. The manual will continue to use $z_{.005} = 2.575$.]

1. σ unknown and population approximately normal:
   If the sample is a simple random sample, use t.
   n = 5; df = 4
   $\alpha = .05$; $\alpha/2 = .025$
   $t_{4,.025} = 2.776$

2. σ unknown and population approximately normal:
   If the sample is a simple random sample, use t.
   n = 10; df = 9
   $\alpha = .05$; $\alpha/2 = .025$
   $t_{9,.025} = 2.262$

3. σ known and the population is not approximately normal
   Since n < 30, neither the z distribution nor the t distribution applies.

4. σ known and n > 30:
   If the sample is a simple random sample, use z.
   $\alpha = .01$; $\alpha/2 = .005$
   $z_{.005} = 2.575$

5. σ unknown and population approximately normal:
   If the sample is a simple random sample, use t.
   n = 92; df = 91 [90]
   $\alpha = .10$; $\alpha/2 = .05$
   $t_{91,.05} = 1.662$

6. σ known and the population is not approximately normal:
   Since n < 30, neither the z distribution nor the t distribution applies.

7. σ known and population approximately normal:
   If the sample is a simple random sample, use z.
   $\alpha = .02$; $\alpha/2 = .01$
   $z_{.01} = 2.236$

8. σ unknown and population approximately normal:
   If the sample is a simple random sample, use t.
   $n = 37$; $df = 36$
   $\alpha = .02$; $\alpha/2 = .01$
   $t_{36,.01} = 2.434$

9. $\alpha = .05$, $\alpha/2 = .025$; $n = 15$, $df = 14$
   a. $E = t_{14,.025} \cdot s/\sqrt{n}$
      $= 2.145 \cdot (108)/\sqrt{15}$
      $= 59.8$
   b. $\bar{x} \pm E$
      496 ± 60
      $436 < \mu < 556$

10. $\alpha = .01$, $\alpha/2 = .005$; $n = 32$, $df = 31$
    a. $E = t_{31,.005} \cdot s/\sqrt{n}$
       $= 2.744 \cdot (.70)/\sqrt{32}$
       $= .3396$ (inches)
    b. $\bar{x} \pm E$
       $14.50 \pm .34$
       $14.16 < \mu < 14.84$ (inches)

11. $112.84 < \mu < 121.56$
    We are 95% confident that the interval from 112.84 to 121.56 contains μ, the true value of the population mean.

12. $77.297 < \mu < 80.453$
    We are 99% confident that the interval from 77.297 to 80.453 contains μ, the true value of the population mean.

13. σ unknown and distribution approximately normal, use t
    $\bar{x} \pm t_{11,.025} \cdot s/\sqrt{n}$
    $26,227 \pm 2.201 \cdot (15,873)/\sqrt{12}$
    $26,227 \pm 10,085$
    $16,142 < \mu < 36,312$ (dollars)
    We are 95% certain that the interval from $16,142 to $36,312 contains the true mean repair cost for repairing Dodge Vipers under the specified conditions.

14. a. σ unknown and distribution approximately normal, use t
       $\bar{x} \pm t_{19,.005} \cdot s/\sqrt{n}$
       $9004 \pm 2.861 \cdot (5629)/\sqrt{20}$
       9004 ± 3601
       $5,403 < \mu < 12,605$ (dollars)
    b. The upper confidence limit of $12,605 would be a reasonable high-end estimate for the long-run average hospital costs of such accident victims.

15. a. σ unknown and $n > 30$, use t
       $\bar{x} \pm t_{30,.005} \cdot s/\sqrt{n}$
       $-.419 \pm 2.750 \cdot (3.704)/\sqrt{31}$
       $-.419 \pm 1.829$
       $-2.248 < \mu < 1.410$ (°F)
    b. Yes, the confidence interval includes 0. No; since the confidence interval includes 0, the true difference could be 0 – i.e., there is not evidence to conclude the three-day forecasts are too high or too low.

16. a. σ unknown and distribution approximately normal, use t
$\bar{x} \pm t_{19,.005} \cdot s/\sqrt{n}$
$4.4 \pm 2.861 \cdot (4.2)/\sqrt{20}$
4.4 ± 2.7
$1.7 < \mu < 7.1$ (inches)
b. No, the confidence interval does not include 0. Yes; since the confidence interval lies entirely above 0, it supports the claim that women tend to marry men who are taller than themselves.
NOTE: In truth, the above interval in part (a) and conclusion in part (b) is not necessarily valid for all women who marry – but only for the population of women from which the sample was taken (viz., women who had children and were presumably within some age range determined by the details of the survey).

17. preliminary values: $n = 7$, $\Sigma x = .85$, $\Sigma x^2 = .1123$
$\bar{x} = (\Sigma x)/n = (.85)/7 = .121$
$s^2 = [n(\Sigma x^2) - (\Sigma x)^2]/[n(n-1)] = [7(.1123) - (.85)^2]/[7(6)] = .00151$
$s = .039$
σ unknown (and assuming the distribution is approximately normal), use t
$\bar{x} \pm t_{6,.01} \cdot s/\sqrt{n}$
$.121 \pm 3.143 \cdot (.039)/\sqrt{7}$
$.121 \pm .046$
$.075 < \mu < .168$ (grams/mile)
No; since the confidence interval includes values greater than .165, there is a reasonable possibility that the requirement is not being met.

18. preliminary values: $n = 6$, $\Sigma x = 9.23$, $\Sigma x^2 = 32.5197$
$\bar{x} = (\Sigma x)/n = (0.23)/6 = 1.538$
$s^2 = [n(\Sigma x^2) - (\Sigma x)^2]/[n(n-1)] = [6(32.5197) - (9.23)^2]/[6(5)] = 3.664$
$s = 1.914$
σ unknown (and assuming the distribution is approximately normal), use t
$\bar{x} \pm t_{5,.025} \cdot s/\sqrt{n}$
$1.538 \pm 2.571 \cdot (1.914)/\sqrt{6}$
1.538 ± 2.009
$-.471 < \mu < 3.547$ (micrograms/cubic meter)
Yes; the fact that 4 of the 5 sample values are below $\bar{x}$ raises a question about whether the data meets the requirement that the underlying population distribution is normal.

19. a. σ unknown (and assuming the distribution is approximately normal), use t
$\bar{x} \pm t_{9,.025} \cdot s/\sqrt{n}$
$175 \pm 2.262 \cdot (15)/\sqrt{10}$
175 ± 11
$164 < \mu < 186$ (beats per minute)
b. σ unknown (and assuming the distribution is approximately normal), use t
$\bar{x} \pm t_{9,.025} \cdot s/\sqrt{n}$
$124 \pm 2.262 \cdot (18)/\sqrt{10}$
124 ± 13
$111 < \mu < 137$ (beats per minute)
c. The maximum likely value for the true mean heart rate for hand-shovelers, 186.
d. Since the two confidence intervals do not overlap, there is evidence that hand-shovelers experience significantly higher heart rates than those using electric snow blowers.

20. a. σ unknown and $n > 30$, use t
$\bar{x} \pm t_{39,.025} \cdot s/\sqrt{n}$
$69.4 \pm 2.024 \cdot (11.3)/\sqrt{40}$
69.4 ± 3.6
$65.8 < \mu < 73.0$ (beats per minute)

b. σ unknown and n > 30, use t
$\bar{x} \pm t_{39,.025} \cdot s/\sqrt{n}$
$76.3 \pm 2.024 \cdot (12.5)/\sqrt{40}$
76.3 ± 4.0
$72.3 < \mu < 80.3$ (beats per minute)

c. No; since the two confidence intervals overlap, we cannot conclude that the population means for males and females are difference – both populations could have the same mean, some value in the region of the overlap.

21. 4000 BC
$n = 12$
$\Sigma x = 1544$
$\Sigma x^2 = 198898$
$\bar{x} = (\Sigma x)/n = (1544)/12 = 128.67$
$s^2 = [n(\Sigma x^2) - (\Sigma x)^2]/[n(n-1)]$
$= [12(198898) - (1544)^2]/[12(11)]$
$= 21.515$
$s = 4.638$
σ unknown (and assuming normality), use t
$\bar{x} \pm t_{11,.025} \cdot s/\sqrt{n}$
$128.67 \pm 2.201 \cdot (4.638)/\sqrt{12}$
128.67 ± 2.95
$125.7 < \mu < 131.6$

150 AD
$n = 12$
$\Sigma x = 1600$
$\Sigma x^2 = 213610$
$\bar{x} = (\Sigma x)/n = (1600)/12 = 133.33$
$s^2 = [n(\Sigma x^2) - (\Sigma x)^2]/[n(n-1)]$
$= [12(213610) - (1600)^2]/[12(11)]$
$= 25.152$
$s = 5.015$
σ unknown (and assuming normality), use t
$\bar{x} \pm t_{11,.025} \cdot s/\sqrt{n}$
$133.33 \pm 2.201 \cdot (5.015)/\sqrt{12}$
133.33 ± 3.19
$130.1 < \mu < 136.5$

Since the two 95% confidence intervals overlap, the two samples could from populations with the same mean. This is not evidence to conclude that the head sizes have changed. NOTE: The level of confidence was not specified. While 95% is the standard level of confidence when no specific value is given, it is possible that using less than 95% confidence would shorten the intervals to the point where there is no overlap.

22. males
$n = 50$
$\Sigma x = 2054.9$
$\Sigma x^2 = 84562.23$
$\bar{x} = (\Sigma x)/n = (2054.9)/50 = 41.098$
$s^2 = [n(\Sigma x^2) - (\Sigma x)^2]/[n(n-1)]$
$= [50(84562.23) - (2054.9)^2]/[50(49)]$
$= 2.244$
$s = 1.498$
σ unknown and n > 30, use t
$\bar{x} \pm t_{49,.025} \cdot s/\sqrt{n}$
$41.098 \pm 2.009 \cdot (1.498)/\sqrt{50}$
$41.098 \pm .426$
$40.67 < \mu < 41.52$ (cm)

females
$n = 50$
$\Sigma x = 2002.4$
$\Sigma x^2 = 80323.82$
$\bar{x} = (\Sigma x)/n = (2002.4)/50 = 40.048$
$s^2 = [n(\Sigma x^2) - (\Sigma x)^2]/[n(n-1)]$
$= [50(80323.82) - (2002.4)^2]/[50(49)]$
$= 2.688$
$s = 1.639$
σ unknown and n > 30, use t
$\bar{x} \pm t_{49,.025} \cdot s/\sqrt{n}$
$40.048 \pm 2.009 \cdot (1.639)/\sqrt{50}$
$40.048 \pm .466$
$39.58 < \mu < 40.51$ (cm)

Since the two 95% confidence intervals do not overlap, the two samples appear to have come from populations with different means. This is evidence to conclude that the males have larger head circumferences.
NOTE: The level of confidence was not specified. While 95% is the standard level of confidence when no specific value is given, it is possible that using more than 95% confidence would lengthen the intervals to the point where they overlap.

23. a. preliminary values: $n = 36$, $\Sigma x = 29.6677$, $\Sigma x^2 = 24.45037155$
$\bar{x} = (\Sigma x)/n = (29.6677)/36 = .82410$
$s^2 = [n(\Sigma x^2) - (\Sigma x)^2]/[n(n-1)] = [36(24.45037155) - (29.6677)^2]/[36(35)] = .00003249$
$s = .005700$

INSTRUCTOR'S SOLUTIONS Chapter 6   S-175

   σ unknown and n > 30, use t
   $\bar{x} \pm t_{35,.025} \cdot s/\sqrt{n}$
   $.82410 \pm 2.032 \cdot (.005700)/\sqrt{36}$
   $.82410 \pm .00193$
   $.82217 < \mu < .82603$ (lbs)
   b. preliminary values: n = 36, $\Sigma x = 28.2189$, $\Sigma x^2 = 22.12028574$
   $\bar{x} = (\Sigma x)/n = (28.2189)/36 = .78386$
   $s^2 = [n(\Sigma x^2) - (\Sigma x)^2]/[n(n-1)] = [36(22.12028574) - (28.2189)^2]/[36(35)] = .00001900$
   $s = .004359$
   σ unknown and n > 30, use t
   $\bar{x} \pm t_{35,.025} \cdot s/\sqrt{n}$
   $.78386 \pm 2.032 \cdot (.004359)/\sqrt{36}$
   $.78386 \pm .00148$
   $.78238 < \mu < .78533$ (lbs)
   c. The confidence intervals do not overlap. Cans of diet Pepsi weigh significantly less than cans of regular Pepsi.

24. a. preliminary values: n = 40, $\Sigma x = 1039.9$, $\Sigma x^2 = 27493.83$
   $\bar{x} = (\Sigma x)/n = (1039.9)/40 = 25.9975$
   $s^2 = [n(\Sigma x^2) - (\Sigma x)^2]/[n(n-1)] = [40(27493.83) - (1039.9)^2]/[40(39)] = 11.7700$
   $s = 3.431$
   σ unknown and n > 30, use t
   $\bar{x} \pm t_{39,.005} \cdot s/\sqrt{n}$
   $25.9975 \pm 2.712 \cdot (3.431)/\sqrt{40}$
   25.9975 ± 1.4711
   $24.53 < \mu < 27.47$
   b. preliminary values: n = 40, $\Sigma x = 1029.6$, $\Sigma x^2 = 27984.46$
   $\bar{x} = (\Sigma x)/n = (1029.6)/40 = 25.7400$
   $s^2 = [n(\Sigma x^2) - (\Sigma x)^2]/[n(n-1)] = [40(27984.46) - (1029.6)^2]/[40(39)] = 38.0143$
   $s = 6.166$
   σ unknown and n > 30, use t
   $\bar{x} \pm t_{39,.005} \cdot s/\sqrt{n}$
   $25.7400 \pm 2.712 \cdot (6.166)/\sqrt{40}$
   25.7400 ± 2.6438
   $23.10 < \mu < 28.38$
   c. The confidence intervals overlap – in fact, the interval for the males is completely contained within the interval for the females. No, men do not appear to have a mean body mass index that is greater than the mean body mass index of women. If there are any body mass index differences, it may be that the female values are more variable.

25. preliminary values: n = 7, $\Sigma x = 60.79$, $\Sigma x^2 = 3600.1087$
   $\bar{x} = (\Sigma x)/n = (60.79)/7 = 8.684$
   $s^2 = [n(\Sigma x^2) - (\Sigma x)^2]/[n(n-1)] = [7(3600.1087) - (60.79)^2]/[7(6)] = 512.0318$
   $s = 22.628$
   σ unknown (and assuming the distribution is approximately normal), use t
   $\bar{x} \pm t_{6,.025} \cdot s/\sqrt{n}$
   $8.684 \pm 2.447 \cdot (22.628)/\sqrt{7}$
   8.684 ± 20.928
   $-12.244 < \mu < 29.613$ (grams/mile)
   This is considerably different from the $.075 < \mu < .168$ interval of exercise #17. Confidence intervals appear to be very sensitive to outliers. Outliers found in sample data should be examined for two reasons: (1) to determine whether they are errors, (2) to determine whether their presence in small samples indicates that the data fail to meet the normal distribution requirement.

26. For any α, the z value is smaller than the corresponding t value (although the difference decreases as n increases). This creates a smaller E and a narrower confidence interval than one is entitled to – i.e., it does not take into consideration the extra uncertainty created by using the sample s instead of the true population σ.

27. Applying the principles from chapter 2, if $y = a \cdot x + b$
then $\bar{y} = a \cdot \bar{x} + b$ and $s_y = a \cdot s_x$
Applying this to C and F, where $C = 5(F-32)/9 = (5/9) \cdot F - 160/9$,
then $\bar{C} = (5/9) \cdot \bar{F} - 160/9$ and $s_C = (5/9) \cdot s_F$

a. $E_C = t_{df,\alpha/2} s_C / \sqrt{n}$
$= t_{df,\alpha/2} (5/9) \cdot s_F / \sqrt{n}$
$= (5/9) \cdot t_{df,\alpha/2} s_F / \sqrt{n}$
$= (5/9) \cdot E_F$

b. $L = \bar{C} - E_C$ $\qquad U = \bar{C} + E_C$
$= [(5/9) \cdot \bar{F} - 160/9] - (5/9) \cdot E_F$ $\qquad = [(5/9) \cdot \bar{F} - 160/9] + (5/9) \cdot E_F$
$= (5/9) \cdot (\bar{F} - E_F) - 160/9$ $\qquad = (5/9) \cdot (\bar{F} + E_F) - 160/9$
$= (5/9) \cdot a - 160/9$ $\qquad = (5/9) \cdot b - 160/9$
$= (5/9) \cdot (a - 32)$ $\qquad = (5/9) \cdot (b - 32)$

c. Yes.

28. While the heights of adult aliens may be normally distributed on their home planet, there would likely be minimum and maximum height requirements to qualify for the space exploration program. This could prevent the height of lone alien landing on earth from being a sample of size n=1 from a normal distribution.

a. In general, one sample value gives no information about the variation of the variable. It is possible, however, that one value plus some other considerations can give some insight. If one knows that 0 is a possible value, for example, then one large sample value would indicate a large variance. [For example: If you sample the snowfall in Nome AK for one day and find 10.0 feet snow fell that day, you would assume that there are days with no snow and that there must be a large variable in the amounts of daily snowfall.]

b. The formula for E requires a value for s and a t score with n-1 degrees of freedom. When n=1, the formula for s fails to produce a value and there is no df=0 row for the t statistic. No confidence interval can be constructed.

c. $x \pm 9.68|x|$
$3.2 \pm 9.68|3.2|$
3.2 ± 31.0
$-27 < \mu < 34.2$ (feet) [see exercise #49 of section 6.1 for comments on
$0 < \mu < 34.2$ (feet) endpoints that are not physically possible]
Yes. Based on this technique it would be unlikely for the <u>mean</u> of such aliens to be 50 feet, but the apparent large variation makes it reasonable that a <u>single</u> alien selected at random could be that tall. If the true mean is 30, for example, the fact that the single randomly selected alien already observed is 3.2 feet tall makes it reasonable that the next such alien would be about 56.8 feet tall – so that their sample mean would be about 30.

6-5 Estimating a Population Variance

1. $\chi_L^2 = \chi_{15,.975}^2 = 6.262$; $\chi_R^2 = \chi_{15,.025}^2 = 27.488$

2. $\chi_L^2 = \chi_{50,.975}^2 = 32.357$; $\chi_R^2 = \chi_{50,.025}^2 = 71.420$

3. $\chi_L^2 = \chi_{79,.995}^2 = 51.172$; $\chi_R^2 = \chi_{79,.005}^2 = 116.321$
NOTE: Use df = 80, the closest entry.

4. $\chi_L^2 = \chi_{39,.95}^2 = 26.509$; $\chi_R^2 = \chi_{39,.05}^2 = 55.758$

5. $(n-1)s^2/\chi^2_{19,.025} < \sigma^2 < (n-1)s^2/\chi^2_{19,.975}$
$(19)(12345)^2/32.852 < \sigma^2 < (19)(12345)^2/8.907$
$88{,}140{,}188 < \sigma^2 < 325{,}090{,}544$
$9{,}388 < \sigma < 18{,}030$ (dollars)

6. $(n-1)s^2/\chi^2_{26,.005} < \sigma^2 < (n-1)s^2/\chi^2_{26,.995}$
$(26)(4.6)^2/48.920 < \sigma^2 < (26)(4.6)^2/11.160$
$12.246 < \sigma^2 < 49.297$
$3.4 < \sigma < 7.0$ (years)

7. $(n-1)s^2/\chi^2_{29,.05} < \sigma^2 < (n-1)s^2/\chi^2_{29,.95}$
$(29)(2.50^2/42.557 < \sigma^2 < (29)(2.50^2/17.708$
$4.2590 < \sigma^2 < 10.2355$
$2.06 < \sigma < 3.20$ (seconds)

8. $(n-1)s^2/\chi^2_{50,.025} < \sigma^2 < (n-1)s^2/\chi^2_{50,.975}$
$(50)(9900)^2/71.420 < \sigma^2 < (50)(9900)^2/32.357$
$68{,}615{,}234 < \sigma^2 < 151{,}451{,}000$
$8{,}283 < \sigma < 12{,}307$ (dollars)

9. From the upper right section of Table 6-2, n = 191.

10. From the upper right section of Table 6-2, n = 20.

11. From the lower left section of Table 6-2, n = 133,448.
No, for most applications this is not a practical sample size.

12. From the upper left section of Table 6-2, n = 210.

NOTE: When raw scores are available, $\bar{x}$ and s should be calculated as the primary descriptive statistics -- but use <u>the unrounded value of s^2</u> in the confidence interval formula. In addition, always make certain that the confidence interval for σ includes the calculated value of s.

13. $(n-1)s^2/\chi^2_{11,.025} < \sigma^2 < (n-1)s^2/\chi^2_{11,.975}$
$(11)(15873)^2/21.920 < \sigma^2 < (11)(15873)^2/3.816$
$126{,}435{,}831 < \sigma^2 < 726{,}277{,}101$
$11{,}244 < \sigma < 26{,}950$ (dollars)
We are 95% confident that the interval from $11,244 to $26,950 contains σ, the true standard deviation for the repair amounts of all such cars.

14. summary information
n = 18 $\bar{x}$ = 3787.0
Σx = 68,166 s^2 = 3066.82
Σx² = 258,196,778   s = 55.4
$(n-1)s^2/\chi^2_{17,.005} < \sigma^2 < (n-1)s^2/\chi^2_{17,.995}$
$(17)(3066.82)/35.718 < \sigma^2 < (17)(3066.82)/5.697$
$1459.65 < \sigma^2 < 9151.47$
$38.2 < \sigma < 95.7$ (mL)
No; the interval indicates 99% confidence that σ > 30.

15. summary information
n = 6 $\bar{x}$ = 1.538
Σx = 9.23 s^2 = 3.664
Σx² = 32.5197   s = 1.914

$(n-1)s^2/\chi^2_{5,.025} < \sigma^2 < (n-1)s^2/\chi^2_{5,.975}$
$(5)(3.664)/12.833 < \sigma^2 < (5)(3.664)/.831$
$1.4276 < \sigma^2 < 22.0468$
$1.195 < \sigma < 4.695$ (micrograms/cubic meter)

Yes; the fact that 4 of the 5 sample values are below $\bar{x}$ raises a question about whether the data meets the requirement that the underlying population distribution is normal.

16. summary information
$n = 12$ $\bar{x} = 3.504$
$\Sigma x = 42.05$ $s^2 = .011899$
$\Sigma x^2 = 147.4811$ $s = .109$
$(n-1)s^2/\chi^2_{11,.025} < \sigma^2 < (n-1)s^2/\chi^2_{11,.975}$
$(11)(.011899)/21.920 < \sigma^2 < (11)(.011899)/3.816$
$.006 < \sigma^2 < .034$
$.077 < \sigma < .185$ (oz)

Yes; he is in trouble. Since .06 is below the above confidence interval, we can be 95% confident that the population standard deviation is greater than .06.

17. a. $(n-1)s^2/\chi^2_{9,.025} < \sigma^2 < (n-1)s^2/\chi^2_{9,.975}$
$(9)(15)^2/19.023 < \sigma^2 < (9)(15)^2/2.700$
$106.5 < \sigma^2 < 750.0$
$10 < \sigma < 27$ (beats per minute)
b. $(n-1)s^2/\chi^2_{9,.025} < \sigma^2 < (n-1)s^2/\chi^2_{9,.975}$
$(9)(18)^2/19.023 < \sigma^2 < (9)(18)^2/2.700$
$153.3 < \sigma^2 < 1080.0$
$12 < \sigma < 33$ (beats per minute)
c. No; the two groups do not appear to differ much in variation.

18. a. $(n-1)s^2/\chi^2_{39,.025} < \sigma^2 < (n-1)s^2/\chi^2_{9,.975}$
$(39)(11.3)^2/59.342 < \sigma^2 < (39)(11.3)^2/24.433$
$83.919 < \sigma^2 < 203.819$
$9.2 < \sigma < 14.3$ (beats per minute)
b. $(n-1)s^2/\chi^2_{39,.025} < \sigma^2 < (n-1)s^2/\chi^2_{39,.975}$
$(39)(12.5)^2/59.342 < \sigma^2 < (39)(12.5)^2/24.433$
$102.689 < \sigma^2 < 249.407$
$10.1 < \sigma < 15.8$ (beats per minute)
c. No; the two groups do not appear to differ much in variation.

19. a. summary information
$n = 10$ $\bar{x} = 7.15$
$\Sigma x = 71.5$ $s^2 = .2272$
$\Sigma x^2 = 513.27$ $s = .48$
$(n-1)s^2/\chi^2_{9,.025} < \sigma^2 < (n-1)s^2/\chi^2_{9,.975}$
$(9)(.2272)/19.023 < \sigma^2 < (9)(.2272)/2.700$
$.1075 < \sigma^2 < .7573$
$.33 < \sigma < .87$ (minutes)
b. summary information
$n = 10$ $\bar{x} = 7.15$
$\Sigma x = 71.5$ $s^2 = 3.3183$
$\Sigma x^2 = 541.09$ $s = 1.82$
$(n-1)s^2/\chi^2_{9,.025} < \sigma^2 < (n-1)s^2/\chi^2_{9,.975}$
$(9)(3.3183)/19.023 < \sigma^2 < (9)(3.3183)/2.700$
$1.5699 < \sigma^2 < 11.0610$
$1.25 < \sigma < 3.33$ (minutes)

c. Yes, there is a difference. The multiple line system exhibits more variability among the waiting times. The greater consistency among the single line waiting times seems fairer to the customers and more professional.

20. a. summary information
$n = 40$ $\quad\quad\quad\quad\quad \bar{x} = 26.00$
$\Sigma x = 1039.9$ $\quad\quad\quad s^2 = 11.7700$
$\Sigma x^2 = 27493.83$ $\quad s = 3.431$
$(n-1)s^2/\chi^2_{39,.005} < \sigma^2 < (n-1)s^2/\chi^2_{39,.995}$
$(39)(11.7700)/66.766 < \sigma^2 < (39)(11.7700)/20.707$
$6.8752 < \sigma^2 < 22.1679$
$2.62 < \sigma < 4.71$ (minutes)

b. summary information
$n = 40$ $\quad\quad\quad\quad\quad \bar{x} = 25.74$
$\Sigma x = 1029.6$ $\quad\quad\quad s^2 = 38.0143$
$\Sigma x^2 = 27984.46$ $\quad s = 6.166$
$(n-1)s^2/\chi^2_{39,.005} < \sigma^2 < (n-1)s^2/\chi^2_{39,.995}$
$(39)(38.0143)/66.766 < \sigma^2 < (39)(38.0143)/20.707$
$22.2053 < \sigma^2 < 71.5969$
$4.71 < \sigma < 8.46$ (minutes)

c. Yes; since the confidence intervals do not overlap, there is a difference – just barely at the 99% confidence level, but more obviously for lower confidence levels. This is evidence that there is more variation in body mass index among females than among males.

21. a. The given interval $2.8 < \sigma < 6.0$
$\quad\quad\quad\quad 7.84 < \sigma^2 < 36.00$
and the usual calculations $(n-1)s^2/\chi^2_{19,\alpha/2} < \sigma^2 < (n-1)s^2/\chi^2_{19,1-\alpha/2}$
$\quad\quad\quad\quad\quad\quad (19)(3.8)^2/\chi^2_{19,\alpha/2} < \sigma^2 < (19)(3.8)^2/\chi^2_{19,1-\alpha/2}$
$\quad\quad\quad\quad\quad\quad 274.36/\chi^2_{19,\alpha/2} < \sigma^2 < 274.36/\chi^2_{19,1-\alpha/2}$
imply that $7.84 = 274.37/\chi^2_{19,\alpha/2}$ $\quad$ and $\quad 36.00 = 274.36/\chi^2_{19,1-\alpha/2}$
$\quad\quad\quad\quad \chi^2_{19,\alpha/2} = 274.36/7.84$ $\quad\quad\quad\quad \chi^2_{19,1-\alpha/2} = 274.36/36.00$
$\quad\quad\quad\quad\quad\quad = 34.99$ $\quad\quad\quad\quad\quad\quad\quad\quad\quad = 7.62$
The closest entries in Table A-4 are $\chi^2_{19,\alpha/2} = 34.805$ $\quad$ and $\quad \chi^2_{19,1-\alpha/2} = 7.633$
which imply $\quad\quad\quad\quad\quad\quad\quad\quad\quad\quad \alpha/2 = .01$ $\quad\quad\quad\quad\quad 1 - \alpha/2 = .99$
$\quad\quad\quad\quad\quad\quad\quad\quad\quad\quad\quad\quad\quad \alpha = .02$ $\quad\quad\quad\quad\quad\quad\quad \alpha/2 = .01$
$\quad\quad\quad\quad\quad\quad\quad\quad\quad\quad\quad\quad\quad\quad\quad\quad\quad\quad\quad\quad\quad\quad \alpha = .02$
The level of confidence is therefore is $1-\alpha = 98\%$.

b. $(n-1)s^2/\chi^2_{11,.025} < \sigma^2 < (n-1)s^2/\chi^2_{11,.975}$
using the lower endpoint $\quad\quad$ OR $\quad\quad$ using the upper endpoint
$(11)s^2/21.920 = (19.1)^2$ $\quad\quad\quad\quad\quad (11)s^2/3.816 = (45.8)^2$
$\quad\quad\quad s^2 = 726.97$ $\quad\quad\quad\quad\quad\quad\quad\quad\quad s^2 = 727.69$
$\quad\quad\quad s = 27.0$ $\quad\quad\quad\quad\quad\quad\quad\quad\quad\quad s = 27.0$

22. $\chi^2 = \frac{1}{2}[\pm z_{.025} + \sqrt{2\cdot df - 1}\,]^2$
$\quad = \frac{1}{2}[\pm 1.960 + \sqrt{2\cdot(771) - 1}\,]^2$
$\quad = \frac{1}{2}[\pm 1.960 + 39.256]^2$
$\quad = \frac{1}{2}[37.296]^2$ $\quad\quad$ or $\quad\quad \frac{1}{2}[41.216]^2$
$\quad = 695.48$ $\quad\quad\quad\quad\quad$ or $\quad\quad\quad 849.36$
$(n-1)s^2/\chi^2_{771,.025} < \sigma^2 < (n-1)s^2/\chi^2_{771,.975}$
$(771)(2.8)^2/849.36 < \sigma^2 < (771)(2.8)^2/695.48$
$\quad\quad\quad 7.117 < \sigma^2 < 8.691$
$\quad\quad\quad\quad 2.7 < \sigma < 2.9$ (inches)

S-180 INSTRUCTOR'S SOLUTIONS Chapter 6

Review Exercises

1. a. $\hat{p} = x/n = 111/1233 = .0900 = 9.00\%$
   b. $\hat{p} \pm z_{.05}\sqrt{\hat{p}\hat{q}/n}$
      $.0900 \pm 1.96\sqrt{(.0900)(.9100)/1233}$
      $.0900 \pm .0160$
      $.0740 < p < .1060$
      $7.40\% < p < 10.60\%$
   c. $\hat{p}$ unknown, use $\hat{p} = .5$
      $n = [(z_{.005})^2\hat{p}\hat{q}]/E^2 = [(2.575)^2(.5)(.5)]/(.025)^2 = 2652.25$, rounded up to 2653

2. a. σ unknown and population normal, use t
      $\bar{x} \pm t_{24,.025} \cdot s/\sqrt{n}$
      $7.01 \pm 2.064 \cdot 3.741/\sqrt{25}$
      7.01 ± 1.54
      $5.47 < \mu < 8.55$ (years)
   b. $(n-1)s^2/\chi^2_{24,.025} < \sigma^2 < (n-1)s^2/\chi^2_{24,.975}$
      $(24)(3.74)^2/39.364 < \sigma^2 < (24)(3.74)^2/12.401$
      $8.528 < \sigma^2 < 27.071$
      $2.92 < \sigma < 5.20$ (years)
   c. $n = [z_{.005} \cdot \sigma/E]^2 = [2.575 \cdot (3.74)/.25]^2 = 1483.94$, rounded up to 1484
   d. No; those who purchased a General Motors car are not necessarily representative of all car owners.

3. a. $\hat{p} = x/n = 308/611 = .504 = 50.4\%$
   b. $\hat{p} \pm z_{.01}\sqrt{\hat{p}\hat{q}/n}$
      $.5041 \pm 2.326\sqrt{(.5041)(.4959)/611}$
      $.5041 \pm .0470$
      $.4570 < p < .5511$
      $45.7\% < p < 55.1\%$
   c. No; since 43% is not within the confidence interval, the survey results are not consistent with the facts. People might not want to admit they voted for a losing candidate, especially if the winner is turning out to be a good president. [If the winner is turning out to be a bad president, the percentage in the survey who said they voted for the winner might actually be lower than the true value.]

4. a. summary information
      $n = 12$ $\bar{x} = 6.500$
      $\Sigma x = 78.0$ $s^2 = 5.9945$
      $\Sigma x^2 = 572.94$ $s = 2.448$
      σ unknown (and assuming normality), use t
      $\bar{x} \pm t_{11,.025} \cdot s/\sqrt{n}$
      $6.500 \pm 2.201 \cdot (2.448)/\sqrt{12}$
      6.500 ± 1.556
      $4.94 < \mu < 8.06$
   b. summary information
      $n = 12$ $\bar{x} = 5.075$
      $\Sigma x = 60.9$ $s^2 = 1.3639$
      $\Sigma x^2 = 324.07$ $s = 1.168$
      σ unknown (and assuming normality), use t
      $\bar{x} \pm t_{11,.025} \cdot s/\sqrt{n}$
      $5.075 \pm 2.201 \cdot (1.168)/\sqrt{12}$
      5.075 ± 0.742
      $4.33 < \mu < 5.82$

c. summary information
   $n = 12$ $\bar{x} = 8.433$
   $\Sigma x = 101.2$ $s^2 = 4.0333$
   $\Sigma x^2 = 897.82$ $s = 2.008$
   σ unknown (and assuming normality), use t
   $\bar{x} \pm t_{11,.025} \cdot s/\sqrt{n}$
   $8.433 \pm 2.201 \cdot (2.008)/\sqrt{12}$
   8.433 ± 1.276
   $7.16 < \mu < 8.71$
d. The mean grade-level rating appears to be significantly higher for Tolstoy than for Clancy or Rowling.

5. $n = [z_{.05} \cdot \sigma/E]^2 = [1.645 \cdot (2.45)/.5]^2 = 64.97$, rounded up to 65

6. $(n-1)s^2/\chi^2_{11,.025} < \sigma^2 < (n-1)s^2/\chi^2_{11,.975}$
   $(11)(1.17)^2/21.920 < \sigma^2 < (11)(1.17)^2/3.816$
   $.687 < \sigma^2 < 3.946$
   $.83 < \sigma < 1.99$

7. $\hat{p}$ unknown, use $\hat{p} = .5$
   $n = [(z_{.015})^2 \hat{p}\hat{q}]/E^2 = [(2.17)^2(.5)(.5)]/(.02)^2 = 2943.06$, rounded up to 2944

8. $\hat{p} = .93$
   $n = [(z_{.01})^2 \hat{p}\hat{q}]/E^2 = [(2.326)^2(.93)(.07)]/(.04)^2 = 220.32$, rounded up to 221

Cumulative Review Exercises

1. Begin by making a stem-and-leaf plot and calculating summary statistics.
   ```
   10 | 5
   11 |
   11 | 599
   12 | 3
   12 | 5788
   ```
   $n = 9$
   $\Sigma x = 1089$
   $\Sigma x^2 = 132223$

   a. $\bar{x} = (\Sigma x)/n = (1089)/9 = 121.0$ lbs
   b. $\tilde{x} = 123.0$ lbs
   c. $M = 119, 128$ (bi-modal)
   d. m.r. $= (105 + 128)/2 = 116.5$ lbs
   e. $R = 128 - 105 = 23$ lbs
   f. $s^2 = [n(\Sigma x^2) - (\Sigma x)^2]/[n(n-1)]$
      $= [9(132223) - (1089)^2]/[9(8)]$
      $= 56.75$ lbs^2
   g. $s = 7.5$ lbs
   h. for $Q_1 = P_{25}$, $L = (25/100)(9) = 2.25$, round up to 3
      $Q_1 = x_3 = 119$ lbs
   i. for $Q_2 = P_{50}$, $L = (50/100)(9) = 4.50$, round up to 5
      $Q_2 = x_5 = 123$ lbs
   j. for $Q_3 = P_{75}$, $L = (75/100)(9) = 6.75$, round up to 7
      $Q_3 = x_7 = 127$ lbs
   k. ratio, since differences are consistent and there is a meaningful zero

l. The boxplot is at the right.

m. σ unknown (and assuming normality), use t
$\bar{x} \pm t_{8,.005} \cdot s/\sqrt{n}$
$121.0 \pm 3.355 \cdot 7.5/\sqrt{9}$
121.0 ± 8.4
$112.6 < \mu < 129.4$ (lbs)

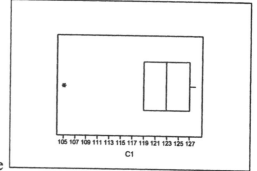

NOTE: Since both the stem-and-leaf plot and the boxplot indicate that the weights do not appear to come from a normal population, the methods of the chapter are not appropriate. It appears the distribution of the weights of supermodels is truncated at some upper limit and skewed to the left. Even though the confidence intervals for μ and σ should not be constructed, the intervals in part (m) above and part (n) below are given as a review of the techniques.

n. $(n-1)s^2/\chi^2_{8,.005} < \sigma^2 < (n-1)s^2/\chi^2_{8,.995}$
$(8)(56.75)/21.955 < \sigma^2 < (8)(56.75)/1.344$
$20.68 < \sigma^2 < 337.80$
$4.5 < \sigma < 18.4$ (lbs)

o. $E = 2$ and $\alpha = .01$
$n = [z_{.005} \cdot \sigma/E]^2$
$= [2.575 \cdot 7.5/2]^2 = 94.07$, rounded up to 95

p. For the general female population, an unusually low weight would be one below 85 lbs (i.e., more than 2σ below μ). Individually, none of the supermodels has an unusually low weight. As a group, however, they are each well below the general population mean weight and appear to weigh substantially less than the general population – as evidenced by their mean weight as given by the point estimate in part (a) and the interval estimate in part (m).

2. a. binomial: $n = 200$ and $p = .25$
a normal approximation is appropriate since
$np = 200(.25) = 50 \geq 5$
$n(1-p) = 200(.75) = 150 \geq 5$
use $\mu = np = 200(.25) = 50$
$\sigma = \sqrt{n(p)(1-p)} = \sqrt{200(.25)(.75)}$
$= 6.214$
$P(x \geq 65)$
$= P_c(x > 64.5)$
$= P(z > 2.37)$
$= 1 - P(z < 2.37)$
$= 1 - .9911$
$= .0089$

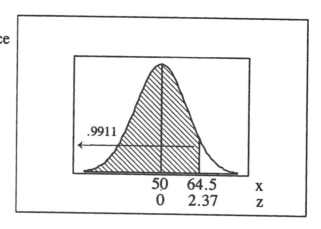

b. $\hat{p} = 65/200 = .325$
$\alpha = .05$
$\hat{p} \pm z_{.025}\sqrt{\hat{p}\hat{q}/n}$
$.325 \pm 1.960\sqrt{(.325)(.675)/200}$
$.325 \pm .065$
$.260 < p < .390$

c. No; the expert's value of .25 does not seem correct for two reasons.
From part (a): if the expert is correct, then the probability of getting the sample obtained is very small – less than 1%.
From part (b): according to the sample obtained, there is 95% confidence that the interval from .26 to .39 includes the true value – that is, there is 95% confidence that the value is not .25.

Chapter 7

Hypothesis Testing

7-2 Basics of Hypothesis testing

1. There is not sufficient evidence to support the claim. Under the assumption that the claim is not true, having 26 girls among 50 babies is the type of result you expect to occur.

2. There is sufficient evidence to support the claim. Under the assumption that the claim is not true, having 49 girls among 50 babies would be a very rare event. Therefore the assumption that the claim is not true is probably not correct.

3. There is sufficient evidence to support the claim. Under the assumption that the claim is not true, having 475 of 500 persons like pizza would be a rare event. Therefore the assumption that the claim is not true is probably not correct.

4. There is not sufficient evidence to support the claim. Under the assumption that the claim is not true, having s=14.99 is the type of result you would expect to occur.

5. original statement: $\mu > 50{,}000$ (does not contain the equality; must be H_1)
   competing idea: $\mu \leq 50{,}000$
   H_o: $\mu = 50{,}000$
   H_1: $\mu > 50{,}000$

6. original statement: $\mu \geq 110$
   competing idea: $\mu < 100$ (does not contain the equality; must be H_1)
   H_o: $\mu = 110$
   H_1: $\mu < 110$

7. original statement: $p > .5$ (does not contain the equality; must be H_1)
   competing idea: $p \leq .5$
   H_o: $p = .5$
   H_1: $p > .5$

8. original statement: $p \neq .70$ (does not contain the equality; must be H_1)
   competing idea: $p = .70$
   H_o: $p = .70$
   H_1: $p \neq .70$

9. original statement: $\sigma < 2.8$ (does not contain the equality; must be H_1)
   competing idea: $\sigma \geq 2.8$
   H_o: $\sigma = 2.8$
   H_1: $\sigma < 2.8$

10. original statement: $p = .24$
    competing idea: $p \neq .24$ (does not contain the equality; must be H_1)
    H_o: $p = .24$
    H_1: $p \neq .24$

S-184 INSTRUCTOR'S SOLUTIONS Chapter 7

11. original statement: $\mu \geq 12$
    competing idea: $\mu < 12$ (does not contain the equality; must be H_1)
    $H_o: \mu = 12$
    $H_1: \mu < 12$

12. original statement: $\sigma > 3000$ (does not contain the equality; must be H_1)
    competing idea: $\sigma \leq 3000$
    $H_o: \sigma = 3000$
    $H_1: \sigma > 3000$

NOTE: Recall that z_α is the z with α <u>above</u> it. By the symmetry of the normal distribution, the z with α <u>below</u> it is $z_{1-\alpha} = -z_\alpha$.

13. Two-tailed test; place $\alpha/2$ in each tail.
    Use $A = 1-\alpha/2 = 1-.0250 = .9750$ and $z = 1.96$.
    critical values are $\pm z_{\alpha/2} = \pm z_{.025} = \pm 1.96$

14. Two-tailed test; place $\alpha/2$ in each tail.
    Use $A = 1-\alpha/2 = 1-.0005 = .9995$ and $z = 2.575$.
    critical values are $\pm z_{\alpha/2} = \pm z_{.005} = \pm 2.575$

15. Right-tailed test; place α in the upper tail.
    Use $A = 1-\alpha = 1-.0100 = .9900$ [closest entry = .9901] and $z = 2.33$.
    critical value is $+z_\alpha = +z_{.01} = +2.33$ [or 2.326 from the "large" row of the t table]

16. Left-tailed test; place α in the lower tail.
    Use $A = \alpha = .0500$ and $z = -1.645$.
    critical value is $z_{1-\alpha} = -z_\alpha = -z_{.05} = -1.645$

17. Two-tailed test; place $\alpha/2$ in each tail.
    Use $A = 1-\alpha/2 = 1-.0500 = .9500$ and $z = 1.645$.
    critical values are $\pm z_{\alpha/2} = \pm z_{.05} = \pm 1.645$

18. Right-tailed test; place α in the upper tail.
    Use $A = 1-\alpha = 1-.1000 = .9000$ [closest entry = .8997] and $z = 1.28$.
    critical value is $z_\alpha = z_{.10} = 1.28$ [or 1.282 from the "large" row of the t table]

19. Left-tailed test; place α in the lower tail.
    Use $A = \alpha = .0200$ [closest entry = .0202] and $z = -2.05$.
    critical value is $-z_\alpha = -z_{.02} = -2.05$

20. Two-tailed test; place $\alpha/2$ in each tail.
    Use $A = 1-\alpha/2 = 1-.0025 = .9975$ and $z = 2.81$.
    critical value is $\pm z_\alpha = \pm z_{.0025} = \pm 2.81$

21. $\hat{p} = x/n = x/1025 = .29$
    $z_{\hat{p}} = (\hat{p} - p)/\sqrt{pq/n} = (.29-.50)/\sqrt{(.50)(.50)/1025} = -.21/.0156 = -13.45$

22. $\hat{p} = x/n = x/580 = .262$
    $z_{\hat{p}} = (\hat{p} - p)/\sqrt{pq/n} = (.262-.25)/\sqrt{(.25)(.75)/580} = .012/.0180 = .67$

23. $\hat{p} = x/n = x/400 = .290$
    $z_{\hat{p}} = (\hat{p} - p)/\sqrt{pq/n} = (.290-.25)/\sqrt{(.25)(.75)/400} = .04/.0217 = 1.85$

24. $\hat{p} = x/n = x/800 = .12$
    $z_{\hat{p}} = (\hat{p} - p)/\sqrt{pq/n} = (.12-.103)/\sqrt{(.103)(.897)/800} = .017/.0107 = 1.58$

25. P-value = P(z > .55)
    = 1 - P(z < .55)
    = 1 - .7088
    = .2912

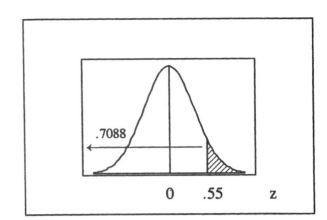

26. P-value = P(z < -1.72)
    = .0427

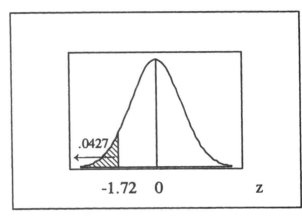

27. P-value = 2·P(z > 1.95)
    = 2·[1 - P(z < 1.95)]
    = 2·[1 - .9744]
    = 2·[.0256]
    = .0512

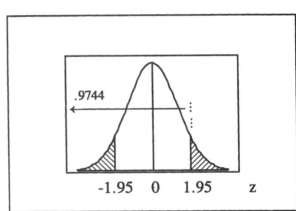

28. P-value = 2·P(z < -1.63)
    = 2·(.0516)
    = .1032

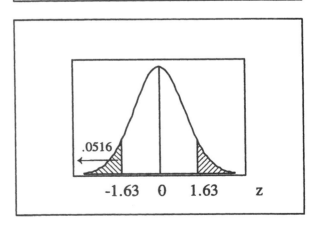

29. P-value = P(z > 1.97)
          = 1 - P(z < 1.97)
          = 1 - .9756
          = .0244

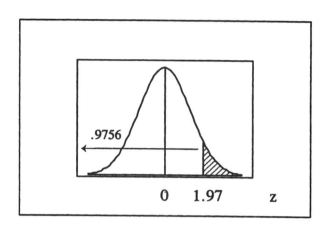

30. P-value = 2 · P(z > 2.44)
          = 2 · [1 - P(z < 2.44)]
          = 2 · [1 - .9927]
          = 2 · [.0073]
          = .0146

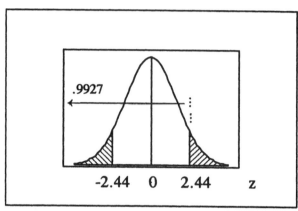

31. P-value = 2 · P(z > .77)
          = 2 · [1 - P(z < .77)]
          = 2 · [1 - .7794]
          = 2 · [.2206]
          = .4412

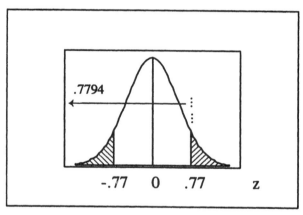

32. P-value = P(z < -1.90)
          = .0287

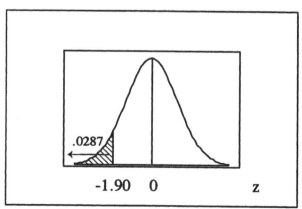

33. original claim: p > .5 (does not contain the equality; must be H_1)
    competing idea: p ≤ .5
       H_o: p = .5
       H_1: p > .5
    initial conclusion: Reject H_o.
    final conclusion: There is sufficient evidence to support the claim that the proportion of married women is greater than .5.

34. original claim: p < .27 (does not contain the equality; must be H_1)
    competing idea: p ≥ .27
       H_o: p = .27
       H_1: p < .27
    initial conclusion: Reject H_o.
    final conclusion: There is sufficient evidence to support the claim that the proportion of college graduates that smoke is less than .27.

35. original claim: p ≠ .038 (does not contain the equality; must be H_1)
    competing idea: p = .038
       H_o: p = .038
       H_1: p ≠ .038
    initial conclusion: Fail to reject H_o.
    final conclusion: There is not sufficient evidence to support the claim that the proportion of commercial aviation crashes that are fatal is different from .038.

36. original claim: p = .10
    competing idea: p ≠ .10 (does not contain the equality; must be H_1)
       H_o: p = .10
       H_1: p ≠ .10
    initial conclusion: Reject H_o.
    final conclusion: There is sufficient evidence to reject the claim that the proportion of M&M's that are blue is equal to .10.

37. type I error: supporting the claim p > .5 when p = .5 is true
    type II error: failing to support the claim p > .5 when p > .5 is true

38. type I error: supporting the claim p < .27 when p = .27 is true
    type II error: failing to support the claim p < .27 when p < .27 is true

39. type I error: supporting the claim p ≠ .038 when p = .038 is true
    type II error: failing to support the claim p ≠ .038 when p ≠ .038 is true

40. type I error: rejecting the claim p = .10 when p = .10 is true
    type II error: failing to reject the claim p = .10 when p ≠ .10 is true

41. $\hat{p}$ = x/n = x/491 = .27
    $z_{\hat{p}}$ = ($\hat{p}$ - p)/$\sqrt{pq/n}$ = (.27-.50)/$\sqrt{(.50)(.50)/491}$ = -.23/.0226 = -10.19
    P-value = P(z > -10.19) = 1 - (z < -10.19) = 1 - .0001 = .9999
    The claim is that p > .5. Only $\hat{p}$ values so much larger than .5 that they are unlikely to occur by chance if p = .5 is true give statistical support to the claim. A $\hat{p}$ smaller than .5 does not give any support to the claim.

42. Not necessarily. Rejection at the .05 level of significance means the result was among the most extreme 5% of the possible results, but not necessarily among the most extreme 1% of the possible results as required for rejection at the .01 level of significance.

43. Prefer .01, the lowest of the given values. The P-value is the probability of getting by chance alone less than or equal to the number of defects observed if the old rate is still in effect. The lower the P-value, the less likely your results could have occurred by chance alone – and the more likely your new process truly reduces the rate of defects.

44. No, since the claim that p=.10 must be the null hypothesis. The two possible conclusions in a statistical hypothesis test are "reject the null hypothesis" or "fail to reject the null hypothesis" – but failing to reject to reject a claim merely means there was not enough evidence to discredit it, not necessarily that the data supported it.
NOTE: In general, a statistical hypothesis test never <u>proves</u> anything – it merely provides a conclusion with a certain level of confidence attached. So long as the entire population is not examined, there will always be a certain probability of making an error because an unusual sample occurs.

45. Mathematically, in order for α to equal 0 the magnitude of the critical value would have to be infinite. Practically, the only way never to make a type I error is to always fail to reject H_o. From either perspective, the only way to achieve $\alpha = 0$ is to never reject H_o no matter how extreme the sample data might be.

46. The basic test of hypothesis is given below and illustrated by the figure at the right.
H_o: $p = .5$
H_1: $p < .5$
$\alpha = .05$
C.R. $z < -z_{.05} = -1.645$

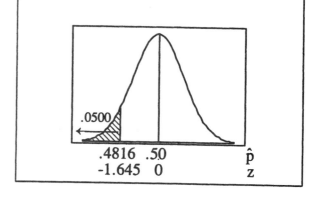

The $\hat{p}$ corresponding to $z = -1.645$ is found by solving $z_{\hat{p}} = (\hat{p} - p)/\sqrt{pq/n}$ for $\hat{p}$ as follows.
$\hat{p} = p + z_{\hat{p}} \cdot \sqrt{pq/n}$
$= .5 + (-1.645) \cdot \sqrt{(.5)(.5)/1998}$
$= .5 + (-1.645) \cdot (.0112)$
$= .5 - .0184$
$= .4816$

a. β = P(not rejecting $H_o | H_o$ is false)
$= P(\hat{p} > .4816 | p = .45)$
$= P(z > 2.84)$
$= 1 - P(z < 2.84)$
$= 1 - .9977$
$= .0023$

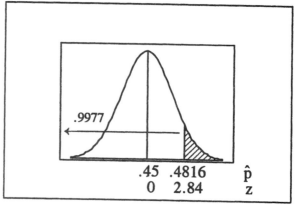

The z corresponding to $\hat{p} = .4816$ is found as follows.
$z_{\hat{p}} = (\hat{p} - p)/\sqrt{pq/n}$
$= (.4816 - .45)/\sqrt{(.45)(.55)/1998}$
$= (.0316)/(.0111)$
$= 2.84$

b. In this particular scenario,
the power is given by $1 - \beta = 1 - .0023 = .9977$.
In general,
β = P(failing to reject $H_o | H_o$ is false)
$1 - \beta$ = P(rejecting $H_o | H_o$ is false)

7-3 Testing a Claim about a Proportion

NOTE: To reinforce the concept that all z scores are standardized rescalings obtained by subtracting the mean and dividing by the standard deviation, the manual uses the "usual" z formula written to apply to $\hat{p}$'s
$$z_{\hat{p}} = (\hat{p} - \mu_{\hat{p}})/\sigma_{\hat{p}}.$$
When the normal approximation to the binomial applies, the $\hat{p}$'s are normally distributed
with $\mu_{\hat{p}} = p$ and $\sigma_{\hat{p}} = \sqrt{pq/n}$.
And so the formula for the z statistic may also be written as
$$z_{\hat{p}} = (\hat{p} - p)/\sqrt{pq/n}.$$

1. a. $z_{\hat{p}} = (\hat{p} - \mu_{\hat{p}})/\sigma_{\hat{p}} = (\hat{p} - p)/\sqrt{pq/n} = (.2494 - .25)/\sqrt{(.25)(.75)/8023} = -.0006/.0048$
   $= -.12$
   b. $z = \pm 1.96$
   c. P-value $= 2 \cdot P(z < -.12) = 2 \cdot (.4522) = .9044$
   d. Do not reject H_o; there is not sufficient evidence to conclude $p \neq .25$.
   e. No, the hypothesis test will either "reject" or "fail to reject" the claim that a population parameter is equal to a specified value.

2. a. $z_{\hat{p}} = (\hat{p} - \mu_{\hat{p}})/\sigma_{\hat{p}} = (\hat{p} - p)/\sqrt{pq/n} = (.62 - .50)/\sqrt{(.50)(.50)/1087} = .12/.0152 = 7.91$
   b. $z = 1.645$
   c. P-value $= P(z > 7.91) = 1 - P(z < 7.91) = 1 - .9999 = .0001$
   d. Reject H_o; there is sufficient evidence to conclude $p > .50$.
   e. No, the value of the test statistic is influenced by the sample size. While the difference between 62% and 50% is highly significant for this relatively large sample of size n=1087, that difference may not be significant for smaller samples – i.e., for smaller samples, that difference might reasonably occur by chance even when p=50% is true.

3. original claim: $p < .62$
   $\hat{p} = x/n = x/2500 = .60$
   H_o: $p = .62$
   H_1: $p < .62$
   $\alpha = .01$
   C.R. $z < -z_{.01} = -2.326$
   calculations:
   $z_{\hat{p}} = (\hat{p} - \mu_{\hat{p}})/\sigma_{\hat{p}}$
   $= (.60 - .62)/\sqrt{(.62)(.38)/2500}$
   $= -.02/.00971$
   $= -2.06$
   P-value $= P(z < -2.06) = .0197$
   conclusion:
   Do not reject H_o; there is not sufficient evidence to conclude that $p < .62$.
   If the data came from self-selected voluntary respondents, the test is not valid.

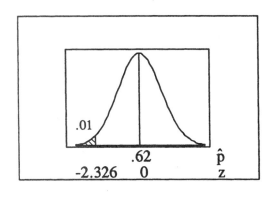

4. original claim: $p = .30$
   $\hat{p} = x/n = x/109,857 = .291$
   H_o: $p = .30$
   H_1: $p \neq .30$
   $\alpha = .01$
   C.R. $z < -z_{.005} = -2.575$
   $z > z_{.005} = 2.575$
   calculations:
   $z_{\hat{p}} = (\hat{p} - \mu_{\hat{p}})/\sigma_{\hat{p}}$
   $= (.291 - .30)/\sqrt{(.30)(.70)/109,857}$
   $= -.009/.00138$
   $= -6.51$

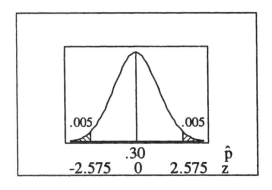

P-value = 2·P(z < -6.51) = 2·(.0001) = .0002
conclusion:
    Reject H_o; there is sufficient evidence reject the claim that p = .30 and to conclude that p ≠ .30 (in fact, that p < .30).
Even though 29.1% is so close to 30%, the sample size was so large that it was able to distinguish between those values. In general, a very large sample allows one to make precise conclusions.

5. original claim: p > .15
    $\hat{p}$ = x/n = 149/880 = .169
    H_o: p = .15
    H_1: p > .15
    α = .05
    C.R. z > $z_{.05}$ = 1.645
    calculations:
        $z_{\hat{p}}$ = ($\hat{p}$ - $\mu_{\hat{p}}$)/$\sigma_{\hat{p}}$
           = (.169 - .15)/√(.15)(.85)/880
           = .019/.0120
           = 1.60
        P-value = P(z > 1.60) = 1 - P(z < 1.60) = 1 - .9452 = .0548
    conclusion:

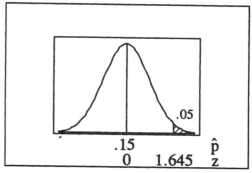

Do not reject H_o; there is not sufficient evidence to conclude that p > .15.
No; since technology and the use of technology is changing so rapidly, any figures from 1997 would no longer be valid today. [By the time you are reading this, e-mail could be so common that essentially 100% of the households use it -- or it could have been replaced by something newer and be so out-dated that essentially 0% of the households still use it!]

6. original claim: p > .35
    $\hat{p}$ = x/n = 4019/4276 = .940
    H_o: p = .35
    H_1: p > .35
    α = .01
    C.R. z > $z_{.01}$ = 2.326
    calculations:
        $z_{\hat{p}}$ = ($\hat{p}$ - $\mu_{\hat{p}}$)/$\sigma_{\hat{p}}$
           = (.940 - .35)/√(.35)(.65)/4276
           = .590/.00729
           = 80.87
        P-value = P(z > 80.87) = 1 - P(z < 80.87) = 1 - .9999 = .0001
    conclusion:

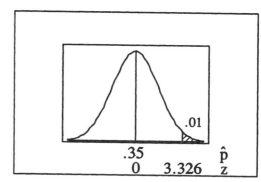

    Reject H_o; there is sufficient evidence to conclude that p > .35.
Yes; what appears to be obvious supporting information really is statistically significant evidence.

7. original claim: p > .50
    $\hat{p}$ = x/n = x/829 = .51
    H_o: p = .50
    H_1: p > .50
    α = .10
    C.R. z > $z_{.10}$ = 1.282
    calculations:
        $z_{\hat{p}}$ = ($\hat{p}$ - $\mu_{\hat{p}}$)/$\sigma_{\hat{p}}$
           = (.51 - .50)/√(.50)(.50)/829
           = .01/.0174
           = .58
        P-value = P(z > .58) = 1 - P(z < .58) = 1 - .7190 = .2810

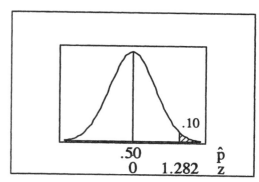

conclusion:
    Do not reject H_o; there is not sufficient evidence to conclude that $p > .50$.
No; since there are state and regional differences (in both circumstances and opinions) the conclusions reached for any one state can not be generalized to the country as a whole.

8. original claim: $p < .10$
    $\hat{p} = x/n = x/1012 = .09$
    H_o: $p = .10$
    H_1: $p < .10$
    $\alpha = .05$
    C.R. $z < -z_{.05} = -1.645$
    calculations:
        $z_{\hat{p}} = (\hat{p} - \mu_{\hat{p}})/\sigma_{\hat{p}}$
        $= (.09 - .10)/\sqrt{(.10)(.90)/1012}$
        $= -.01/.00943$
        $= -1.06$
    P-value $= P(z<-1.06) = .1446$

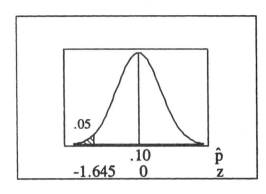

conclusion:
    Do not reject H_o; there is not sufficient evidence to conclude that $p < .10$.
If a newspaper ran such a headline, it would be doing so with P(error) = .1446 – for even when p equals 10%, there is a 14.46% of getting results as "convincing" as those above.

9. original claim: $p = .01$
    $\hat{p} = x/n = 20/1234 = .016$
    H_o: $p = .01$
    H_1: $p \neq .01$
    $\alpha = .05$
    C.R. $z < -z_{.025} = -1.96$
         $z > z_{.025} = 1.96$
    calculations:
        $z_{\hat{p}} = (\hat{p} - \mu_{\hat{p}})/\sigma_{\hat{p}}$
        $= (.016 - .01)/\sqrt{(.01)(.99)/1234}$
        $= .006/.00283$
        $= 2.19$
    P-value $= 2 \cdot P(z>2.19) = 2 \cdot [1 - P(z<2.19)] = 2 \cdot [1 - .9857] = 2 \cdot [.0143] = .0286$

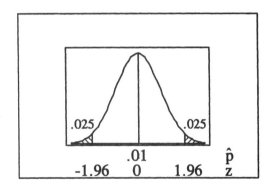

conclusion:
    Reject H_o; there is sufficient evidence reject the claim that $p = .01$ and to conclude that $p \neq .01$ (in fact, that $p > .01$).
No; based on these results, consumers appear to be subjected to more overcharges than under the old pre-scanner system.

10. original claim: $p < .058$
    $\hat{p} = x/n = 58/1520 = .038$
    H_o: $p = .058$
    H_1: $p < .058$
    $\alpha = .01$
    C.R. $z < -z_{.01} = -2.326$
    calculations:
        $z_{\hat{p}} = (\hat{p} - \mu_{\hat{p}})/\sigma_{\hat{p}}$
        $= (.038 - .058)/\sqrt{(.058)(.942)/1520}$
        $= -.020/.00600$
        $= -3.31$
    P-value $= P(z<-3.31) = .0005$

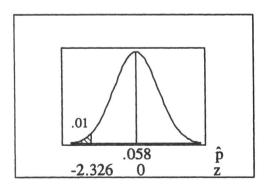

conclusion:
    Reject H_o; there is sufficient evidence to conclude that $p < .058$
Yes, the result suggests that fewer job applicants now use drugs.

11. original claim: p > .610
    $\hat{p} = x/n = 2231/3581 = .623$
    H_o: p = .610
    H_1: p > .610
    $\alpha = .05$
    C.R. $z > z_{.05} = 1.645$
    calculations:
    $z_{\hat{p}} = (\hat{p} - \mu_{\hat{p}})/\sigma_{\hat{p}}$
      $= (.623 - .610)/\sqrt{(.610)(.390)/3581}$
      $= .013/.00815$
      $= 1.60$
    P-value = P(z > 1.60) = 1 - P(z < 1.60) = 1 - .9452 = .0548
    conclusion:
    Do not reject H_o; there is not sufficient evidence to conclude that p > .610.

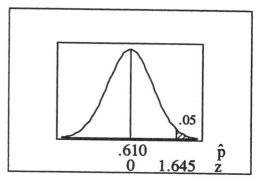

12. original claim: p > .019
    $\hat{p} = x/n = 19/863 = .022$
    H_o: p = .019
    H_1: p > .019
    $\alpha = .01$
    C.R. $z > z_{.01} = 2.326$
    calculations:
    $z_{\hat{p}} = (\hat{p} - \mu_{\hat{p}})/\sigma_{\hat{p}}$
      $= (.022 - .019)/\sqrt{(.019)(.981)/863}$
      $= .003/.00465$
      $= .65$
    P-value = P(z > .65) = 1 - P(z < .65) = 1 - .7422 = .2578
    conclusion:
    Do not reject H_o; there is not sufficient evidence to conclude that p > .019.
    No, flu symptoms do not appear to be an adverse reaction of the treatment.

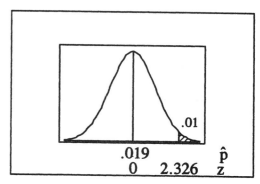

13. original claim: p = .000340
    $\hat{p} = x/n = 135/420,095 = .000321$
    H_o: p = .000340
    H_1: p ≠ .000340
    $\alpha = .005$
    C.R. $z < -z_{.0025} = -2.81$
        $z > z_{.0025} = 2.81$
    calculations:
    $z_{\hat{p}} = (\hat{p} - \mu_{\hat{p}})/\sigma_{\hat{p}}$
      $= (.000321 - .000340)/\sqrt{(.000340)(.999660)/420,095}$
      $= -.000019/.0000284$
      $= -.66$
    P-value = 2·P(z < -.66) = 2·(.2546) = .5092
    conclusion:
    Do not reject H_o; there is not sufficient evidence reject the claim that p = .000340.
    No; based on these results, cell phone users have no reason for such concern.

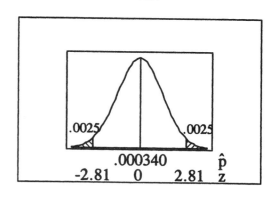

14. original claim: p > .50
    $\hat{p}$ = x/n = 39/(39+32) = 39/71 = .549
    H_o: p = .50
    H_1: p > .50
    α = .10
    C.R. z > $z_{.10}$ = 1.282
    calculations:
    $z_{\hat{p}} = (\hat{p} - \mu_{\hat{p}})/\sigma_{\hat{p}}$
    $= (.549 - .50)/\sqrt{(.50)(.50)/71}$
    = .049/.0593
    = .83

    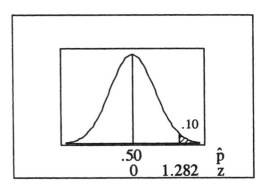

    P-value = P(z > .83) = 1 - P(z < .83) = 1 - .7967 = .2033
    conclusion:
    Do not reject H_o; there is not sufficient evidence to conclude that p > .50.
    Not necessarily. To test whether the therapy is effective, the results need to be compared to those of a control group – i.e., a group that tried to quit smoking without the therapy.

15. original claim: p < .27
    $\hat{p}$ = x/n = 144/785 = .183
    H_o: p = .27
    H_1: p < .27
    α = .01
    C.R. z < -$z_{.01}$ = -2.326
    calculations:
    $z_{\hat{p}} = (\hat{p} - \mu_{\hat{p}})/\sigma_{\hat{p}}$
    $= (.183 - .27)/\sqrt{(.27)(.73)/785}$
    = -.087/.0158
    = -5.46
    P-value = P(z < -5.46) = .0001
    conclusion:
    Reject H_o; there is sufficient evidence to conclude that p < .27.
    Since not smoking is the better decision from both a health and a financial perspective, one would expect it to be the choice of the better decision-makers – and presumably the additional education college graduates have makes them better decision-makers than the general population.

16. original claim: p > .20
    $\hat{p}$ = x/n = 1024/(1024+3836) = 1024/4860 = .211
    H_o: p = .20
    H_1: p > .20
    α = .025
    C.R. z > $z_{.025}$ = 1.96
    calculations:
    $z_{\hat{p}} = (\hat{p} - \mu_{\hat{p}})/\sigma_{\hat{p}}$
    $= (.211 - .20)/\sqrt{(.20)(.80)/4860}$
    = .011/.00574
    = 1.86

    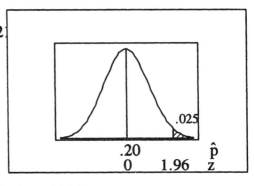

    P-value = P(z > 1.86) = 1 - P(z < 1.86) = 1 - .9686 = .0314
    conclusion:
    Do not reject H_o; there is not sufficient evidence to conclude that p > .20.
    Argue that even though $\hat{p}$ = .211 happens to be greater than 20%, that is not enough evidence to be 97.5% certain that the true population proportion is greater than 20%.

S-194 INSTRUCTOR'S SOLUTIONS Chapter 7

17. original claim: $p > .75$
   $\hat{p} = x/n = x/500 = .91$
   $H_o: p = .75$
   $H_1: p > .75$
   $\alpha = .01$
   C.R. $z > z_{.01} = 2.326$
   calculations:
   $z_{\hat{p}} = (\hat{p} - \mu_{\hat{p}})/\sigma_{\hat{p}}$
   $= (.91 - .75)/\sqrt{(.75)(.25)/500}$
   $= .16/.0194$
   $= 8.26$
   P-value $= P(z > 8.26) = 7.22 \times 10^{-17} = .0000000000000000722$ [from TI-83 Plus]
   conclusion:
   Reject H_o; there is sufficient evidence to conclude that $p > .75$.
   The TI-83 Plus values agree with those calculated above. Yes; based on these sample results, the funding will be approved.

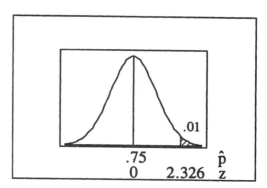

18. original claim: $p < .50$
   $\hat{p} = x/n = x/1998 = .48$
   $H_o: p = .50$
   $H_1: p < .50$
   $\alpha = .05$
   C.R. $z < -z_{.05} = -1.645$
   calculations:
   $z_{\hat{p}} = (\hat{p} - \mu_{\hat{p}})/\sigma_{\hat{p}}$
   $= (.48 - .50)/\sqrt{(.50)(.50)/1998}$
   $= -.02/.0112$
   $= -1.79$
   P-value $= P(z < -1.79) = .0367$
   conclusion:
   Reject H_o; there is sufficient evidence to conclude that $p < .50$.
   The TI-83 Plus values agree with those calculated above. Yes; based on these sample results, the executive's claim is supported.

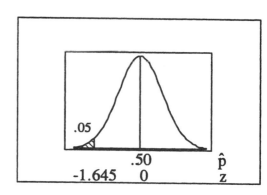

19. There are 100 M&M candies listed in Data Set 19, and 5 of them are blue.
   original claim: $p = .10$
   $\hat{p} = x/n = 5/100 = .05$
   $H_o: p = .10$
   $H_1: p \neq .10$
   $\alpha = .05$ [assumed]
   C.R. $z < -z_{.025} = -1.96$
   $z > z_{.025} = 1.96$
   calculations:
   $z_{\hat{p}} = (\hat{p} - \mu_{\hat{p}})/\sigma_{\hat{p}}$
   $= (.05 - .10)/\sqrt{(.10)(.90)/100}$
   $= -.05/.03$
   $= -1.67$
   P-value $= 2 \cdot P(z < -1.67) = 2 \cdot (.0475) = .0950$
   conclusion:
   Do not reject H_o; there is not sufficient evidence reject the claim that $p = .10$.

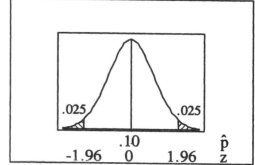

20. There are 50 movies listed in Data Set 7; 16 of them show no [0 seconds] alcohol or tobacco use, 34 show some use of alcohol or tobacco (or both).
original claim: $p > .50$
$\hat{p} = x/n = 34/50 = .68$
$H_o: p = .50$
$H_1: p > .50$
$\alpha = .05$
C.R. $z > z_{.05} = 1.645$
calculations:
$z_{\hat{p}} = (\hat{p} - \mu_{\hat{p}})/\sigma_{\hat{p}}$
$= (.68 - .50)/\sqrt{(.50)(.50)/50}$
$= .18/.0707$
$= 2.55$
P-value $= P(z>2.55) = 1 - P(z<2.55) = 1 - .9946 = .0054$
conclusion:
Reject H_o; there is sufficient evidence to conclude that $p > .50$.

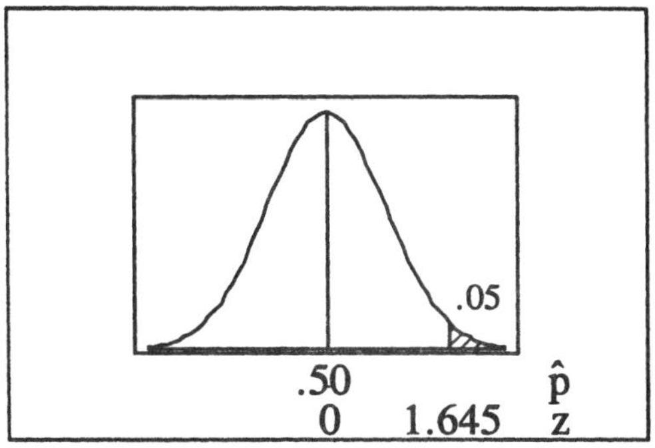

21. original claim: $p = .10$
$\hat{p} = x/n = 119/1000 = .119$
$H_o: p = .10$
$H_1: p \neq .10$
$\alpha = .05$
C.R. $z < -z_{.025} = -1.96$
$z > z_{.025} = 1.96$
calculations:
$z_{\hat{p}} = (\hat{p} - \mu_{\hat{p}})/\sigma_{\hat{p}}$
$= (.119 - .10)/\sqrt{(.10)(.90)/1000}$
$= .019/.00949$
$= 2.00$
P-value $= 2 \cdot P(z>2.00) = 2 \cdot [1 - P(z<2.00)] = 2 \cdot [1 - .9772] = 2 \cdot [.0228] = .0456$
conclusion:
Reject H_o; there is sufficient evidence reject the claim that $p = .10$ and conclude that $p \neq .10$ (in fact, that $p > .10$).

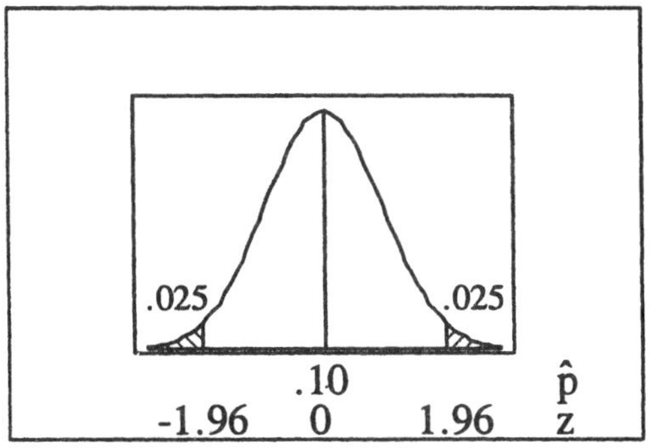

a. As seen above, the traditional method leads to rejection of the claim that $p=.10$ because the calculated $z=2.00$ is greater than the critical value of 1.96.
b. As seen above, the P-value method leads to rejection of the claim that $p=.10$ because the calculated P-value$=.0456$ is less than the level of significance of .05.
c. $\alpha = .05$ and $\hat{p} = x/n = 119/1000 = .119$
$\hat{p} \pm z_{.025}\sqrt{\hat{p}\hat{q}/n}$
$.119 \pm 1.96\sqrt{(.119)(.881)/1000}$
$.119 \pm .020$
$.099 < p < .139$
Since .10 is inside the confidence interval, this suggests that $p=.10$ is a reasonable claim that should not be rejected.
d. The traditional method and the P-value method are mathematically equivalent and will always agree. As seen by this example, the confidence interval method does not always lead to the same conclusion as the other two methods.

22. There are 50 movies listed in Data Set 7; 16 of them show no [0 seconds] alcohol or tobacco use, 34 show some use of alcohol or tobacco (or both).
original claim: $p > .50$
$\hat{p} = x/n = 34/50 = .68$
The test of hypothesis considers the probability of obtaining a result as extreme as or more extreme than the one obtained. In this problem, that would be $P(x \geq 34) = P_c(x > 33.5)$. The value of the sample proportion corrected for continuity is therefore
$\hat{p} = x/n = 33.5/50 = .67$.

H_o: p = .50
H_1: p > .50
α = .05
C.R. z > $z_{.05}$ = 1.645
calculations:
$z_{\hat{p}}$ = $(\hat{p} - \mu_{\hat{p}})/\sigma_{\hat{p}}$
  = $(.67 - .50)/\sqrt{(.50)(.50)/50}$
  = .17/.0707
  = 2.40
P-value = P(z>2.40)
  = 1 - P(z<2.40) = 1 - .9918 = .0082

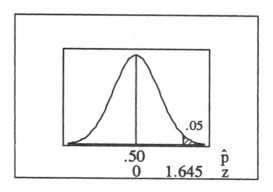

conclusion:
   Reject H_o; there is sufficient evidence to conclude that p > .50.
The calculated test statistic is slightly smaller than the 2.55 in exercise #20, and so the P-value is slightly larger than the .0054 in exercise #20. The change is minor and does not affect the conclusion.

23. original claim: p ≤ c
competing claim: p > c (does not contain the equality; must be H_1)
H_o: p = c
H_1: p > c
The possible conclusions are "reject p=c, and conclude p>c" or "fail to reject p=c, and say there is not enough evidence to conclude p>c" – but the test cannot conclude in favor of or support the null hypothesis or the original claim.

24. a. For the claim p > .0025, we have H_o:p = .0025.
Since np = (80)(.0025) = 0.2 < 5, the normal approximation to the binomial is not appropriate and the methods of this section cannot be used.

b. binomial: n = 80, p = .0025
$P(x) = [n!/x!(n-x)!]p^x(1-p)^{n-x}$
$P(x = 0) = [80!/0!80!](.0025)^0(.9975)^{80}$ = .818525754
$P(x = 1) = [80!/1!79!](.0025)^1(.9975)^{79}$ = .164115439
$P(x = 2) = [80!/2!78!](.0025)^2(.9975)^{78}$ = .016247017
$P(x = 3) = [80!/3!77!](.0025)^3(.9975)^{77}$ = .001058703
$P(x = 4) = [80!/4!76!](.0025)^4(.9975)^{76}$ = .000051078
$P(x = 5) = [80!/5!75!](.0025)^5(.9975)^{75}$ = .000001946   P(x ≥ 7) = 1 - P(x ≤ 6)
$P(x = 6) = [80!/6!74!](.0025)^6(.9975)^{74}$ = .000000061   = 1 - .999999998
                                   .999999998   = .000000002

c. Based on the result from part (b), getting 7 or more males with this color blindness if p = .0025 would be an extremely rare event. This leads one to reject the claim that p = .0025. This is essentially the P-value approach to hypothesis testing – for in a one-tailed test of H_o: p = .0025 vs. H_1: p > .0025, the observed value x = 7 produces:
P-value = P(getting a result as extreme as or more extreme than x = 7)
        = P(x ≥ 7)
        = .000000002

NOTE: This problem may be worked using the classical approach to hypothesis testing using the following binomial distribution (for n=80 and p = .0025) and rationale:

x	P(x)
0	.818526
1	.164115
2	.016247
3	.001059
4	.000051
5	.000002
6	0^+
7	0^+
⋮	⋮
	1.000000

We want to place .01 (or as close to it as possible without going over) in the upper tail of this one-tailed test.
P(x > 1) = P(x = 2) + ...
       = .016247 + ...
       > .01
P(x > 2) = 1 - [P(x = 0) + P(x = 1) + P(x = 2)]
       = 1 - [.818526 + .164115 + .016247]
       = 1 - .998888
       = .001112 ≤ .01; hence this is the desired critical region

This produces the following classical test of hypothesis [accompanied by a non-appropriate normal figure for illustrative purposes only].
original claim: $p > .0025$ [$np = (80)(.0025) = 0.2 < 5$; use binomial distribution]
$\quad x = 7$
H_o: $p = .0025$
H_1: $p > .0025$
$\alpha = .01$
C.R. $x > 2$
calculations
$\quad x = 7$
conclusion:
$\quad$ Reject H_o; there is sufficient evidence to conclude $p > .0025$.

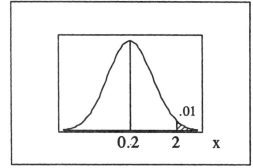

25. original claim: $p = .10$ [normal approximation to the binomial, use z]
$\hat{p} = x/n = 0/50 = 0$
Yes, the methods of this section can be used. In general, the appropriateness of a test depends on the design of the experiment and not the particular results. In particular, for this problem the normal approximation applies because
$\quad np = (50)(.1) = 5 \geq 5$
$\quad n(1-p) = (50)(.9) = 45 \geq 5$
H_o: $p = .10$
H_1: $p \neq .10$
$\alpha = .01$
C.R. $z < -z_{.005} = -2.575$
$\quad z > z_{.005} = 2.575$
calculations:
$\quad z_{\hat{p}} = (\hat{p} - \mu_{\hat{p}})/\sigma_{\hat{p}}$
$\quad\quad = (0 - .10)/\sqrt{(.10)(.90)/50}$
$\quad\quad = -2.357$

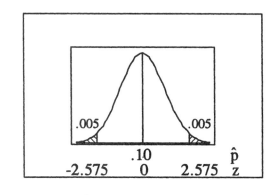

P-value = $2 \cdot P(z < -2.36) = 2 \cdot (.0091) = .0182$
conclusion:
$\quad$ Do no reject H_o; there is not sufficient evidence reject the claim that $p = .10$.

26. original claim: $p \leq d$, a rate to be determined
$\hat{p} = x/n = 15/500 = .03$
H_o: $p = d$
H_1: $p > d$
$\alpha = .05$
C.R. $z > z_{.05} = 1.645$
calculations:
$\quad z_{\hat{p}} = (\hat{p} - \mu_{\hat{p}})/\sigma_{\hat{p}}$
$\quad\quad 1.645 = (.03 - d)/\sqrt{(d)(1-d)/500}$
$\quad 1.645\sqrt{d(1-d)} = \sqrt{500}(.03 - d)$
$\quad 2.706(d - d^2) = 500(.0009 - .06d + d^2)$
$\quad\quad d - d^2 = 184.7729(.0009 - .06d + d^2)$
$\quad\quad d - d^2 = .1663 - 11.0864d + 184.7729\, d^2$

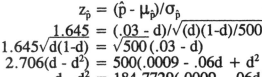

$185.7729\, d^2 - 12.0864d + .1163 = 0$
$\quad\quad d = [12.0864 \pm \sqrt{(-12.0864)^2 - 4(185.7729)(.1163)}]/2(185.7729)$
$\quad\quad\quad = [12.0864 \pm 4.7442]/371.5457$
$\quad\quad\quad = .01976$ or $.04530$
$\quad\quad$ [with .04530 being the d for which .03 has a z score of -1.645]
The lowest defective rate he can claim and still have $\hat{p} = x/n = 15/500 = .03$ <u>not</u> cause the claim to be rejected is .01977 (rounded up).

27. Assuming that each mouse is either a success or not a success, the number of successes must be a whole number between 0 and 20 inclusive. The possible success rates are all multiples of 5%: 0/20 = 0%, 1/20 = 5%, 2/20 = 10%, 3/20 = 15%, etc. A success rate of 47% is not a possibility.

28. The basic test of hypothesis is given below and illustrated by the figure at the right.
H_o: p = .5
H_1: p > .5
α = .05
C.R. z > $z_{.05}$ = 1.645

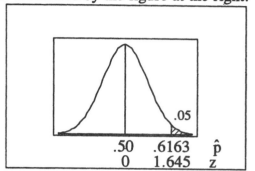

The $\hat{p}$ corresponding to z = 1.645 is found by solving $z_{\hat{p}} = (\hat{p} - p)/\sqrt{pq/n}$ for $\hat{p}$ as follows.
$\hat{p} = p + z_{\hat{p}} \cdot \sqrt{pq/n}$
   = .5 + 1.645·√(.5)(.5)/50
   = .5 + 1.645·(.0707)
   = .5 + .1163
   = .6163

β = P(not rejecting H_o | H_o is false)
   = P($\hat{p}$ < .6163 | p = .45)
   = P(z < 2.36)
   = .9909

The z corresponding to $\hat{p}$ = .6163 is found as follows.
$z_{\hat{p}} = (\hat{p} - p)/\sqrt{pq/n}$
   = (.6163 - .45)/√(.45)(.55)/50
   = (.1663)/(.0704)
   = 2.36

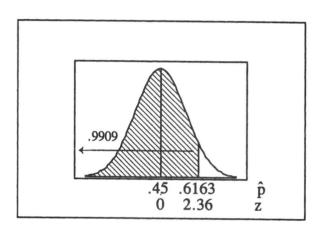

7-4 Testing a Claim about a Mean: σ Known

1. Yes, since σ is known and the original population is normally distributed.

2. No, since σ is unknown.

3. No, since σ is unknown.

4. Yes, since σ is known and n > 30.

5. original claim: μ > 118
   $\bar{x}$ = 120
   H_o: μ = 118
   H_1: μ > 118
   α = .05
   C.R. z > $z_{.05}$ = 1.645
   calculations:
   $z_{\bar{x}} = (\bar{x} - \mu)/\sigma_{\bar{x}}$
      = (120 - 118)/(12/√50)
      = 2/1.697
      = 1.18
   P-value = P(z > 1.18) = 1 - P(z < 1.18) = 1 - .8810 = .1190

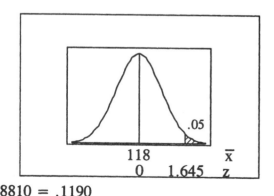

conclusion:
Do not reject H₀; there is not sufficient evidence to conclude that $\mu > 118$.

6. original claim: $\mu < 98.6$
   $\bar{x} = 98.20$
   $H_o: \mu = 98.6$
   $H_1: \mu < 98.6$
   $\alpha = .01$
   C.R. $z < -z_{.01} = -2.326$
   calculations:
   $z_{\bar{x}} = (\bar{x} - \mu)/\sigma_{\bar{x}}$
   $= (98.20 - 98.6)/(.62/\sqrt{106})$
   $= -.40/.0602$
   $= -6.64$
   P-value $= P(z<-6.64) = .0001$
   conclusion:
   Reject H₀; there is sufficient evidence to conclude that $\mu < 98.6$.

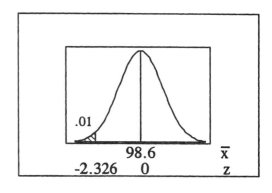

7. original claim: $\mu = 5.00$
   $\bar{x} = 5.25$
   $H_o: \mu = 5.00$
   $H_1: \mu \neq 5.00$
   $\alpha = .01$
   C.R. $z < -z_{.005} = -2.575$
   $z > z_{.005} = 2.575$
   calculations:
   $z_{\bar{x}} = (\bar{x} - \mu)/\sigma_{\bar{x}}$
   $= (5.25 - 5.00)/(2.50/\sqrt{80})$
   $= .25/.2795$
   $= .89$
   P-value $= 2 \cdot P(z>.89) = 2 \cdot [1 - P(z<.89)] = 2 \cdot [1 - .8133] = 2 \cdot [.1867] = .3734$
   conclusion:
   Do not reject H₀; there is not sufficient evidence reject the claim that $\mu = .10$.

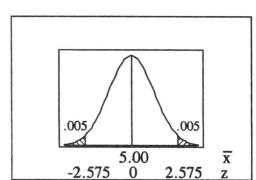

8. original claim: $\mu = 46,000$
   $\bar{x} = 45,678$
   $H_o: \mu = 46,000$
   $H_1: \mu \neq 46,000$
   $\alpha = .05$
   C.R. $z < -z_{.025} = -1.96$
   $z > z_{.025} = 1.96$
   calculations:
   $z_{\bar{x}} = (\bar{x} - \mu)/\sigma_{\bar{x}}$
   $= (45,678 - 46,000)/(9900/\sqrt{65})$
   $= -322/1227.944$
   $= -.26$
   P-value $= 2 \cdot P(z<-.26) = 2 \cdot (.3974) = .7948$
   conclusion:
   Do not reject H₀; there is not sufficient evidence reject the claim that $\mu = 46,000$.

S-200 INSTRUCTOR'S SOLUTIONS Chapter 7

9. original claim: $\mu > 30.0$
$\bar{x} = 30.4$
$H_o: \mu = 30.0$
$H_1: \mu > 30.0$
$\alpha = .05$
C.R. $z > z_{.05} = 1.645$
calculations:
$z_{\bar{x}} = (\bar{x} - \mu)/\sigma_{\bar{x}}$
$= (30.4 - 30.0)/(1.7/\sqrt{61})$
$= .4/.218$
$= 1.84$
P-value $= P(z>1.84) = 1 - P(z<1.84) = 1 - .9671 = .0329$
conclusion:
Reject H_o; there is sufficient evidence to conclude that $\mu > 30.0$.

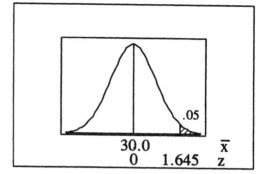

10. original claim: $\mu < 200$
$\bar{x} = 182.9$
$H_o: \mu = 200$
$H_1: \mu < 200$
$\alpha = .10$
C.R. $z < -z_{.10} = -1.282$
calculations:
$z_{\bar{x}} = (\bar{x} - \mu)/\sigma_{\bar{x}}$
$= (182.9 - 200)/(121.8/\sqrt{54})$
$= -17.1/16.575$
$= -1.03$
P-value $= P(z<-1.03) = .1515$
conclusion:
Do not reject H_o; there is not sufficient evidence to conclude that $\mu < 200$.

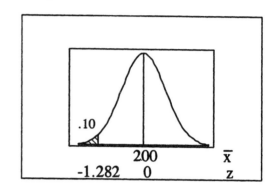

11. original claim: $\mu = 200.0$
$\bar{x} = 172.5$
$H_o: \mu = 200.0$
$H_1: \mu \neq 200.0$
$\alpha = .01$
C.R. $z < -z_{.005} = -2.575$
$z > z_{.005} = 2.575$
calculations:
$z_{\bar{x}} = (\bar{x} - \mu)/\sigma_{\bar{x}}$
$= (172.5 - 200.0)/(119.5/\sqrt{40})$
$= -27.5/18.895$
$= -1.46$
P-value $= 2 \cdot P(z<-1.46) = 2 \cdot (.0721) = .1442$
conclusion:
Do not reject H_o; there is not sufficient evidence reject the claim that $\mu = 200.0$.

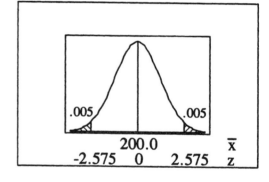

12. original claim: $\mu = 40.0$; $\bar{x} = 40.6$
$H_o: \mu = 40.0$
$H_1: \mu \neq 40.0$
$\alpha = .05$
C.R. $z < -z_{.025} = -1.96$
$z > z_{.025} = 1.96$
calculations:
$z_{\bar{x}} = (\bar{x} - \mu)/\sigma_{\bar{x}}$
$= (40.6 - 40.0)/(1.6/\sqrt{100})$
$= .6/.16 = 3.75$

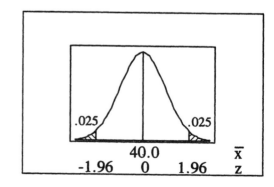

INSTRUCTOR'S SOLUTIONS Chapter 7   S-201

P-value = $2 \cdot P(z > 3.75) = 2 \cdot [1 - P(z < 3.75)] = 2 \cdot [1 - .9999] = 2 \cdot (.0001) = .0002$
conclusion:
   Reject H_o; there is sufficient evidence reject the claim that $\mu = 40.0$ and conclude that
   $\mu \neq 40.0$ (in fact, that $\mu > 40.0$).

13. original claim: $\mu \neq .9085$
    $\overline{x} = .91470$
    $H_o: \mu = .9085$
    $H_1: \mu \neq .9085$
    $\alpha = .05$ [assumed]
    C.R. $z < -z_{.025} = -1.96$
         $z > z_{.025} = 1.96$
    calculations:
    $z_{\overline{x}} = (\overline{x} - \mu)/\sigma_{\overline{x}}$
         $= (.91470 - .9085)/(.03691/\sqrt{100})$
         $= .0062/.003691$
         $= 1.68$

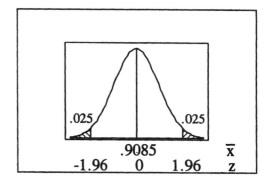

   P-value = $2 \cdot P(z > 1.68) = 2 \cdot [1 - P(z < 1.68)] = 2 \cdot [1 - .9535] = 2 \cdot (.0465) = .0930$
conclusion:
   Do not reject H_o; there is not sufficient evidence to conclude that $\mu \neq .9085$
Both the calculated z and the P-value agree with the Minitab output. No; since $\overline{x} >$
.9085, it appears that any deviance is in favor of the consumer. No; since the company
presumably figures its costs based on what is actually produced rather than on what is
advertised, they are not losing money by making the candy slightly heavier than necessary.

14. original claim: $\mu = 4.50$
    $\overline{x} = 1.753$
    $H_o: \mu = 4.50$
    $H_1: \mu \neq 4.50$
    $\alpha = .05$
    C.R. $z < -z_{.025} = -1.96$
         $z > z_{.025} = 1.96$
    calculations:
    $z_{\overline{x}} = (\overline{x} - \mu)/\sigma_{\overline{x}}$
         $= (1.753 - 4.50)/(2.87/\sqrt{73})$
         $= -2.747/.3359$
         $= -8.18$

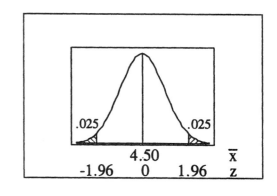

   P-value = $2 \cdot P(z < -8.18) = 2 \cdot (.0001) = .0002$
conclusion:
   Reject H_o; there is sufficient evidence reject the claim that $\mu = 4.50$ and conclude that
   $\mu \neq 4.50$ (in fact, that $\mu < 4.50$).
Both the calculated z and the P-value agree with the Minitab output. No; it appears that
the distances were not accurately measured.

15. original claim: $\mu = 0$
    $\overline{x} = -.419$
    $H_o: \mu = 0$
    $H_1: \mu \neq 0$
    $\alpha = .05$ [assumed]
    C.R. $z < -z_{.025} = -1.96$
         $z > z_{.025} = 1.96$
    calculations:
    $z_{\overline{x}} = (\overline{x} - \mu)/\sigma_{\overline{x}}$
         $= (-.419 - 0)/(3.704/\sqrt{31})$
         $= -.419/.6653$
         $= -.63$

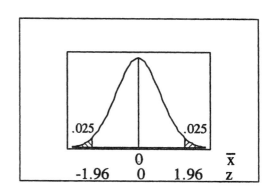

P-value = 2·P(z<-.63) = 2·(.2643) = .5286
conclusion:
  Do not reject H₀; there is not sufficient evidence reject the claim that $\mu = 0$.
Both the calculated z and the P-value agree with the TI-83 Plus output. The mean difference does not appear to be significantly different form 0. This suggests that the 3-day forecast high temperatures are correct <u>on the average</u> – but that does not necessarily instill confidence in any one single prediction, as the $\sigma=3.7$ suggests that it's not unusual for the predictions to be off by 2·(3.7) = 7.4 degrees in either direction.

16. original claim: $\mu < 281.81$
    $\bar{x} = 267.11$
    $H_0: \mu = 281.81$
    $H_1: \mu < 281.81$
    $\alpha = .01$
    C.R. $z < -z_{.01} = -2.326$
    calculations:
    $z_{\bar{x}} = (\bar{x} - \mu)/\sigma_{\bar{x}}$
    $= (267.11 - 281.81)/(22.11/\sqrt{175})$
    $= -14.7/1.6714$
    $= -8.80$
    P-value = P(z<-8.80) = .0001

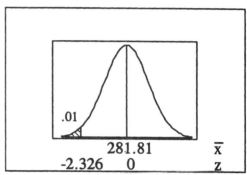

conclusion:
    Reject H₀; there is sufficient evidence to conclude that $\mu < 281.81$.
Both the caluclated z and the P-value agree with the TI-83 Plus output. Yes, the thinner cans do appear to have a mean axial load less than 281.82 lbs.

17. a. It is unrealistic that the value of σ would be known.
    b. original claim: $\mu \neq 98.6$; $\bar{x} = 98.20$
       $H_0: \mu = 98.6$
       $H_1: \mu \neq 98.6$
       $\alpha = .05$
       C.R. $z < -z_{.025} = -1.96$
            $z > z_{.025} = 1.96$
       calculations:
       $z_{\bar{x}} = (\bar{x} - \mu)/\sigma_{\bar{x}}$
       $-1.96 < (98.20 - 98.6)/(\sigma/\sqrt{106})$
            $\sigma < (98.20 - 98.6)/(-1.96/\sqrt{106})$
            $\sigma < (-.4)/(-.1904)$
            $\sigma < 2.101$; The largest possible standard deviation is 2.10

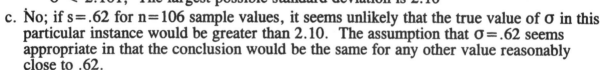

    c. No; if s=.62 for n=106 sample values, it seems unlikely that the true value of σ in this particular instance would be greater than 2.10. The assumption that σ=.62 seems appropriate in that the conclusion would be the same for any other value reasonably close to .62.

18. original claim: $\mu = 100$; $\bar{x} = 103.6$
    $H_0: \mu = 100$
    $H_1: \mu \neq 100$
    $\alpha = .01$
    C.R. $z < -z_{.005} = -2.575$
         $z > z_{.005} = 2.575$
    calculations:
    $z_{\bar{x}} = (\bar{x} - \mu)/\sigma_{\bar{x}}$
    $2.575 < (103.6 - 100)/(\sigma/\sqrt{62})$
         $\sigma < (103.6 - 100)/(2.575/\sqrt{62})$
         $\sigma < 3.6/.327$
         $\sigma < 11.008$; The largest possible standard deviation is 11.0.

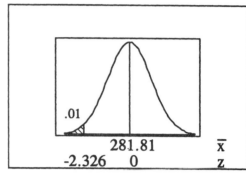

19. original claim: $\mu \neq 98.6$
    H_o: $\mu = 98.6$
    H_1: $\mu \neq 98.6$
    $\alpha = .05$
    C.R. $z < -z_{.025} = -1.960$
    $\quad\;\;\, z > z_{.025} = 1.960$

    in terms of $\bar{x}$, as seen by the calculations below,
    C.R. $\bar{x} < 98.482$
    $\quad\;\;\, \bar{x} > 98.718$

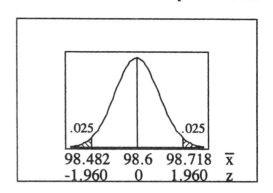

calculations:
$z_{\bar{x}} = (\bar{x} - \mu)/\sigma_{\bar{x}}$ $\qquad$ $z_{\bar{x}} = (\bar{x} - \mu)/\sigma_{\bar{x}}$
$-1.960 = (\bar{x} - 98.6)/(.62/\sqrt{106})$ $\qquad$ $1.960 = (\bar{x} - 98.6)/(.62/\sqrt{106})$
$\bar{x} = 98.482$ $\qquad\qquad\qquad\qquad$ $\bar{x} = 98.718$

a. $\beta = P(98.482 < \bar{x} < 98.718 | \mu = 98.7)$
   $\quad = P(-3.62 < z < .30)$
   $\quad = P(z < .30) - P(z < -3.62)$
   $\quad = .6179 - .0001$
   $\quad = .6178$

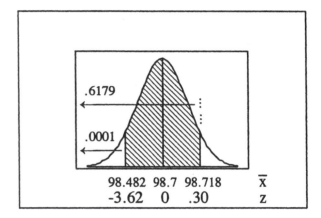

b. $\beta = P(98.482 < \bar{x} < 98.718 | \mu = 98.4)$
   $\quad = P(1.36 < z < 5.28)$
   $\quad = P(5.28 < z < 1.36)$
   $\quad = .9999 - .9131$
   $\quad = .0868$

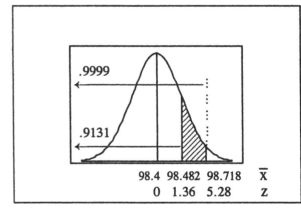

7-5 Testing a Claim About a Mean: σ Not Known

1. σ unknown and distribution approximately normal, use t

2. distribution not normal and n≤30, neither

3. σ known and distribution approximately normal, use z

4. σ unknown and n>30, use t

5. $t_{11,.01} = 2.718 < 2.998 < 3.106 = t_{11,.005}$
   .01 > P-value < .005
   .005 < P-value < .01

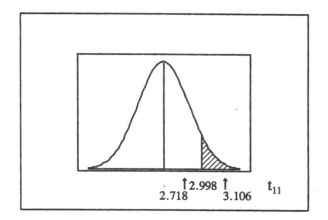

6. $-t_{11,.10} = -1.363 < -.855$
   .10 < P-value
   P-value > .10

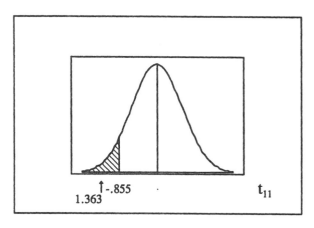

7. $t_{15,.005} = 2.947 < 4.629$
   .005 > ½(P-value)
   P-value < .01

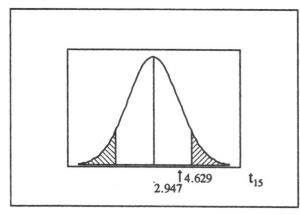

8. $t_{8,.05} = -1.860 < -1.577 < -1.397 = -t_{8,.10}$
   .05 < ½(P-value) < .10
   .10 < P-value < .20

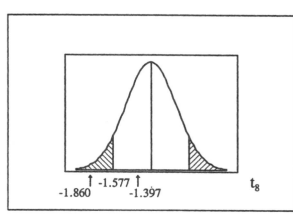

9. original claim: $\mu > 118$
   H_o: $\mu = 118$
   H_1: $\mu > 118$
   $\alpha = .05$
   C.R. $t > t_{19,.05} = 1.729$
   calculations:
   $t_{\bar{x}} = (\bar{x} - \mu)/s_{\bar{x}}$
   $= (120 - 118)/(12/\sqrt{20})$
   $= 2/2.683$
   $= .745$
   P-value $= P(t_{19} > .745) > .10$
   conclusion:
   Do not reject H_o; there is not sufficient evidence to conclude that $\mu > 118$.

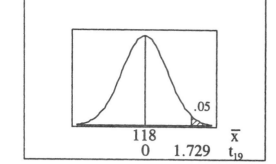

10. original claim: $\mu < 98.6$
    H_o: $\mu = 98.6$
    H_1: $\mu < 98.6$
    $\alpha = .01$
    C.R. $t < -t_{34,.01} = -2.441$
    calculations:
    $t_{\bar{x}} = (\bar{x} - \mu)/s_{\bar{x}}$
    $= (98.20 - 98.6)/(.62/\sqrt{35})$
    $= -.40/.1047$
    $= -3.817$
    P-value $= P(t_{34} < -3.817) < .005$
    conclusion:
    Reject H_o; there is sufficient evidence to conclude that $\mu < 98.6$.

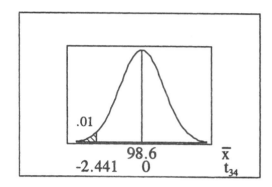

11. original claim: $\mu = 5.00$
    H_o: $\mu = 5.00$
    H_1: $\mu \neq 5.00$
    $\alpha = .01$
    C.R. $t < -t_{80,.005} = -2.639$
        $t > t_{80,.005} = 2.639$
    calculations:
    $t_{\bar{x}} = (\bar{x} - \mu)/t_{\bar{x}}$
    $= (5.25 - 5.00)/(2.50/\sqrt{81})$
    $= .25/.2778$
    $= .900$
    P-value $= 2 \cdot P(t_{19} > .900) > .20$
    conclusion:
    Do not reject H_o; there is not sufficient evidence reject the claim that $\mu = .10$.

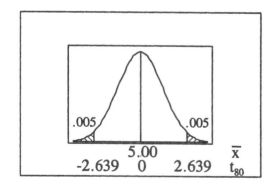

12. original claim: $\mu = 46,000$
    H_o: $\mu = 46,000$
    H_1: $\mu \neq 46,000$
    $\alpha = .05$
    C.R. $t < -t_{26,.025} = -2.056$
        $t > t_{26,.025} = 2.056$
    calculations:
    $t_{\bar{x}} = (\bar{x} - \mu)/s_{\bar{x}}$
    $= (45,678 - 46,000)/(9900/\sqrt{27})$
    $= -322/1905.256 = -.169$
    P-value $= 2 \cdot P(t_{26} < -.169) > .20$
    conclusion:
    Do not reject H_o; there is not sufficient evidence reject the claim that $\mu = 46,000$.

13. original claim: $\mu > 4$
    H_0: $\mu = 4$
    H_1: $\mu > 4$
    $\alpha = .05$
    C.R. $t > t_{11,.05} = 1.796$
    calculations:
    $t_{\bar{x}} = (\bar{x} - \mu)/s_{\bar{x}}$
    $= (5.075 - 4)/(1.168/\sqrt{12})$
    $= 1.075/.337$
    $= 3.188$
    P-value $= P(t_{11} > 3.188) < .005$
    conclusion:
    Reject H_0; there is sufficient evidence to conclude that $\mu > 4$.
    Yes; based on this sample, the teachers will use the book.

    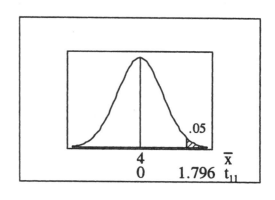

14. original claim: $\mu < .3$
    H_0: $\mu = .3$
    H_1: $\mu < .3$
    $\alpha = .05$
    C.R. $t < -t_{15,.05} = -1.753$
    calculations:
    $t_{\bar{x}} = (\bar{x} - \mu)/s_{\bar{x}}$
    $= (.295 - .3)/(.168/\sqrt{16})$
    $= -.005/.042$
    $= -.119$
    P-value $= P(t_{15} < -.119) > .10$
    conclusion:
    Do not reject H_0; there is not sufficient evidence to conclude that $\mu < .3$.

    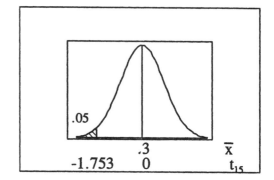

15. original claim: $\mu \neq 0$
    H_0: $\mu = 0$
    H_1: $\mu \neq 0$
    $\alpha = .05$
    C.R. $t < -t_{30,.025} = -2.042$
    $t > t_{30,.025} = 2.042$
    calculations:
    $t_{\bar{x}} = (\bar{x} - \mu)/s_{\bar{x}}$
    $= (-.419 - 0)/(3.704/\sqrt{31})$
    $= -.419/.665$
    $= -.630$
    P-value $= 2 \cdot P(t_{30} < -.630) > .20$
    conclusion:
    Do not reject H_0; there is not sufficient evidence to reject the claim that $\mu = 0$.
    Yes; based on this result, the forecasts do seem to be reasonably accurate.

    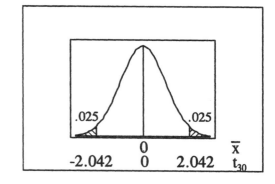

16. original claim: $\mu > 0$
    H_0: $\mu = 0$
    H_1: $\mu > 0$
    $\alpha = .01$
    C.R. $t > t_{19,.01} = 2.539$
    calculations:
    $t_{\bar{x}} = (\bar{x} - \mu)/s_{\bar{x}}$
    $= (4.4 - 0)/(4.2/\sqrt{20})$
    $= 4.4/.939$
    $= 4.685$
    P-value $= P(t_{19} > 4.685) < .005$

    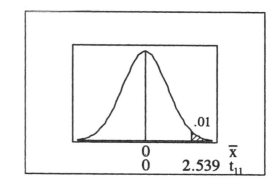

conclusion:
Reject H_o; there is sufficient evidence to conclude that $\mu > 0$.
Yes; the results support the claim that women tend to marry men taller than themselves.
NOTE: Since the women were not selected at random from among all married women, the results do not apply to the general population of married women. The women in the survey were those who had children (presumably by choice) – and most likely whose children were in a certain age range, which places them in a certain age range.

17. original claim: $\mu = 0$
H_o: $\mu = 0$
H_1: $\mu \neq 0$
$\alpha = .01$
C.R. $t < -t_{39,.005} = -2.712$
$t > t_{39,.005} = 2.712$
calculations:
$t_{\bar{x}} = (\bar{x} - \mu)/s_{\bar{x}}$
$= (117.3 - 0)/(185.0/\sqrt{40})$
$= 117.3/29.251$
$= 4.010$
P-value $= 2 \cdot P(t_{39} > 4.010) < .01$

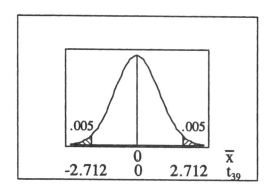

conclusion:
Reject H_o; there is sufficient evidence to reject the claim that $\mu = 0$ and to conclude that $\mu \neq 0$ (in fact, that $\mu > 0$).
One conclude that on the average, people's wrist watches tend to be almost two minutes fast – which is probably by deliberate choice.

18. original claim: $\mu < 75$
H_o: $\mu = 75$
H_1: $\mu < 75$
$\alpha = .05$ [assumed]
C.R. $t < -t_{15,.05} = -1.753$
calculations:
$t_{\bar{x}} = (\bar{x} - \mu)/s_{\bar{x}}$
$= (70.41 - 75)/(19.70/\sqrt{16})$
$= -4.59/4.925$
$= -.932$
P-value $= P(t_{15} < -.932) > .10$

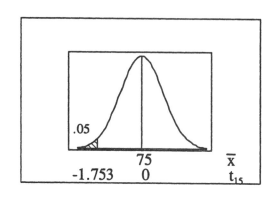

conclusion:
Do not reject H_o; there is not sufficient evidence to conclude that $\mu < 75$.
Since the sample mean was less than 75, these are certainly not results that would cause one to reject the college's claim. But since the sample mean was not far enough below 75 to rule out getting such a result by chance alone when $\mu < 75$ is not true, there is not enough evidence to conclude with 95% certainty that the college's claim is correct.
NOTE: It is possible that the student and the college are using two different definitions of "the mean price of a textbook." The student appears to have selected at random from among the available titles, regardless of the quantities sold. The college may be basing their claim on the total paid for textbooks divided by the number of books sold – which would be a weighted average. Suppose, for example, the college offers only two classes – a large one of 95 students with a textbook that costs $10, and a small one of 5 students with a textbook that costs $100. If each student buys the text, what would you consider "the mean price of a textbook" that semester? Is it $(10+100)/2 = $55, or is it $(950+500)/100 = $1450/100 = $14.50?

19. original claim: $\mu > 69.5$
    H_o: $\mu = 69.5$
    H_1: $\mu > 69.5$
    $\alpha = .05$
    C.R. $t > t_{34,.05} = 1.691$
    calculations:
    $t_{\bar{x}} = (\bar{x} - \mu)/s_{\bar{x}}$
    $= (73.4 - 69.5)/(8.7/\sqrt{35})$
    $= 3.9/1.471$
    $= 2.652$
    $.005 < $ P-value $= P(t_{34} > 2.652) < .01$

    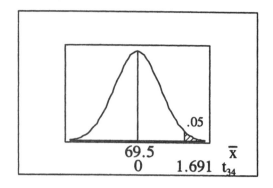

    conclusion:
    Reject H_o; there is sufficient evidence to conclude that $\mu > 69.5$.
    Yes, it does appear that male symphony conductors live longer than males form the general population – but in one sense that is like saying "male centenarians live longer than males from the general population." Male conductors should not be compared to "males from the general population" – but to "males from general population who have reached, say, age 35." Males who die in their teens, for example, are included in the general male population but (since we have no foreknowledge about what vocation they might have chosen had they lived) not in the population of male symphony orchestra conductors.

20. original claim: $\mu \neq 98.24$
    H_o: $\mu = 98.24$
    H_1: $\mu \neq 98.24$
    $\alpha = .05$
    C.R. $t < -t_{39,.025} = -2.024$
    $t > t_{39,.025} = 2.024$
    calculations:
    $t_{\bar{x}} = (\bar{x} - \mu)/s_{\bar{x}}$
    $= (92.67 - 92.84)/(1.79/\sqrt{40})$
    $= -.17/.283$
    $= -.601$
    P-value $= 2 \cdot P(t_{39} < -.601) > .20$

    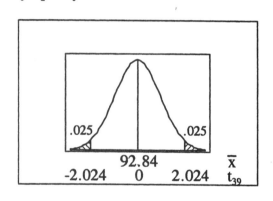

    conclusion:
    Do not reject H_o; there is not sufficient evidence to conclude that $\mu \neq 98.24$.
    No, it does not appear that the new baseballs are significantly different in this respect.

21. original claim: $\mu < 1000$
    H_o: $\mu = 1000$
    H_1: $\mu < 1000$
    $\alpha = .05$ [assumed]
    C.R. $t < -t_{4,.05} = -2.132$
    calculations:
    $t_{\bar{x}} = (\bar{x} - \mu)/s_{\bar{x}}$
    $= (767 - 1000)/(285/\sqrt{5})$
    $= -233/127.46$
    $= -1.828$
    P-value $= .071$

    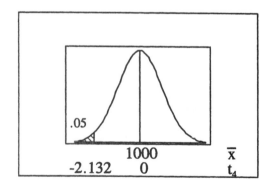

    conclusion:
    Do not reject H_o; there is not sufficient evidence to conclude that $\mu < 1000$.
    No; while the sample mean was less than $1000, the company cannot be 95% certain that the population mean is less than that amount.

22. original claim: $\mu > 420$
H_o: $\mu = 420$
H_1: $\mu > 420$
$\alpha = .05$ [assumed]
C.R. $t > t_{14,.05} = 1.761$
calculations:
$t_{\bar{x}} = (\bar{x} - \mu)/s_{\bar{x}}$
$= (442.2 - 420)/(44.0/\sqrt{15})$
$= 22.2/11.361$
$= 1.954$
P-value = .035
conclusion:
Reject H_o; there is sufficient evidence to conclude that $\mu > 420$.
Yes, appears (at the .05 level of significance) that the modifications improved reliability.

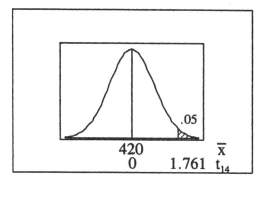

23. original claim: $\mu = 3.39$
H_o: $\mu = 3.39$
H_1: $\mu \neq 3.39$
$\alpha = .05$ [assumed]
C.R. $t < -t_{15,.025} = -2.131$
    $t > t_{15,.025} = 2.131$
calculations:
$t_{\bar{x}} = (\bar{x} - \mu)/s_{\bar{x}}$
$= (3.675 - 3.39)/(.6573/\sqrt{16})$
$= .285/.1643$
$= 1.734$
P-value = .1034
conclusion:
Do not reject H_o; there is not sufficient evidence to conclude that $\mu \neq 3.39$.
No; based on these results the supplement does not appear to have a significant effect on birth weight.

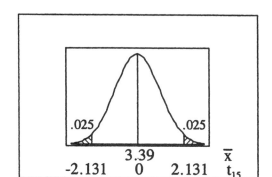

24. original claim: $\mu > 60$
H_o: $\mu = 60$
H_1: $\mu > 60$
$\alpha = .05$ [assumed]
C.R. $t > t_{19,.05} = 1.729$
calculations:
$t_{\bar{x}} = (\bar{x} - \mu)/s_{\bar{x}}$
$= (74.35 - 60)/(10.038/\sqrt{20})$
$= 14.35/2.2446$
$= 6.393$
P-value = 1.968×10^{-6} = .000001968
conclusion:
Reject H_o; there is sufficient evidence to conclude that $\mu > 60$.
Yes, the evidence supports the claim that the mean student rate is greater than 60.

25. summary statistics: n = 6, Σx = 9.23, Σx^2 = 32.5197, $\bar{x}$ = 1.538, s = 1.914
    original claim: $\mu > 1.5$
    H_o: $\mu = 1.5$
    H_1: $\mu > 1.5$
    $\alpha = .05$
    C.R. t > $t_{5,.05}$ = 2.015
    calculations:
    $t_{\bar{x}} = (\bar{x} - \mu)/s_{\bar{x}}$
    $= (1.538 - 1.5)/(1.914/\sqrt{6})$
    $= .0383/.78147$
    $= .049$

    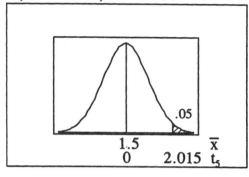

    P-value = $P(t_5 > .049) > .10$
    conclusion:
    Do not reject H_o; there is not sufficient evidence to conclude that $\mu > 1.5$.
    Yes; since 5 of the 6 values are below the sample mean, there is reason to doubt that the population values are normally distributed.

26. summary statistics: n = 21, Σx = 84, Σx^2 = 430, $\bar{x}$ = 4.000, s = 2.1679
    original claim: $\mu > 0$
    H_o: $\mu = 0$
    H_1: $\mu > 0$
    $\alpha = .01$
    C.R. t > $t_{20,.01}$ = 2.528
    calculations:
    $t_{\bar{x}} = (\bar{x} - \mu)/s_{\bar{x}}$
    $= (4.000 - 0)/(2.1679/\sqrt{21})$
    $= 4.000/.4731$
    $= 8.455$

    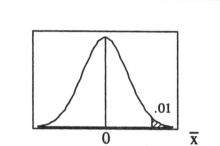

    P-value = $P(t_{20} > 8.455) < .005$
    conclusion:
    Reject H_o; there is sufficient evidence to conclude that $\mu > 0$.
    Yes, the treatment appears to be effective.

27. summary statistics: n = 23, Σx = 241.27, Σx^2 = 2536.5762, $\bar{x}$ = 10.490 s = .507
    original claim: $\mu < 10.5$
    H_o: $\mu = 10.5$
    H_1: $\mu < 10.5$
    $\alpha = .05$ [assumed]
    C.R. t < $-t_{22,.05}$ = -1.717
    calculations:
    $t_{\bar{x}} = (\bar{x} - \mu)/s_{\bar{x}}$
    $= (10.490 - 10.5)/(.507/\sqrt{23})$
    $= -.01/.1057$
    $= -.095$
    P-value = $P(t_{22} < -.095) > .10$
    conclusion:
    Do not reject H_o; there is not sufficient evidence to conclude that $\mu < 10.5$.
    The precision of the numbers changed from 1 to 2 decimal accuracy, presumably as improved technology produced more accurate timing devices. The test does not consider the sequential nature of the data, and would give the same results no matter what the order of the data. Because the test fails to consider the significant order effect, it only summarizes the past and cannot accurately project into the future. No; the test does not address the issue about future times being close to 10.5, and such a conclusion would not be valid.

28. summary statistics: n = 10, $\Sigma x = 433.0$, $\Sigma x^2 = 18{,}878.90$, $\bar{x} = 43.3$, s = 3.801
original claim: $\mu > 40$
H_o: $\mu = 40$
H_1: $\mu > 40$
$\alpha = .01$
C.R. t > $t_{9,.01}$ = 2.821
calculations:
$t_{\bar{x}} = (\bar{x} - \mu)/s_{\bar{x}}$
   = (43.3 - 40)/(3.801/$\sqrt{10}$)
   = 3.3/1.202
   = 2.746
.01 < P-value = P(t_9 > 2.746) < .025
conclusion:
Do not reject H_o; there is not sufficient evidence to conclude that $\mu > 40$.

29. summary statistics: n = 12, $\Sigma x = 78.0$, $\Sigma x^2 = 572.94$, $\bar{x} = 6.50$, s = 2.448
original claim: $\mu > 6$
H_o: $\mu = 6$
H_1: $\mu > 6$
$\alpha = .05$ [assumed]
C.R. t > $t_{11,.05}$ = 1.796
calculations:
$t_{\bar{x}} = (\bar{x} - \mu)/s_{\bar{x}}$
   = (6.50 - 6)/(2.448/$\sqrt{12}$)
   = .50/.7068
   = .707
P-value = P(t_{11} > .707) > .10
conclusion:
Do not reject H_o; there is not sufficient evidence to conclude that $\mu > 6$.
While the sample mean is above 6, the teacher cannot be 95% certain that the population mean is above 6.

30. Of the 50 films, 28 show the use of tobacco and are included in the sample.
summary statistics: n = 28, $\Sigma x = 2872$, $\Sigma x^2 = 694{,}918$, $\bar{x} = 102.57$, s = 121.77
original claim: $\mu = 120$ seconds
H_o: $\mu = 120$
H_1: $\mu \neq 120$
$\alpha = .05$ [assumed]
C.R. t < $-t_{27,.025}$ = -2.052
    t > $t_{27,.025}$ = 2.052
calculations:
$t_{\bar{x}} = (\bar{x} - \mu)/s_{\bar{x}}$
   = (102.57 - 120)/(121.77/$\sqrt{28}$)
   = -17.429/23.012
   = -.757
P-value = 2·P(t_{27} < -.757) > .20
conclusion:
Do not reject H_o; there is not sufficient evidence to conclude that $\mu \neq 120$.
Whether the claim is deceptive depends on what is meant by "given the sample data." Considering the sample mean, the claim is not deceptive. If $\mu = 120$, one would not expect to find $\bar{x} = 120$ in every sample – but values relatively near 120. The above $\bar{x} = 102.57$ is near enough that we expect values that far or farther from 120 more than 20% of the time when $\mu = 120$ is true. The claim could be considered deceptive, however, in that it refers to "movies that show the use of tobacco" without indicating whether that is 1% or 75% of the population – and most casual readers will mentally generalize the 2 minute figure to include all films [instead of the approximately 56% suggested by the data].

31. summary statistics: $n = 36$, $\Sigma x = 439.0$, $\Sigma x^2 = 5353.82$, $\bar{x} = 12.194$, $s = .1145$
    original claim: $\mu > 12$
    H_o: $\mu = 12$
    H_1: $\mu > 12$
    $\alpha = .01$
    C.R. $t > t_{35,.01} = 2.441$
    calculations:
    $t_{\bar{x}} = (\bar{x} - \mu)/s_{\bar{x}}$
    $= (12.194 - 12)/(.1145/\sqrt{36})$
    $= .1944/.01908$
    $= 10.189$
    P-value $= P(t_{35} > 10.189) < .005$
    conclusion:
    Reject H_o; there is sufficient evidence to conclude that $\mu > 12$.
    While the population mean appears to be above 12, the volume should not be reduced unless the standard deviation can be lowered. If $\mu \approx 12.194$ and $\sigma \approx .1145$ as the data suggests, then the limits $\mu \pm 2\sigma$ for "usual" occurrences already include values below 12 – i.e., it would not be unusual to get a can with less than 12 ounces. Reducing the volume would further increase the number of sub-standard cans, possibly to the point of serious legal or public relations problems.

32. summary statistics: $n = 16$, $\Sigma x = 104.8$, $\Sigma x^2 = 724.74$, $\bar{x} = 6.55$, $s = 1.598$
    original claim: $\mu > 6$
    H_o: $\mu = 6$
    H_1: $\mu > 6$
    $\alpha = .05$ [assumed]
    C.R. $t > t_{15,.05} = 1.753$
    calculations:
    $t_{\bar{x}} = (\bar{x} - \mu)/s_{\bar{x}}$
    $= (6.55 - 6)/(1.598/\sqrt{16})$
    $= .55/.39947$
    $= 1.377$
    $.05 < $ P-value $= P(t_{15} > 1.377) < .10$
    conclusion:
    Do not reject H_o; there is not sufficient evidence to conclude that $\mu > 6$.
    No, we cannot be 95% certain that cereal has excessive sodium content.

33. The P-value is changed. The summary values n, $\bar{x}$ and s do not change. The standard error of the mean, which is $s_{\bar{x}} = s/\sqrt{n}$, does not change. The calculated t, which is $t_{\bar{x}} = (\bar{x} - \mu)/s_{\bar{x}}$, does not change. Since the test is now two-tailed, the P-value doubles and is now $2 \cdot P(t_{14} > 1.954) = 2 \cdot (.035) = .070$. In this instance, changing the P-value changes the conclusion and (at the assumed .05 level of significance) there is not enough evidence to reject H_o: $\mu = 420$.

34. Because the z distribution has less spread than a t distribution, $z_\alpha < t_\alpha$ for any α. Since a smaller z score (in absolute value) is needed to reject H_o, rejection is more likely with z than with t.

35. The new test of hypothesis is given below. In this instance the presence of a drastic outlier outlier drastically changed the summery statistics but did not have much of an impact on the overall test of hypothesis. In general, however, the effects of an outlier can be substantial. One could argue that the effect is not so great in this exercise because the original test already involved an outlier: by the $\bar{x} \pm s$ standard, 5.40 is an "unusual" result – and it is so much larger than the others that it is the only value above $\bar{x}$.

summary statistics: n=6, $\Sigma x = 543.83$, $\Sigma x^2 = 291,603.3597$, $\bar{x} = 90.638$, s = 220.142
original claim: $\mu > 1.5$
H_o: $\mu = 1.5$
H_1: $\mu > 1.5$
$\alpha = .05$
C.R. $t > t_{5,.05} = 2.015$
calculations:
$t_{\bar{x}} = (\bar{x} - \mu)/s_{\bar{x}}$
  = $(90.638 - 1.5)/(220.142/\sqrt{6})$
  = $89.138/89.872$
  = .992

P-value = $P(t_5 > .992) > .10$
conclusion:
  Do not reject H_o; there is not sufficient evidence to conclude that $\mu > 1.5$.

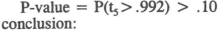

36. $A = z_{.05} \cdot [8 \cdot df + 3)/(8 \cdot df + 1)]$ $t = \sqrt{df \cdot (e^{A \cdot A/df} - 1)}$
    = $1.645 \cdot [8 \cdot 9 + 3)/(8 \cdot 9 + 1)]$ = $\sqrt{9 \cdot (e^{(1.960)(1.960)/9} - 1)}$
    = $1.645 \cdot [75/73]$ = $\sqrt{9 \cdot (e^{.317} - 1)}$
    = 1.690   = $\sqrt{9 \cdot (.3735)}$
                                                                = 1.833

    This agrees exactly with $t_{9,.05} = 1.833$ given in Table A-3.

37. original claim: $\mu > 40$
    H_o: $\mu = 40$
    H_1: $\mu > 40$
    $\alpha = .01$
    C.R. $t > t_{9,.01} = 2.821$
    calculations:
    $t_{\bar{x}} = (\bar{x} - \mu)/s_{\bar{x}}$
    $2.821 = (\bar{x} - 40)/(3.801/\sqrt{10})$
    $3.390 = \bar{x} - 40$
    $\bar{x} = 43.3899$

    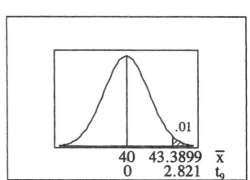

    β = P(failing to reject $H_o | H_o$ is false)
       = $P(\bar{x} < 43.3899 | \mu = 45.0518)$
       = $P(t_9 < -1.383)$
       = .10 [since $-t_{9,.10} = -1.383$]

    in the above calculation of β,
    $t_{\bar{x}} = (\bar{x} - \mu)/s_{\bar{x}}$
       = $(43.3899 - 45.0518)/(3.801/\sqrt{10})$
       = $-1.6619/1.202$
       = -1.383

    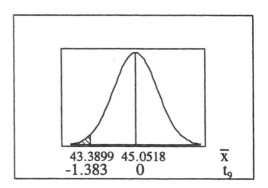

7-6 Testing a Claim about a Standard Deviation or Variance

NOTE: Following the pattern used with the z and t distributions, this manual uses the closest entry from Table A-4 for χ^2 as if it were the precise value necessary and does not use interpolation. This procedure sacrifices very little accuracy – and even interpolation does not yield precise values. When extreme accuracy is needed in practice, statisticians refer either to more accurate tables or to computer-produced values.

ADDITIONAL NOTE: The χ^2 distribution depends upon n, and it "bunches up" around df = n-1. In addition, the formula $\chi^2 = (n-1)s^2/\sigma^2$ used in the calculations contains df = n-1. When the exact df needed in the problem does not appear in the table and the closest χ^2 value is used to determine the critical region, some instructors recommend using the same df in the calculations that were used to determine the C.R. This manual typically uses the closest entry to determine the C.R. and the n from the problem in the calculations, even though this introduces a slight discrepancy.

1. H_o: $\sigma = 15$
    H_1: $\sigma \neq 15$
    $\alpha = .05$
    C.R. $\chi^2 < \chi^2_{19,.975} = 8.907$
    $\chi^2 > \chi^2_{19,.025} = 32.852$
    calculations:
    $\chi^2 = (n-1)s^2/\sigma^2$
    $ = (19)(10)^2/(15)^2$
    $ = 8.444$
    $.02 <$ P-value $< .05$

    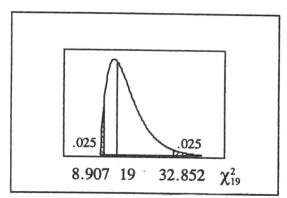

    conclusion:
    Reject H_o; there is sufficient evidence to conclude that $\sigma \neq 15$ (in fact, that $\sigma < 15$).

GRAPHICS NOTE: To illustrate χ^2 tests of hypotheses, this manual uses a "generic" figure, resembling a chi-squared distribution with approximately 4 degrees of freedom – chi-squared distributions with 1 and 2 degrees of freedom actually have no upper limit and approach the y axis asymptotically, while chi-squared distributions with more than 30 degrees of freedom are essentially symmetric and normal-looking. The expected value of the chi-squared distribution is the degrees of freedom for the problem, typically n-1. Since the distribution is positively skewed, this manual indicates that value slightly to the right of the figure's "peak." Loosely speaking, the distribution "bunches up" around the degrees of freedom.

2. H_o: $\sigma = 12$
    H_1: $\sigma > 12$
    $\alpha = .01$
    C.R. $\chi^2 > \chi^2_{4,.01} = 13.277$
    calculations:
    $\chi^2 = (n-1)s^2/\sigma^2$
    $ = (4)(18)^2/(12)^2$
    $ = 9.000$
    $.05 <$ P-value $< .10$

    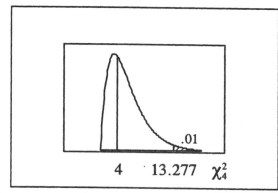

    conclusion:
    Do not reject H_o; there is not sufficient evidence to conclude that $\sigma > 12$.

3. H_o: $\sigma = 50$
    H_1: $\sigma < 50$
    $\alpha = .01$
    C.R. $\chi^2 < \chi^2_{29,.99} = 14.257$
    calculations:
    $\chi^2 = (n-1)s^2/\sigma^2$
    $ = (29)(30)^2/(50)^2$
    $ = 10.440$
    P-value $< .005$

    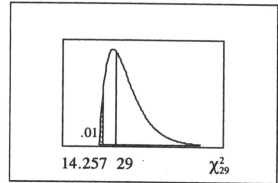

    conclusion:
    Reject H_o; there is sufficient evidence to conclude that $\sigma < 50$.

4. H_o: $\sigma = 4$
   H_1: $\sigma \neq 4$
   $\alpha = .05$
   C.R. $\chi^2 < \chi^2_{80,.975} = 57.153$
   $\chi^2 > \chi^2_{80,.025} = 106.629$
   calculations:
   $\chi^2 = (n-1)s^2/\sigma^2$
   $= (80)(4.7)^2/(4)^2$
   $= 110.450$
   $.02 < \text{P-value} < .05$

   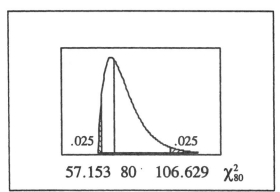

   conclusion:
   Reject H_o; there is sufficient evidence to conclude that $\sigma \neq 4$ (in fact, that $\sigma > 4$).

5. original claim: $\sigma > .04$
   H_o: $\sigma = .04$
   H_1: $\sigma > .04$
   $\alpha = .01$
   C.R. $\chi^2 > \chi^2_{39,.01} = 63.691$
   calculations:
   $\chi^2 = (n-1)s^2/\sigma^2$
   $= (39)(.31)^2/(.04)^2$
   $= 2342.4375$
   P-value $< .005$

   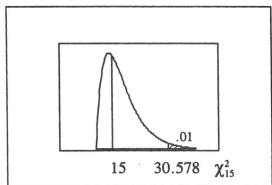

   conclusion:
   Reject H_o; there is sufficient evidence to conclude that $\sigma > .04$.
   Peanut M&M's vary more than plain M&M's since lack of uniformity in peanut sizes introduces a significant additional source of variation. Plain M&M's are produced to specifications with little variation from piece to piece. Peanut M&M's start with a peanut – not produced to specifications, but occurring with varying sizes. After rejecting very large and very small peanuts, there is still more variation than in a controlled process.

6. original claim: $\sigma < .0005$
   H_o: $\sigma = .0005$
   H_1: $\sigma < .0005$
   $\alpha = .05$
   C.R. $\chi^2 < \chi^2_{11,.95} = 4.575$
   calculations:
   $\chi^2 = (n-1)s^2/\sigma^2$
   $= (11)(.00047)^2/(.0005)^2$
   $= 9.7196$
   $.10 < \text{P-value} < .90$

   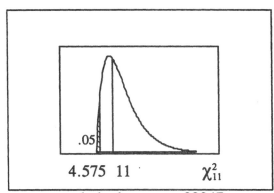

   conclusion:
   Do not reject H_o; there is not sufficient evidence to conclude that $\sigma < .00047$.

7. original claim: $\sigma \neq 43.7$
   H_o: $\sigma = 43.7$
   H_1: $\sigma \neq 43.7$
   $\alpha = .05$
   C.R. $\chi^2 < \chi^2_{80,.975} = 57.153$
   $\chi^2 > \chi^2_{80,.025} = 106.629$
   calculations:
   $\chi^2 = (n-1)s^2/\sigma^2$
   $= (80)(52.3)^2/(43.7)^2$
   $= 114.586$
   $.01 < \text{P-value} < .02$

   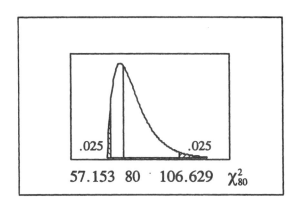

conclusion:
   Reject H_o; there is sufficient evidence to conclude that $\sigma \neq 43.7$ (in fact, that $\sigma > 43.7$). Since it appears that the standard deviation has increased, the new production method seems to be worse (at least so far as product consistency is concerned) than the old one.

8. original claim: $\sigma < 14.1$
   H_o: $\sigma = 14.1$
   H_1: $\sigma < 14.1$
   $\alpha = .01$
   C.R. $\chi^2 < \chi^2_{26,.99} = 12.198$
   calculations:
      $\chi^2 = (n-1)s^2/\sigma^2$
         $= (26)(9.3)^2/(14.1)^2$
         $= 11.311$
      $.005 <$ P-value $< .01$

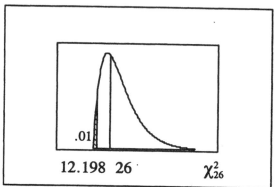

   conclusion:
      Reject H_o; there is sufficient evidence to conclude that $\sigma < 14.1$.
   No; a lower standard deviation does not suggest that the current class is doing better, but that they are more similar to each other and show less variation. The standard deviation speaks only about the spread of the scores and not their location.

9. original claim: $\sigma < 6.2$
   H_o: $\sigma = 6.2$
   H_1: $\sigma < 6.2$
   $\alpha = .05$
   C.R. $\chi^2 < \chi^2_{24,.95} = 13.848$
   calculations:
      $\chi^2 = (n-1)s^2/\sigma^2$
         $= (24)(3.8)^2/(6.2)^2$
         $= 9.016$
      P-value $< .005$

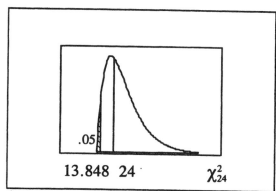

   conclusion:
      Reject H_o; there is sufficient evidence to conclude that $\sigma < 6.2$.
   Customers prefer waiting times with less variation because eliminating very long or very short waits depending only on which line is chosen creates a more predictable and more professional business climate. No; the smaller variability does not mean a smaller mean time.

10. original claim: $\sigma < 2.11$
    H_o: $\sigma = 2.11$
    H_1: $\sigma < 2.11$
    $\alpha = .005$
    C.R. $\chi^2 < \chi^2_{105,.005} = 67.328$
    calculations:
       $\chi^2 = (n-1)s^2/\sigma^2$
          $= (105)(.62)^2/(2.11)^2$
          $= 9.066$
       P-value $< .005$

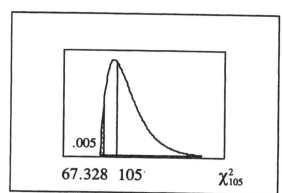

    conclusion:
       Reject H_o; there is sufficient evidence to conclude that $\sigma < 2.11$.
    It appears we can be very confident that $\sigma < 2.11$ and that the previous conclusion about the mean is valid.

NOTE: When dealing with raw data, use the calculated value of s^2 directly in the formula $\chi^2 = (n-1)s^2/\sigma^2$. It is good to take the square root to find the value of s to see whether it is larger or smaller than the hypothesized σ, but do not introduce unnecessary rounding errors by entering the value for s and then squaring it.

11. summary statistics: n = 9, Σx = 1089, Σx^2 = 132223, $\bar{x}$ = 121.0, s^2 = 56.75
    original claim: $\sigma < 29$
    H_o: $\sigma = 29$
    H_1: $\sigma < 29$
    $\alpha = .01$
    C.R. $\chi^2 < \chi^2_{8,.99} = 1.646$
    calculations:
    $\chi^2 = (n-1)s^2/\sigma^2$
    $= (8)(56.75)/(29)^2$
    $= .540$
    P-value < .005

    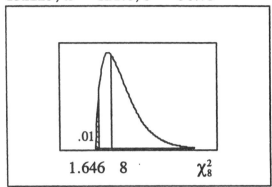

    conclusion:
    Reject H_o; there is sufficient evidence to conclude that $\sigma < 29$.

12. summary statistics: n = 18, Σx = 1259.0, Σx^2 = 88084.00, $\bar{x}$ = 69.944, s^2 = 1.408
    original claim: $\sigma < 2.5$
    H_o: $\sigma = 2.5$
    H_1: $\sigma < 2.5$
    $\alpha = .05$
    C.R. $\chi^2 < \chi^2_{17,.95} = 8.672$
    calculations:
    $\chi^2 = (n-1)s^2/\sigma^2$
    $= (17)(1.408)/(2.5)^2$
    $= 3.831$
    P-value < .005

    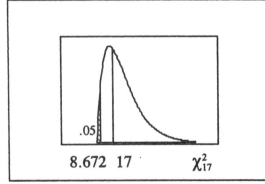

    conclusion:
    Reject H_o; there is sufficient evidence to conclude that $\sigma < 2.5$.

13. Usually we use the closest table entry -- or, when the desired df is exactly halfway between the closest two tabled values, we act conservatively and choose the smaller df. For this exercise only (and related exercise #17), use interpolation to estimate
    $\chi^2_{35,.95} = (18.493 + 26.509)/2 = 22.501$.
    summary statistics: n = 36, Σx = 442.5000, Σx^2 = 5439.35004672
    $\bar{x}$ = 12.2917, s^2 = .008216 [s = .0907]
    original claim: $\sigma < .10$
    H_o: $\sigma = .10$
    H_1: $\sigma < .10$
    $\alpha = .05$
    C.R. $\chi^2 < \chi^2_{35,.95} = 22.501$
    calculations:
    $\chi^2 = (n-1)s^2/\sigma^2$
    $= (35)(.008216)/(.10)^2$
    $= 28.755$
    .10 < P-value < .90

    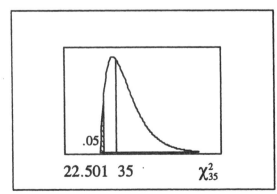

    conclusion:
    Do not reject H_o; there is not sufficient evidence to conclude that $\sigma < .10$.
    If the mean is less than 12 ounces, then the company is guilty of cheating the consumers. If the mean is greater than 12 ounces but the standard deviation is high, there still may be a significant number of individual cans containing less than 12 ounces.

14. summary statistics: $n = 40$, $\Sigma x = 4432$, $\Sigma x^2 = 502{,}488$, $\bar{x} = 111.0$, $s^2 = 292.882$
    original claim: $\sigma = 23.4$
    H_o: $\sigma = 23.4$
    H_1: $\sigma \neq 23.4$
    $\alpha = .05$ [assumed]
    C.R. $\chi^2 < \chi^2_{39,.975} = 24.433$
    $\chi^2 > \chi^2_{39,.025} = 59.342$
    calculations:
    $\chi^2 = (n-1)s^2/\sigma^2$
    $= (39)(292.882)/(23.4)^2$
    $= 20.861$
    $.01 < \text{P-value} < .02$

    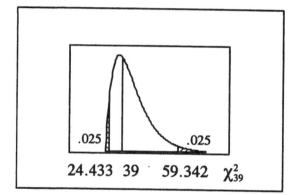

    conclusion:
      Reject H_o; there is sufficient evidence to conclude that $\sigma \neq 23.4$ (in fact, that $\sigma < 23.4$).

15. summary statistics: $n = 40$, $\Sigma x = 6902.0$, $\Sigma x^2 = 1{,}217{,}971.76$
    $\bar{x} = 172.55$, $s^2 = 693.119$ [$s = 26.327$]
    original claim: $\sigma = 28.7$
    H_o: $\sigma = 28.7$
    H_1: $\sigma \neq 28.7$
    $\alpha = .05$
    C.R. $\chi^2 < \chi^2_{39,.975} = 24.433$
    $\chi^2 > \chi^2_{39,.025} = 59.342$
    calculations:
    $\chi^2 = (n-1)s^2/\sigma^2$
    $= (39)(693.110)/(28.7)^2$
    $= 32.818$
    P-value $> .20$

    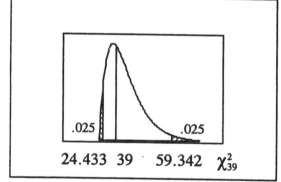

    conclusion:
      Do not reject H_o; there is not sufficient evidence to conclude that $\sigma \neq 28.7$. Underestimating the variation in the weight could lead to designing an elevator that handled the average expected weight per load, but that had a relatively high risk of being overloaded.

16. summary statistics: $n = 40$, $\Sigma x = 2527.8$, $\Sigma x^2 = 160{,}037.38$, $\bar{x} = 63.195$, $s^2 = 7.5413$
    original claim: $\sigma = 2.52$
    H_o: $\sigma = 2.52$
    H_1: $\sigma \neq 2.52$
    $\alpha = .05$
    C.R. $\chi^2 < \chi^2_{39,.975} = 24.433$
    $\chi^2 > \chi^2_{39,.025} = 59.342$
    calculations:
    $\chi^2 = (n-1)s^2/\sigma^2$
    $= (39)(7.5413)/(2.52)^2$
    $= 46.148$
    P-value $> .20$
    conclusion:
      Do not reject H_o; there is not sufficient evidence to conclude that $\sigma \neq 2.52$. Underestimating the variation in height could lead to designing car seats unsuitable for too many women at the extremes – i.e., the proportion of women too short or to tall for the seat could be unacceptably large.

17. Usually we use the closest table entry – or, when the desired df is exactly halfway between the closest two tabled values, we act conservatively and choose the smaller df. For this exercise only (and related exercise #13), use interpolation to estimate
    $\chi^2_{35,.95} = (18.493 + 26.509)/2 = 22.501$.

original claim: $\sigma < .10$
H_o: $\sigma = .10$
H_1: $\sigma < .10$
$\alpha = .05$
C.R. $\chi^2 < \chi^2_{35,.95} = 22.501$
calculations:
$(n-1)s^2/\sigma^2 < 22.501$
$(35)s^2/(.10)^2 < 22.501$
$s^2 < .006428$
$s < .0802$
The largest s that would lead to rejection of H_o (and support of $\sigma < .10$) is $s = .08$

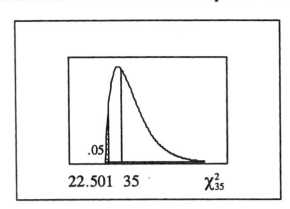

18. a. lower $\chi^2 = \frac{1}{2}(-z_{.025} + \sqrt{2 \cdot df - 1})^2$
    $= \frac{1}{2}(-1.96 + \sqrt{2 \cdot (100) - 1})^2$
    $= \frac{1}{2}(-1.96 + \sqrt{199})^2$
    $= \frac{1}{2}(147.543)$
    $= 73.7722$ [compare to $\chi^2_{100,.975} = 74.222$ from Table A-4]
    upper $\chi^2 = \frac{1}{2}(z_{.025} + \sqrt{2 \cdot df - 1})^2$
    $= \frac{1}{2}(1.96 + \sqrt{2 \cdot (100) - 1})^2$
    $= \frac{1}{2}(1.96 + \sqrt{199})^2$
    $= \frac{1}{2}(258.140)$
    $= 129.070$ [compare to $\chi^2_{100,.025} = 129.561$ from Table A-4]
   b. lower $\chi^2 = \frac{1}{2}(-z_{.025} + \sqrt{2 \cdot df - 1})^2$
    $= \frac{1}{2}(-1.96 + \sqrt{2 \cdot (149) - 1})^2$
    $= \frac{1}{2}(-1.96 + \sqrt{297})^2$
    $= \frac{1}{2}(233.286)$
    $= 116.643$
    upper $\chi^2 = \frac{1}{2}(z_{.025} + \sqrt{2 \cdot df - 1})^2$
    $= \frac{1}{2}(1.96 + \sqrt{2 \cdot (149) - 1})^2$
    $= \frac{1}{2}(1.96 + \sqrt{297})^2$
    $= \frac{1}{2}(368.398)$
    $= 184.199$

19. a. lower $\chi^2 = df \cdot [1 - 2/(9 \cdot df) - z_{.025}\sqrt{2/(9 \cdot df)}]^3$
    $= 100 \cdot [1 - 2/(9 \cdot 100) - 1.96\sqrt{2/(9 \cdot 100)}]^3$
    $= 100 \cdot [.9054]^3$
    $= 74.216$ [compare to $\chi^2_{100,.975} = 74.222$ from Table A-4]
    upper $\chi^2 = df \cdot [1 - 2/(9 \cdot df) + z_{.025}\sqrt{2/(9 \cdot df)}]^3$
    $= 100 \cdot [1 - 2/(9 \cdot 100) + 1.96\sqrt{2/(9 \cdot 100)}]^3$
    $= 100 \cdot [1.0902]^3$
    $= 129.565$ [compare to $\chi^2_{100,.025} = 129.561$ from Table A-4]
   b. lower $\chi^2 = df \cdot [1 - 2/(9 \cdot df) - z_{.025}\sqrt{2/(9 \cdot df)}]^3$
    $= 149 \cdot [1 - 2/(9 \cdot 149) - 1.96\sqrt{2/(9 \cdot 149)}]^3$
    $= 149 \cdot [.9228]^3$
    $= 117.093$

upper χ^2 = df·[1 - 2/(9·df) - $z_{.025}\sqrt{2/(9 \cdot df)}$]³
 = 149·[1 - 2/(9·149) - 1.96$\sqrt{2/(9 \cdot 149)}$]³
 = 149·[1.0742]³
 = 184.690

20. Yes, although the effect may not be "dramatic" if the sample is extremely large. Because this section tests the "spread" of scores, and by definition an outlier is one that is "spread" away from the others, the presence of an outlier directly affects the property being measured and tested.

21. a. If there are disproportionately more 0's and 5's, there would be an overabundance of digits less than 6 – i.e., at the lower end of the scale. This would reduce the mean to something less than 4.5 and (since most of the digits are bunched within these narrower limits) reduce the standard deviation to less than 3.
 b. If H_o were true and $\sigma = 3$, we would expect a uniform distribution of digits; if H_1 were true and $\sigma < 3$, we would expect a distribution of digits that peaks at 0 and/or 5 and is positively skewed. In either case, the data would not be coming from a normal distribution – which is one of the assumptions of the test.

22. original claim: $\sigma < 6.2$
H_o: $\sigma = 6.2$
H_1: $\sigma < 6.2$
$\alpha = .05$
C.R. $\chi^2 < \chi^2_{24,.95} = 13.848$
calculations:
$\chi^2 = (n-1)s^2/\sigma^2$
13.848 = $(24)s^2/(6.2)^2$
532.317 = $(24)s^2$
22.180 = s^2

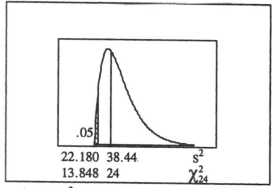

In terms of s^2, then, the expected value of the sampling distribution is $(6.2)^2 = 38.44$ critical region is $s^2 < 22.180$.

β = P(not rejecting $H_o | H_o$ is false)
 = P($s^2 > 22.180 | \sigma = 4$)
 = P($\chi^2_{24} > 33.270$)
 = .10 [from Table A-4, closest entry =33.196]

The above value was obtained as follows:
If $\sigma = 4$ is true, the sampling distribution at the right applies and $\chi^2 = (n-1)s^2/\sigma^2$
 = (24)(22.180)/(16)
 = 33.270

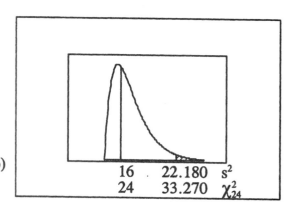

Review Exercises

1. a. No; because the sample consisted only of those who chose to reply, the respondents are not necessarily representative of the general population.
 b. No; While the 0.2 lb loss is statistically significant, it is not of practical significance. In truth, there is essentially no chance that any H_o is <u>exactly</u> true to an infinite number of decimal places – and so given a large enough sample, virtually any H_o will be rejected. When the sample is so large that it is able to detect minute differences of no useful value, some statisticians call this "an insignificant significant difference."

c. Choose .001, the smallest of the suggested P-values. This corresponds to the smallest chance that such results could have occurred by chance alone, and hence gives the most support to the conclusion that the cure was effective.

d. Do not reject $H_o: \mu = 12$, there is not sufficient evidence to conclude that $\mu > 12$.

e. "...rejecting the null hypothesis when it is true."

2. a. original claim: $\mu < 10,000 \Rightarrow H_o: \mu = 10,000$ and $H_1: \mu < 10,000$
   asks about μ, with σ unknown and $n > 30$: use the t distribution [with df=749]

   b. original claim: $\sigma > 1.8 \Rightarrow H_o: \sigma = 1.8$ and $H_1: \sigma < 1.8$
   asks about σ: assuming the population is approximately normal, use the χ^2 distribution

   c. original claim: $p > .50 \Rightarrow H_o: p = .50$ and $H_1: p > .50$
   asks about p, with $np \geq 5$ and $n(1-p) \geq 5$: use the z distribution

   d. original claim: $\mu = 100 \Rightarrow H_o: \mu = 100$ and $H_1: \mu \neq 100$
   asks about μ, with σ known and $n > 30$: use the z distribution

3. a. original claim: $\mu = 100$
   $\bar{x} = 98.4$
   $H_o: \mu = 100$
   $H_1: \mu \neq 100$
   $\alpha = .10$
   C.R. $z < -z_{.05} = -1.645$
   $z > z_{.05} = 1.645$
   calculations:
   $z_{\bar{x}} = (\bar{x} - \mu)/\sigma_{\bar{x}}$
   $= (98.4 - 100)/(15/\sqrt{50})$
   $= -1.6/2.121 = -.75$
   P-value $= 2 \cdot P(z < -.75) = 2 \cdot (.2266) = .4532$
   conclusion:
   Do not reject H_o; there is not sufficient evidence reject the claim that $\mu = 100$.

   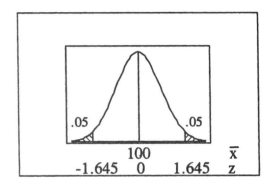

   b. original claim: $\mu = 100$
   $H_o: \mu = 100$
   $H_1: \mu \neq 100$
   $\alpha = .10$
   C.R. $t < -t_{49,.05} = -1.676$
   $t > t_{49,.05} = 1.676$
   calculations:
   $t_{\bar{x}} = (\bar{x} - \mu)/s_{\bar{x}}$
   $= (98.4 - 100)/(16.3/\sqrt{50})$
   $= -1.6/2.305 = -.694$
   P-value $> .20$
   conclusion:
   Do not reject H_o; there is not sufficient evidence to reject the claim that $\mu = 100$.

   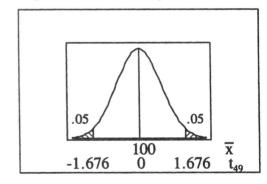

   c. original claim: $\sigma = 15$
   $H_o: \sigma = 15$
   $H_1: \sigma \neq 15$
   $\alpha = .10$
   C.R. $\chi^2 < \chi^2_{49,.95} = 34.764$
   $\chi^2 > \chi^2_{49,.05} = 67.505$
   calculations:
   $\chi^2 = (n-1)s^2/\sigma^2$
   $= (49)(16.3)^2/(15)^2$
   $= 57.861$
   P-value $> .20$
   conclusion:
   Do not reject H_o; there is not sufficient evidence to reject the claim that $\sigma = 15$.

   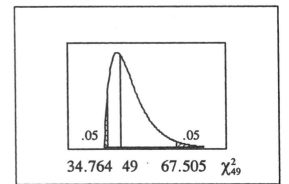

d. Yes; based on the preceding results, the random number generator appears to be working correctly.

4. original claim: p < .50
$\hat{p} = x/n = x/15 = = .44$
H_o: p = .50
H_1: p < .50
$\alpha = .05$
C.R. $z < -z_{.05} = -1.645$
calculations:
$z_{\hat{p}} = (\hat{p} - \mu_{\hat{p}})/\sigma_{\hat{p}}$
$= (.44 - .50)/\sqrt{(.50)(.50)/150}$
$= -.06/.0408$
$= -1.47$
P-value = P(z < -1.47) = .0708

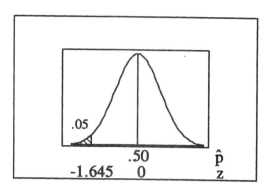

conclusion:
Do not reject H_o; there is not sufficient evidence to conclude that p < .50.

5. original claim: $\mu = 5.670$
H_o: $\mu = 5.670$
H_1: $\mu \neq 5.670$
$\alpha = .01$
C.R. $t < -t_{49,.005} = -2.678$
   $t > t_{49,.005} = 2.678$
calculations:
$t_{\bar{x}} = (\bar{x} - \mu)/s_{\bar{x}}$
$= (5.622 - 5.670)/(.068/\sqrt{50})$
$= -.048/.009617$
$= -4.991$
P-value < .01

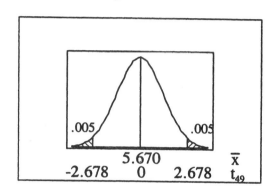

conclusion:
Reject H_o; there is sufficient evidence to reject the claim that $\mu = 5.670$ and conclude that $\mu \neq 5.670$ (in fact, that $\mu < 5.670$).
The quarters may have had a mean weight of 5.670 g when they were minted, but wear from being in circulation will reduce the weight of coins.

6. There are n=5 blue M%M's listed in Data Set 19.
original claim: $\mu \geq .9085$
H_o: $\mu = .9085$
H_1: $\mu < .9085$
$\alpha = .05$
C.R. $t < -t_{4,.05} = -2.132$
calculations:
$t_{\bar{x}} = (\bar{x} - \mu)/s_{\bar{x}}$
$= (.9014 - .9085)/(.0573/\sqrt{5})$
$= -.0071/.0256$
$= -.277$
P-value > .10

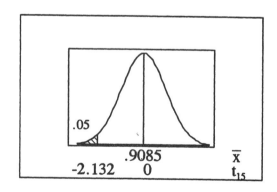

conclusion:
Do not reject H_o; there is not sufficient evidence to reject the claim that $\mu \geq .9085$.
No, this result does not provide evidence of such a discrepancy.

7. original claim: $p < .10$
   $\hat{p} = x/n = 111/1233 = .090$
   H_o: $p = .10$
   H_1: $p < .10$
   $\alpha = .05$
   C.R. $z < -z_{.05} = -1.645$
   calculations:
   $z_{\hat{p}} = (\hat{p} - \mu_{\hat{p}})/\sigma_{\hat{p}}$
   $= (.090 - .10)/\sqrt{(.10)(.90)/1233}$
   $= -.010/.00854$
   $= -1.17$
   P-value $= P(z < -1.17) = .1210$
   conclusion:
   Do not reject H_o; there is not sufficient evidence to conclude that $p < .10$.
   She can use that claim if she is willing to run a 12.10% risk of being wrong.

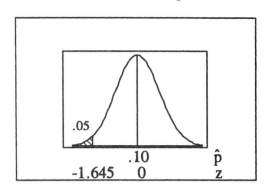

8. original claim: $p = .43$
   $\hat{p} = x/n = 308/611 = .504$
   H_o: $p = .43$
   H_1: $p \neq .43$
   $\alpha = .04$
   C.R. $z < -z_{.02} = -2.05$
   $z > z_{.02} = 2.05$
   calculations:
   $z_{\hat{p}} = (\hat{p} - \mu_{\hat{p}})/\sigma_{\hat{p}}$
   $= (.0504 - .43)/\sqrt{(.43)(.57)/611}$
   $= .074/.0200$
   $= 3.70$
   P-value $= 2 \cdot P(z > 3.70) = 2 \cdot [1 - P(z < 3.70)] = 2 \cdot [1 - .9999] = 2 \cdot [.0001] = .0002$
   conclusion:
   Reject H_o; there is sufficient evidence to reject the claim that $p = .43$ and conclude that $p \neq .43$ (in fact, that $p > .43$).
   The result indicates that the proportion who say they voted for the winner is greater than proportion that actually did so vote. It appears that people are not responding honestly to the question – and this should raise a concern about the honesty of respondents in general, in situations where there is no independent reliable knowledge about the true value.

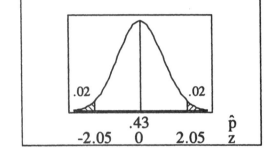

9. original claim: $\mu < 12$
   H_o: $\mu = 12$
   H_1: $\mu < 12$
   $\alpha = .05$ [assumed]
   C.R. $t < -t_{23,.05} = -1.714$
   calculations:
   $t_{\bar{x}} = (\bar{x} - \mu)/s_{\bar{x}}$
   $= (11.4 - 12)/(.62/\sqrt{24})$
   $= -.6/.1266$
   $= -4.741$
   P-value $< .005$
   conclusion:
   Reject H_o; there is sufficient evidence to conclude that $\mu < 12$.

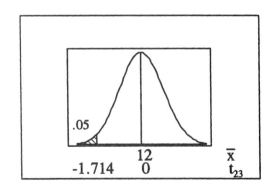

No; assuming that proper random sampling techniques were used, his claim that the sample is too small to be meaningful is not valid. The t distribution adjusts for the various sample sizes to produce a valid test for any sample size n.

10. original claim: p < .10
    $\hat{p} = x/n = x/1248 = .08$
    H_o: p = .10
    H_1: p < .10
    $\alpha = .01$
    C.R. $z < -z_{.01} = -2.326$
    calculations:
     $z_{\hat{p}} = (\hat{p} - \mu_{\hat{p}})/\sigma_{\hat{p}}$
       $= (.08 - .10)/\sqrt{(.10)(.90)/1248}$
       $= -.02/.00849$
       $= -2.36$
    P-value = P(z < -2.36) = .0091

    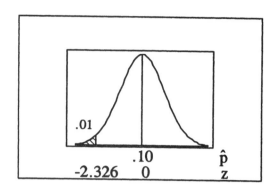

    conclusion:
     Reject H_o; there is sufficient evidence to conclude that p < .10.
    Yes; based on this result, the phrase "almost 1 out of 10" is justified. While "almost" is not well-defined, 10% is the closest round number to 8%. The test indicates we are confident that p < .10, but does not specify how much less.
    NOTE: x=(.08)(1248)=99.84; any 94≤x≤106 rounds to 8%. Using the upper possibility $\hat{p} = x/n = 106/1248 = .0849$ $z_{\hat{p}} = -1.774$, which is not enough evidence to reject H_o: p=.10. This not only supports the opinion that "almost 1 out of 10" is justified, but it also shows how sensitive tests may be to rounding procedures – and that in this case the results are actually not reported with sufficient accuracy to conduct the test.

11. original claim: $\sigma < .15$
    H_o: $\sigma = .15$
    H_1: $\sigma < .15$
    $\alpha = .05$
    C.R. $\chi^2 < \chi^2_{70,.95} = 51.739$
    calculations:
     $\chi^2 = (n-1)s^2/\sigma^2$
       $= (70)(.12)^2/(.15)^2$
       $= 44.800$
     $.005 < $ P-value $ < .01$

    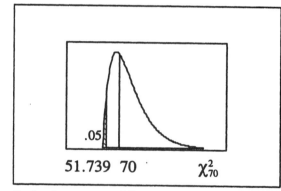

    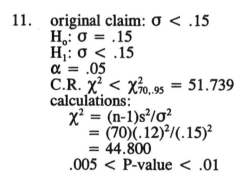

    conclusion:
     Reject H_o; there is sufficient evidence to conclude that $\sigma < .15$.
    Yes; assuming the costs and any other factors do not suggest otherwise, the company should consider purchasing the machine.

12. summary statistics: n = 70, $\Sigma x = 251.023$, $\Sigma x^2 = 900.557440$, $\bar{x} = 3.586$, s = .0740
    original claim: $\mu = 3.5$
    H_o: $\mu = 3.5$
    H_1: $\mu \neq 3.5$
    $\alpha = .05$ [assumed]
    C.R. $t < -t_{69,.025} = -1.994$
        $t > t_{69,.025} = 1.994$
    calculations:
     $t_{\bar{x}} = (\bar{x} - \mu)/s_{\bar{x}}$
       $= (3.586 - 3.5)/(.0740/\sqrt{70})$
       $= .086/.008849$
       $= 9.724$
    P-value < .01

    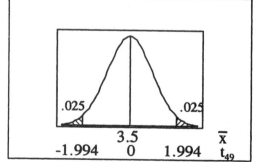

    conclusion:
     Reject H_o; there is sufficient evidence to reject the claim that $\mu = 3.5$ and conclude that $\mu \neq 3.5$ (in fact, that $\mu > 3.5$).
    It appears the true mean weight is greater that the 3.5 g claimed on the label.

Cumulative Review Exercises

1. scores in order of magnitude: .018 .0268 .0281 .0320 .0440 .0524 .161 .175 .176
    summary statistics: n = 9, Σx = .7133, Σx^2 = .09505961
    a. $\bar{x} = (\Sigma x)/n = .7133/9 = .0793$
    b. $\tilde{x} = .044$
    c. $s^2 = [n(\Sigma x^2) - (\Sigma x)^2]/[n(n-1)] = [9(.09505961) - (.7133)^2]/[9(8)] = .004816$
       $s = .0694$
    d. $s^2 = .004816$
    e. R = .176 - .018 = .158
    f. $\bar{x} = \pm t_{8,.025} s_{\bar{x}}$
       .0793 $\pm$ 2.306·(.0694/$\sqrt{9}$)
       .0793 $\pm$ .0533
       .0259 < μ < .1326
    g. original claim: μ < .16
       H_o: μ = .16
       H_1: μ < .16
       α = .05
       C.R. t < $-t_{8,.05}$ = -1.860
       calculations:
       $t_{\bar{x}} = (\bar{x} - \mu)/s_{\bar{x}}$
           = (.0793 - .16)/(.0694/$\sqrt{9}$)
           = -.0807/.0231 = -3.491
       P-value < .005

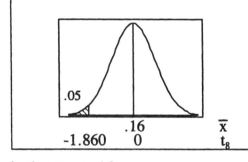

       conclusion:
       Reject H_o; there is sufficient evidence to conclude that μ < .16.
    h. Yes; the values were listed in chronological order, and there seems to be a dereasing trend over time.

2. a. normal distribution
       μ = 496
       σ = 108
       P(x > 500)
           = P(z > .04)
           = 1 - P(z < .04)
           = 1 - .5160
           = .4840

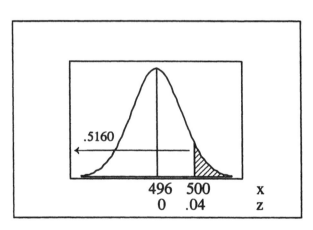

    b. let A = a selected score is above 500
       P(A) = .4840, for each selection
       P(A_1 and A_2 and A_3 and A_4 and A_5)
           = P(A_1)·P(A_2)·P(A_3)·P(A_4)·P(A_5)
           = $(.4840)^5$ = .0266

    c. normal distribution,
           since the original distribution is so
       $\mu_{\bar{x}} = \mu = 496$
       $\sigma_{\bar{x}} = \sigma/\sqrt{n} = 108/\sqrt{5} = 48.30$
       P($\bar{x}$ > 500)
           = P(z > .08)
           = 1 - P(z < .08)
           = 1 - .5319
           = .4681

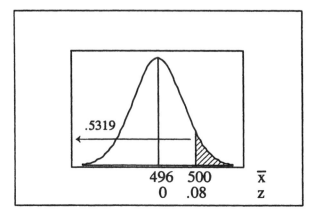

d. For P_{90}, $A = .9000$ [.8997] and $z = 1.28$.
   [or, $z_{.10} = t_{large,.10} = 1.282$]
   $x = \mu + z \cdot \sigma$
   $= 496 + (1.282)(108)$
   $= 496 + 138$
   $= 634$

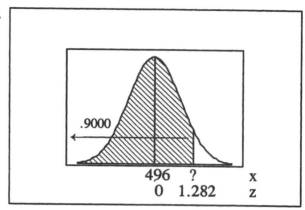

3. binomial distribution: $n = 25$ and $p = .25$
   a. $\mu = np = 25(.25) = 6.25$
   b. $\sigma = \sqrt{np(1-p)} = \sqrt{25(.25)(.75)} = 2.165$
   c. normal approximation appropriate since
      $np = 25(.25) = 6.25 \geq 5$
      $n(1-p) = 25(.75) = 18.75 \geq 5$
      $P(x > 12) = P_c(x > 12.5)$
      $= P(z > 2.89)$
      $= 1 - P(z < 2.89)$
      $= 1 - .9981$
      $= .0019$

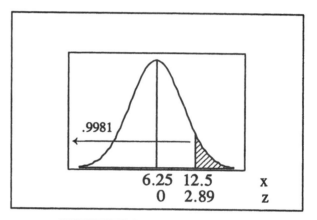

   d. original claim: $p = .25$
      $\hat{p} = x/n = (\text{more than } 12)/25 = 13/25 = .52$
      $H_o: p = .25$
      $H_1: p \neq .25$
      $\alpha = .05$
      C.R. $z < -z_{.025} = -1.96$
           $z > z_{.025} = 1.96$
      calculations:
      $z_{\hat{p}} = (\hat{p} - \mu_{\hat{p}})/\sigma_{\hat{p}}$
      $= (.52 - .25)/\sqrt{(.25)(.75)/25}$
      $= .27/.0866$
      $= 3.12$

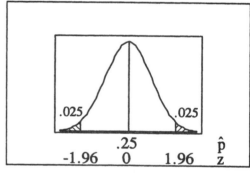

      P-value $= 2 \cdot P(z > 3.12) = 2 \cdot [1 - P(z < 3.12)] = 2 \cdot [1 - .9991] = 2 \cdot [.0009] = .0018$
      conclusion:
         Reject H_o; there is sufficient evidence to reject the claim that $p = .25$ and conclude that $p \neq .25$ (in fact, that $p > .25$).
      NOTE: The above test of hypothesis follows the form and notation used in this chapter. One could also consider the .0019 of part (c) above, calculate P-value $= 2 \cdot (.0019) = .0038$, and reject H_o because $.0038 < .05$. The difference in P-values arises because (c) uses the correction for continuity while the section on testing hypotheses about p and part (d) ignore the correction.
   e. $\hat{p}$ unknown, use $\hat{p} = .5$
      $n = [(z_{.05})^2 \hat{p}\hat{q}]/E^2 = [(1.645)^2(.5)(.5)]/(.04)^2 = 422.8$, rounded up to 423

Chapter 8

Inferences from Two Samples

8-2 Inferences about Two Proportions

NOTE: To be consistent with the notation of the previous chapters, reinforcing the patterns and concepts presented there, the manual uses the "usual" z formula written to apply to $\hat{p}_1-\hat{p}_2$'s:

$z_{\hat{p}_1-\hat{p}_2} = (\hat{p}_1-\hat{p}_2 - \mu_{\hat{p}_1-\hat{p}_2})/\sigma_{\hat{p}_1-\hat{p}_2}$

with $\mu_{\hat{p}_1-\hat{p}_2} = p_1 - p_2$ and $\sigma_{\hat{p}_1-\hat{p}_2} = \sqrt{\bar{p}\cdot\bar{q}/n_1 + \bar{p}\cdot\bar{q}/n_2}$ [when H_o includes $p_1=p_2$]
where $\bar{p} = (x_1 + x_2)/(n_1 + n_2)$

And so the formula for the z statistic may also be written as

$z_{\hat{p}_1-\hat{p}_2} = ((\hat{p}_1-\hat{p}_2) - (p_1-p_2))/\sqrt{\bar{p}\cdot\bar{q}/n_1 + \bar{p}\cdot\bar{q}/n_2}$

1. $x = \hat{p}\cdot n = .81\cdot 37 = 29.97$, rounded to 30

2. $x = \hat{p}\cdot n = .63\cdot 240 = 151.2$, rounded to 151 [although 152/240 also rounds to 63%]

3. $x = \hat{p}\cdot n = .289\cdot 294 = 84.966$, rounded to 85

4. $x = \hat{p}\cdot n = .24\cdot 205 = 49.2$, rounded to 49 [although 50/205 also rounds to 24%]

5. $\hat{p}_1 = x_1/n_1 = 192/436 = .440$ $\hat{p}_2 = x_2/n_2 = 40/121 = .331$
   $\hat{p}_1-\hat{p}_2 = .440 - .331 = .110$
   a. $\bar{p} = (x_1 + x_2)/(n_1 + n_2) = (192 + 40)/(436 + 121) = 232/557 = .417$
   b. $z_{\hat{p}_1-\hat{p}_2} = (\hat{p}_1-\hat{p}_2 - \mu_{\hat{p}_1-\hat{p}_2})/\sigma_{\hat{p}_1-\hat{p}_2}$
      $= (.110 - 0)/\sqrt{(.417)(.583)/436 + (.417)(.583)/121}$
      $= .100/.0507 = 2.17$
   c. $\pm z_{.025} = \pm 1.960$
   d. P-value $= 2\cdot P(z>2.17) = 2\cdot[1 - P(z<2.17)] = 2\cdot[1 - .9850] = 2\cdot[.0150] = .0300$

NOTE: Since $\bar{p}$ is the weighted average of $\hat{p}_1$ and $\hat{p}_2$, it must always fall between those two values. If it does not, then an error has been made that must be corrected before proceeding. Calculation of $\sigma_{\hat{p}_1-\hat{p}_2} = \sqrt{\bar{p}\cdot\bar{q}/n_1 + \bar{p}\cdot\bar{q}/n_2}$ can be accomplished with no round-off loss on most calculators by calculating $\bar{p}$ and proceeding as follows: STORE 1-RECALL = * RECALL = STORE RECALL ÷ n_1 + RECALL ÷ n_2 = √. The quantity $\sigma_{\hat{p}_1-\hat{p}_2}$ may then be STORED for future use. Each calculator is different -- learn how your calculator works, and do the homework on the same calculator you will use for the exam. If you have any questions about performing/storing calculations on your calculator, check with your instructor or class assistant.

6. $\hat{p}_1 = x_1/n_1 = 101/10239 = .00986$ $\hat{p}_2 = x_2/n_2 = 56/9877 = .00567$
   $\hat{p}_1-\hat{p}_2 = .00986 - .00567 = .00419$
   a. $\bar{p} = (x_1 + x_2)/(n_1 + n_2) = (101 + 56)/(10239 + 9877) = 157/20116 = .00780$
   b. $z_{\hat{p}_1-\hat{p}_2} = (\hat{p}_1-\hat{p}_2 - \mu_{\hat{p}_1-\hat{p}_2})/\sigma_{\hat{p}_1-\hat{p}_2}$
      $= (.00419 - 0)/\sqrt{(.00780)(.99220)/10239 + (.00780)(.99220)/9877}$
      $= .00419/.00124 = 3.38$
   c. $\pm z_{.025} = \pm 1.960$
   d. P-value $= 2\cdot P(z>3.38) = 2\cdot[1 - P(z<3.38)] = 2\cdot[1 - .9996] = 2\cdot[.0004] = .0008$

7. Let the employees be group 1.
   original claim: $p_1-p_2 > 0$
   $\hat{p}_1 = x_1/n_1 = 192/436 = .440$
   $\hat{p}_2 = x_2/n_2 = 40/121 = .331$
   $\hat{p}_1-\hat{p}_2 = .440 - .331 = .110$
   $\bar{p} = (x_1 + x_2)/(n_1 + n_2)$
   $\phantom{\bar{p}} = (192 + 40)/(436 + 121)$
   $\phantom{\bar{p}} = 232/557 = .417$
   H_o: $p_1-p_2 = 0$
   H_1: $p_1-p_2 > 0$
   $\alpha = .05$
   C.R. $z > z_{.05} = 1.645$
   calculations:
   $z_{\hat{p}_1-\hat{p}_2} = (\hat{p}_1-\hat{p}_2 - \mu_{\hat{p}_1-\hat{p}_2})/\sigma_{\hat{p}_1-\hat{p}_2}$
   $\phantom{z_{\hat{p}_1-\hat{p}_2}} = (.110 - 0)/\sqrt{(.417)(.583)/436 + (.417)(.583)/121}$
   $\phantom{z_{\hat{p}_1-\hat{p}_2}} = .110/.0507 = 2.17$
   P-value = $P(z>2.17) = 1 - P(z<2.17) = 1 - .9850 = .0150$
   conclusion:
   Reject H_o; there is sufficient evidence to conclude that $p_1-p_2 > 0$.

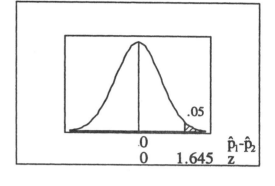

8. Refer to the data and calculations in exercise #7.
   $(\hat{p}_1-\hat{p}_2) \pm z_{.05}\sqrt{\hat{p}_1\hat{q}_1/n_1 + \hat{p}_2\hat{q}_2/n_2}$
   $.1098 \pm 1.645\cdot\sqrt{(.440)(.560)/436 + (.331)(.669)/121}$
   $.1098 \pm .0805$
   $.029 < p_1-p_2 < .190$
   Yes; since the interval does not contain zero, there is a significant difference between the employees and their bosses.

9. Let the women with low activity be group 1.
   $\hat{p}_1 = x_1/n_1 = 101/10239 = .00986$ $\hat{p}_2 = x_2/n_2 = 56/9877 = .00567$
   $\hat{p}_1-\hat{p}_2 = .00986 - .00567 = .00419$
   $(\hat{p}_1-\hat{p}_2) \pm z_{.05}\sqrt{\hat{p}_1\hat{q}_1/n_1 + \hat{p}_2\hat{q}_2/n_2}$
   $.00419 \pm 1.645\cdot\sqrt{(.00986)(.99014)/10239 + (.00567)(.99433)/9877}$
   $.00419 \pm .00203$
   $.00216 < p_1-p_2 < .00623$
   Yes; since the confidence interval does not include 0, there does seem to be a significant difference – and that physical activity corresponds to a lower rate of the disease. Whether the difference is "substantial" is subjective, as the difference is only about 0.4%.

10. Refer to the data and calculations in exercise #9.
    original claim: $p_1-p_2 > 0$
    $\bar{p} = (x_1 + x_2)/(n_1 + n_2)$
    $\phantom{\bar{p}} = (101 + 56)/(10239 + 9877)$
    $\phantom{\bar{p}} = 157/20116 = .00780$
    H_o: $p_1-p_2 = 0$
    H_1: $p_1-p_2 > 0$
    $\alpha = .05$
    C.R. $z > z_{.05} = 1.645$
    calculations:
    $z_{\hat{p}_1-\hat{p}_2} = (\hat{p}_1-\hat{p}_2 - \mu_{\hat{p}_1-\hat{p}_2})/\sigma_{\hat{p}_1-\hat{p}_2}$
    $\phantom{z_{\hat{p}_1-\hat{p}_2}} = (.00419 - 0)/\sqrt{(.00780)(.99220)/10239 + (.00780)(.99220)/9877}$
    $\phantom{z_{\hat{p}_1-\hat{p}_2}} = .00419/.00124 = 3.38$
    P-value = $P(z>3.38) = 1 - P(z<3.38) = 1 - .9996 = .0004$
    conclusion:
    Reject H_o; there is sufficient evidence to conclude that $p_1-p_2 > 0$.

11. Let the 2000 season be group 1.
original claim: $p_1-p_2 = 0$
$\hat{p}_1 = x_1/n_1 = 83/247 = .336$
$\hat{p}_2 = x_2/n_2 = 89/258 = .345$
$\hat{p}_1-\hat{p}_2 = .336 - .345 = -.00893$
$\bar{p} = (x_1 + x_2)/(n_1 + n_2)$
$= (83 + 89)/(247 + 258)$
$= 172/505 = .341$
H_o: $p_1-p_2 = 0$
H_1: $p_1-p_2 \neq 0$
$\alpha = .05$ [assumed]
C.R. $z < -z_{.025} = -1.96$
$z > z_{.025} = 1.96$

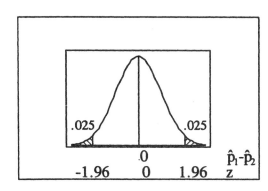

calculations:
$z_{\hat{p}_1-\hat{p}_2} = (\hat{p}_1-\hat{p}_2 - \mu_{\hat{p}_1-\hat{p}_2})/\sigma_{\hat{p}_1-\hat{p}_2}$
$= (-.00893 - 0)/\sqrt{(.341)(.659)/247 + (.341)(.659)/258}$
$= -.00893/.0422 = -.21$
P-value $= 2 \cdot P(z<-.21) = 2 \cdot (.4168) = .8336$
conclusion:
Do not reject H_o; there is not sufficient evidence to reject the claim that $p_1-p_2 = 0$.
Yes, the reversal rate appears to be the same for both years.

12. Let the employees of hospitals with the smoking ban be group 1.
original claim: $p_1-p_2 = 0$
$\hat{p}_1 = x_1/n_1 = 56/843 = .0664$
$\hat{p}_2 = x_2/n_2 = 27/703 = .0384$
$\hat{p}_1-\hat{p}_2 = .0664 - .0384 = .0280$
$\bar{p} = (x_1 + x_2)/(n_1 + n_2)$
$= (56 + 27)/(843 + 703)$
$= 83/1546 = .0537$
H_o: $p_1-p_2 = 0$
H_1: $p_1-p_2 \neq 0$
$\alpha = .05$
C.R. $z < -z_{.025} = -1.96$
$z > z_{.025} = 1.96$

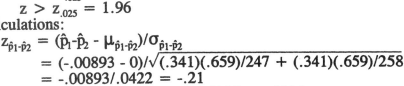

calculations:
$z_{\hat{p}_1-\hat{p}_2} = (\hat{p}_1-\hat{p}_2 - \mu_{\hat{p}_1-\hat{p}_2})/\sigma_{\hat{p}_1-\hat{p}_2}$
$= (.0280 - 0)/\sqrt{(.0537)(.9463)/843 + (.0537)(.9463)/703}$
$= .0280/.0115 = 2.43$
P-value $= 2 \cdot P(z>2.43) = 2 \cdot [1 - P(z<2.43)] = 2 \cdot [1 - .9925] = 2 \cdot [.0075] = .0150$
conclusion:
Reject H_o; there is sufficient evidence to reject the claim that $p_1-p_2 = 0$ and conclude that $p_1-p_2 \neq 0$ (in fact, that $p_1-p_2 > 0$).
Since the calculated $z=2.43<z_{.005}<2.575$ [and the calculated P-value$=.0150>.01$], the difference is not significant at the .01 level of significance. Even though one cannot be 99% certain, it does appear that the ban had an effect on the smoking quit rate.

13. Let those who received the vaccine be group 1.
original claim: $p_1-p_2 < 0$
$\hat{p}_1 = x_1/n_1 = 14/1070 = .013$
$\hat{p}_2 = x_2/n_2 = 95/532 = .179$
$\hat{p}_1-\hat{p}_2 = .013 - .179 = -.165$
$\bar{p} = (x_1 + x_2)/(n_1 + n_2)$
$= (14 + 95)/(1070 + 532)$
$= 109/1602 = .068$

S-230   INSTRUCTOR'S SOLUTIONS Chapter 8

H_o: $p_1-p_2 = 0$
H_1: $p_1-p_2 < 0$
$\alpha = .05$ [assumed]
C.R. $z < -z_{.05} = -1.645$

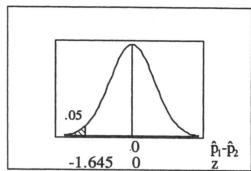

calculations:
$z_{\hat{p}_1-\hat{p}_2} = (\hat{p}_1-\hat{p}_2 - \mu_{\hat{p}_1-\hat{p}_2})/\sigma_{\hat{p}_1-\hat{p}_2}$
$= (-.165 - 0)/\sqrt{(.068)(.932)/1070 + (.068)(.932)/532}$
$= -.165/.0134 = -12.39$
P-value = $P(z<-12.39) = .0001$
conclusion:
Reject H_o; there is sufficient evidence to conclude that $p_1-p_2 < 0$.

14. Let the men group 1.
   a. original claim: $p_1-p_2 > 0$
   $\hat{p}_1 = x_1/n_1 = 45/500 = .09000$
   $\hat{p}_2 = x_2/n_2 = 6/2100 = .00286$
   $\hat{p}_1-\hat{p}_2 = .09000 - .00286 = .08714$
   $\bar{p} = (x_1 + x_2)/(n_1 + n_2)$
   $= (45 + 6)/(500 + 2100)$
   $= 51/2600 = .0196$
   H_o: $p_1-p_2 = 0$
   H_1: $p_1-p_2 > 0$
   $\alpha = .01$
   C.R. $z > z_{.01} = 2.326$
   calculations:

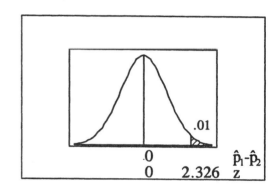

   $z_{\hat{p}_1-\hat{p}_2} = (\hat{p}_1-\hat{p}_2 - \mu_{\hat{p}_1-\hat{p}_2})/\sigma_{\hat{p}_1-\hat{p}_2}$
   $= (.08714 - 0)/\sqrt{(.0196)(.9804)/500 + (.0196)(.9804)/2100}$
   $= .08714/.00690 = 12.63$
   P-value = $P(z>12.63) = 1 - P(z<12.63) = 1 - .9999 = .0001$
   conclusion:
   Reject H_o; there is sufficient evidence to conclude that $p_1-p_2 > 0$.

   b. $(\hat{p}_1-\hat{p}_2) \pm z_{.01}\sqrt{\hat{p}_1\hat{q}_1/n_1 + \hat{p}_2\hat{q}_2/n_2}$
   $.08714 \pm 2.326 \cdot \sqrt{(.09)(.91)/500 + (.00286)(.99714)/2100}$
   $.08714 \pm .02989$
   $.057 < p_1-p_2 < .117$
   Yes; the interval does not contain zero, indicating a significant difference between the two groups.

   c. The proportion of women with this form of color blindness is very small. In general, the sample should be large enough to guarantee the occurrence of some persons with the desired characteristic so that a meaningful point estimate can be obtained. In particular, in order to use the normal approximation it must be true that $n_2p_2 \geq 5$ – i.e., if $p \approx .003$ we need $n_2 \geq 5/(.003) = 1667$.

15. Let those not wearing seat belts be group 1.
   original claim: $p_1-p_2 > 0$
   $\hat{p}_1 = x_1/n_1 = 50/290 = .1724$
   $\hat{p}_2 = x_2/n_2 = 16/123 = .1301$
   $\hat{p}_1-\hat{p}_2 = .1724 - .1301 = .0423$
   $\bar{p} = (x_1 + x_2)/(n_1 + n_2) = (50 + 16)/(290 + 123) = 66/413 = .160$

H_o: $p_1-p_2 = 0$
H_1: $p_1-p_2 > 0$
$\alpha = .05$
C.R. $z > z_{.05} = 1.645$

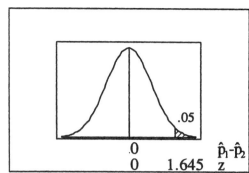

calculations:
$z_{\hat{p}_1-\hat{p}_2} = (\hat{p}_1-\hat{p}_2 - \mu_{\hat{p}_1-\hat{p}_2})/\sigma_{\hat{p}_1-\hat{p}_2}$
$= (.0423 - 0)/\sqrt{(.160)(.840)/290 + (.160)(.840)/123}$
$= .0423/.0394 = 1.07$
P-value $= P(z > 1.07) = 1 - P(z < 1.07) = 1 - .8577 = .1423$
conclusion:
  Do not reject H_o; there is not sufficient evidence to conclude that $p_1-p_2 > 0$.
Based on these results, no specific action should be taken.
NOTE: This test involves only children who were hospitalized – and perhaps a greater percentage of children wearing seat belts avoided being hospitalized in the first place.

16. Let the convicted arsonists be group 1.
  original claim: $p_1-p_2 > 0$
  $\hat{p}_1 = x_1/n_1 = 50/(50 + 43) = 50/93 = .538$
  $\hat{p}_2 = x_2/n_2 = 63/(63 + 144) = 63/207 = .304$
  $\hat{p}_1-\hat{p}_2 = .538 - .304 = .233$
  $\bar{p} = (x_1 + x_2)/(n_1 + n_2)$
  $= (50 + 63)/(93 + 207)$
  $= 113/300 = .377$
H_o: $p_1-p_2 = 0$
H_1: $p_1-p_2 > 0$
$\alpha = .01$
C.R. $z > z_{.01} = 2.326$

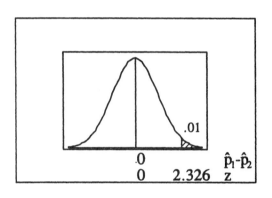

calculations:
$z_{\hat{p}_1-\hat{p}_2} = (\hat{p}_1-\hat{p}_2 - \mu_{\hat{p}_1-\hat{p}_2})/\sigma_{\hat{p}_1-\hat{p}_2}$
$= (.233 - 0)/\sqrt{(.377)(.623)/93 + (.377)(.623)/207}$
$= .233/.0605 = 3.86$
P-value $= P(z > 3.86) = 1 - P(z < 3.86) = 1 - .9999 = .0001$
conclusion:
  Reject H_o; there is sufficient evidence to conclude that $p_1-p_2 > 0$.
Yes; it does seem reasonable that drinking might be related to the type of crime. The kinds of problems and personalities associated with drinking may well be more likely to be associated with some crimes more than others.

17. Let the husband defendants be group 1.
  original claim: $p_1-p_2 > 0$
  $\hat{p}_1 = x_1/n_1 = 277/318 = .871$
  $\hat{p}_2 = x_2/n_2 = 155/222 = .698$
  $\hat{p}_1-\hat{p}_2 = .871 - .692 = .173$
  $\bar{p} = (x_1 + x_2)/(n_1 + n_2)$
  $= (277 + 155)/(318 + 222)$
  $= 432/540 = .80$
H_o: $p_1-p_2 = 0$
H_1: $p_1-p_2 > 0$
$\alpha = .05$ [assumed]
C.R. $z > z_{.05} = 1.645$

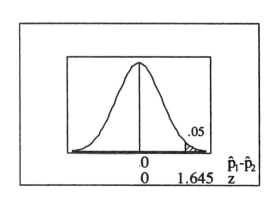

calculations:
$$z_{\hat{p}_1-\hat{p}_2} = (\hat{p}_1-\hat{p}_2 - \mu_{\hat{p}_1-\hat{p}_2})/\sigma_{\hat{p}_1-\hat{p}_2}$$
$$= (.173 - 0)/\sqrt{(.80)(.20)/318 + (.80)(.20)/222}$$
$$= .173/.0350 = 4.941$$
P-value = $P(z>4.94) = 1 - P(z<4.94) = 1 - .9999 = .0001$
conclusion:
Reject H_o; there is sufficient evidence to conclude that $p_1-p_2 > 0$.
In spouse murder cases the question is usually not about who committed the murder, but whether the murder was committed in self-defense. Since women are usually physically smaller and weaker than men, a self-defense plea would be more believable from a wife defendant.

18. Let those receiving the Salk vaccine be group 1.
original claim: $p_1-p_2 < 0$
$\hat{p}_1 = x_1/n_1 = 33/200,000 = .000165$
$\hat{p}_2 = x_2/n_2 = 115/200,000 = .000575$
$\hat{p}_1-\hat{p}_2 = .000165 - .000575 = -.000410$
$\bar{p} = (x_1 + x_2)/(n_1 + n_2)$
$= (33 + 115)/(200,000 + 200,000)$
$= 148/400,000 = .00037$
H_o: $p_1-p_2 = 0$
H_1: $p_1-p_2 < 0$
$\alpha = .01$
C.R. $z < -z_{.01} = -2.326$

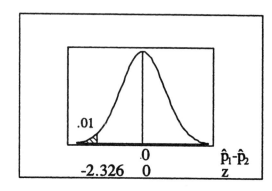

calculations:
$$z_{\hat{p}_1-\hat{p}_2} = (\hat{p}_1-\hat{p}_2 - \mu_{\hat{p}_1-\hat{p}_2})/\sigma_{\hat{p}_1-\hat{p}_2}$$
$$= (-.000410 - 0)/\sqrt{(.00037)(.99963)/200,000 + (.00037)(.99963)/200,000}$$
$$= -.000410/.0000608 = -6.7416 \text{ [from TI-83 Plus]}$$
P-value = $P(z < -6.74) = 7.88E-12 = .00000000000788$ [from TI-83 Plus]
conclusion:
Reject H_o; there is sufficient evidence to conclude that $p_1-p_2 < 0$.
Yes; assuming the group assignments were random, the vaccine appears to be effective.

19. Let the Autozone stores be group 1.
original claim: $p_1-p_2 = 0$
$\hat{p}_1 = x_1/n_1 = 63/100 = .630$
$\hat{p}_2 = x_2/n_2 = 30/37 = .811$
$\hat{p}_1-\hat{p}_2 = .630 - .811 = -.181$
$\bar{p} = (x_1 + x_2)/(n_1 + n_2)$
$= (63 + 30)/(100 + 37)$
$= 93/137 = .679$
H_o: $p_1-p_2 = 0$
H_1: $p_1-p_2 \neq 0$
$\alpha = .05$
C.R. $z < -z_{.025} = -1.96$
$z > z_{.025} = 1.96$

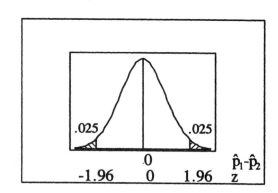

calculations:
$$z_{\hat{p}_1-\hat{p}_2} = (\hat{p}_1-\hat{p}_2 - \mu_{\hat{p}_1-\hat{p}_2})/\sigma_{\hat{p}_1-\hat{p}_2}$$
$$= (-.181 - 0)/\sqrt{(.679)(.321)/100 + (.679)(.321)/37}$$
$$= -.181/.0898 = -2.01$$
P-value = $2 \cdot P(z<-2.01) = 2 \cdot (.0222) = .0444$
conclusion:
Reject H_o; there is sufficient evidence to reject the claim that $p_1-p_2 = 0$ and conclude that $p_1-p_2 \neq 0$ (in fact, that $p_1-p_2 < 0$).
Yes; since Autozone has the lower failure rate, it appears to be the better choice.

20. Let Southwest be group 1.
  a. original claim: $p_1-p_2 = 0$
    $\hat{p}_1 = x_1/n_1 = 2181604/3131727 = .696613$
    $\hat{p}_2 = x_2/n_2 = 1448255/2091859 = .692329$
    $\hat{p}_1-\hat{p}_2 = .696613 - .692329 = .004285$
    $\bar{p} = (x_1 + x_2)/(n_1 + n_2)$
      $= 3629859/5223586$
      $= .695$
    H_o: $p_1-p_2 = 0$
    H_1: $p_1-p_2 \neq 0$
    $\alpha = .05$ [assumed]
    C.R. $z < -z_{.025} = -1.96$
       $z > z_{.025} = 1.96$

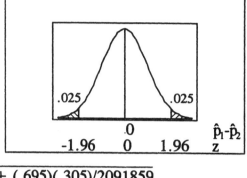

    calculations:
      $z_{\hat{p}_1-\hat{p}_2} = (\hat{p}_1-\hat{p}_2 - \mu_{\hat{p}_1-\hat{p}_2})/\sigma_{\hat{p}_1-\hat{p}_2}$
      $= (.004285 - 0)/\sqrt{(.695)(.305)/3131727 + (.695)(.305)/2091859}$
      $= .004285/.0004112 = 10.42$
      P-value $= 2 \cdot P(z>10.42) = 2 \cdot [1 - P(z<10.42)] = 2 \cdot [1 - .9999] = 2 \cdot [.0001] = .0002$
    conclusion:
      Reject H_o; there is sufficient evidence to reject the claim that $p_1-p_2 = 0$ and conclude that $p_1-p_2 \neq 0$ (in fact, that $p_1-p_2 > 0$).

  b. The very large sample sizes provide so much precision that it is possible to distinguish between the two values. In general, no two measurable values are <u>exactly</u> equal – i.e., given sufficiently large sample sizes and/or sufficiently precise measuring tools, the difference can always be detected. This illustrates the difference between statistical significance and practical significance. Some statisticians refer to this as "the insignificant significant difference."

  c. "...can be statistically significant."

21. Let the single women be group 1. Since there is not enough information to determine the exact values of x_1 and x_2, the problem is limited to precision given below.
  $\hat{p}_1 = x_1/n_1 = [49 \text{ or } 50]/205 = .24$ $\hat{p}_2 = x_2/n_2 = [70 \text{ or } 71]/260 = .27$
  $\hat{p}_1-\hat{p}_2 = .24 - .27 = -.03$
  $(\hat{p}_1-\hat{p}_2) \pm z_{.005}\sqrt{\hat{p}_1\hat{q}_1/n_1 + \hat{p}_2\hat{q}_2/n_2}$
  $-.03 \pm 2.575 \cdot \sqrt{(.24)(.76)/205 + (.27)(.73)/260}$
  $-.03 \pm .10$
  $-.13 < p_1-p_2 < .07$
  No; since the confidence interval includes 0, there does not appear to be a gender gap on this issue.

22. Refer to the data and calculations in exercise #22. Since the x's cannot be determined with certainty, one must accept the accuracy limitations imposed by the reported percents.
  original claim: $p_1-p_2 \neq 0$
    $\bar{p} = [n_1\hat{p}_1 - n_2\hat{p}_2]/(n_1 + n_2)$
      $= [205(.24) + 260(.27)]/(205 + 260)$
      $= 119.4/465 = .2568$
    H_o: $p_1-p_2 = 0$
    H_1: $p_1-p_2 \neq 0$
    $\alpha = .01$
    C.R. $z < -z_{.005} = -2.575$
       $z > z_{.005} = 2.575$

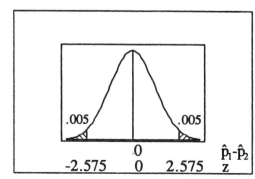

calculations:
$$z_{\hat{p}_1-\hat{p}_2} = (\hat{p}_1-\hat{p}_2 - \mu_{\hat{p}_1-\hat{p}_2})/\sigma_{\hat{p}_1-\hat{p}_2}$$
$= (-.03 - 0)/\sqrt{(.2568)(.7432)/205 + (.2568)(.7432)/260}$
$= -.03/.0408 = -.74$
P-value $= 2 \cdot P(z < -.74) = 2 \cdot (.2296) = .4592$
conclusion:
  Do not reject H_o; there is not sufficient evidence to conclude that $p_1-p_2 \neq 0$.
  No, there does not appear to be a difference in the proportions.

23. Let those younger than 21 be group 1.
$\hat{p}_1 = x_1/n_1 = .0425$ $\hat{p}_2 = x_2/n_2 = .0455$
$\hat{p}_1-\hat{p}_2 = .0425 - .0455 = -.0029$
$(\hat{p}_1-\hat{p}_2) \pm z_{.025}\sqrt{\hat{p}_1\hat{q}_1/n_1 + \hat{p}_2\hat{q}_2/n_2}$
$-.0029 \pm 1.96 \cdot \sqrt{(.0425)(.9575)/2750 + (.0455)(.9545)/2200}$
$-.0029 \pm .0115$
$-.0144 < p_1-p_2 < .0086$
Yes; the interval contains zero, indicating no significant difference between the two rates of violent crimes.

24. Let the vinyl gloves be group 1. Since the x's cannot be determined with certainty, one must accept the accuracy limitations imposed by the reported percents.
original claim: $p_1-p_2 > 0$
$\hat{p}_1 = x_1/n_1 = x_1/240 = .63$ [note: $x_1/240 = 63\%$ for $151 \leq x_1 \leq 152$]
$\hat{p}_2 = x_2/n_2 = x_2/240 = .07$ [note: $x_2/240 = 7\%$ for $16 \leq x_2 \leq 17$]
$\hat{p}_1-\hat{p}_2 = .63 - .07 = .56$
$\bar{p} = (x_1 + x_2)/(n_1 + n_2)$
$\phantom{\bar{p}} = (.63 \cdot 240 + .07 \cdot 240)/(240 + 240)$
$\phantom{\bar{p}} = 168/480 = .350$
H_o: $p_1-p_2 \leq 0$
H_1: $p_1-p_2 > 0$
$\alpha = .005$
C.R. $z > z_{.005} = 2.575$

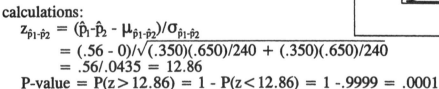

calculations:
$$z_{\hat{p}_1-\hat{p}_2} = (\hat{p}_1-\hat{p}_2 - \mu_{\hat{p}_1-\hat{p}_2})/\sigma_{\hat{p}_1-\hat{p}_2}$$
$= (.56 - 0)/\sqrt{(.350)(.650)/240 + (.350)(.650)/240}$
$= .56/.0435 = 12.86$
P-value $= P(z > 12.86) = 1 - P(z < 12.86) = 1 - .9999 = .0001$
conclusion:
  Reject H_o; there is sufficient evidence to conclude that $p_1-p_2 > 0$.

25. Let those given the written survey be group 1.
$\hat{p}_1 = x_1/n_1 = 67/850 = .079$ [note: $x_1 = (.079)(850) = 67$]
$\hat{p}_2 = x_2/n_2 = 105/850 = .124$ [note: $x_2 = (.124)(850) = 105$]
$\hat{p}_1-\hat{p}_2 = .079 - .124 = -.0447$
$\bar{p} = (x_1 + x_2)/(n_1 + n_2)$
$\phantom{\bar{p}} = (67 + 105)/(850 + 850)$
$\phantom{\bar{p}} = 172/1700 = .1012$

a. original claim: $p_1-p_2 = 0$
H_o: $p_1-p_2 = 0$
H_1: $p_1-p_2 \neq 0$
$\alpha = .05$ [assumed]
C.R. $z < -z_{.025} = -1.96$
$\phantom{\text{C.R. }} z > z_{.025} = 1.96$

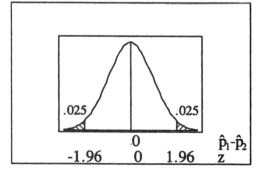

calculations:
$z_{\hat{p}_1-\hat{p}_2} = (\hat{p}_1-\hat{p}_2 - \mu_{\hat{p}_1-\hat{p}_2})/\sigma_{\hat{p}_1-\hat{p}_2}$
$= (-.0447 - 0)/\sqrt{(.1012)(.8988)/850 + (.1012)(.8988)/850}$
$= -.0447/.0146 = -3.06$
P-value = $2 \cdot P(z<-3.06) = 2 \cdot (.0011) = .0022$
conclusion:
Reject H_o; there is sufficient evidence to reject the claim that $p_1 - p_2 = 0$ and to conclude that $p_1-p_2 \ne 0$ (in fact, that $p_1 - p_2 < 0$ – i.e., that fewer students receiving the written test admit carrying a gun.)
Yes; based on this result the difference between 7.9% and 12.4% is significant.
b. $(\hat{p}_1-\hat{p}_2) \pm z_{.005}\sqrt{\hat{p}_1\hat{q}_1/n_1 + \hat{p}_2\hat{q}_2/n_2}$
$-.0447 \pm 2.575 \cdot \sqrt{(.079)(.921)/850 + (.124)(.876)/850}$
$-.0447 \pm .0376$
$-.0823 < p_1-p_2 < -.0071$
The interval does not contain zero, indicating a significant difference between the two response rates. We are 99% confident the interval from -8.3% to -0.7% contains the true difference between the population percentages.

26. Let the Viagra users be group 1. Since the x's cannot be determined with certainty, one must accept the accuracy limitations imposed by the reported percents.
a. original claim: $p_1-p_2 > 0$
$\hat{p}_1 = x_1/n_1 = x_1/734 = .16$ [note: $x_1/734 = 16\%$ for $114 \le x_1 \le 121$]
$\hat{p}_2 = x_2/n_2 = x_2/725 = .04$ [note: $x_2/725 = 4\%$ for $26 \le x_2 \le 32$]
$\hat{p}_1-\hat{p}_2 = .16 - .04 = .12$
$\bar{p} = (x_1 + x_2)/(n_1 + n_2)$
$= (.16 \cdot 734 + .04 \cdot 725)/(734 + 725)$
$= 146.44/1459 = .1004$
H_o: $p_1-p_2 = 0$
H_1: $p_1-p_2 > 0$
$\alpha = .01$
C.R. $z > z_{.01} = 2.326$

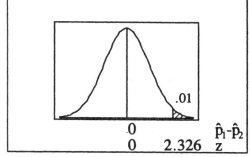

calculations:
$z_{\hat{p}_1-\hat{p}_2} = (\hat{p}_1-\hat{p}_2 - \mu_{\hat{p}_1-\hat{p}_2})/\sigma_{\hat{p}_1-\hat{p}_2}$
$= (.12 - 0)/\sqrt{(.1004)(.8996)/734 + (.1004)(.8996)/725}$
$= .12/.0157 = 7.63$
P-value = $P(z>7.63) = 1 - P(z<7.63) = 1 - .9999 = .0001$
conclusion:
Reject H_o; there is sufficient evidence to conclude that $p_1-p_2 > 0$.
Yes, men taking Viagra seem to experience more headaches.
b. $(\hat{p}_1-\hat{p}_2) \pm z_{.005}\sqrt{\hat{p}_1\hat{q}_1/n_1 + \hat{p}_2\hat{q}_2/n_2}$
$.12 \pm 2.575 \cdot \sqrt{(.16)(.86)/734 + (.04)(.96)/725}$
$.12 \pm .04$
$.08 < p_1-p_2 < .16$
Since the interval does not contain zero, the headache rate was significantly larger for the Viagra group – i.e., headaches appear to be a side effect of Viagra.

27. Let the central city be group 1.
original claim: $p_1-p_2 = 0$
$\hat{p}_1 = x_1/n_1 = 85/294 = .289$ [note: $x_1 = (.289)(294) = 85$]
$\hat{p}_2 = x_2/n_2 = 174/1015 = .171$ [note: $x_2 = (.171)(1015) = 174$]
$\hat{p}_1-\hat{p}_2 = .289 - .171 = .118$
$\bar{p} = (x_1 + x_2)/(n_1 + n_2)$
$= (84 + 174)/(294 + 1015)$
$= 259/1309 = .198$

H_o: $p_1-p_2 = 0$
H_1: $p_1-p_2 \neq 0$
$\alpha = .01$
C.R. $z < -z_{.005} = -2.575$
     $z > z_{.005} = 2.575$

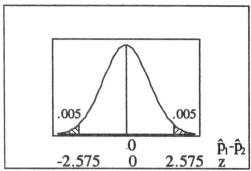

calculations:
$z_{\hat{p}_1-\hat{p}_2} = (\hat{p}_1-\hat{p}_2 - \mu_{\hat{p}_1-\hat{p}_2})/\sigma_{\hat{p}_1-\hat{p}_2}$
$= (.118 - 0)/\sqrt{(.198)(.802)/294 + (.198)(.802)/1015}$
$= .118/.0264 = 4.47$
P-value = $2 \cdot P(z>4.47) = 2 \cdot [1 - P(z<4.47)] = 2 \cdot [1 - .9999] = 2 \cdot [.0001] = .0002$
conclusion:
  Reject H_o; there is sufficient evidence to reject the claim that $p_1-p_2 = 0$ and to conclude that $p_1-p_2 \neq 0$ (in fact, $p_1-p_2 > 0$).

28. Let the basketball results be group 1.
    $\hat{p}_1 = x_1/n_1 = 127/198 = .6414$ $\hat{p}_2 = x_2/n_2 = 57/99 = .5758$
    $\hat{p}_1-\hat{p}_2 = .6414 - .5758 = .0657$
    $(\hat{p}_1-\hat{p}_2) \pm z_{.025}\sqrt{\hat{p}_1\hat{q}_1/n_1 + \hat{p}_2\hat{q}_2/n_2}$
    $.0657 \pm 1.96 \cdot \sqrt{(.6414)(.3586)/198 + (.5758)(.4242)/99}$
    $.0657 \pm .1181$
    $-.052 < p_1-p_2 < .184$
    No; since the confidence interval includes zero, the home court advantage for basketball is not significantly different from the home field advantage in football. To decide whether there is a home advantage at all, one would have to see whether individual confidence intervals for p include the value .50.

29. Let the movies tested for alcohol use be group 1.
    original claim: $p_1-p_2 < 0$
    $\hat{p}_1 = x_1/n_1 = 25/50 = .50$
    $\hat{p}_2 = x_2/n_2 = 28/50 = .56$
    $\hat{p}_1-\hat{p}_2 = .50 - .56 = -.06$
    $\bar{p} = (x_1 + x_2)/(n_1 + n_2)$
       $= (25 + 28)/(50 + 50)$
       $= 53/100 = .53$
    H_o: $p_1-p_2 = 0$
    H_1: $p_1-p_2 < 0$
    $\alpha = .05$ [assumed]
    C.R. $z < -z_{.01} = -1.645$
    calculations:
    $z_{\hat{p}_1-\hat{p}_2} = (\hat{p}_1-\hat{p}_2 - \mu_{\hat{p}_1-\hat{p}_2})/\sigma_{\hat{p}_1-\hat{p}_2}$
    $= (-.06 - 0)/\sqrt{(.53)(.47)/50 + (.53)(.47)/50}$
    $= -.06/.0998 = -.60$
    P-value = $P(z<-.60) = .2743$
conclusion:
    Do not reject H_o; there is not sufficient evidence to conclude that $p_1-p_2 < 0$.
No; even though the data match those in Data Set 7, the results do not apply. One requirement for this section is that the two samples be independent. The data in Data Set 7 are not from independent samples, but from the same sample – analogous to studying whether the proportion of people who declare bankruptcy is different from the proportion of people with more than $10,000 of credit card debt, and using the same 50 people.

30. Let the men be group 1.
original claim: $p_1 - p_2 = 0$
$\hat{p}_1 = x_1/n_1 = 23/40 = .575$
$\hat{p}_2 = x_2/n_2 = 22/40 = .550$
$\hat{p}_1 - \hat{p}_2 = .575 - .550 = .025$
$\bar{p} = (x_1 + x_2)/(n_1 + n_2)$
$\phantom{\bar{p}} = (23 + 22)/(40 + 40)$
$\phantom{\bar{p}} = 45/80 = .5625$
$H_o: p_1 - p_2 = 0$
$H_1: p_1 - p_2 \neq 0$
$\alpha = .05$ [assumed]
C.R. $z < -z_{.025} = -1.96$
$\ z > z_{.025} = 1.96$
calculations:
$z_{\hat{p}_1-\hat{p}_2} = (\hat{p}_1-\hat{p}_2 - \mu_{\hat{p}_1-\hat{p}_2})/\sigma_{\hat{p}_1-\hat{p}_2}$
$\phantom{z_{\hat{p}_1-\hat{p}_2}} = (.025 - 0)/\sqrt{(.5625)(.4375)/40 + (.5625)(.4375)/40}$
$\phantom{z_{\hat{p}_1-\hat{p}_2}} = .025/.1109 = .23$
P-value $= 2 \cdot P(z > .23) = 2 \cdot [1 - P(z < .23)] = 2 \cdot [1 - .5910] = 2 \cdot [.4090] = .8180$
conclusion:
Do not reject H_o; there is not sufficient evidence to reject the claim that $p_1 - p_2 = 0$.

31. For all parts of this exercise.
$\hat{p}_1 = x_1/n_1 = 112/200 = .56$ $\hat{p}_2 = x_2/n_2 = 88/200 = .44$
$\hat{p}_1 - \hat{p}_2 = .56 - .44 = .12$ $\bar{p} = (x_1 + x_2)/(n_1 + n_2) = 200/400 = .50$

a. $(\hat{p}_1 - \hat{p}_2) \pm z_{.025}\sqrt{\hat{p}_1\hat{q}_1/n_1 + \hat{p}_2\hat{q}_2/n_2}$
$.120 \pm 1.96 \cdot \sqrt{(.56)(.44)/200 + (.44)(.56)/200}$
$.120 \pm .097$
$.023 < p_1 - p_2 < .217$
Since the interval does not include zero, the implication is that p_1 and p_2 are different.
Since the interval lies entirely above, conclude that $p_1 - p_2 > 0$.

b. for group 1   for group 2
$\hat{p} \pm z_{.025}\sqrt{\hat{p}\hat{q}/n}$ $\hat{p} \pm z_{.025}\sqrt{\hat{p}\hat{q}/n}$
$.560 \pm 1.96\sqrt{(.56)(.44)/200}$ $.440 \pm 1.96\sqrt{(.44)(.56)/200}$
$.560 \pm .069$ $.440 \pm .069$
$.491 < p < .629$ $.371 < p < .509$
Since the intervals overlap, the implication is that p_1 and p_2 could be the same.

c. original claim: $p_1 - p_2 = 0$
$H_o: p_1 - p_2 = 0$
$H_1: p_1 - p_2 \neq 0$
$\alpha = .05$
C.R. $z < -z_{.025} = -1.96$
$\ z > z_{.025} = 1.96$

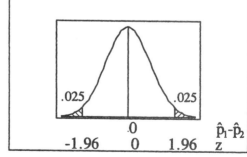

calculations:
$z_{\hat{p}_1-\hat{p}_2} = (\hat{p}_1-\hat{p}_2 - \mu_{\hat{p}_1-\hat{p}_2})/\sigma_{\hat{p}_1-\hat{p}_2}$
$\phantom{z_{\hat{p}_1-\hat{p}_2}} = (.12 - 0)/\sqrt{(.5)(.5)/200 + (.5)(.5)/200}$
$\phantom{z_{\hat{p}_1-\hat{p}_2}} = .12/.05 = 2.40$
P-value $= 2 \cdot P(z > 2.40) = 2 \cdot [1 - P(z < 2.40)] = 2 \cdot [1 - .9918] = 2 \cdot [.0082] = .0164$
conclusion:
Reject H_o; there is sufficient evidence to reject the claim that $p_1 - p_2 = 0$ and conclude that $p_1 - p_2 \neq 0$ (in fact, that $p_1 - p_2 > 0$).

d. Based on parts (a)-(c), conclude that p_1 and p_2 are unequal, and that $p_1 > p_2$. The overlapping interval method of part (b) appears to be the least effective method for comparing two populations.

32. For all parts of this exercise.
$\hat{p}_1 = x_1/n_1 = 10/20 = .500$ $\hat{p}_2 = x_2/n_2 = 1404/2000 = .702$
$\hat{p}_1 - \hat{p}_2 = .500 - .702 = -.202$ $\bar{p} = (x_1+x_2)/(n_1+n_2) = 1414/2020 = .70$

The test of hypothesis is as follows.
  original claim: $p_1 - p_2 = 0$
  H_o: $p_1 - p_2 = 0$
  H_1: $p_1 - p_2 \neq 0$
  $\alpha = .05$
  C.R. $z < -z_{.025} = -1.96$
        $z > z_{.025} = 1.96$

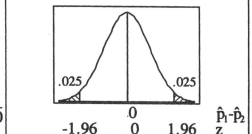

  calculations:
    $z_{\hat{p}_1-\hat{p}_2} = (\hat{p}_1-\hat{p}_2 - \mu_{\hat{p}_1-\hat{p}_2})/\sigma_{\hat{p}_1-\hat{p}_2}$
    $= (-.202 - 0)/\sqrt{(.7)(.3)/20 + (.7)(.3)/2000}$
    $= -.202/.10298 = -1.9615$
    P-value $= 2 \cdot P(z<-1.96) = 2 \cdot (.0250) = .0500$ [actually, slightly less than .0500]
  conclusion:
    Reject H_o; there is sufficient evidence to reject the claim that $p_1-p_2 = 0$ and conclude that $p_1-p_2 \neq 0$ (in fact, that $p_1-p_2 > 0$).

The confidence interval is as follows.
$(\hat{p}_1-\hat{p}_2) \pm z_{.025}\sqrt{\hat{p}_1\hat{q}_1/n_1 + \hat{p}_2\hat{q}_2/n_2}$
$-.202 \pm 1.96 \cdot \sqrt{(.500)(.500)/20 + (.702)(.298)/2000}$
$-.202 \pm .220$
$-.422 < p_1-p_2 < .018$
Since the interval include zero, the implication is that p_1 and p_2 could be the same that one should not reject the claim $p_1-p_2 = 0$.
The test of hypothesis and the confidence interval lead to different conclusions. In this instance, they are not equivalent.

33. Let the black drivers be group 1.
The new test of hypothesis is as follows.
  original claim: $p_1 - p_2 > 0$
    $\hat{p}_1 = x_1/n_1 = 240/2000 = .120$
    $\hat{p}_2 = x_2/n_2 = 1470/14,000 = .105$
    $\hat{p}_1 - \hat{p}_2 = .120 - .105 = .015$
    $\bar{p} = (x_1 + x_2)/(n_1 + n_2)$
       $= (240 + 1470)/(2000 + 14000)$
       $= 1710/16,000 = .107$
  H_o: $p_1-p_2 = 0$
  H_1: $p_1-p_2 > 0$
  $\alpha = .05$
  C.R. $z > z_{.05} = 1.645$

  calculations:
    $z_{\hat{p}_1-\hat{p}_2} = (\hat{p}_1-\hat{p}_2 - \mu_{\hat{p}_1-\hat{p}_2})/\sigma_{\hat{p}_1-\hat{p}_2}$
    $= (.015 - 0)/\sqrt{(.107)(.893)/2000 + (.107)(.893)/14000}$
    $= .015/.007385 = 2.03$
    P-value $= P(z>2.03) = 1 - P(z<2.03) = 1 - .9788 = .0212$
  conclusion:
    Reject H_o; there is sufficient evidence to conclude that $p_1-p_2 > 0$.

The new confidence interval is as follows.
$(\hat{p}_1-\hat{p}_2) \pm z_{.05}\sqrt{\hat{p}_1\hat{q}_1/n_1 + \hat{p}_2\hat{q}_2/n_2}$
$.015 \pm 1.645 \cdot \sqrt{(.120)(.880)/2000 + (.105)(.895)/14000}$
$.015 \pm .013$
$.002 < p_1-p_2 < .028$

The test statistic increased from .64 to 2.03, now allowing one to conclude $p_1-p_2 > 0$. The confidence interval grew narrower to not include zero, now allowing one to conclude $p_1-p_2 > 0$. With the same point estimates, the increased sample size provides enough evidence to support the original claim.

34. Let the Viagra users be group 1. Since the x's cannot be determined with certainty, one must accept the accuracy limitations imposed by the reported percents.
    original claim: $p_1-p_2 = .10$
    $\hat{p}_1 = x_1/n_1 = x_1/734 = .16$ [note: $x_1/734 = 16\%$ for $114 \le x_1 \le 121$]
    $\hat{p}_2 = x_2/n_2 = x_2/725 = .04$ [note: $x_2/725 = 4\%$ for $26 \le x_2 \le 32$]
    $\hat{p}_1 - \hat{p}_2 = .16 - .04 = .12$
    H_0: $p_1-p_2 = .10$
    H_1: $p_1-p_2 \ne .10$
    $\alpha = .05$
    C.R. $z < -z_{.025} = -1.96$
    $z > z_{.025} = 1.96$

    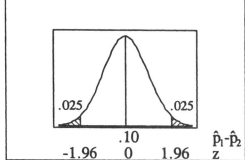

    calculations:
    $z_{\hat{p}_1-\hat{p}_2} = (\hat{p}_1-\hat{p}_2 - \mu_{\hat{p}_1-\hat{p}_2})/\sigma_{\hat{p}_1-\hat{p}_2}$
    $= (.12 - .10)/\sqrt{(.16)(.84)/734 + (.04)(.96)/725}$
    $= .02/.0154 = 1.30$
    P-value $= 2 \cdot P(z > 1.30) = 2 \cdot [1 - P(z < 1.30)] = 2 \cdot [1 - .9032] = 2 \cdot [.0968] = .1936$
    conclusion:
    Do not reject H_0; there is not sufficient evidence to reject the claim that $p_1-p_2 = .10$.

35. $\hat{p}_1 = x_1/n_1 = 40/100 = .40$ for groups 1 and 2, $\bar{p} = 70/200 = .35$
    $\hat{p}_2 = x_2/n_2 = 30/100 = .30$ for groups 2 and 3, $\bar{p} = 50/200 = .25$
    $\hat{p}_3 = x_3/n_3 = 20/100 = .20$ for groups 1 and 3, $\bar{p} = 50/200 = .30$

    a. H_0: $p_1-p_2 = 0$
       H_1: $p_1-p_2 \ne 0$
       $\alpha = .05$
       C.R. $z < -z_{.025} = -1.96$
       $z > z_{.025} = 1.96$

       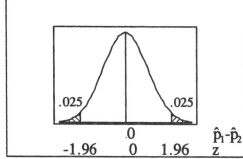

       calculations:
       $z_{\hat{p}_1-\hat{p}_2} = (\hat{p}_1-\hat{p}_2 - \mu_{\hat{p}_1-\hat{p}_2})/\sigma_{\hat{p}_1-\hat{p}_2}$
       $= (.10 - 0)/\sqrt{(.35)(.65)/100 + (.35)(.65)/100}$
       $= .10/.0675 = 1.482$
       conclusion:
       Do not reject H_0; there is not sufficient evidence to reject the claim that $p_1-p_2 = 0$.

    b. H_0: $p_2-p_3 = 0$
       H_1: $p_2-p_3 \ne 0$
       $\alpha = .05$
       C.R. $z < -z_{.025} = -1.96$
       $z > z_{.025} = 1.96$

       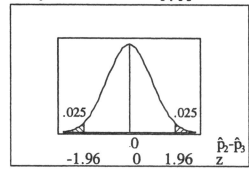

S-240 INSTRUCTOR'S SOLUTIONS Chapter 8

    calculations:
$$z_{\hat{p}_2-\hat{p}_3} = (\hat{p}_2-\hat{p}_3 - \mu_{\hat{p}_2-\hat{p}_3})/\sigma_{\hat{p}_2-\hat{p}_3}$$
$$= (.10 - 0)/\sqrt{(.25)(.75)/100 + (.25)(.75)/100}$$
$$= .10/.0612 = 1.633$$
    conclusion:
        Do not reject H_o; there is not sufficient evidence to reject the claim that $p_2-p_3 = 0$.

c. H_o: $p_1-p_3 = 0$
   H_1: $p_1-p_3 \neq 0$
   $\alpha = .05$
   C.R. $z < -z_{.025} = -1.96$
       $z > z_{.025} = 1.96$

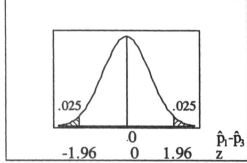

   calculations:
$$z_{\hat{p}_1-\hat{p}_3} = (\hat{p}_1-\hat{p}_3 - \mu_{\hat{p}_1-\hat{p}_3})/\sigma_{\hat{p}_1-\hat{p}_3}$$
$$= (.20 - 0)/\sqrt{(.30)(.70)/100 + (.30)(.70)/100}$$
$$= .20/.0648 = 3.086$$
   conclusion:
      Reject H_o; there is sufficient evidence to reject the claim that $p_1-p_2 = 0$ and to conclude that $p_1-p_3 \neq 0$ (in fact, $p_1-p_3 > 0$).

d. No; failing to find a difference between population 1 and population 2, and between population 2 and population 3, does not necessarily mean one will fail to find a difference between population 1 and population 3. The fact that adjacent values may be equal does not necessarily mean that the possibility of equality extends to non-adjacent values.

36. $E^2 = (z_{\alpha/2})^2(p_1q_1/n_1 + p_2q_2/n_2)$ squaring the original equation
    $E^2 = (z_{\alpha/2})^2(.25/n_1 + .25/n_2)$ setting the unknown proportions to .5
    $E^2 = (z_{\alpha/2})^2(.25/n + .25/n)$ requiring $n_1 = n_2 = n$
    $E^2 = (z_{\alpha/2})^2(.50/n)$ addition
    $n = (z_{\alpha/2})^2(.50)/E^2$ solving for n
    $= (1.96)^2(.50)/(.03)^2$ for $\alpha = .05$ and $E = .03$
    $= 2134.2$, rounded up to 2135

37. a. No; for the placebo group, $np \approx (144)(.018) = 2.592 < 5$. The normal approximation to the binomial does not apply.
   b. For the 144 people in the placebo group, 1.8% is not a possible sample result since $2/144 = 1.4\%$ and $3/144 = 2.1\%$

38. a.

#	sample	$\hat{p}$	$\hat{p}-\mu_{\hat{p}}$	$(\hat{p}-\mu_{\hat{p}})^2$
1	HH	1.0	0.5	.25
2	HT	0.5	0.0	.00
3	TH	0.5	0.0	.00
4	TT	0.0	-.5	.25
		2.0		.50

$\mu_{\hat{p}} = \Sigma\hat{p}/N = 2.0/4 = .5$

$\sigma_{\hat{p}} = \Sigma(\hat{p}-\mu_{\hat{p}})^2/N = .50/4 = .125$

b. The 16 possible $\hat{p}_D-\hat{p}_Q$ values are given below, identified by the sample numbers used to generate them.

#'s	1-1	1-2	1-3	1-4	2-1	2-2	2-3	2-4	3-1	3-2	3-3	3-4	4-1	4-2	4-3	4-4	sum
A	0	.5	.5	1	-.5	0	0	.5	-.5	0	0	.5	-1	-.5	-.5	0	0
B	0	.25	.25	1	.25	0	0	.25	.25	0	0	.25	1	.25	.25	0	4

The A row gives the $\hat{p}_D-\hat{p}_Q$ value for each outcome, $\mu_{D-Q} = (\Sigma A_i)/N = 0/16 = 0$.
The B row gives $[(\hat{p}_D-\hat{p}_Q) - \mu]^2 = [(\hat{p}_D-\hat{p}_Q) - 0]^2$, $\sigma^2_{D-Q} = (\Sigma B_i)/N = 4/16 = .25$.

c. $\sigma^2_{D-Q} = \sigma^2_D + \sigma^2_Q = .125 + .125 = .25$

8-2 Inferences about Two Means: Independent Samples

NOTE: To be consistent with the previous notation, reinforcing the patterns and concepts presented in those sections, the manual uses the "usual" t formula written to apply to $\bar{x}_1 - \bar{x}_2$'s
$$t_{\bar{x}_1-\bar{x}_2} = (\bar{x}_1 - \bar{x}_2 - \mu_{\bar{x}_1-\bar{x}_2})/s_{\bar{x}_1-\bar{x}_2}$$
with $\mu_{\bar{x}_1-\bar{x}_2} = \mu_1 - \mu_2$ and $s_{\bar{x}_1-\bar{x}_2} = \sqrt{s_1^2/n_1 + s_2^2/n_2}$
And so the formula for the t statistic <u>may also be written</u> as
$$t_{\bar{x}_1-\bar{x}_2} = ((\bar{x}_1 - \bar{x}_2) - (\mu_1-\mu_2))/\sqrt{s_1^2/n_1 + s_2^2/n_2}$$

1. Independent samples, since two groups are selected and evaluated separately.

2. Matched pairs, since each "before" measurement is paired with its "after" measurement.

3. Matched pairs, since each reported weight is paired with its measured weight.

4. Independent samples, since the two groups are selected and analyzed separately.

5. Let the light users be group 1.
   original claim: $\mu_1 - \mu_2 > 0$
   $\bar{x}_1 - \bar{x}_2 = 53.3 - 51.3 = 2.0$
   $H_o: \mu_1 - \mu_2 = 0$
   $H_1: \mu_1 - \mu_2 > 0$
   $\alpha = .01$
   C.R. $t > t_{63,.01} = 2.385$
   calculations:
   $t_{\bar{x}_1-\bar{x}_2} = (\bar{x}_1 - \bar{x}_2 - \mu_{\bar{x}_1-\bar{x}_2})/s_{\bar{x}_1-\bar{x}_2}$
   $= (2.0 - 0)/\sqrt{(3.6)^2/64 + (4.5)^2/65}$
   $= 2.0/.7170$
   $= 2.790$
   P-value $= P(t_{63} > 2.790) < .005$
   conclusion:
   Reject H_o; there is sufficient evidence to conclude that $\mu_1 - \mu_2 > 0$.
   Yes; based on these results, heavy marijuana use appears to impede performance.

6. Refer to the data and calculations of exercise #5.

   $(\bar{x}_1 - \bar{x}_2) \pm t_{63,.01} \sqrt{s_1^2/n_1 + s_2^2/n_2}$
   $2.0 \pm 2.385 \cdot \sqrt{(3.6)^2/64 + (4.5)^2/65}$
   2.0 ± 1.7
   $.3 < \mu_1 - \mu_2 < 3.7$
   Since the confidence interval does not include zero, it suggests that the two population means are not equal – and that $\mu_1 > \mu_2$.

7. Let the placebo users be group 1.
   $\bar{x}_1 - \bar{x}_2 = 21.57 - 20.38 = 1.19$
   $(\bar{x}_1 - \bar{x}_2) \pm t_{32,.025} \sqrt{s_1^2/n_1 + s_2^2/n_2}$
   $1.19 \pm 2.037 \cdot \sqrt{(3.87)^2/43 + (3.91)^2/33}$
   1.19 ± 1.84
   $-.65 < \mu_1 - \mu_2 < 3.03$
   No; based on these results, we cannot be 95% certain that the two populations have different means. There is not enough evidence to make this the generally recommended treatment for bipolar depression.

8. Refer to the data and calculations of exercise #7.
original claim: $\mu_1-\mu_2 = 0$
H_o: $\mu_1-\mu_2 = 0$
H_1: $\mu_1-\mu_2 \neq 0$
$\alpha = .05$
C.R. $t < -t_{32,.025} = -2.037$
$t > t_{32,.025} = 2.037$
calculations:
$t_{\bar{x}_1-\bar{x}_2} = (\bar{x}_1-\bar{x}_2 - \mu_{\bar{x}_1-\bar{x}_2})/s_{\bar{x}_1-\bar{x}_2}$
$= (1.19 - 0)/\sqrt{(3.87)^2/43 + (3.91)^2/33}$
$= 1.19/.9009 = 1.321$
$.10 < $ P-value $= 2 \cdot P(t_{32} > 1.321) < .20$
conclusion:
Do not reject H_o; there is not sufficient evidence to reject the claim that $\mu_1-\mu_2 = 0$. This result indicates that there is not sufficient evidence to be 95% certain that the drug does any better than no drug at all.

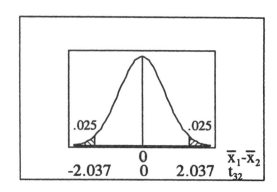

9. Let those receiving the magnet treatment be group 1.
original claim: $\mu_1-\mu_2 > 0$
$\bar{x}_1-\bar{x}_2 = .49 - .44 = .05$
H_o: $\mu_1-\mu_2 = 0$
H_1: $\mu_1-\mu_2 > 0$
$\alpha = .05$
C.R. $t > t_{19,.05} = 1.729$
calculations:
$t_{\bar{x}_1-\bar{x}_2} = (\bar{x}_1-\bar{x}_2 - \mu_{\bar{x}_1-\bar{x}_2})/s_{\bar{x}_1-\bar{x}_2}$
$= (.05 - 0)/\sqrt{(.96)^2/20 + (1.4)^2/20}$
$= .05/.3796 = .132$
P-value $= P(t_{19} > .132) > .10$

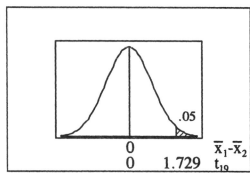

conclusion:
Do not reject H_o; there is not sufficient evidence to conclude that $\mu_1-\mu_2 > 0$.
No, it does not appear that magnets are effective in treating back pain. If much larger sample sizes achieved these same results, the calculated t could fall in the critical region and appear to provide evidence that the treatment is effective – but the observed difference would still be .05, and one would have to decide whether that statistically significant difference is of practical significance.

10. Refer to the data and calculations of exercise #9.
$(\bar{x}_1-\bar{x}_2) \pm t_{19,.05} \sqrt{s_1^2/n_1 + s_2^2/n_2}$
$.05 \pm 1.729 \cdot \sqrt{(.96)^2/20 + (1.4)^2/20}$
$.05 \pm .66$
$-.61 < \mu_1-\mu_2 < .71$
Since the interval includes zero, that the treatment has no effect is a reasonable possibility.

11. a. original claim: $\mu_1-\mu_2 = 0$
$\bar{x}_1-\bar{x}_2 = .81682 - .78479 = .03203$
H_o: $\mu_1-\mu_2 = 0$
H_1: $\mu_1-\mu_2 \neq 0$
$\alpha = .01$
C.R. $t < -t_{35,.005} = -2.728$
$t > t_{35,.005} = 2.728$
calculations:
$t_{\bar{x}_1-\bar{x}_2} = (\bar{x}_1-\bar{x}_2 - \mu_{\bar{x}_1-\bar{x}_2})/s_{\bar{x}_1-\bar{x}_2}$

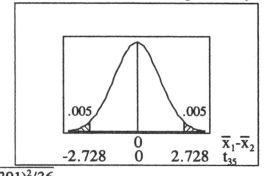

$= (.03203 - 0)/\sqrt{(.007507)^2/36 + (.004391)^2/36}$
$= .03203/.0014495 = 22.098$

INSTRUCTOR'S SOLUTIONS Chapter 8   S-243

P-value = $2 \cdot P(t_{35} > 22.098) < .01$
conclusion:
  Reject H_o; there is sufficient evidence to reject the claim that $\mu_1 - \mu_2 = 0$ and conclude that $\mu_1 - \mu_2 \neq 0$ (in fact, that $\mu_1 - \mu_2 > 0$).
Diet Coke probably weighs less because it uses an artificial sweetener that weighs less than real sugar – or perhaps because it simply uses less sweetener, and sweeteners in general are more dense than the rest of the product.

b. $(\bar{x}_1 - \bar{x}_2) \pm t_{35,.005} \sqrt{s_1^2/n_1 + s_2^2/n_2}$
  $.03203 \pm 2.728 \cdot \sqrt{(.007507)^2/36 + (.004391)^2/36}$
  $.03203 \pm .00395$
  $.02808 < \mu_1 - \mu_2 < .03598$

12. a. original claim: $\mu_1 - \mu_2 < 0$
    $\bar{x}_1 - \bar{x}_2 = .94 - 1.65 = -.71$
    $H_o: \mu_1 - \mu_2 = 0$
    $H_1: \mu_1 - \mu_2 < 0$
    $\alpha = .05$
    C.R. $t < -t_{7,.05} = -1.895$
    calculations:
    $t_{\bar{x}_1-\bar{x}_2} = (\bar{x}_1 - \bar{x}_2 - \mu_{\bar{x}_1-\bar{x}_2})/s_{\bar{x}_1-\bar{x}_2}$
      $= (-.71 - 0)/\sqrt{(.31)^2/21 + (.16)^2/8}$
      $= -.71/.08818 = -8.051$
    P-value = $P(t_7 < -8.051) < .005$

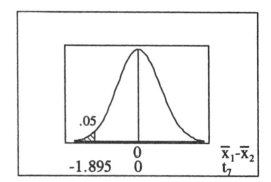

conclusion:
  Reject H_o; there is sufficient evidence to conclude that $\mu_1 - \mu_2 < 0$.

b. $(\bar{x}_1 - \bar{x}_2) \pm t_{7,.05} \sqrt{s_1^2/n_1 + s_2^2/n_2}$
  $-.71 \pm 1.895 \cdot \sqrt{(.31)^2/21 + (.16)^2/8}$
  $-.71 \pm .17$
  $-.88 < \mu_1 - \mu_2 < -.54$

c. Yes, cigarette filters appear to be effective in reducing the amount of nicotine transferred from the cigarette to the user. There is, of course, a fool-proof way to eliminate completely the amount of such nicotine taken into the body.

13. Let the control subjects be group 1.
    $\bar{x}_1 - \bar{x}_2 = .45 - .34 = .11$
    $(\bar{x}_1 - \bar{x}_2) \pm t_{9,.005} \sqrt{s_1^2/n_1 + s_2^2/n_2}$
    $.11 \pm 3.250 \cdot \sqrt{(.08)^2/10 + (.08)^2/10}$
    $.11 \pm .12$
    $-.01 < \mu_1 - \mu_2 < .23$
    The confidence intervals suggests that the two samples could come from populations with the same mean. Based on this result, one cannot be 99% certain that there is such a biological basis for the disorders.

14. Refer to the data and calculations of exercise #13.
    original claim: $\mu_1 - \mu_2 = 0$
    $H_o: \mu_1 - \mu_2 = 0$
    $H_1: \mu_1 - \mu_2 \neq 0$
    $\alpha = .05$
    C.R. $t < -t_{9,.005} = -3.250$
        $t > t_{9,.005} = 3.250$
    calculations:
    $t_{\bar{x}_1-\bar{x}_2} = (\bar{x}_1 - \bar{x}_2 - \mu_{\bar{x}_1-\bar{x}_2})/s_{\bar{x}_1-\bar{x}_2}$
      $= (.11 - 0)/\sqrt{(.08)^2/10 + (.08)^2/10}$
      $= .11/.0358 = 3.075$

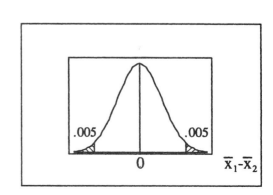

.01 < P-value = 2·P(t_9 > 3.075) < .02
conclusion:
   Do not reject H_o; there is not sufficient evidence to reject the claim that $\mu_1-\mu_2 = 0$. The test of hypothesis suggests that the two samples could come from populations with the same mean. Based on this result, one cannot be 99% certain that there is such a biological basis for the disorders.

15. Let the treatment subjects be group 1.
   $\bar{x}_1-\bar{x}_2 = 4.20 - 1.71 = 2.49$
   $(\bar{x}_1-\bar{x}_2) \pm t_{21,.025} \sqrt{s_1^2/n_1 + s_2^2/n_2}$
   $2.49 \pm 2.080 \cdot \sqrt{(2.20)^2/22 + (.72)^2/22}$
   2.49 ± 1.03
   $1.46 < \mu_1-\mu_2 < 3.52$
   Yes, since high scores indicate the presence of more errors and the confidence interval falls entirely above zero, the results support the common belief that drinking alcohol is hazardous for operators of passenger vehicles – and for their passengers.

16. Refer to the data and calculations of exercise #15.
   original claim: $\mu_1-\mu_2 = 0$
   H_o: $\mu_1-\mu_2 = 0$
   H_1: $\mu_1-\mu_2 \neq 0$
   $\alpha = .05$
   C.R. $t < -t_{21,.025} = -2.080$
         $t > t_{21,.025} = 2.080$
   calculations:
   $t_{\bar{x}_1-\bar{x}_2} = (\bar{x}_1-\bar{x}_2 - \mu_{\bar{x}_1-\bar{x}_2})/s_{\bar{x}_1-\bar{x}_2}$
      $= (2.49 - 0)/\sqrt{(2.20)^2/22 + (.72)^2/22}$
      $= 2.49/.4935 = 5.045$
   P-value = 2·P(t_{21} > 5.045) < .01

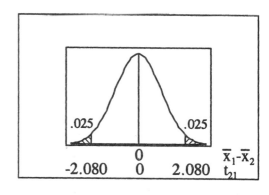

   conclusion:
      Reject H_o; there is sufficient evidence to reject the claim that $\mu_1-\mu_2 = 0$ and conclude that $\mu_1-\mu_2 \neq 0$ (in fact, that $\mu_1-\mu_2 > 0$).
   Yes; if there was no difference other than the treatment between the two groups, we can be 95% certain that the treatment was responsible for the decrease in visual and motor skills.

17. original claim: $\mu_1-\mu_2 > 0$
   $\bar{x}_1-\bar{x}_2 = 53.3 - 45.3 = 8.0$
   H_o: $\mu_1-\mu_2 = 0$
   H_1: $\mu_1-\mu_2 > 0$
   $\alpha = .01$
   C.R. $t > t_{39,.01} = 2.429$
   calculations:
   $t_{\bar{x}_1-\bar{x}_2} = (\bar{x}_1-\bar{x}_2 - \mu_{\bar{x}_1-\bar{x}_2})/s_{\bar{x}_1-\bar{x}_2}$
      $= (8.0 - 0)/\sqrt{(11.6)^2/40 + (13.2)^2/40}$
      $= 8.0/2.778 = 2.879$
   P-value = P(t_{39} > 2.879) < .005

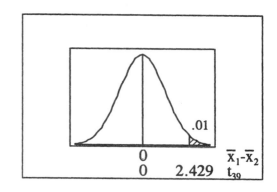

   conclusion:
      Reject H_o; there is sufficient evidence to conclude that $\mu_1-\mu_2 > 0$.

18. Refer to the data and calculations of exercise #17.
   $(\bar{x}_1-\bar{x}_2) \pm t_{39,.01} \sqrt{s_1^2/n_1 + s_2^2/n_2}$
   $8.0 \pm 2.429 \cdot \sqrt{(11.6)^2/40 + (13.2)^2/40}$
   8.0 ± 6.7
   $1.3 < \mu_1-\mu_2 < 14.7$

zero, the results support the claim that stress decreased the amount recalled.

19. Let the westbound stowaways be group 1.
original claim: $\mu_1-\mu_2 = 0$
$\overline{x}_1-\overline{x}_2 = 26.71 - 24.84 = 1.87$
H_o: $\mu_1-\mu_2 = 0$
H_1: $\mu_1-\mu_2 \neq 0$
$\alpha = .05$
C.R. $t < -t_{104,.025} = -1.983$ [from Excel]
$\quad\quad t > t_{104,.025} = 1.983$
calculations:
$t_{\overline{x}_1-\overline{x}_2} = (\overline{x}_1-\overline{x}_2 - \mu_{\overline{x}_1-\overline{x}_2})/s_{\overline{x}_1-\overline{x}_2}$
$= (1.87 - 0)/\sqrt{(103.30)/56 + (67.81)/75}$
$= 1.130$ [from Excel]
P-value $= 2 \cdot P(t_{104} > 1.130) = .2609$ [from Excel]

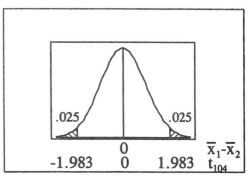

conclusion:
Do not reject H_o; there is not sufficient evidence to reject the claim that $\mu_1-\mu_2 = 0$.
No, there appears to be no significant difference between the ages.

20. Let the Clancy pages be group 1.
original claim: $\mu_1-\mu_2 = 0$
$\overline{x}_1-\overline{x}_2 = 70.73 - 80.75 = -10.02$
H_o: $\mu_1-\mu_2 = 0$
H_1: $\mu_1-\mu_2 \neq 0$
$\alpha = .05$ [assumed]
C.R. $t < -t_{14.65,.025} = $ [unstated by TI-83 Plus]
$\quad\quad t > t_{14.65,.025} = $ [unstated by TI-83 Plus]
calculations:
$t_{\overline{x}_1-\overline{x}_2} = (\overline{x}_1-\overline{x}_2 - \mu_{\overline{x}_1-\overline{x}_2})/s_{\overline{x}_1-\overline{x}_2}$
$= (-10.02 - 0)/\sqrt{(s_1)^2/12 + (s_2)^2/12}$
$= -2.831$ [from TI-83 Plus]

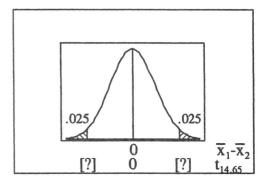

P-value $= 2 \cdot P(t_{14.65} < -2.831) = .0127$ [from TI-83 Plus]
conclusion:
Reject H_o; there is sufficient evidence to reject the claim that $\mu_1-\mu_2 = 0$ and conclude $\mu_1-\mu_2 \neq 0$ (in fact, $\mu_1-\mu_2 < 0$).
Yes, there is sufficient reason to believe the reading ease scores are different. Since a higher score represents greater reading ease, conclude that the Rowling material is easier reading that the Clancy material.

21. Let the filtered cigarettes be group 1
group 1: filtered (n = 21)   group 2: unfiltered (n = 8)
$\Sigma x = 279$ $\Sigma x = 192$
$\Sigma x^2 = 3987$ $\Sigma x^2 = 4628$
$\overline{x} = 13.286$ $\overline{x} = 24.000$
$s = 3.744$ $s = 1.690$
original claim: $\mu_1-\mu_2 < 0$
$\overline{x}_1-\overline{x}_2 = 13.286 - 24.000 = -10.714$
H_o: $\mu_1-\mu_2 = 0$
H_1: $\mu_1-\mu_2 < 0$
$\alpha = .05$
C.R. $t < -t_{7,.05} = -1.895$
calculations:
$t_{\overline{x}_1-\overline{x}_2} = (\overline{x}_1-\overline{x}_2 - \mu_{\overline{x}_1-\overline{x}_2})/s_{\overline{x}_1-\overline{x}_2}$
$= (-10.714 - 0)/\sqrt{(3.744)^2/21 + (1.690)^2/8}$
$= -.10.714/1.0122 = -10.585$

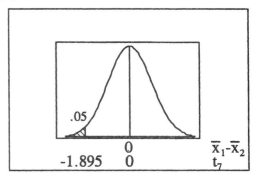

P-value = P(t_7 < -10.585) < .005
conclusion:
Reject H_o; there is sufficient evidence to conclude that $\mu_1 - \mu_2 < 0$.

22. Let the easy to difficult items be group 1
    group 1: E to D (n = 25)   group 2: D to E (n = 16)
    Σx = 677.88 Σx = 507.65
    Σx^2 = 19,509.3278 Σx^2 = 16,379.0161
    $\bar{x}$ = 27.115 $\bar{x}$ = 31.728
    s^2 = 47.0198 s^2 = 18.1489
    original claim: $\mu_1 - \mu_2 = 0$
    $\bar{x}_1 - \bar{x}_2$ = 27.115 - 31.728 = -4.613
    H_o: $\mu_1 - \mu_2 = 0$
    H_1: $\mu_1 - \mu_2 \neq 0$
    α = .05 [assumed]
    C.R. t < -$t_{15,.025}$ = -2.131
         t > $t_{15,.025}$ = 2.131
    calculations:
    $t_{\bar{x}_1 - \bar{x}_2}$ = ($\bar{x}_1 - \bar{x}_2 - \mu_{\bar{x}_1-\bar{x}_2}$)/$s_{\bar{x}_1-\bar{x}_2}$
          = (-4.613 - 0)/$\sqrt{47.0198/25 + 18.1489/16}$
          = -4.613/1.736 = -2.657

    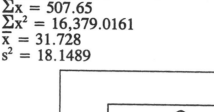

    .01 < P-value = 2·P(t_{15} < -2.657) < .02
    conclusion:
    Reject H_o; there is sufficient evidence to reject the claim that $\mu_1 - \mu_2 = 0$ and conclude $\mu_1 - \mu_2 \neq 0$ (in fact, $\mu_1 - \mu_2 < 0$).
    Yes; since high scores correspond to high anxiety, it appears that the easy to difficult arrangement is associated with less anxiety.

23. Let the men be group 1
    group 1: men (n = 40)   group 2: women (n = 40)
    Σx = 1039.9 Σx = 1029.6
    Σx^2 = 27,493.83 Σx^2 = 27,984.46
    $\bar{x}$ = 25.9975 $\bar{x}$ = 25.7400
    s = 3.43   s = 6.17
    original claim: $\mu_1 - \mu_2 = 0$
    $\bar{x}_1 - \bar{x}_2$ = 25.9975 - 25.7400 = .2575
    H_o: $\mu_1 - \mu_2 = 0$
    H_1: $\mu_1 - \mu_2 \neq 0$
    α = .05 [assumed]
    C.R. t < -$t_{39,.025}$ = -2.024
         t > $t_{39,.025}$ = 2.024
    calculations:
    $t_{\bar{x}_1 - \bar{x}_2}$ = ($\bar{x}_1 - \bar{x}_2 - \mu_{\bar{x}_1-\bar{x}_2}$)/$s_{\bar{x}_1-\bar{x}_2}$
          = (.2575 - 0)/$\sqrt{(3.43)^2/40 + (6.17)^2/40}$
          = .2575/1.1156 = .231

    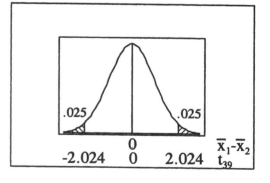

    P-value = 2·P(t_{39} > .231) > .20
    conclusion:
    Do not reject H_o; there is not sufficient evidence to reject the claim that $\mu_1 - \mu_2 = 0$.

24. Let the males be group 1
    group 1: males (n = 111)   group 2: females (n = 39)
    Σx = 4397 Σx = 1433
    Σx^2 = 184,083 Σx^2 = 57,841
    $\bar{x}$ = 39.613 $\bar{x}$ = 36.744
    s = 9.49   s = 11.68

P-value = 2·P(t_{39} > .231) > .20
conclusion:
Do not reject H_o; there is not sufficient evidence to reject the claim that $\mu_1-\mu_2 = 0$.

24. Let the males be group 1
   group 1: males (n = 111)   group 2: females (n=39)
   $\Sigma x = 4397$ $\Sigma x = 1433$
   $\Sigma x^2 = 184,083$ $\Sigma x^2 = 57,841$
   $\bar{x} = 39.613$ $\bar{x} = 36.744$
   s = 9.49   s = 11.68
   original claim: $\mu_1-\mu_2 = 0$
   $\bar{x}_1 - \bar{x}_2 = 39.613 - 36.744 = 2.869$
   $H_o: \mu_1-\mu_2 = 0$
   $H_1: \mu_1-\mu_2 \neq 0$
   $\alpha = .05$ [assumed]
   C.R. $t < -t_{38,.025} = -2.024$
        $t > t_{38,.025} = 2.024$
   calculations:
   $t_{\bar{x}_1-\bar{x}_2} = (\bar{x}_1-\bar{x}_2 - \mu_{\bar{x}_1-\bar{x}_2})/s_{\bar{x}_1-\bar{x}_2}$
       = $(2.869 - 0)/\sqrt{(9.49)^2/111 + (11.68)^2/39}$
       = 2.869/2.076 = 1.382

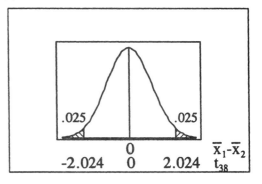

   .10 < P-value = 2·P(t_{38} > 1.382) < .20
conclusion:
   Do not reject H_o; there is not sufficient evidence to reject the claim that $\mu_1-\mu_2 = 0$.

25. Let the placebo users be group 1.
   $\bar{x}_1 - \bar{x}_2 = 21.57 - 20.38 = 1.19$
   $s_p^2 = [(n_1-1) \cdot s_1^2 + (n_2-1) \cdot s_2^2]/(n_1 + n_2 - 2)$
       = $[42(3.87)^2 + 32(3.91)^2]/74$
       = 15.111
   $(\bar{x}_1-\bar{x}_2) \pm t_{74,.025} \sqrt{s_p^2/n_1 + s_p^2/n_2}$
   $1.19 \pm 1.992 \cdot \sqrt{15.111/43 + 15.111/33}$
   1.19 ± 1.79
   $-.60 < \mu_1-\mu_2 < 2.98$

   df = df_1 + df_2
      = 42 + 32
      = 74

No; based on these results, we cannot be 95% certain that the two populations have different means. There is not enough evidence to make this the generally recommended treatment for bipolar depression.
▸The conclusion is the same, but the willingness to accept another assumption allows the confidence interval to be slightly narrower.

26. Refer to the data and calculations of exercise #25.
   original claim: $\mu_1-\mu_2 = 0$
   $H_o: \mu_1-\mu_2 = 0$
   $H_1: \mu_1-\mu_2 \neq 0$
   $\alpha = .05$
   C.R. $t < -t_{74,.025} = -1.992$
        $t > t_{74,.025} = 1.992$
   calculations:
   $t_{\bar{x}_1-\bar{x}_2} = (\bar{x}_1-\bar{x}_2 - \mu_{\bar{x}_1-\bar{x}_2})/s_{\bar{x}_1-\bar{x}_2}$
       = $(1.19 - 0)/\sqrt{15.111/43 + 15.111/33}$
       = 1.19/.8996 = 1.323
   .10 < P-value = 2·P(t_{74} > 1.323) < .20

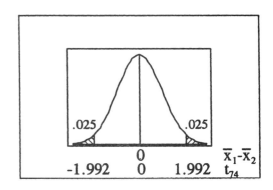

conclusion:
   Do not reject H_o; there is not sufficient evidence to reject the claim that $\mu_1-\mu_2 = 0$.
   This result indicates that there is not sufficient evidence to be 95% certain that the drug

S-248   INSTRUCTOR'S SOLUTIONS Chapter 8

does any better than no drug at all.
▶The conclusion is the same, but the willingness to accept another assumption makes rejection slightly more likely – the absolute value of the critical value is slightly smaller and the absolute value of the calculated test statistic is slightly larger.

27. Let those receiving the magnet treatment be group 1.
original claim: $\mu_1 - \mu_2 > 0$
$\bar{x}_1 - \bar{x}_2 = .49 - .44 = .05$
$s_p^2 = [(n_1-1) \cdot s_1^2 + (n_2-1) \cdot s_2^2]/(n_1 + n_2 - 2)$
$= [19(.96)^2 + 19(1.41)^2]/38 = 1.4408$

$df = df_1 + df_2$
$= 19 + 19 = 38$

$H_0: \mu_1 - \mu_2 = 0$
$H_1: \mu_1 - \mu_2 > 0$
$\alpha = .05$
C.R. $t > t_{38,.05} = 1.686$
calculations:
$t_{\bar{x}_1-\bar{x}_2} = (\bar{x}_1-\bar{x}_2 - \mu_{\bar{x}_1-\bar{x}_2})/s_{\bar{x}_1-\bar{x}_2}$
$= (.05 - 0)/\sqrt{1.4408/20 + 1.4408/20}$
$= .05/.3796 = .132$
P-value = $P(t_{38} > .132) > .10$

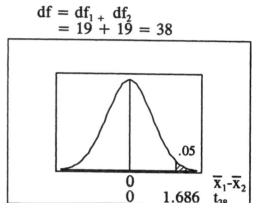

conclusion:
Do not reject H_0; there is not sufficient evidence to conclude that $\mu_1 - \mu_2 > 0$. No, it does not appear that magnets are effective in treating back pain. If much larger sample sizes achieved these same results, the calculated t could fall in the critical region and appear to provide evidence that the treatment is effective – but the observed difference would still be .05, and one would have to decide whether that statistically significant difference is of practical significance.
▶The conclusion is the same, but the willingness to accept another assumption makes rejection slightly more likely – the absolute value of the critical value is slightly smaller and the absolute value of the calculated test statistic is (since $n_1 = n_2$) the same.

28. Refer to the data and calculations of exercise #27.
$(\bar{x}_1-\bar{x}_2) \pm t_{38,.05} \sqrt{s_p^2/n_1 + s_p^2/n_2}$
$.05 \pm 1.686 \cdot \sqrt{1.4408/20 + 1.4408/20}$
$.05 \pm .64$
$-.59 < \mu_1 - \mu_2 < .69$
Since the interval includes zero, that the treatment has no effect is a reasonable possibility.
▶The conclusion is the same, but the willingness to accept another assumption allows the confidence interval to be slightly narrower.

29. NOTE: Exercise #19 used df=104 as determined by Excel, to produce CV = ±1.983. Even though changing the sample size and the variance of one of the groups will change these values, this manual re-uses them unchanged – both to avoid a discussion of Excel's algorithm for determining df, and to better compare the new result to the original one.
a. Let the westbound stowaways be group 1.
new group 1 values based on n = 57
$\Sigma x = 1496 + 90 = 1586$
$\Sigma x^2 = 45646 + 8100 = 53746$
$\bar{x} = 27.825$ $s^2 = 171.72$
original claim: $\mu_1 - \mu_2 = 0$
$\bar{x}_1 - \bar{x}_2 = 27.825 - 24.840 = 2.985$
$H_0: \mu_1 - \mu_2 = 0$
$H_1: \mu_1 - \mu_2 \neq 0$
$\alpha = .05$
C.R. $t < -t_{104,.025} = -1.983$ [from Excel]
$t > t_{104,.025} = 1.983$

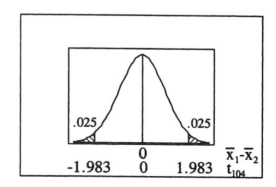

calculations:
$t_{\bar{x}_1-\bar{x}_2} = (\bar{x}_1-\bar{x}_2 - \mu_{\bar{x}_1-\bar{x}_2})/s_{\bar{x}_1-\bar{x}_2}$
= (2.985 - 0)/√(171.72)/57 + (67.81)/75
= 2.985/1.979 = 1.508
.10 < P-value = 2·P(t_{104} > 1.508) < .20

conclusion:
Do not reject H_o; there is not sufficient evidence to reject the claim that $\mu_1-\mu_2 = 0$. No, there appears to be no significant difference between the ages.
▶The conclusion is the same, but the P-value is smaller because the calculated t increased from 1.130 to 1.508.

b. Let the westbound stowaways be group 1.
new group 1 values based on n = 57
Σx = 1496+5000 = 6496
Σx^2 = 45,646+25,000,000 = 25,045,646
$\bar{x}$ = 27.825 s^2 = 434023
original claim: $\mu_1-\mu_2 = 0$
$\bar{x}_1-\bar{x}_2$ = 113.965 - 24.840 = 89.125
H_o: $\mu_1-\mu_2 = 0$
H_1: $\mu_1-\mu_2 \neq 0$
α = .05
C.R. t < $-t_{104,.025}$ = -1.983 [from Excel]
t > $t_{104,.025}$ = 1.983

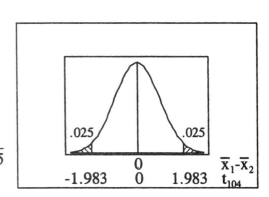

calculations:
$t_{\bar{x}_1-\bar{x}_2} = (\bar{x}_1-\bar{x}_2 - \mu_{\bar{x}_1-\bar{x}_2})/s_{\bar{x}_1-\bar{x}_2}$
= (89.125 - 0)/√434023/57 + 67.81/75
= 89.125/87.266 = 1.021
P-value = 2·P(t_{104} > 1.021) > .20

conclusion:
Do not reject H_o; there is not sufficient evidence to reject the claim that $\mu_1-\mu_2 = 0$. No, there appears to be no significant difference between the ages.
▶The conclusion is the same, but the calculated t statistic actually decreased from 1.130 to 1.021. Even though the difference between the sample means increased, the variability in the problem (which in a sense is a measure of uncertainty) increased even more.

30. Since $t_{\bar{x}_1-\bar{x}_2} = (\bar{x}_1-\bar{x}_2 - \mu_{\bar{x}_1-\bar{x}_2})/s_{\bar{x}_1-\bar{x}_2}$, the numerator and denominator have the same units – which means the units cancel out and the t statistic is unit free. The calculated t statistic in a test of hypothesis will not change. Changing the units will change the confidence interval accordingly – but it will still contain exactly the same points, only now described by their counterparts in terms of the new units.

31. a. x = 5,10,15
$\mu = \Sigma x/n$ = 30/3 = 10
$\sigma^2 = \Sigma(x-\mu)^2/n$ = [$(-5)^2 + (0)^2 + (5)^2$]/3 = 50/3

b. y = 1,2,3
$\mu = \Sigma y/n$ = 6/3 = 2
$\sigma^2 = \Sigma(y-\mu)^2/n$ = [$(-1)^2 + (0)^2 + (1)^2$]/3 = 2/3

c. z = x-y = 4,3,2,9,8,7,14,13,12
$\mu = \Sigma z/n$ = 72/9 = 8
$\sigma^2 = \Sigma(z-\mu)^2/n$
= [$(-4)^2 + (-5)^2 + (-6)^2 + (1)^2 + (0)^2 + (-1)^2 + (6)^2 + (5)^2 + (4)^2$]/9
= 156/9 = 52/3

d. $\sigma^2_{x-y} = \sigma^2_x + \sigma^2_y$
52/3 = 50/3 + 2/3
52/3 = 52/3

S-250 INSTRUCTOR'S SOLUTIONS Chapter 8

e. Let R stand for range.
R_{x-y} = highest$_{x-y}$ - lowest$_{x-y}$
= (highest x - lowest y) - (lowest x - highest y)
= highest x - lowest y - lowest x + highest y
= (highest x - lowest x) + (highest y - lowest y)
= $R_x + R_y$
The range of all possible x-y values is the sum of the individual ranges of x and y.
NOTE: The problem refers to all possible x-y differences (where n_x and n_y might even be different) and not to x-y differences for paired data.

32. original claim: $\mu_1-\mu_2 = 0$
$\bar{x}_1-\bar{x}_2 = .049 - .000 = .049$
$H_o: \mu_1-\mu_2 = 0$
$H_1: \mu_1-\mu_2 \neq 0$
$\alpha = .05$
C.R. $t < -t_{21,.025} = -2.080$
$t > t_{21,.025} = 2.080$
calculations:
$t_{\bar{x}_1-\bar{x}_2} = (\bar{x}_1-\bar{x}_2 - \mu_{\bar{x}_1-\bar{x}_2})/s_{\bar{x}_1-\bar{x}_2}$
$= (.049 - 0)/\sqrt{(.015)^2/22 + (0)^2/22}$
$= .049/.003198 = 15.322$
P-value = $2 \cdot P(t_{21} > 15.322) < .01$

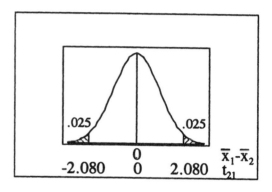

conclusion:
Reject H_o; there is sufficient evidence to reject the claim that $\mu_1-\mu_2 = 0$ and conclude $\mu_1-\mu_2 \neq 0$ (in fact, that $\mu_1-\mu_2 > 0$).
The fact that there was no variation in the second sample did not affect the calculations or present any special problems. Since there is no variation in x_2, it is really equivalent to the constant value zero – and the test is equivalent to the one-sample test $H_o: \mu_1 = 0$, for which $t = (\bar{x}_1 - 0)/s_{\bar{x}_1}$.

33. $A = s_1^2/n_1 = .0064/10 = .00064$ $df = (A + B)^2/(A^2/df_1 + B^2/df_2)$
$B = s_2^2/n_2 = .0064/10 = .00064$ $= (.00064 + .00064)^2/(.00064^2/9 + .00064^2/9)$
     $= .000001638/.000000091$
     $= 18$

When $s_1^2=s_2^2$ and $n_1=n_2$, formula 8-1 yields df = $df_1 + df_2$.
In this exercise the tabled t value changes from $t_{9,.005} = 3.250$ to $t_{18,.005} = 2.878$.
In general, the larger df signifies a "tighter" t distribution that is closer to the z distribution. In the test of hypothesis, the sampling distribution will be "tighter" and the critical t value will be smaller – since the calculated t is not affected, the P-value will be smaller. In the confidence interval, the tabled t value will be smaller – since the other values are not affected, the interval will be narrower. Using df = (smaller of df_1 or df_2) is more conservative in that will not reject H_o as often and will lead to wider confidence intervals – i.e., it will allows for a wider range of possible values for the parameter.

8-4 Inferences from Matched Pairs

NOTE: To be consistent with the notation of the previous sections, and thereby reinforcing the patterns and concepts presented in those sections, the manual uses the "usual" t formula written to apply to d's
$t_{\bar{d}} = (\bar{d} - \mu_{\bar{d}})/s_{\bar{d}}$, with $\mu_{\bar{d}} = \mu_d$ and $s_{\bar{d}} = s_d/\sqrt{n}$
And so the formula for the t statistic may also be written as
$t_{\bar{d}} = (\bar{d} - \mu_d)/(s_d/\sqrt{n})$

INSTRUCTOR'S SOLUTIONS Chapter 8   S-251

1. d = x - y: 1 -1 -2 -3 4
    summary: n = 5, $\Sigma d = -1$, $\Sigma d^2 = 31$
    a. $\bar{d} = (\Sigma d)/n = -1/5 = -.2$
    b. $s_d^2 = [n \cdot \Sigma d^2 - (\Sigma d)^2]/[n(n-1)]$
        $= [5 \cdot 31-(-1)^2]/[5(4)] = 154/20 = 7.7$
        $s_d = 2.8$
    c. $t_{\bar{d}} = (\bar{d} - \mu_{\bar{d}})/s_{\bar{d}}$
        $= (-.2 - 0)/(2.774/\sqrt{5})$
        $= -.2/1.241 = -.161$
    d. $\pm t_{4,.025} = \pm 2.776$

2. d = x - y: 0 2 5 3 -4 1
    summary: n = 6, $\Sigma d = 7$, $\Sigma d^2 = 55$
    a. $\bar{d} = (\Sigma d)/n = 7/6 = 1.2$
    b. $s_d^2 = [n \cdot \Sigma d^2 - (\Sigma d)^2]/[n(n-1)]$
        $= [6 \cdot 55-(7)^2]/[6(5)] = 281/30 = 9.367$
        $s_d = 3.1$
    c. $t_{\bar{d}} = (\bar{d} - \mu_{\bar{d}})/s_{\bar{d}}$
        $= (1.167 - 0)/(3.061/\sqrt{6})$
        $= 1.167/1.249 = .934$
    d. $\pm t_{5,.025} = \pm 2.571$

3. $\bar{d} \pm t_{4,.025} \cdot s_d/\sqrt{n}$
    $-.2 \pm 2.776 \cdot 2.775/\sqrt{5}$
    $-.2 \pm 3.4$
    $-3.6 < \mu_d < 3.2$

4. $\bar{d} \pm t_{5,.005} \cdot s_d/\sqrt{n}$
    $1.167 \pm 4.032 \cdot 3.061/\sqrt{6}$
    1.167 ± 5.037
    $-3.9 < \mu_d < 6.2$

5. d = x - y: -5.1 1.3 -0.1 1.2 0.8 -1.4 0.4 -0.5 -2.8 0.1 1.9 -1.5
    n = 12
    $\Sigma d = -5.7$ $\bar{d} = -.475$
    $\Sigma d^2 = 45.87$ $s_d = 1.981$
    a. original claim: $\mu_d = 0$
        $H_o: \mu_d = 0$
        $H_1: \mu_d \neq 0$
        $\alpha = .05$
        C.R. $t < -t_{11,.025} = -2.201$
             $t > t_{11,.025} = 2.201$
        calculations:
        $t_{\bar{d}} = (\bar{d} - \mu_{\bar{d}})/s_{\bar{d}}$
            $= (-.475 - 0)/(1.981/\sqrt{12})$
            $= -.475/.5718 = -.831$
        P-value = $2 \cdot P(t_{11} < -.831) > .20$

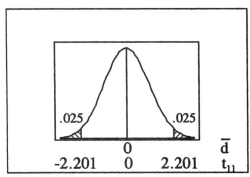

    conclusion:
        Do not reject H_o; there is not sufficient evidence to reject the claim that $\mu_d = 0$.
    b. $\bar{d} \pm t_{11,.025} \cdot s_d/\sqrt{n}$
       $-.475 \pm 2.201 \cdot 1.981/\sqrt{12}$
       $-.475 \pm 1.259$
       $-1.73 < \mu_d < .78$
    Since the confidence interval contains 0, there is no significant difference between the reported and measured heights.

6. d = x - y: 0.1 1.1 -1.9 1.7 0.7 -0.6 0.5 -3.0 -1.6 -11.2 1.0 1.2
    n = 12
    $\Sigma d = -12.0$ $\bar{d} = -1.0$
    $\Sigma d^2 = 148.26$ $s_d = 3.520$
    a. original claim: $\mu_d = 0$
        $H_o: \mu_d = 0$
        $H_1: \mu_d \neq 0$
        $\alpha = .05$
        C.R. $t < -t_{11,.025} = -2.201$
             $t > t_{11,.025} = 2.201$
        calculations:
        $t_{\bar{d}} = (\bar{d} - \mu_{\bar{d}})/s_{\bar{d}}$
            $= (-1.0 - 0)/(3.520/\sqrt{12})$
            $= -1.0/1.016 = -.984$
        P-value = $2 \cdot P(t_{11} < -.984) > .20$

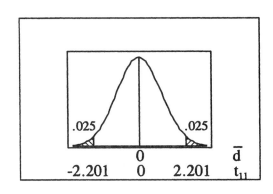

conclusion:
    Do not reject H_o; there is not sufficient evidence to reject the claim that $\mu_d = 0$.
b. $\bar{d} \pm t_{11,.025} \cdot s_d/\sqrt{n}$
   -1.0 $\pm$ 2.201·3.520/$\sqrt{12}$
   -1.0 $\pm$ 2.2
   -3.2 < μ_d < 1.2
Since the confidence interval contains 0, there is no significant difference between the reported and measured heights.

NOTE: The one person who understated his height by 11.2 inches (63 vs. 74.2) deserves special consideration. This almost certainly represents someone who meant to report 6'3" – and either mis-converted and entered 63", or entered 6'3" and it was mis-read as 63". The practicing statistician encounters data like this frequently and has 4 options: (1) go back to the original source and/or data sheet for clarification, (2) change the entry to 75", (3) throw out that pair of data, (4) accept the entry as is.

7. $d = x - y$: -20   0   10   -40   -30   -10   30   -20   -20   -10
   $n = 10$
   $\Sigma d = -110$ $\bar{d} = -11.0$
   $\Sigma d^2 = 4900$ $s_d = 20.248$
  a. original claim: $\mu_d < 0$
    $H_o: \mu_d = 0$
    $H_1: \mu_d < 0$
    $\alpha = .05$
    C.R. $t < -t_{9,.05} = -1.833$
    calculations:
      $t_{\bar{d}} = (\bar{d} - \mu_{\bar{d}})/s_{\bar{d}}$
         $= (-11.0 - 0)/(20.248/\sqrt{10})$
         $= -11.0/6.40$
         $= -1.718$
    .05 < P-value = $P(t_9 < -1.718)$ < .10

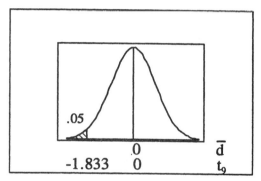

    conclusion:
      Do not reject H_o; there is not sufficient evidence to conclude that $\mu_d < 0$.
  b. $\bar{d} \pm t_{9,.025} \cdot s_d/\sqrt{n}$
   -11.0 $\pm$ 2.262·20.248/$\sqrt{10}$
   -11.0 $\pm$ 14.5
   -25.5 < μ_d < 3.5
We have 95% confidence that the interval from -25.5 to 3.5 contains the true mean population difference. Since this interval includes 0, the mean before and after scores are not significantly different, and there is not enough evidence to say that the course has any effect.

8. $d = x_B - x_A$: 9   4   21   3   20   31   17   26   26   10   23   33
   $n = 12$
   $\Sigma d = 223$ $\bar{d} = 18.58$
   $\Sigma d^2 = 5267$ $s_d = 10.104$
  a. $\bar{d} \pm t_{11,.005} \cdot s_d/\sqrt{n}$
   18.58 $\pm$ 3.106·10.104/$\sqrt{12}$
   18.58 $\pm$ 9.06
   9.5 < μ_d < 27.6
  b. original claim: $\mu_d > 0$
    $H_o: \mu_d \leq 0$
    $H_1: \mu_d > 0$
    $\alpha = .05$ [assumed]
    C.R. $t > t_{11,.05} = 1.796$

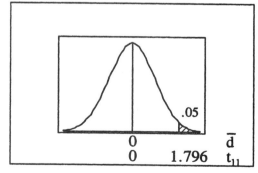

calculations:
$t_{\bar{d}} = (\bar{d} - \mu_{\bar{d}})/s_{\bar{d}}$
$= (18.58 - 0)/(10.104/\sqrt{12})$
$= 6.371$
P-value $= P(t_{11} > 6.371) < .005$
conclusion:
Reject H_o; there is sufficient evidence to conclude that $\mu_d > 0$.

9. $d = x_B - x_A$: -.2 4.1 1.6 1.8 3.2 2.0 2.9 9.6
$n = 8$
$\Sigma d = 25.0$ $\bar{d} = 3.125$
$\Sigma d^2 = 137.46$ $s_d = 2.9114$

a. $\bar{d} \pm t_{7,.025} \cdot s_d/\sqrt{n}$
$3.125 \pm 2.365 \cdot 2.9114/\sqrt{8}$
3.125 ± 2.434
$.69 < \mu_d < 5.56$

b. original claim: $\mu_d > 0$
$H_o: \mu_d = 0$
$H_1: \mu_d > 0$
$\alpha = .05$
C.R. $t > t_{7,.05} = 1.895$
calculations:
$t_{\bar{d}} = (\bar{d} - \mu_{\bar{d}})/s_{\bar{d}}$
$= (3.125 - 0)/(2.9114/\sqrt{8})$
$= 3.125/1.029 = 3.036$
$.005 < $ P-value $= P(t_7 > 3.036) < .01$

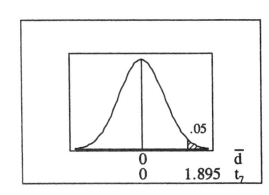

conclusion:
Reject H_o; there is sufficient evidence to conclude that $\mu_d > 0$.

c. Yes; hypnotism appears to be effective in reducing pain.

10. $d = x_1 - x_2$: 0 13 5 15 15 126 28 -2 -5 31 3 51 3 14 37
$n = 15$
$\Sigma d = 334$ $\bar{d} = 22.27$
$\Sigma d^2 = 22478$ $s_d = 32.777$

a. original claim: $\mu_d = 0$
$H_o: \mu_d = 0$
$H_1: \mu_d \neq 0$
$\alpha = .01$
C.R. $t < -t_{14,.005} = -2.977$
$t > t_{14,.005} = 2.977$
calculations:
$t_{\bar{d}} = (\bar{d} - \mu_{\bar{d}})/s_{\bar{d}}$
$= (22.27 - 0)/(32.777/\sqrt{15})$
$= 22.27/8.463 = 2.631$
$.01 < $ P-value $= 2 \cdot P(t_{14} > 2.631) < .02$

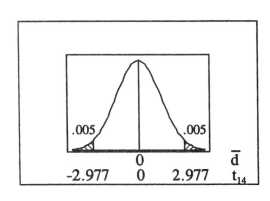

conclusion:
Do not reject H_o; there is not sufficient evidence to reject the claim that $\mu_d = 0$.

b. $\bar{d} \pm t_{14,.005} \cdot s_d/\sqrt{n}$
$22.27 \pm 2.977 \cdot 32.777/\sqrt{15}$
22.27 ± 25.19
$-2.9 < \mu_d < 47.5$
Yes; the confidence interval limits contain 0, thus indicating we cannot be 99% certain there is a difference between the first and second trials.

S-254 INSTRUCTOR'S SOLUTIONS Chapter 8

11. d = x - y: -106 20 -101 33 -72 36 -62 -38 70 -127 -24
    n = 11
    $\Sigma d = -371$ $\bar{d} = -33.73$
    $\Sigma d^2 = 56299$ $s_d = 66.171$
    a. original claim: $\mu_d = 0$
       H_o: $\mu_d = 0$
       H_1: $\mu_d \neq 0$
       $\alpha = .05$
       C.R. $t < -t_{10,.025} = -2.228$
            $t > t_{10,.025} = 2.228$
       calculations:
          $t_{\bar{d}} = (\bar{d} - \mu_{\bar{d}})/s_{\bar{d}}$
               $= (-33.73 - 0)/(66.171/\sqrt{11})$
               $= -33.73/19.951 = -1.690$
          $.10 < \text{P-value} = 2 \cdot P(t_{11} < -1.690) < .20$
       conclusion:
          Do not reject H_o; there is not sufficient evidence to reject the claim that $\mu_d = 0$.
    b. $\bar{d} \pm t_{10,.025} \cdot s_d/\sqrt{n}$
       $-33.73 \pm 2.228 \cdot 66.171/\sqrt{11}$
       -33.73 ± 44.45
       $-78.2 < \mu_d < 10.7$
    c. We cannot be 95% certain that either type of seed is better. If there are no differences in cost, or any other considerations, choose the kiln dried – even though we can't be sure that it's generally better, it did have the higher yield in this particular trial.

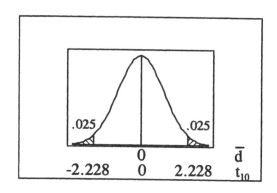

12. d = x - y: -4 -11 2 -3 -4 -4 -8 -10 2 -4 -3 -4 -10 3 -10 -1 -8 -1 -6 -4
    n = 20
    $\Sigma d = -88$ $\bar{d} = -4.4$
    $\Sigma d^2 = 718$ $s_d = 4.173$
    a. original claim: $\mu_d < 0$
       H_o: $\mu_d = 0$
       H_1: $\mu_d < 0$
       $\alpha = .01$
       C.R. $t < -t_{19,.01} = -2.539$
       calculations:
          $t_{\bar{d}} = (\bar{d} - \mu_{\bar{d}})/s_{\bar{d}}$
               $= (-4.4 - 0)/(4.173/\sqrt{20})$
               $= -4.4/.933 = -4.716$
          P-value = $P(t_{19} < -4.716) < .005$
       conclusion:
          Reject H_o; there is sufficient evidence to conclude that $\mu_d < 0$.
    b. $\bar{d} \pm t_{19,.01} \cdot s_d/\sqrt{n}$
       $-4.4 \pm 2.539 \cdot 4.173/\sqrt{20}$
       -4.4 ± 2.4
       $-6.8 < \mu_d < -2.0$

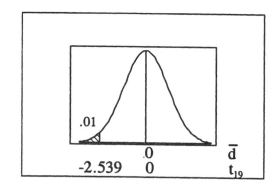

13. a. original claim: $\mu_d \neq 0$
       H_o: $\mu_d = 0$
       H_1: $\mu_d \neq 0$
       $\alpha = .05$
       C.R. $t < -t_{9,.025} = -2.262$
            $t > t_{9,.025} = 2.262$
       calculations:
          $t_{\bar{d}} = (\bar{d} - \mu_{\bar{d}})/s_{\bar{d}}$
               $= (-7.5 - 0)/(57.7/\sqrt{10})$
               $= -7.5/18.2 = -.41$ [from Minitab]

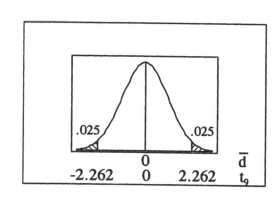

P-value = 2·P(t_9 < -.41) = .691 [from Minitab]
conclusion:
    Do not reject H_o; there is not sufficient evidence to conclude that $\mu_d \neq 0$.
No; based on this result, do not spend the money for the drug.

  b. original claim: $\mu_d < 0$
    H_o: $\mu_d = 0$
    H_1: $\mu_d < 0$
    $\alpha = .05$
    C.R. $t < -t_{9,.05} = -1.833$
    calculations:
      $t_{\bar{d}} = (\bar{d} - \mu_{\bar{d}})/s_{\bar{d}}$
         = (-7.5 - 0)/(57.7/$\sqrt{10}$)
         = -7.5/18.2 = -.41 [from Minitab]
    P-value = P(t_9 < -.41) = ½(.691) = .3455

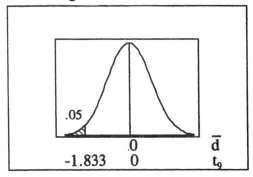

conclusion:
    Do not reject H_o; there is not sufficient evidence to conclude that $\mu_d < 0$.
All the calculations remain the same. The alternative hypothesis and critical region change to reflect the one-tailed test. The P-value is ½ the .691 given by Minitab for the two-tailed test.

14. a. original claim: $\mu_d > 0$
    H_o: $\mu_d = 0$
    H_1: $\mu_d > 0$
    $\alpha = .05$
    C.R. $t > t_{6,.05} = 1.943$
    calculations:
      $t_{\bar{d}} = (\bar{d} - \mu_{\bar{d}})/s_{\bar{d}}$
         = (5.142 - 0)/(2.116/$\sqrt{7}$)
         = 6.431 [from TI-83 Plus]
    P-value = P(t_6 > 6.431) = 3.34x10^{-4} = .000334

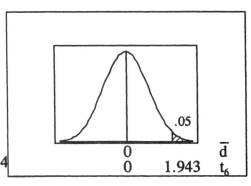

conclusion:
    Reject H_o; there is sufficient evidence to conclude that $\mu_d > 0$.
Yes; only .0334% of the time would such a weight loss occur by chance if the diet had no effect.

  b. $\bar{d} = 5.142$ lbs.
    Yes; the weight loss is large enough to make the diet practical

  c. $\bar{d} \pm t_{6,.025} \cdot s_d/\sqrt{n}$
    5.14 ± 2.447·2.116/$\sqrt{7}$
    5.14 ± 1.96
    3.2 < μ_d < 7.1

15. original claim: $\mu_d \neq 0$
    H_o: $\mu_d = 0$
    H_1: $\mu_d \neq 0$
    $\alpha = .05$ [assumed]
    C.R. $t < -t_{11,.025} = -2.201$
         $t > t_{11,.025} = 2.201$
    calculations:
      $t_{\bar{d}} = (\bar{d} - \mu_{\bar{d}})/s_{\bar{d}}$
         = -.501 [from Excel]
    P-value = 2·P(t_{11} < -.51) = .626 [from Excel]

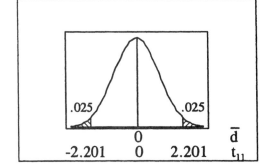

conclusion:
    Do not reject H_o; there is not sufficient evidence to conclude $\mu_d \neq 0$.

16. original claim: $\mu_d > 0$
$H_0: \mu_d = 0$
$H_1: \mu_d > 0$
$\alpha = .05$
C.R. $t > t_{10,.05} = 1.812$
calculations:
$t_{\bar{d}} = (\bar{d} - \mu_{\bar{d}})/s_{\bar{d}}$
$= (.6727 - 0)/(.8259/\sqrt{11})$
$= .6727/.2490 = 2.701$ [from Statdisk]
P-value $= P(t_{10} > 2.701) = .011$ [from Statdisk]

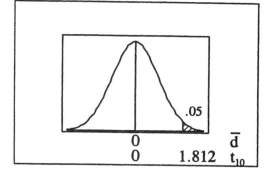

conclusion:
Reject H_0; there is sufficient evidence to conclude that $\mu_d > 0$.
Yes, these is enough evidence to support the claim that male statistics students exaggerate their heights.

17. $d = x_8 - x_{12}$: given at the right
$n = 11$
$\Sigma d = -8.6$ $\bar{d} = -.782$
$\Sigma d^2 = 15.06$ $s_d = .913$

sbj#	8am	12am	diff
19	97.0	97.7	-0.7
20	98.0	98.8	-0.8
22	96.4	98.0	-1.6
26	98.2	98.7	-0.5
71	98.8	98.0	0.8
78	98.6	98.5	0.1
80	97.8	98.3	-0.5
81	98.7	98.7	0.0
83	97.8	99.1	-1.3
98	96.4	98.2	-1.8
99	96.9	99.2	-2.3

a. $\bar{d} \pm t_{10,.025} \cdot s_d/\sqrt{n}$
$-.782 \pm 2.228 \cdot .913/\sqrt{11}$
$-.782 \pm .613$
$-1.40 < \mu_d < -.17$

b. original claim: $\mu_d = 0$
$H_0: \mu_d = 0$
$H_1: \mu_d \neq 0$
$\alpha = .05$
C.R. $t < -t_{10,.025} = -2.228$
    $t > t_{10,.025} = 2.228$
calculations:
$t_{\bar{d}} = (\bar{d} - \mu_{\bar{d}})/s_{\bar{d}}$
$= (-.782 - 0)/(.913/\sqrt{11})$
$= -.782/.275 = -2.840$
$.01 < $ P-value $= 2 \cdot P(t_{10} < -2.840) < .02$

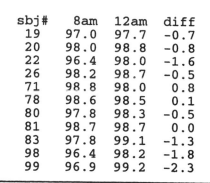

conclusion:
Reject H_0; there is sufficient evidence to reject the claim that $\mu_d = 0$ and conclude that $\mu_d \neq 0$ (in fact, that $\mu_d < 0$).
No; based on this result. morning and night body temperatures do not appear to be about the same. The 8 am (morning) temperatures are significantly lower.

18. The Minitab output is given below.
MTB > let c6 = c4-c5
MTB > ttest c6   TEST OF MU = 0.0 VS MU N.E. 0.0
         N   MEAN   STDEV   SE MEAN   T   P VALUE
   C6   50   25.0   109.2   15.4   1.62   0.11

a. original claim: $\mu_d = 0$
$H_0: \mu_d = 0$
$H_1: \mu_d \neq 0$
$\alpha = .05$ [assumed]
C.R. $t < -t_{49,.025} = -2.009$
    $t > t_{49,.025} = 2.009$
calculations:
$t_{\bar{d}} = (\bar{d} - \mu_{\bar{d}})/s_{\bar{d}}$
$= (25.0 - 0)/(109.2/\sqrt{50})$
$= 25.0/15.44 = 1.62$ [Minitab]

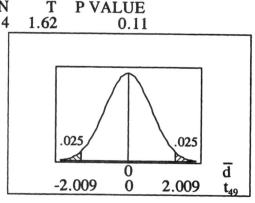

P-value = $2 \cdot P(t_{49} > 1.62) = .11$ [Minitab]
conclusion:
Do not reject H_o; there is not sufficient evidence to reject the claim that $\mu_d = 0$.

b. $\bar{d} \pm t_{49,.005} \cdot s_d/\sqrt{n}$
25.0 $\pm 2.678 \cdot 109.2/\sqrt{50}$
25.0 $\pm$ 41.4
$-16.4 < \mu_d < 66.4$
No; since the interval includes zero, there does not appear to be a significant difference in the times.

19. The Minitab output is given below.
   MTB > let c11=c3-c9
   MTB > ttest c11   TEST OF MU = 0.00 VS MU N.E. 0.00
     N MEAN STDEV SE MEAN T P VALUE
   C11 31 -6.00 11.26   2.02 -2.97 0.0059

   a. original claim: $\mu_d = 0$
   H_o: $\mu_d = 0$
   H_1: $\mu_d \ne 0$
   $\alpha = .05$
   C.R. $t < -t_{30,.025} = -2.042$
     $t > t_{30,.025} = 2.042$
   calculations:
    $t_{\bar{d}} = (\bar{d} - \mu_{\bar{d}})/s_{\bar{d}}$
     $= (-6.0 - 0)/(11.26/\sqrt{31})$
     $= -6.0/2.022 = -2.97$ [Minitab]
   P-value = $2 \cdot P(t_{30} < -2.97) = .0059$ [Minitab]

   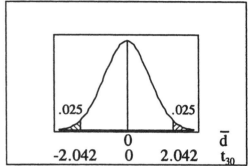

   conclusion:
    Reject H_o; there is sufficient evidence to reject the claim that $\mu_d = 0$ and conlcude that $\mu_d \ne 0$ (in fact, that $\mu_d < 0$).

   b. $\bar{d} \pm t_{30,.025} \cdot s_d/\sqrt{n}$
   $-6.0 \pm 2.042 \cdot 11.26/\sqrt{31}$
   -6.0 ± 4.1
   $-10.1 < \mu_d < -1.9$

   c. Even though the conclusions are different for the example in the chapter (using the first 5 values) and this exercise (using all the data), the results are not contradictory. With n=5, there was almost (but not quite) enough evidence to declare a difference; with n=31 (even though the mean difference is smaller), there is enough evidence to declare a difference. It appears that the five-day low forecasts tend to be higher than the low temperatures actually reached.

20. The Minitab output is given below.
   MTB > let c11 = c3-c5
   MTB > ttest c11   TEST OF MU = 0.00 VS MU N.E. 0.00
     N MEAN STDEV SE MEAN T P VALUE
   C11 31 -3.52 11.81   2.12 -1.66 0.11

   a. original claim: $\mu_d = 0$
   H_o: $\mu_d = 0$
   H_1: $\mu_d \ne 0$
   $\alpha = .05$
   C.R. $t < -t_{30,.025} = -2.042$
     $t > t_{30,.025} = 2.042$
   calculations:
    $t_{\bar{d}} = (\bar{d} - \mu_{\bar{d}})/s_{\bar{d}}$
     $= (-3.52 - 0)/(11.81/\sqrt{31})$
     $= -3.52/2.121 = -1.66$ [Minitab]

   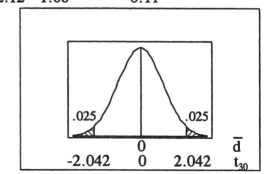

P-value = $2 \cdot P(t_{30} < -1.66) = .11$ [Minitab]
conclusion:
    Do not reject H_o; there is not sufficient evidence to reject the claim that $\mu_d = 0$.

b. $\bar{d} \pm t_{30,.025} \cdot s_d/\sqrt{n}$
  $-3.52 \pm 2.042 \cdot 11.81/\sqrt{31}$
  -3.52 ± 4.331
  $-7.9 < \mu_d < 0.8$

c. In exercise #19 there was a significant difference between the five-day forecasts and the actual temperatures. In this exercise there is not a significant difference between the one-day forecasts and the actual temperatures. Yes, the one-day forecasts appear more accurate than the five-day forecasts.

21. a. Yes, an outlier can have a drastic effect on the test of hypothesis and/or the confidence interval. Depending on the data an outlier could increase the mean difference to create a significant difference where none exists, or an outlier could increase the variability to hide a significant different that does exist.

    b. Since the test statistic is unit free, it will be same no matter what units are used. The confidence interval will be stated in the new units, but it will include precisely the same climactic conditions no matter how those conditions are described in the various units.

22. If the lower limit of the 95% confidence interval is 0, then
$$\bar{d} - t_{df,.025} \cdot s_d/\sqrt{n} = 0$$
$$\bar{d} = t_{df,.025} \cdot s_d/\sqrt{n}$$
$$\bar{d} - 0 = t_{df,.025} \cdot s_d/\sqrt{n}$$
$$(\bar{d} - 0)/(s_d/\sqrt{n}) = t_{df,.025}$$
Now consider the following test of hypothesis.
H_o: $\mu_d = 0$
H_1: $\mu_d > 0$
$\alpha =$
C.R. $t > t_{df,\alpha}$
calculations
$t_{\bar{d}} = (\bar{d} - \mu_{\bar{d}})/s_{\bar{d}}$
    $= (\bar{d} - 0)/(s_d/\sqrt{n})$
    $= t_{df,.025}$ [from the algebraic rearrangement above]
This means that the calculated t will fall right on the boundary of the critical region when testing at the $\alpha = .025$ level of significance. For $\alpha > .025$, $t_{df,\alpha}$ will be smaller and H_o is rejected. For $\alpha < .025$, $t_{df,\alpha}$ will be larger and H_o is not rejected. The smallest possible level of significance at which H_1: $\mu_d > 0$ will be supported is .025.

23. The following table of summary statistics applies to all parts of this exercise.

	values										n	Σv	Σv^2	$\bar{v}$	s_v
x:	1	3	2	2	1	2	3	3	2	1	10	20	46	2.0	.816
y:	1	2	1	2	1	2	1	2	1	2	10	15	25	1.5	.527
d=x-y:	0	1	1	0	0	0	2	1	1	-1	10	5	9	.5	.850

a. original claim: $\mu_d > 0$
H_o: $\mu_d = 0$
H_1: $\mu_d > 0$
$\alpha = .05$
C.R. $t > t_{9,.05} = 1.833$
calculations:
$t_{\bar{d}} = (\bar{d} - \mu_{\bar{d}})/s_{\bar{d}}$
    $= (.5 - 0)/(.850/\sqrt{10})$
    $= .5/.2687 = 1.861$
$.025 < \text{P-value} = P(t_9 > 1.861) < .05$

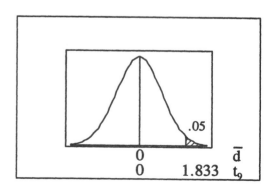

conclusion:
   Reject H_o; there is sufficient evidence to conclude that $\mu_d > 0$.

b. original claim: $\mu_1 - \mu_2 > 0$
   $\bar{x}_1 - \bar{x}_2 = 2.0 - 1.5 = .5$
   $H_o: \mu_1 - \mu_2 = 0$
   $H_1: \mu_1 - \mu_2 > 0$
   $\alpha = .01$
   C.R. $t > t_{9,.05} = 1.833$
   calculations:
   $t_{\bar{x}_1-\bar{x}_2} = (\bar{x}_1 - \bar{x}_2 - \mu_{\bar{x}_1-\bar{x}_2})/s_{\bar{x}_1-\bar{x}_2}$
   $= (.5 - 0)/\sqrt{(.816)^2/10 + (.527)^2/10}$
   $= .5/.3073 = 1.627$
   $.05 < $ P-value $= P(t_9 > 1.627) < .10$

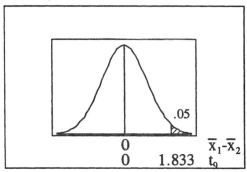

conclusion:
   Do not reject H_o; there is not sufficient evidence to conclude that $\mu_1 - \mu_2 > 0$.

c. Yes; since different methods can give different results, it is important that the correct method is used.

8-5 Comparing Variation in Two Samples

NOTE: The following convention are used in this manual regarding the F test.
* The set if scores with the larger sample variance is designated group 1.
* Even though always designating the scores with the larger sample variance as group 1 makes lower critical unnecessary in two-tailed tests, the lower critical value are calculated (using the method given in exercise #19) and included for completeness and consistency with the other tests. The F distribution always "bunches up" with expected value 1.0, regardless of the df.
* The df for group 1 (numerator) and group 2 (denominator) are given with the F as a superscript and subscript respectively.
* If the desired df does not appear in Table A-5, the closest entry is used. If the desired entry is exactly halfway between two tabled values, the conservative approach of using the smaller df is employed. Since any finite number is closer to 120 than ∞, 120 is used for all df larger than 120.
* Since all hypotheses in the text question the equality of σ_1^2 and σ_2^2, the calculation of F [which is statistically defined to be $F = (s_1^2/\sigma_1^2)/(s_2^2/\sigma_2^2)$] is shortened to $F = s_1^2/s_2^2$.
* Some problems are stated in terms of variance, and some are stated in terms of standard deviation. Since $\sigma_1^2 = \sigma_2^2$ is equivalent to $\sigma_1^2 = \sigma_2^2$, the manual simply states all claims and hypotheses and conclusions using the variance.

1. Let the treatment population be group 1.
   original claim: $\sigma_1^2 \neq \sigma_2^2$
   $H_o: \sigma_1^2 = \sigma_2^2$
   $H_1: \sigma_1^2 \neq \sigma_2^2$
   $\alpha = .05$
   C.R. $F < F^{24}_{29,.975} = .4527$
   $F > F^{24}_{29,.025} = 2.1540$
   calculations:
   $F = s_1^2/s_2^2$
   $= (.78)^2/(.52)^2$
   $= 2.25$

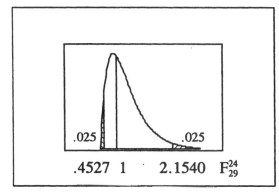

conclusion:
   Reject H_o; there is sufficient evidence to conclude that $\sigma_1^2 \neq \sigma_2^2$ (in fact, that $\sigma_1^2 > \sigma_2^2$).

2. Let the males be group 1.
   original claim: $\sigma_1^2 > \sigma_2^2$
   H_o: $\sigma_1^2 = \sigma_2^2$
   H_1: $\sigma_1^2 > \sigma_2^2$
   $\alpha = .05$
   C.R. $F > F_{11,.05}^{15} = 2.7186$
   calculations:
   $F = s_1^2/s_2^2$
     $= (.54)^2/(.39)^2$
     $= 1.9172$

   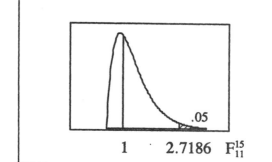

   conclusion:
   Do not reject H_o; there is not sufficient evidence to conclude that $\sigma_1^2 > \sigma_2^2$.

3. Let the sham population be group 1.
   original claim: $\sigma_1^2 > \sigma_2^2$
   H_o: $\sigma_1^2 = \sigma_2^2$
   H_1: $\sigma_1^2 > \sigma_2^2$
   $\alpha = .05$
   C.R. $F > F_{19,.05}^{19} = 2.1555$
   calculations:
   $F = s_1^2/s_2^2$
     $= (1.4)^2/(.96)^2$
     $= 2.1267$

   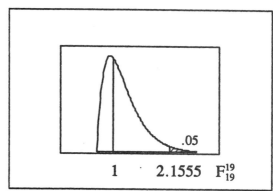

   conclusion:
   Do not reject H_o; there is not sufficient evidence to conclude that $\sigma_1^2 > \sigma_2^2$.

4. Let the heavy users be group 1.
   original claim: $\sigma_1^2 \neq \sigma_2^2$
   H_o: $\sigma_1^2 = \sigma_2^2$
   H_1: $\sigma_1^2 \neq \sigma_2^2$
   $\alpha = .05$
   C.R. $F < F_{63,.975}^{64} = .6000$
       $F > F_{63,.025}^{64} = 1.6668$
   calculations:
   $F = s_1^2/s_2^2$
     $= (4.5)^2/(3.6)^2$
     $= 1.5625$

   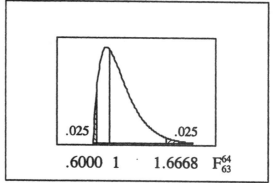

   conclusion:
   Do not reject H_o; there is not sufficient evidence to conclude that $\sigma_1^2 \neq \sigma_2^2$.

5. Let the regular Coke be group 1.
   original claim: $\sigma_1^2 \neq \sigma_2^2$
   H_o: $\sigma_1^2 = \sigma_2^2$
   H_1: $\sigma_1^2 \neq \sigma_2^2$
   $\alpha = .05$
   C.R. $F < F_{35,.975}^{35} = .4822$
       $F > F_{35,.025}^{35} = 2.0739$
   calculations:
   $F = s_1^2/s_2^2$
     $= (.007507)^2/(.004391)^2$
     $= 2.9228$
   conclusion:
   Reject H_o; there is sufficient evidence to conclude that $\sigma_1^2 \neq \sigma_2^2$ (in fact, that $\sigma_1^2 > \sigma_2^2$). The significant difference in variability between the weights of regular and diet Coke is not necessarily due to differences in quality control efforts. Regular Coke may contain a

relatively heavy ingredient (e.g., sugar) that is difficult to dispense with the same level of consistency as the other ingredients.

6. Let the .111" cans be group 1.
   original claim: $\sigma_1^2 \neq \sigma_2^2$
   $H_o: \sigma_1^2 = \sigma_2^2$
   $H_1: \sigma_1^2 \neq \sigma_2^2$
   $\alpha = .05$
   C.R. $F < F^{174}_{174,.975} = .6980$
   $F > F^{174}_{174,.025} = 1.4327$
   calculations:
   $F = s_1^2/s_2^2$
   $= (27.8)^2/(22.1)^2$
   $= 1.5824$
   conclusion:
   Reject H_o; there is sufficient evidence to conclude that $\sigma_1^2 \neq \sigma_2^2$ (in fact, that $\sigma_1^2 > \sigma_2^2$).

   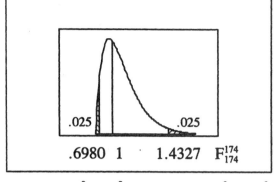

7. Let the filtered cigarettes be group 1.
   original claim: $\sigma_1^2 > \sigma_2^2$
   $H_o: \sigma_1^2 = \sigma_2^2$
   $H_1: \sigma_1^2 > \sigma_2^2$
   $\alpha = .05$
   C.R. $F > F^{20}_{7,.05} = 3.4445$
   calculations:
   $F = s_1^2/s_2^2$
   $= (.31)^2/(.16)^2$
   $= 3.7539$
   conclusion:
   Reject H_o; there is sufficient evidence to conclude that $\sigma_1^2 > \sigma_2^2$.

   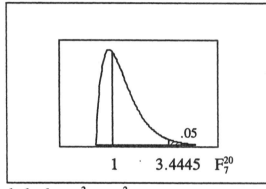

8. Let the treatment population be group 1.
   original claim: $\sigma_1^2 > \sigma_2^2$
   $H_o: \sigma_1^2 = \sigma_2^2$
   $H_1: \sigma_1^2 > \sigma_2^2$
   $\alpha = .05$
   C.R. $F > F^{21}_{21,.05} = 2.0960$
   calculations:
   $F = s_1^2/s_2^2$
   $= (2.20)^2/(.72)^2$
   $= 9.3364$
   conclusion:
   Reject H_o; there is sufficient evidence to conclude that $\sigma_1^2 > \sigma_2^2$.

   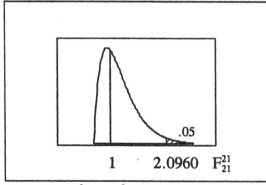

9. Let the students be group 1.
   original claim: $\sigma_1^2 > \sigma_2^2$
   $H_o: \sigma_1^2 = \sigma_2^2$
   $H_1: \sigma_1^2 > \sigma_2^2$
   $\alpha = .05$ [assumed]
   C.R. $F > F^{216}_{151,.05} = 1.3519$
   calculations:
   $F = s_1^2/s_2^2$
   $= (3.67)^2/(3.65)^2$
   $= 1.0110$
   conclusion:
   Do not reject H_o; there is not sufficient evidence to conclude that $\sigma_1^2 > \sigma_2^2$.

   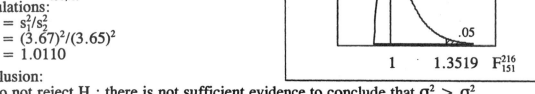

S-262 INSTRUCTOR'S SOLUTIONS Chapter 8

10. Let the placebo population be group 1.
original claim: $\sigma_1^2 > \sigma_2^2$
H_o: $\sigma_1^2 = \sigma_2^2$
H_1: $\sigma_1^2 > \sigma_2^2$
$\alpha = .05$
C.R. $F > F^{285}_{293,.05} = 1.3519$
calculations:
$F = s_1^2/s_2^2$
$= (728)^2/(669)^2$
$= 1.1842$

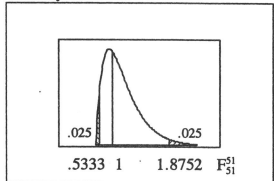

conclusion:
Do not reject H_o; there is not sufficient evidence to conclude that $\sigma_1^2 > \sigma_2^2$.

11. NOTE: While Data Set 11 covers a full year of 365 days with 53 Wednesdays and 52 observations for each of the other days, part (a) chooses to use equal sample sizes by dropping December 31 and using the first 52 Wednesdays.

a. Let the Sundays be group 1.
original claim: $\sigma_1^2 = \sigma_2^2$
H_o: $\sigma_1^2 = \sigma_2^2$
H_1: $\sigma_1^2 \neq \sigma_2^2$
$\alpha = .05$
C.R. $F < F^{51}_{51,.975} = .5333$
$F > F^{51}_{51,.025} = 1.8752$
calculations:
$F = s_1^2/s_2^2$
$= (.2000)^2/(.1357)^2$
$= 2.1722$

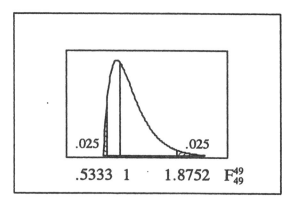

conclusion:
Reject H_o; there is sufficient evidence to reject the claim that $\sigma_1^2 = \sigma_2^2$ and conclude that $\sigma_1^2 \neq \sigma_2^2$ (in fact, that $\sigma_1^2 > \sigma_2^2$).

b. For Sundays, 37 of the 52 observations are 0; for Wednesdays, 37 of the 53 [or 36 of the first 52] observations are 0. The data for both days are very positively skewed. The rainfall amounts do not come from populations with normal distributions.

c. Because the original populations are not normally distributed, the test in part (a) is not valid – which is good news, since there is no reason why rainfall amounts should be more variable on Sundays than on Wednesdays.

12. a. Let the tobacco times be group 1.
original claim: $\sigma_1^2 \neq \sigma_2^2$
H_o: $\sigma_1^2 = \sigma_2^2$
H_1: $\sigma_1^2 \neq \sigma_2^2$
$\alpha = .05$
C.R. $F < F^{49}_{49,.975} = .5333$
$F > F^{49}_{49,.025} = 1.8752$
calculations:
$F = s_1^2/s_2^2$
$= (104.0)^2/(66.3)^2$
$= 2.4606$
conclusion:
Reject H_o; there is sufficient evidence to conclude that $\sigma_1^2 \neq \sigma_2^2$ (in fact, that $\sigma_1^2 > \sigma_2^2$).

b. For tobacco, 22 of the 50 observations are 0; for alcohol, 25 of the 50 observations are 0. The data for both usages are very positively skewed. The times do not come from populations with normal distributions.

c. Because the original populations are not normally distributed, the test in part (a) is not valid. **NOTE:** There is a second reason why the test in part (a) is not valid. One of the requirements for the F test is that the samples be independent. Since both times came from the same 50 movies, they were not 50 tobacco times from randomly selected films and then 50 independently selected alcohol times.

13. The summary statistics are as follows.
   placebo: $n=13$ $\Sigma x=1490.5$ $\Sigma x^2=171965.47$ $\bar{x}=114.65$ $s^2=89.493$
   calcium: $n=15$ $\Sigma x=1740.4$ $\Sigma x^2=202936.92$ $\bar{x}=116.03$ $s^2=71.722$

Let the placebo population be group 1.
original claim: $\sigma_1^2 = \sigma_2^2$
H_o: $\sigma_1^2 = \sigma_2^2$
H_1: $\sigma_1^2 \neq \sigma_2^2$
$\alpha = .05$
C.R. $F < F^{12}_{14,.975} = .3147$
    $F > F^{12}_{14,.025} = 3.0502$
calculations:
    $F = s_1^2/s_2^2$
      $= 89.493/71.722$
      $= 1.2478$ [agrees with TI-83 Plus]

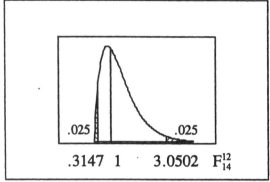

P-value $= 2 \cdot P(F^{12}_{14} > 1.2478) = .6852$ [from TI-83 Plus]
conclusion:
    Do not reject H_o; there is not sufficient evidence to reject the claim that $\sigma_1^2 = \sigma_2^2$.
    Yes, these two groups appear acceptable for an experiment that requires $\sigma_1^2 = \sigma_2^2$.

14. Let the "easy to difficult" values be group 1.
original claim: $\sigma_1^2 = \sigma_2^2$
H_o: $\sigma_1^2 = \sigma_2^2$
H_1: $\sigma_1^2 \neq \sigma_2^2$
$\alpha = .05$
C.R. $F < F^{24}_{15,.975} = .4103$
    $F > F^{24}_{15,.025} = 2.7006$
calculations:
    $F = s_1^2/s_2^2$
      $= 47.0198/18.1489$
      $= 2.5908$ [from Excel]

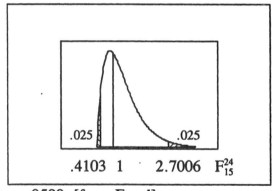

P-value $= 2 \cdot P(F^{24}_{15} > 2.5908) = 2 \cdot (.029928) = .0599$ [from Excel]
conclusion:
    Do not reject H_o; there is not sufficient evidence to reject the claim that $\sigma_1^2 = \sigma_2^2$.

15. The Minirab output is given below.
    MTB > describe c1-c2

	N	MEAN	MEDIAN	TRMEAN	STDEV	SEMEAN	MIN	MAX	Q1	Q3
RowlFRE	12	80.75	81.35	81.19	4.68	1.35	70.90	86.20	78.75	84.53
TolsFRE	12	66.15	68.95	66.75	7.86	2.27	51.90	74.40	59.85	72.05

Let the Tolstoy pages be group 1.
original claim: $\sigma_1^2 = \sigma_2^2$
H_o: $\sigma_1^2 = \sigma_2^2$
H_1: $\sigma_1^2 \neq \sigma_2^2$
$\alpha = .05$
C.R. $F < F^{11}_{11,.975} = .2836$
    $F > F^{11}_{11,.025} = 3.5257$
calculations:
    $F = s_1^2/s_2^2$
      $= (7.86)^2/(4.68)^2$
      $= 2.8208$

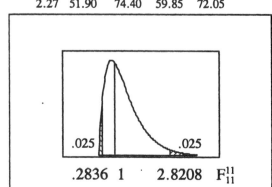

conclusion:
    Do not reject H_o; there is not sufficient evidence to reject the claim that $\sigma_1^2 = \sigma_2^2$.

S-264 INSTRUCTOR'S SOLUTIONS Chapter 8

16. The Minitab output is given below.
```
MTB > describe c1-c2
            N    MEAN   MEDIAN  TRMEAN  STDEV  SEMEAN   MIN     MAX      Q1      Q3
agemen    111  39.613   39.000  39.455  9.490   0.901  21.000  65.000  32.000  46.000
agewomen   39  36.74    32.00   36.06   11.68   1.87   19.00   68.00   29.00   47.00
```
Let the women be group 1.
original claim: $\sigma_1^2 \neq \sigma_2^2$
$H_o: \sigma_1^2 = \sigma_2^2$
$H_1: \sigma_1^2 \neq \sigma_2^2$
$\alpha = .05$
C.R. $F < F^{38}_{110,.975} = .5800$
     $F > F^{38}_{110,.025} = 1.6141$
calculations:
$F = s_1^2/s_2^2$
  $= (11.68)^2/(9.490)^2$
  $= 1.5148$

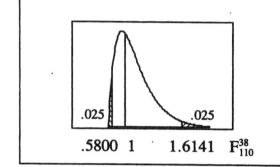

conclusion:
Do not reject H_o; there is not sufficient evidence to conclude that $\sigma_1^2 \neq \sigma_2^2$.

17. Let the .111" cans be group 1.
original claim: $\sigma_1^2 \neq \sigma_2^2$
$H_o: \sigma_1^2 = \sigma_2^2$
$H_1: \sigma_1^2 \neq \sigma_2^2$
$\alpha = .05$
C.R. $F < F^{173}_{174,.975} = .6980$
     $F > F^{174}_{173,.025} = 1.4327$
calculations:
$F = s_1^2/s_2^2$
  $= (22.1)^2/(22.1)^2$
  $= 1.0000$

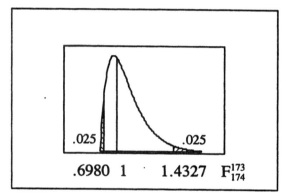

conclusion:
Do not reject H_o; there is not sufficient evidence to conclude that $\sigma_1^2 \neq \sigma_2^2$.
Yes, eliminating the outlier changed the decision.

18. With $n_1 = n_2 = 36$, the calculated F value for the two-tailed test in exercise #5 was 2.9228. Since $2.9228 > F^{35}_{35,.025} = 2.0739$, P-value $< 2 \cdot (.025) = .05$.

19. a. $F_L = F^9_{9,.975} = 1/F^9_{9,.025} = 1/4.0260 = .2484$
       $F_R = F^9_{9,.025} = 4.0260$
    b. $F_L = F^9_{6,.975} = 1/F^6_{9,.025} = 1/4.3197 = .2315$
       $F_R = F^9_{6,.025} = 5.5234$
    c. $F_L = F^6_{9,.975} = 1/F^9_{6,.025} = 1/5.5234 = .1810$
       $F_R = F^6_{9,.025} = 4.3197$

20. Let those who took the placebo be group 1, and refer to the calculations for exercise #13.
$F_L = F^{12}_{14,.975} = 1/F^{14}_{12,.025} = 1/3.1772 = .3147$
$F_R = F^{12}_{14,.025} = 3.0502$
$$(s_1^2/s_2^2) \cdot (1/F_R) < \sigma_1^2/\sigma_2^2 < (s_1^2/s_2^2) \cdot (1/F_L)$$
$$(89.493/71.722) \cdot (1/3.0502) < \sigma_1^2/\sigma_2^2 < (89.493/71.222) \cdot (1/.3147)$$
$$.409 < \sigma_1^2/\sigma_2^2 < 3.964$$

Review Exercises

1. Let those that were warmed be group 1.
   a. original claim: $p_1 - p_2 < 0$
   $\hat{p}_1 = x_1/n_1 = 6/104 = .0577$
   $\hat{p}_2 = x_2/n_2 = 18/96 = .1875$
   $\hat{p}_1 - \hat{p}_2 = .0577 - .1875 = -.1298$
   $\bar{p} = (x_1 + x_2)/(n_1 + n_2)$
   $= (6 + 18)/(104 + 96)$
   $= 24/200 = .12$
   $H_o: p_1 - p_2 = 0$
   $H_1: p_1 - p_2 < 0$
   $\alpha = .05$
   C.R. $z < -z_{.05} = -1.645$

   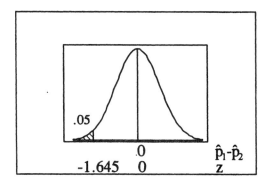

   calculations:
   $z_{\hat{p}_1-\hat{p}_2} = (\hat{p}_1 - \hat{p}_2 - \mu_{\hat{p}_1-\hat{p}_2})/\sigma_{\hat{p}_1-\hat{p}_2}$
   $= (-.1298 - 0)/\sqrt{(.12)(.88)/104 + (.12)(.88)/96}$
   $= -.1298/.0460 = -2.82$
   P-value = $P(z < -2.82) = .0024$
   conclusion:
   Reject H_o; there is sufficient evidence to conclude that $p_1 - p_2 < 0$.
   Yes; if these results are verified, surgical patients should be routinely warmed.

   b. The test in part (a) places 5% in the lower tail. To correspond with this test, use a 90% two-sided confidence interval [or a 95% one-sided confidence interval].

   c. $(\hat{p}_1 - \hat{p}_2) \pm z_{.05}\sqrt{\hat{p}_1\hat{q}_1/n_1 + \hat{p}_2\hat{q}_2/n_2}$
   $-.1298 \pm 1.645 \cdot \sqrt{(.0577)(.9423)/104 + (.1875)(.8125)/96}$
   $-.1298 \pm .0756$
   $-.205 < p_1 - p_2 < -.054$

   d. Since the test of hypothesis and the confidence interval use different estimates for $\sigma_{\hat{p}_1-\hat{p}_2}$, they are not mathematically equivalent and it is possible that they may might lead to different conclusions.

2. d = x - y: -5.75 -1.25 -1.00 -5.00 0.00 0.25 2.25 -0.50 0.75 -1.50 -0.25
   n = 11
   $\Sigma d = -12.00$ $\bar{d} = -1.091$
   $\Sigma d^2 = 68.8750$ $s_d = 2.3619$

   a. original claim: $\mu_d = 0$
   $H_o: \mu_d = 0$
   $H_1: \mu_d \neq 0$
   $\alpha = .05$
   C.R. $t < -t_{10,.025} = -2.228$
   $t > t_{10,.025} = 2.228$

   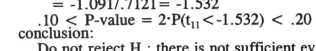

   calculations:
   $t_{\bar{d}} = (\bar{d} - \mu_{\bar{d}})/s_{\bar{d}}$
   $= (-1.091 - 0)/(2.3619/\sqrt{11})$
   $= -1.091/.7121 = -1.532$
   $.10 < $ P-value $= 2 \cdot P(t_{11} < -1.532) < .20$
   conclusion:
   Do not reject H_o; there is not sufficient evidence to reject the claim that $\mu_d = 0$.

   b. $\bar{d} \pm t_{10,.025} \cdot s_d/\sqrt{n}$
   $-1.091 \pm 2.228 \cdot 2.3619/\sqrt{11}$
   -1.091 ± 1.587
   $-2.67 < \mu_d < .50$

   c. We cannot be 95% certain that either type of seed is better. If there are no differences in cost, or any other considerations, choose the kiln dried – even though we can't be sure that it's generally better, it did have the higher yield in this particular trial.

3. Let the obsessive-compulsive patients be group 1.
$\bar{x}_1 - \bar{x}_2 = 1390.03 - 1268.41 = 121.62$

a. $(\bar{x}_1 - \bar{x}_2) \pm t_{9,.025} \sqrt{s_1^2/n_1 + s_2^2/n_2}$
$121.62 \pm 2.262 \cdot \sqrt{(156.84)^2/10 + (137.97)^2/10}$
121.62 ± 149.42
$-27.80 < \mu_1 - \mu_2 < 271.04$

b. original claim: $\mu_1 - \mu_2 = 0$
$H_o: \mu_1 - \mu_2 = 0$
$H_1: \mu_1 - \mu_2 \neq 0$
$\alpha = .05$
C.R. $t < -t_{9,.025} = -2.262$
$t > t_{9,.025} = 2.262$
calculations:
$t_{\bar{x}_1 - \bar{x}_2} = (\bar{x}_1 - \bar{x}_2 - \mu_{\bar{x}_1 - \bar{x}_2})/s_{\bar{x}_1 - \bar{x}_2}$
$= (121.62 - 0)/\sqrt{(156.84)^2/10 + (137.97)^2/10}$
$= 121.62/66.056 = 1.841$
$.05 < $ P-value $= 2 \cdot P(t_9 > 1.841) < .10$

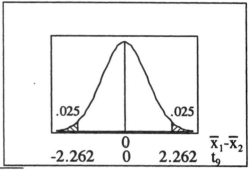

conclusion:
Do not reject H_o; there is not sufficient evidence to reject the claim that $\mu_1 - \mu_2 = 0$.

c. No, it does not appear that the total brain volume can be used as a reliable indicator.

4. Let the obsessive-compulsive patients be group 1.
original claim: $\sigma_1^2 \neq \sigma_2^2$
$H_o: \sigma_1^2 = \sigma_2^2$
$H_1: \sigma_1^2 \neq \sigma_2^2$
$\alpha = .05$
C.R. $F < F^9_{9,.975} = .2484$
$F > F^9_{9,.025} = 4.0260$
calculations:
$F = s_1^2/s_2^2$
$= (156.84)^2/(137.97)^2$
$= 1.2922$

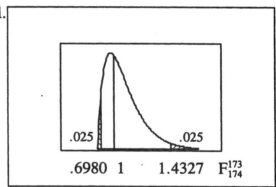

conclusion:
Do not reject H_o; there is not sufficient evidence to conclude that $\sigma_1^2 \neq \sigma_2^2$.

5. Let the filtered cigarettes be group 1

group 1: filtered (n = 21)
$\Sigma x = 270$
$\Sigma x^2 = 3660$
$\bar{x} = 12.857$
$s = 3.071$

group 2: unfiltered (n = 8)
$\Sigma x = 125$
$\Sigma x^2 = 1963$
$\bar{x} = 15.625$
$s = 1.188$

original claim: $\mu_1 - \mu_2 = 0$
$\bar{x}_1 - \bar{x}_2 = 12.857 - 15.625 = -2.768$
$H_o: \mu_1 - \mu_2 = 0$
$H_1: \mu_1 - \mu_2 \neq 0$
$\alpha = .05$
C.R. $t < -t_{7,.025} = -2.365$
$t > t_{7,.025} = 2.365$
calculations:
$t_{\bar{x}_1 - \bar{x}_2} = (\bar{x}_1 - \bar{x}_2 - \mu_{\bar{x}_1 - \bar{x}_2})/s_{\bar{x}_1 - \bar{x}_2}$
$= (-2.768 - 0)/\sqrt{(3.071)^2/21 + (1.188)^2/8}$
$= -2.768/.7908 = -3.500$

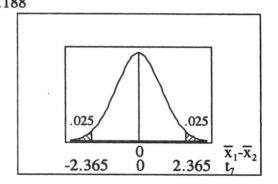

INSTRUCTOR'S SOLUTIONS Chapter 8   S-267

P-value = $2 \cdot P(t_7 < -3.500) \approx .01$
conclusion:
   Reject H_o; there is sufficient evidence to reject the claim that $\mu_1-\mu_2 = 0$ and conclude
   that $\mu_1-\mu_2 \neq 0$ (in fact, that $\mu_1-\mu_2 < 0$).
Yes; based on this result, cigarette filters are effective in reducing the amount of carbon monoxide.

6. Let those receiving the zinc supplement be group 1.
   original claim: $\mu_1-\mu_2 > 0$
   $\bar{x}_1-\bar{x}_2 = 3214 - 3088 = 126$
   H_o: $\mu_1-\mu_2 = 0$
   H_1: $\mu_1-\mu_2 > 0$
   $\alpha = .05$
   C.R. $t > t_{285,.05} = 1.650$
   calculations:
   $t_{\bar{x}_1-\bar{x}_2} = (\bar{x}_1-\bar{x}_2 - \mu_{\bar{x}_1-\bar{x}_2})/s_{\bar{x}_1-\bar{x}_2}$
   $= (126 - 0)/\sqrt{(669)^2/294 + (728)^2/286}$
   $= 126/58.098 = 2.169$
   $.01 < $ P-value $= P(t_{285} > 2.169) < .025$

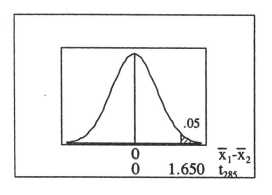

conclusion:
   Reject H_o; there is sufficient evidence to conclude that $\mu_1-\mu_2 > 0$.
Yes, there is sufficient evidence to support the claim that the zinc supplement results in increased birth weights.

7. Let those who saw the first woman be group 1.
   original claim: $p_1-p_2 > 0$
   $\hat{p}_1 = x_1/n_1 = 58/2000 = .0290$
   $\hat{p}_2 = x_2/n_2 = 35/2000 = .0175$
   $\hat{p}_1-\hat{p}_2 = .0290 - .0175 = .0115$
   $\bar{p} = (x_1 + x_2)/(n_1 + n_2)$
      $= (58 + 35)/(2000 + 2000)$
      $= 93/4000 = .02325$
   H_o: $p_1-p_2 = 0$
   H_1: $p_1-p_2 > 0$
   $\alpha = .05$
   C.R. $z > z_{.05} = 1.645$
   calculations:
   $z_{\hat{p}_1-\hat{p}_2} = (\hat{p}_1-\hat{p}_2 - \mu_{\hat{p}_1-\hat{p}_2})/\sigma_{\hat{p}_1-\hat{p}_2}$
      $= (.0115 - 0)/\sqrt{(.02325)(.97675)/2000 + (.02325)(.97576)/2000}$
      $= .0115/.004765 = 2.41$
   P-value $= P(z>2.41) = 1 - P(z<2.41) = 1 - .9920 = .0080$

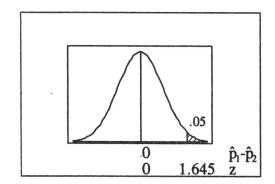

conclusion:
   Reject H_o; there is sufficient evidence to conclude that $p_1-p_2 > 0$.

8. The summary statistics are as follows.
   d = $x_1 - x_2$: 5 0 0 0 8 1 1 4 0 1
   n = 10
   $\Sigma d = 20$ $\bar{d} = 2.00$
   $\Sigma d^2 = 108$ $s_d = 2.749$
   a. original claim: $\mu_d = 0$
      H_o: $\mu_d = 0$
      H_1: $\mu_d \neq 0$
      $\alpha = .05$ [assumed]
      C.R. $t < -t_{9,.025} = -2.262$
         $t > t_{9,.025} = 2.262$

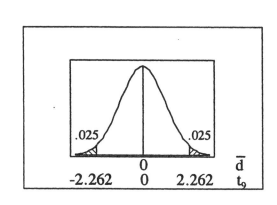

calculations:
$t_{\bar{d}} = (\bar{d} - \mu_{\bar{d}})/s_{\bar{d}}$
$= (2.00 - 0)/(2.749/\sqrt{10})$
$= 2.00/.869 = 2.301$
$.02 < $ P-value $ < .05$
conclusion:
   Reject H_o; there is sufficient evidence to reject the claim that $\mu_d = 0$ and to conclude that $\mu_d \neq 0$ (in fact, that $\mu_d > 0$ – i.e., that training reduces the weights).
b. $\bar{d} \pm t_{9,.025} \cdot s_d/\sqrt{n}$
   $2.00 \pm 2.262 \cdot 2.749/\sqrt{10}$
   2.00 ± 1.97
   $0.0 < \mu_d < 4.0$

Cumulative Review Exercises

1. Refer to the summary table at the right.
   a. $P(Y) = 53/750 = .071$
   b. $P(M \text{ or } Y) = P(M) + P(Y) - P(M \text{ and } Y)$
      $= 250/750 + 53/750 - 26/750$
      $= 277/750 = .369$
   c. $P(Y|M) = 26/250 = .104$
   d. $P(Y|F) = 27/500 = .054$
   e. Let the males be group 1.
      original claim: $p_1 - p_2 > 0$
      $\hat{p}_1 = x_1/n_1 = 26/250 = .104$
      $\hat{p}_2 = x_2/n_2 = 27/500 = .054$
      $\hat{p}_1 - \hat{p}_2 = .104 - .054 = .050$
      $\bar{p} = (x_1 + x_2)/(n_1 + n_2)$
      $= (26 + 27)/(250 + 500)$
      $= 53/750 = .0707$
   H_o: $p_1 - p_2 = 0$
   H_1: $p_1 - p_2 > 0$
   $\alpha = .05$
   C.R. $z > z_{.05} = 1.645$
   calculations:
   $z_{\hat{p}_1 - \hat{p}_2} = (\hat{p}_1 - \hat{p}_2 - \mu_{\hat{p}_1 - \hat{p}_2})/\sigma_{\hat{p}_1 - \hat{p}_2}$
   $= (.050 - 0)/\sqrt{(.0707)(.9293)/250 + (.0707)(.9293)/500}$
   $= .050/.0199 = 2.52$
   P-value $= P(z > 2.52) = 1 - P(z < 2.52) = 1 - .9941 = .0059$
   conclusion:
      Reject H_o; there is sufficient evidence to conclude that $p_1 - p_2 > 0$.
   No; while we can conclude that men receive more speeding tickets than women, we cannot conclude that men speed more often. That conclusion would be valid only if there were a perfect correspondence between speeding and receiving a ticket. It could be that women speed just as often as men but get fewer tickets because they (a) tend to exceed the posted limit by smaller amounts than men do or (b) are more likely to be let off with only a warning than men are. It could also be that men drive more, and so they would be expected to get more tickets even if they did not speed more – consider a population that is ½ male and ½ female in which all people speed equally: if the men do 80% of the driving, for example, then they would receive 80% of the tickets.

   TICKET?
	Y	N	
SEX M	26	224	250
SEX F	27	473	500
	53	697	750

2. There is a problem with the reported results, and no statistical analysis would be appropriate. Since there were 100 drivers in each group, and the number in each group owning a cell phone must be a whole number between 0 and 100 inclusive, the sample

proportion for each group must be a whole percent. The reported values of 13.7% and 10.6% are not mathematical possibilities for the sample success rates of groups of 100.

3. Let the Viagra users be group 1. The following summary statistics apply to all parts [and, as usual, the unrounded values are used in all subsequent calculations.]
$\hat{p}_1 = x_1/n_1 = 29/734 = .040$
$\hat{p}_2 = x_2/n_2 = 15/725 = .021$
$\hat{p}_1 - \hat{p}_2 = .040 - .021 = .019$

a. $\hat{p}_1 \pm z_{.025}\sqrt{\hat{p}_1\hat{q}_1/n_1}$
.040 ± 1.96 $\sqrt{(.040)(.960)/734}$
.040 ± .014
.025 < p_1 < .054

b. $\hat{p}_2 \pm z_{.025}\sqrt{\hat{p}_2\hat{q}_2/n_2}$
.021 ± 1.96 $\sqrt{(.021)(.979)/725}$
.021 ± .010
.010 < p_1 < .031

c. $(\hat{p}_1-\hat{p}_2) \pm z_{.025}\sqrt{\hat{p}_1\hat{q}_1/n_1 + \hat{p}_2\hat{q}_2/n_2}$
.019 ± 1.96 $\sqrt{(.040)(.960)/734 + (.021)(.979)/725}$
.019 ± .018
.001 < p_1-p_2 < .036

d. The method in part (iii) is best. It uses all the information at once and asks whether the data is consistent with the claim that $p_1 = p_2$. Method (ii) constructs an interval with no specific claim in mind. Method (i) does not use all the information at once – in general, methods that incorporate all the information into a single statistical procedure are superior to ones that use smaller sample sizes to give partial answers that need to be combined and/or compared afterward.

4. a. Let p = the proportion of runners who finished the race that were females.
original claim: p < .50
$\hat{p} = x/n = 39/150 = .26$
$H_o: p = .50$
$H_1: p < .50$
α = .05 [assumed]
C.R. z < $-z_{.05}$ = -1.645
calculations:
$z_{\hat{p}} = (\hat{p} - \mu_{\hat{p}})/\sigma_{\hat{p}}$
    = $(.26 - .50)/\sqrt{(.50)(.50)/150}$
    = -.24/.0408
    = -5.88
P-value = P(z<-5.88) = .0001

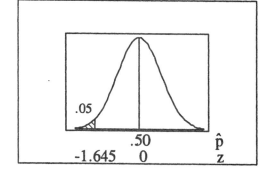

conclusion:
Reject H_o; there is sufficient evidence to conclude that p < .50.

b. The ordered 39 times are as follows.
12047 12289 13593 13704 13854 14216 14235 14721 15036 15077
15326 15357 15402 16013 16297 16352 16401 16758 16771 16792
16871 16991 17211 17260 17286 17636 17726 17799 18469 18580
18647 10177 20084 20675 20891 21911 21983 25399 25858

The summary statistics are as follows.
n = 39   Σx = 670735   Σx² = 11,902,396,416
$\bar{x}$ = (Σx)/n = 670735/39 = 17198.3
$\tilde{x}$ = x_{20} = 17792
s^2 = [n(Σx²) - (Σx)²]/[n(n-1)] = [19(11902396416) - (670735)²]/[39(38)]
s = 3107.2

The following frequency distribution indicates that the times appear to be approximately normally distributed.

time (seconds)	frequency
11000 - 12999	2
13000 - 14999	6
15000 - 16999	14
17000 - 18999	9
19000 - 20999	4
21000 - 22999	2
23000 - 24999	0
25000 - 26999	2
	39

There is no absolute definition of an outlier. The section on boxplots suggests that any score more than (1.5)(IQR) from the median is an outlier.

$IQR = Q_3 - Q_1 = x_{30} - x_{10} = 18580-15077 = 3503$
$16792 \pm 2(3503)$
16792 ± 7006
9786 to 23798

By this definition, the 25399 and 25989 may be considered outliers.

c. original claim: $\mu < 5$ hours = 18000 seconds
$H_o: \mu = 18000$
$H_1: \mu < 18000$
$\alpha = .05$
C.R. $t < -t_{38,.05} = -1.686$
calculations:
$t_{\bar{x}} = (\bar{x} - \mu)/s_{\bar{x}}$
$= (17198.3 - 18000)/(3107.2/\sqrt{39})$
$= -801.7/497.6$
$= -1.611$
$.05 < $ P-value $ < .10$

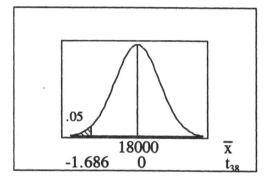

conclusion:
Do not reject H_o; there is not sufficient evidence to conclude that $\mu < 5$ hours.

d. Let the females be group 1.
original claim: $\mu_1-\mu_2 \neq 0$
$\bar{x}_1-\bar{x}_2 = 17198.3 - 15415.2 = 1783.1$
$H_o: \mu_1-\mu_2 = 0$
$H_1: \mu_1-\mu_2 \neq 0$
$\alpha = .05$
C.R. $t < -t_{38,.025} = -2.024$
$t > t_{38,.025} = 2.024$
calculations:
$t_{\bar{x}_1-\bar{x}_2} = (\bar{x}_1-\bar{x}_2 - \mu_{\bar{x}_1-\bar{x}_2})/s_{\bar{x}_1-\bar{x}_2}$
$= (1783.1 - 0)/\sqrt{(3107)^2/39+(3037)^2/111}$
$= 1783.1/572.0 = 3.101$
P-value $= 2 \cdot P(t_{38} > 3.101) < .01$

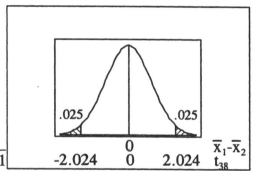

conclusion:
Reject H_o; there is sufficient evidence to conclude that $\mu_1-\mu_2 \neq 0$ (in fact, that $\mu_1-\mu_2 < 0$).

e. Let the females be group 1.
$\hat{p}_1 = x_1/n_1 = 39/150 = .26$
$\hat{p}_2 = x_2/n_2 = 111/150 = .74$

The methods of this chapter cannot be used to make inferences about $p_1 - p_2$ because the data is not from two independent samples. The "n_2" above is really the same 150 that was n_1. Whatever the value of $\hat{p}_1$ is, the value of $\hat{p}_2$ will have to be $1-\hat{p}_1$. To test whether p_1 and p_2 are equal (which means each must be .5), use only the data for one of the genders – for example, test $H_o: p_1 = .50$.

Chapter 9

Correlation and Regression

9-2 Correlation

1. a. From Table A-6, CV = ±.707; therefore r = .993 indicates a significant (positive) linear correlation.
   b. The proportion of the variation in weight that can be explained in terms of the variation in chest size is $r^2 = (.993)^2 = .986$, or 98.6%.

2. a. From Table A-6, CV = ±.707; therefore r = .885 indicates a significant (positive) linear correlation.
   b. The proportion of the variation in the murder rate that can be explained in terms of the variation in registered automatic weapons is $r^2 = (.885)^2 = .783$, or 78.3%.

3. a. From Table A-6, CV = ±.444; therefore r = -.133 does not indicate a significant linear correlation.
   b. The proportion of the variation in Super Bowl points that can be explained in terms of the variation in the DJIA high values is $r^2 = (-.133)^2 = .017$, or 1.7%.

4. a. From Table A-6, CV = ±.444; therefore r = -.284 does not indicate a significant linear correlation.
   b. The proportion of the variation in US car sales that can be explained in terms of the variation in sunspot number is $r^2 = (-.284)^2 = .081$, or 8.1%.

NOTE: In addition to the value of n, calculation of r requires five sums: Σx, Σy, Σx^2, Σy^2 and Σxy. The next problem shows the chart prepared to find these sums. As the sums can usually be found conveniently using a calculator and without constructing the chart, subsequent problems typically give only the values of the sums and do not show a chart.

Also, calculation of r also involves three subcalculations.
(1) $n(\Sigma xy) - (\Sigma x)(\Sigma y)$ determines the sign of r. If large values of x are associated with large values of y, it will be positive. If large values of x are associated with small values of y, it will be negative. If not, a mistake has been made.
(2) $n(\Sigma x^2) - (\Sigma x)^2$ cannot be negative. If it is, a mistake has been made.
(3) $n(\Sigma y^2) - (\Sigma y)^2$ cannot be negative. If it is, a mistake has been made.
Finally, r must be between -1 and 1 inclusive. If not, a mistake has been made. If this or any of the previous mistakes occurs, stop immediately and find the error; continuing will be a fruitless waste of effort.

5.
x	y	xy	x²	y²
0	4	0	0	16
1	1	1	1	1
2	0	0	4	0
3	1	3	9	1
4	4	16	16	16
10	10	20	30	34

$n(\Sigma xy) - (\Sigma x)(\Sigma y) = 5(20) - (10)(10) = 0$
$n(\Sigma x^2) - (\Sigma x)^2 = 5(30) - (10)^2 = 50$
$n(\Sigma y^2) - (\Sigma y)^2 = 5(34) - (10)^2 = 70$

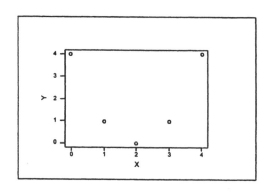

S-272 INSTRUCTOR'S SOLUTIONS Chapter 9

a. According to the scatter diagram, there appears to be a significant "U-shaped" pattern, but no <u>linear</u> relationship between x and y. Expect a value for r close to 0.

b. $r = [n(\Sigma xy) - (\Sigma x)(\Sigma y)]/[\sqrt{n(\Sigma x^2) - (\Sigma x)^2} \cdot \sqrt{n(\Sigma y^2) - (\Sigma y)^2}]$
  $= 0/[\sqrt{50} \cdot \sqrt{70}] = 0$
From Table A-6, assuming $\alpha = .05$, CV = $\pm .878$; therefore r = 0 does not indicate a significant linear correlation. This agrees with the interpretation of the scatter diagram in part (a).

6.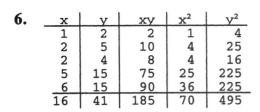

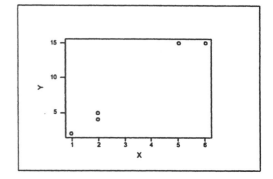

$n(\Sigma xy) - (\Sigma x)(\Sigma y) = 5(185) - (16)(41) = 269$
$n(\Sigma x^2) - (\Sigma x)^2 = 5(70) - (16)^2 = 94$
$n(\Sigma y^2) - (\Sigma y)^2 = 5(495) - (41)^2 = 794$

a. According to the scatter diagram, there appears to be a significant linear relationship between x and y. Expect a positive value for r fairly near 1.00.

b. $r = [n(\Sigma xy) - (\Sigma x)(\Sigma y)]/[\sqrt{n(\Sigma x^2) - (\Sigma x)^2} \cdot \sqrt{n(\Sigma y^2) - (\Sigma y)^2}]$
  $= 269/[\sqrt{94} \cdot \sqrt{794}] = .985$
From Table A-6, assuming $\alpha = .05$, CV = $\pm .878$; therefore r = .976 indicates a significant (positive) linear correlation. This agrees with the interpretation of the scatter diagram in part (a).

7. The following table and summary statistics apply to all parts of this exercise.
   x: 1 1 1 2 2 2 3 3 3 10
   y: 1 2 3 1 2 3 1 2 3 10
   using all the points: n=10 $\Sigma x=28$ $\Sigma y=28$ $\Sigma xy=136$ $\Sigma x^2=142$ $\Sigma y^2=142$
   without the outlier: n=9 $\Sigma x=18$ $\Sigma y=18$ $\Sigma xy=36$ $\Sigma x^2=42$ $\Sigma y^2=42$

a. There appears to be a strong positive linear correlation, with r close to 1.

b. $n(\Sigma xy) - (\Sigma x)(\Sigma y) = 10(136) - (28)(28) = 576$
  $n(\Sigma x^2) - (\Sigma x)^2 = 10(142) - (28)^2 = 636$
  $n(\Sigma y^2) - (\Sigma y)^2 = 10(142) - (28)^2 = 636$
  $r = [n(\Sigma xy) - (\Sigma x)(\Sigma y)]/[\sqrt{n(\Sigma x^2) - (\Sigma x)^2} \cdot \sqrt{n(\Sigma y^2) - (\Sigma y)^2}]$
  $= 576/[\sqrt{636} \cdot \sqrt{636}] = .906$
From Table A-6, assuming $\alpha = .05$, CV = $\pm .632$; therefore r = .906 indicates a significant (positive) linear correlation. This agrees with the interpretation of the scatter diagram in part (a).

c. There appears to be no linear correlation, with r close to 0.
  $n(\Sigma xy) - (\Sigma x)(\Sigma y) = 9(36) - (18)(18) = 0$
  $n(\Sigma x^2) - (\Sigma x)^2 = 9(42) - (18)^2 = 54$
  $n(\Sigma y^2) - (\Sigma y)^2 = 9(42) - (18)^2 = 54$
  $r = [n(\Sigma xy) - (\Sigma x)(\Sigma y)]/[\sqrt{n(\Sigma x^2) - (\Sigma x)^2} \cdot \sqrt{n(\Sigma y^2) - (\Sigma y)^2}]$
  $= 0/[\sqrt{54} \cdot \sqrt{54}] = 0$
From Table A-6, assuming $\alpha = .05$, CV = $\pm .666$; therefore r = 0 does not indicate a significant linear correlation. This agrees with the interpretation of the scatter diagram.

d. The effect of a single pair of values can be dramatic, changing the conclusion entirely.

NOTE: In each of exercises 8-14, the first variable listed is designated x, and the second variable listed is designated y. In correlation problems, the designation of x and y is arbitrary – so long as a person remains consistent after making the designation. For use in the next section, the following summary statistics should be saved for each exercise: n, Σx, Σy, Σx^2, Σy^2, Σxy.

8. a. n = 11
    Σx = 677
    Σy = 33.4
    Σx^2 = 43105
    Σy^2 = 126.04
    Σxy = 2153.0

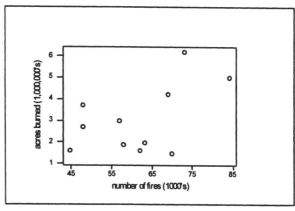

   b. $n(\Sigma xy) - (\Sigma x)(\Sigma y) = 11(2153.0) - (677)(33.4) = 1071.2$
      $n(\Sigma x^2) - (\Sigma x)^2 = 11(43105) - (677)^2 = 15826$
      $n(\Sigma y^2) - (\Sigma y)^2 = 11(126.04) - (33.4)^2 = 270.88$
      $r = [n(\Sigma xy) - (\Sigma x)(\Sigma y)]/[\sqrt{n(\Sigma x^2) - (\Sigma x)^2} \cdot \sqrt{n(\Sigma y^2) - (\Sigma y)^2}]$
        $= 1071.2/[\sqrt{15826} \cdot \sqrt{270.88}] = .517$

   c. H_o: $\rho = 0$
      H_1: $\rho \ne 0$
      $\alpha = .05$
      C.R. r < -.602   OR   C.R. t < $-t_{9,.025}$ = -2.262
           r > .602   t > $t_{9,.025}$ = 2.262

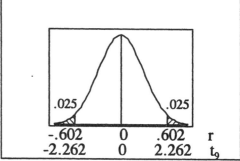

      calculations: calculations:
          r = .517 $t_r = (r - \mu_r)/s_r$
                                   $= (.517 - 0)/\sqrt{(1-(.517)^2)/9}$
                                   $= .517/.2853 = 1.814$
      conclusion:
          Do not reject H_o; there is not sufficient evidence to reject the claim that $\rho = 0$.
          No; there is no significant correlation between the number of fires and the number of acres burned.
          No; the results do not support any conclusion about the removal of trees affecting the risk of fires, because none of the variables addresses the removal/density of trees.

9. a. n = 8
   Σx = 188.2
   Σy = 52.0
   Σx^2 = 11850.04
   Σy^2 = 402.90
   Σxy = 1141.22

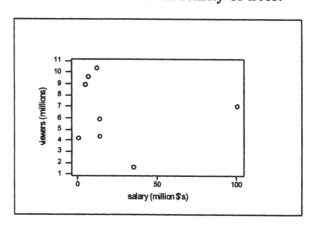

b. $n(\Sigma xy) - (\Sigma x)(\Sigma y) = 8(1141.22) - (188.2)(52.0) = -656.64$
   $n(\Sigma x^2) - (\Sigma x)^2 = 8(11850.04) - (188.2)^2 = 59381.08$
   $n(\Sigma y^2) - (\Sigma y)^2 = 8(402.90) - (52.0)^2 = 519.20$
   $r = [n(\Sigma xy) - (\Sigma x)(\Sigma y)]/[\sqrt{n(\Sigma x^2) - (\Sigma x)^2} \cdot \sqrt{n(\Sigma y^2) - (\Sigma y)^2}]$
   $= -656.64/[\sqrt{59381.08} \cdot \sqrt{519.20}] = -.118$

c. $H_o: \rho = 0$
   $H_1: \rho \neq 0$
   $\alpha = .05$
   C.R. $r < -.707$ OR   C.R. $t < -t_{6,.025} = -2.447$
   $r > .707$ $t > t_{6,.025} = 2.447$

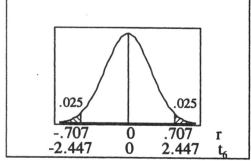

calculations: calculations:
$r = -.118$ $t_r = (r - \mu_r)/s_r$
                        $= (-.118 - 0)/\sqrt{(1-(-.118)^2)/6}$
conclusion: $= -.118/.4054 = -.292$

Do not reject H_o; there is not sufficient evidence to reject the claim that $\rho = 0$.
No; there is no significant correlation between salary and number of viewers.
Susan Lucci (1/4.2 = $0.24 per viewer) has the lowest cost per viewer, and Kelsey
Grammar (35.2/1.6 = $22.00 per viewer) has the highest cost per viewer.

10. a. $n = 9$
    $\Sigma x = 632.0$
    $\Sigma y = 1089$
    $\Sigma x^2 = 44399.50$
    $\Sigma y^2 = 132223$
    $\Sigma xy = 76546.0$

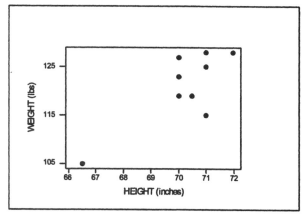

b. $n(\Sigma xy) - (\Sigma x)(\Sigma y) = 9(76546) - (632)(1089) = 666$
   $n(\Sigma x^2) - (\Sigma x)^2 = 9(44399.5) - (632)^2 = 171.5$
   $n(\Sigma y^2) - (\Sigma y)^2 = 9(132223) - (1089)^2 = 4086$
   $r = [n(\Sigma xy) - (\Sigma x)(\Sigma y)]/[\sqrt{n(\Sigma x^2) - (\Sigma x)^2} \cdot \sqrt{n(\Sigma y^2) - (\Sigma y)^2}]$
   $= 666/[\sqrt{171.5} \cdot \sqrt{4086}] = .796$

c. $H_o: \rho = 0$
   $H_1: \rho \neq 0$
   $\alpha = .05$
   C.R. $r < -.666$ OR   C.R. $t < -t_{7,.025} = -2.365$
   $r > .666$ $t > t_{7,.025} = 2.365$

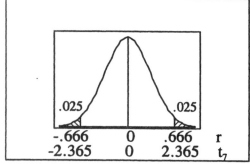

calculations: calculations:
$r = .796$ $t_r = (r - \mu_r)/s_r$
                        $= (.796 - 0)/\sqrt{(1-(.796)^2)/7}$
conclusion: $= .796/.229 = 3.475$

Reject H_o; there is sufficient evidence to reject the claim that $\rho = 0$ and to conclude that $\rho \neq 0$ (in fact, that $\rho > 0$).

Yes; there is a correlation between height and weight for the supermodels.
No; since the supermodels are not representative of all adult women, these results cannot be used to make any conclusions about that population.

11. a. n = 14
    Σx = 1875
    Σy = 1241
    Σx² = 252179
    Σy² = 111459
    Σxy = 167023

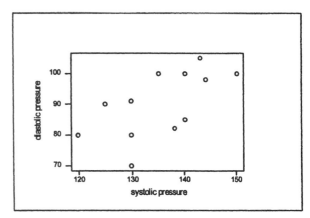

b. n(Σxy) - (Σx)(Σy) = 14(167023) - (1875)(1241) = 11447
   n(Σx²) - (Σx)² = 14(252179) - (1875)² = 14881
   n(Σy²) - (Σy)² = 14(111459) - (1241)² = 20345
   r = [n(Σxy) - (Σx)(Σy)]/[√n(Σx²) - (Σx)² ·√n(Σy²) - (Σy)²]
     = 11447/[√14881·√20345] = .658

c. H_o: ρ = 0
   H_1: ρ ≠ 0
   α = .05
   C.R. r < -.532   OR   C.R. t < $-t_{12,.025}$ = -2.179
        r > .532   t > $t_{12,.025}$ = 2.179

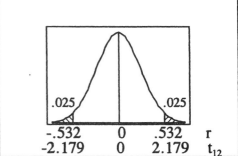

   calculations: calculations:
      r = .658 t_r = (r - $μ_r$)/s_r
                                  = (.658 - 0)/√(1-(.658)²)/12
                                  = .658/.2174 = 3.026
   conclusion:
      Reject H_o; there is sufficient evidence to reject the claim that ρ = 0 and to conclude that ρ ≠ 0 (in fact, ρ > 0).
   Yes; there is a correlation between systolic and diastolic pressure.
   Yes; since the measurements were on the same patient, one could study the variability in a person's blood pressure – or in the variability of medical student readings?

12. a. n = 8
    Σx = 478
    Σy = 1176.617
    Σx² = 29070
    Σy² = 173068.664832
    Σxy = 70318.992

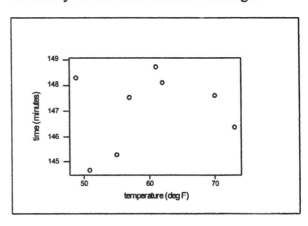

b. $n(\Sigma xy) - (\Sigma x)(\Sigma y) = 8(70318.992) - = (478)(1176.617) = 129.010$
   $n(\Sigma x^2) - (\Sigma x)^2 = 8(29070) - (478)^2 = 4076$
   $n(\Sigma y^2) - (\Sigma y)^2 = 8(173068.664832) - (1176.617)^2 = 121.75371$
   $r = [n(\Sigma xy) - (\Sigma x)(\Sigma y)]/[\sqrt{n(\Sigma x^2) - (\Sigma x)^2} \cdot \sqrt{n(\Sigma y^2) - (\Sigma y)^2}]$
   $= 129.010/[\sqrt{4076} \cdot \sqrt{121.75371}] = .183$

c. $H_o: \rho = 0$
   $H_1: \rho \neq 0$
   $\alpha = .05$
   C.R. $r < -.707$ OR C.R. $t < -t_{6,.025} = -2.447$
   $r > .707$ $t > t_{6,.025} = 2.447$

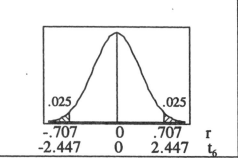

calculations: calculations:
   $r = .183$ $t_r = (r - \mu_r)/s_r$
                           $= (.183 - 0)/\sqrt{(1-(.183)^2)/6}$
conclusion: $= .183/.4013 = .456$

Do not reject H_o; there is not sufficient evidence to reject the claim that $\rho = 0$.
No; there is no correlation between temperature and winning time.
No; winning times do not appear to be affected by temperature.

13. a. $n = 12$
    $\Sigma x = 175$
    $\Sigma y = 2102.46$
    $\Sigma x^2 = 5155$
    $\Sigma y^2 = 601709.8240$
    $\Sigma xy = 37111.51$

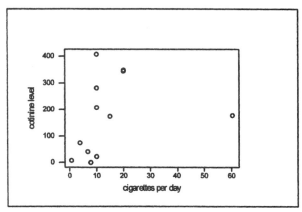

b. $n(\Sigma xy) - (\Sigma x)(\Sigma y) = 12(37111.51) - (175)(2102.46) = 77407.62$
   $n(\Sigma x^2) - (\Sigma x)^2 = 12(5155) - (175)^2 = 31235$
   $n(\Sigma y^2) - (\Sigma y)^2 = 12(601709.8240) - (2102.46)^2 = 2800179.836$
   $r = [n(\Sigma xy) - (\Sigma x)(\Sigma y)]/[\sqrt{n(\Sigma x^2) - (\Sigma x)^2} \cdot \sqrt{n(\Sigma y^2) - (\Sigma y)^2}]$
   $= 77407.62/[\sqrt{31235} \cdot \sqrt{2800179.836}] = .262$

c. $H_o: \rho = 0$
   $H_1: \rho \neq 0$
   $\alpha = .05$
   C.R. $r < -.576$ OR C.R. $t < -t_{10,.025} = -2.228$
   $r > .576$ $t > t_{10,.025} = 2.228$

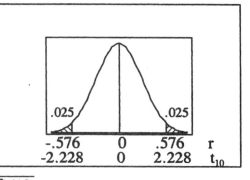

calculations: calculations:
   $r = .262$ $t_r = (r - \mu_r)/s_r$
                           $= (.262 - 0)/\sqrt{(1-(.262)^2)/10}$
conclusion: $= .262/.3052 = .858$

Do not reject H_o; there is not sufficient evidence to reject the claim that $\rho = 0$.
No; based on these results, there is not a correlation between number of cigarettes smoked and the amount of cotinine in the body. Possible explanations for not

identifying the expected correlation are (1) the subjects may not have accurately self-reported their number of cigarettes smoked, (2) the survey did not consider whether or not the cigarettes smoked had filters (which would affect the amounts nicotine reaching the body), and (3) the results could be affected by exposure to second-hand smoke.

14. a. n = 14
     Σx = 66.4
     Σy = 669.1
     Σx^2 = 444.54
     Σy^2 = 37365.71
     Σxy = 3865.67

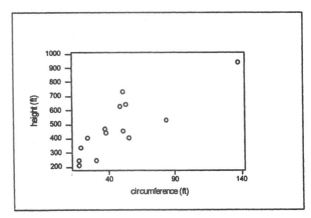

b. $n(\Sigma xy) - (\Sigma x)(\Sigma y) = 14(3865.67) - (66.4)(669.1) = 9691.14$
   $n(\Sigma x^2) - (\Sigma x)^2 = 14(444.54) - (66.4)^2 = 1814.60$
   $n(\Sigma y^2) - (\Sigma y)^2 = 14(37365.71) - (669.1)^2 = 75425.13$
   $r = [n(\Sigma xy) - (\Sigma x)(\Sigma y)]/[\sqrt{n(\Sigma x^2) - (\Sigma x)^2} \cdot \sqrt{n(\Sigma y^2) - (\Sigma y)^2}]$
   $= 9691.14/[\sqrt{1814.60} \cdot \sqrt{75425.13}] = .828$

c. H_o: $\rho = 0$
   H_1: $\rho \neq 0$
   $\alpha = .05$
   C.R. r < -.532 OR C.R. t < $-t_{12,.025}$ = -2.179
        r > .532   t > $t_{12,.025}$ = 2.179

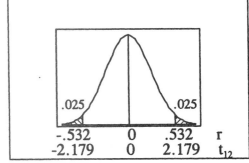

   calculations: calculations:
   r = .828 $t_r = (r - \mu_r)/s_r$
                        $= (.828 - 0)/\sqrt{(1-(.828)^2)/12}$
                        $= .828/.1617 = 5.123$
   conclusion:
     Reject H_o; there is sufficient evidence to reject the claim that $\rho = 0$ and to conclude that $\rho \neq 0$ (in fact, $\rho > 0$).
   Yes; there is a correlation between trunk circumference and height.
   There should be such a correlation because both circumference and height increase with age.

15. a. n = 16
    Σx = .35
    Σy = 60.2
    Σx^2 = .0143
    Σy^2 = 227.24
    Σxy = 1.342

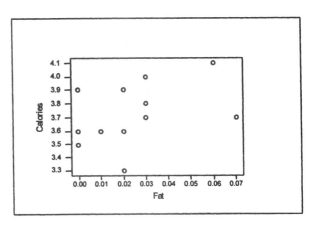

S-278   INSTRUCTOR'S SOLUTIONS Chapter 9

b. $n(\Sigma xy) - (\Sigma x)(\Sigma y) = 16(1.342) - (.35)(60.2) = .402$
$n(\Sigma x^2) - (\Sigma x)^2 = 16(.0143) - (.35)^2 = .1063$
$n(\Sigma y^2) - (\Sigma y)^2 = 16(227.24) - (60.2)^2 = 11.80$
$r = [n(\Sigma xy) - (\Sigma x)(\Sigma y)]/[\sqrt{n(\Sigma x^2) - (\Sigma x)^2} \cdot \sqrt{n(\Sigma y^2) - (\Sigma y)^2}\,]$
$= .402/[\sqrt{.1063} \cdot \sqrt{11.80}] = .359$

c. H_o: $\rho = 0$
H_1: $\rho \neq 0$
$\alpha = .05$
C.R. $r < -.497$ OR   C.R. $t < -t_{14,.025} = -2.145$
$r > .497$ $t > t_{14,.025} = 2.145$

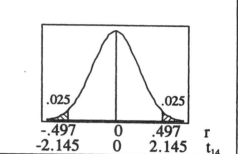

calculations: calculations:
$r = .359$ $t_r = (r - \mu_r)/s_r$
                       $= (.359 - 0)/\sqrt{(1-(.359)^2)/14}$
                       $= .359/.2495 = 1.439$

conclusion:
Do not reject H_o; there is not sufficient evidence to reject the claim that $\rho = 0$.

16. a. $n = 50$
$\Sigma x = 2872$
$\Sigma y = 1623$
$\Sigma x^2 = 694918$
$\Sigma y^2 = 268331$
$\Sigma xy = 173771$

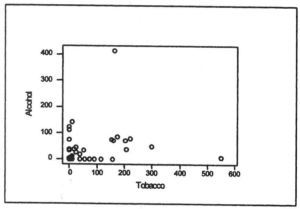

b. $n(\Sigma xy) - (\Sigma x)(\Sigma y) = 50(173771) - (2872)(1623) = 4027294$
$n(\Sigma x^2) - (\Sigma x)^2 = 50(694918) - (2872)^2 = 26497516$
$n(\Sigma y^2) - (\Sigma y)^2 = 50(268331) - (1623)^2 = 10782421$
$r = [n(\Sigma xy) - (\Sigma x)(\Sigma y)]/[\sqrt{n(\Sigma x^2) - (\Sigma x)^2} \cdot \sqrt{n(\Sigma y^2) - (\Sigma y)^2}\,]$
$= 4027294/[\sqrt{26497516} \cdot \sqrt{10782421}] = .238$

c. H_o: $\rho = 0$
H_1: $\rho \neq 0$
$\alpha = .05$
C.R. $r < -.279$ OR   C.R. $t < -t_{48,.025} = -2.009$
$r > .279$ $t > t_{48,.025} = 2.009$

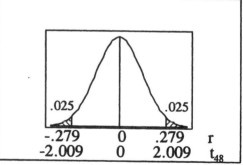

calculations: calculations:
$r = .238$ $t_r = (r - \mu_r)/s_r$
                       $= (.238 - 0)/\sqrt{(1-(.238)^2)/48}$
conclusion: $= .238/.1402 = 1.700$
Do not reject H_o; there is not sufficient evidence to reject the claim that $\rho = 0$.

17. a. n = 40
    Σx = 9635
    Σy = 1029.6
    Σx² = 3669819
    Σy² = 27984.46
    Σxy = 269551.6

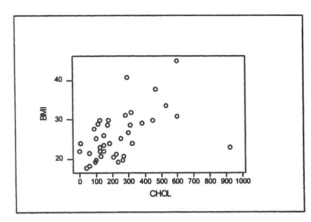

b. n(Σxy) - (Σx)(Σy) = 40(269551.6) - (9635)(1029.6) = 861868.0
   n(Σx²) - (Σx)² = 40(3669819) - (9635)² = 53959535
   n(Σy²) - (Σy)² = 40(27984.46) - (1029.6)² = 59302.24
   r = [n(Σxy) - (Σx)(Σy)]/[√n(Σx²) - (Σx)² · √n(Σy²) - (Σy)²]
     = 861868.0/[√53959535 · √59302.24] = .482

c. H_o: ρ = 0
   H_1: ρ ≠ 0
   α = .05
   C.R. r < -.312   OR   C.R. t < $-t_{38,.025}$ = -2.024
        r > .312   t > $t_{38,.025}$ = 2.024

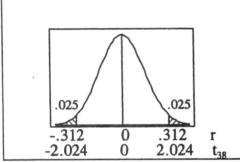

   calculations: calculations:
        r = .482 t_r = (r - μ_r)/s_r
                              = (.482 - 0)/√(1-(.482)²)/38
   conclusion: = .482/.1422 = 3.389
        Reject H_o; there is sufficient evidence to reject the claim that ρ = 0 and to conclude
        that ρ ≠ 0 (in fact, ρ > 0).

18. a. n = 12
    Σx = 848.8
    Σy = 78.0
    Σx² = 61449.54
    Σy² = 572.94
    Σxy = 5238.66

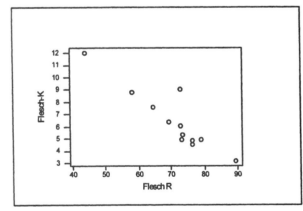

b. n(Σxy) - (Σx)(Σy) = 12(5238.66) - (848.8)(78.0) = -3342.48
   n(Σx²) - (Σx)² = 12(61449.54) - (848.8)² = 16933.04
   n(Σy²) - (Σy)² = 12(572.94) - (78.0)² = 791.28
   r = [n(Σxy) - (Σx)(Σy)]/[√n(Σx²) - (Σx)² · √n(Σy²) - (Σy)²]
     = -3342.48/[√16933.04 · √791.28] = -.913

c. $H_o: \rho = 0$
   $H_1: \rho \neq 0$
   $\alpha = .05$
   C.R. $r < -.576$ OR C.R. $t < -t_{10,.025} = -2.228$
        $r > .576$ $t > t_{10,.025} = 2.228$

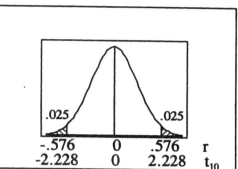

   calculations: calculations:
   $r = -.913$ $t_r = (r - \mu_r)/s_r$
                         $= (-.913 - 0)/\sqrt{(1-(-.913)^2)/10}$
                         $= -.913/.1289 = -7.084$
   conclusion:
      Reject H_o; there is sufficient evidence to reject the claim that $\rho = 0$ and to conclude that $\rho \neq 0$ (in fact, $\rho < 0$).
   There is a negative correlation because the Reading Ease scale assigns high values to the easier material while the Grade Level scale assigns high values to the more difficult material.

19. A. Selling price-list price relationship?
    a. $n = 50$
       $\Sigma x = 8917$
       $\Sigma y = 8517.0$
       $\Sigma x^2 = 1899037$
       $\Sigma y^2 = 1710311.50$
       $\Sigma xy = 1801273.0$

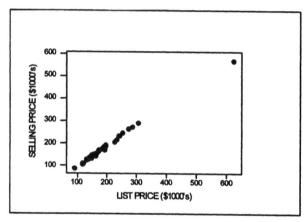

   b. $n(\Sigma xy) - (\Sigma x)(\Sigma y) = 50(1801273.0) - (8917)(8517.0) = 14117561$
      $n(\Sigma x^2) - (\Sigma x)^2 = 50(1899037) - (8917)^2 = 15438961$
      $n(\Sigma y^2) - (\Sigma y)^2 = 50(1710311.50) - (8517.0)^2 = 12976286$
      $r = [n(\Sigma xy) - (\Sigma x)(\Sigma y)]/[\sqrt{n(\Sigma x^2) - (\Sigma x)^2} \cdot \sqrt{n(\Sigma y^2) - (\Sigma y)^2}]$
        $= 14117561/[\sqrt{15438961} \cdot \sqrt{12976286}] = .9974$

   c. $H_o: \rho = 0$
      $H_1: \rho \neq 0$
      $\alpha = .05$
      C.R. $r < -.279$ OR C.R. $t < -t_{48,.025} = -1.960$
           $r > .279$ $t > t_{48,.025} = 1.960$

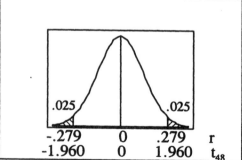

      calculations: calculations:
      $r = .9974$ $t_r = (r - \mu_r)/s_r$
                            $= (.9974 - 0)/\sqrt{(1-(.9974)^2)/48}$
                            $= .9974/.0101 = 96.149$
      conclusion:
         Reject H_o; there is sufficient evidence to reject the claim that $\rho = 0$ and to conclude that $\rho \neq 0$ (in fact, that $\rho > 0$). The data support the given expectation.

B. Selling price-taxes relationship?

a. n = 50
Σx = 185305
Σy = 8517.0
Σx² = 809069344
Σy² = 1710311.50
Σxy = 36631916.5

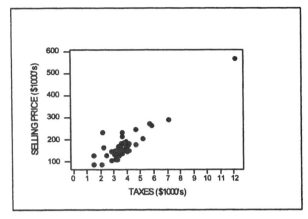

b. $n(\Sigma xy) - (\Sigma x)(\Sigma y) = 50(36631916.5) - (185305)(8517.0) = 253353140$
$n(\Sigma x^2) - (\Sigma x)^2 = 50(809069344) - (185305)^2 = 6115524175$
$n(\Sigma y^2) - (\Sigma y)^2 = 50(1710311.50) - (8517.0)^2 = 12976286$
$r = [n(\Sigma xy) - (\Sigma x)(\Sigma y)]/[\sqrt{n(\Sigma x^2) - (\Sigma x)^2} \cdot \sqrt{n(\Sigma y^2) - (\Sigma y)^2}]$
$= 253353140/[\sqrt{6115524175} \cdot \sqrt{12976286}] = .8994$

c. $H_o: \rho = 0$
$H_1: \rho \neq 0$
$\alpha = .05$
C.R. r < -.279   OR   C.R. t < $-t_{48,.025}$ = -1.960
r > .279   t > $t_{48,.025}$ = 1.960

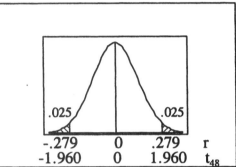

calculations: calculations:
r = .8994 $t_r = (r - \mu_r)/s_r$
                      $= (.8994 - 0)/\sqrt{(1-(.8994)^2)/48}$
conclusion: $= .8994/.0631 = 14.252$

Reject H_o; there is sufficient evidence to reject the claim that $\rho = 0$ and to conclude that $\rho \neq 0$ (in fact, that $\rho > 0$).

Yes; the tax bill appears to be based on the value of the house – since the selling price is assumed to represent the true value of the house, and there is a strong correlation between selling price and the amount of the taxes. The tax assessments appear to be valid.

20. A. Nicotine-tar relationship?

a. n = 29
Σx = 351
Σy = 27.3
Σx² = 4849
Σy² = 28.45
Σxy = 369.5

b. $n(\Sigma xy) - (\Sigma x)(\Sigma y) = 29(369.5) - (351)(27.3) = 1133.2$
$n(\Sigma x^2) - (\Sigma x)^2 = 29(4849) - (351)^2 = 17420$
$n(\Sigma y^2) - (\Sigma y)^2 = 29(28.45) - (27.3)^2 = 79.76$

$r = [n(\Sigma xy) - (\Sigma x)(\Sigma y)]/[\sqrt{n(\Sigma x^2) - (\Sigma x)^2} \cdot \sqrt{n(\Sigma y^2) - (\Sigma y)^2}]$
$= 1133.2/[\sqrt{17420} \cdot \sqrt{79.76}] = .961$

c. H_o: $\rho = 0$
   H_1: $\rho \neq 0$
   $\alpha = .05$
   C.R. $r < -.361$ OR   C.R. $t < -t_{27,.025} = -2.052$
        $r > .361$ $t > t_{27,.025} = 2.052$

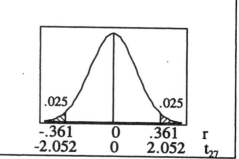

calculations: calculations:
  $r = .961$ $t_r = (r - \mu_r)/s_r$
                            $= (.961 - 0)/\sqrt{(1-(.961)^2)/27}$
                            $= .961/.0530 = 18.148$

conclusion:
  Reject H_o; there is sufficient evidence to reject the claim that $\rho = 0$ and to conclude that $\rho \neq 0$ (in fact, that $\rho > 0$).

Yes; based on this result there is a significant positive linear correlation, suggesting that researchers might be able to reduce their laboratory expenses by measuring only one of these two variables.

B. Nicotine-CO relationship?

a. n = 29
   Σx = 359
   Σy = 27.3
   Σx^2 = 5003
   Σy^2 = 28.45
   Σxy = 371.8

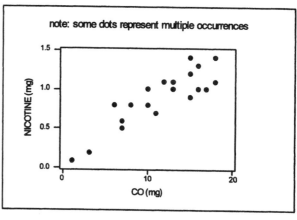

b. $n(\Sigma xy) - (\Sigma x)(\Sigma y) = 29(371.8) - (359)(27.3) = 981.5$
   $n(\Sigma x^2) - (\Sigma x)^2 = 29(5003) - (359)^2 = 16206$
   $n(\Sigma y^2) - (\Sigma y)^2 = 29(28.45) - (27.3)^2 = 79.76$
   $r = [n(\Sigma xy) - (\Sigma x)(\Sigma y)]/[\sqrt{n(\Sigma x^2) - (\Sigma x)^2} \cdot \sqrt{n(\Sigma y^2) - (\Sigma y)^2}]$
   $= 981.5/[\sqrt{16206} \cdot \sqrt{79.76}] = .863$

c. H_o: $\rho = 0$
   H_1: $\rho \neq 0$
   $\alpha = .05$
   C.R. $r < -.361$ OR   C.R. $t < -t_{27,.025} = -2.052$
        $r > .361$ $t > t_{27,.025} = 2.052$

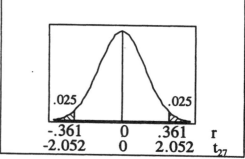

calculations: calculations:
  $r = .863$ $t_r = (r - \mu_r)/s_r$
                            $= (.863 - 0)/\sqrt{(1-(.863)^2)/27}$
                            $= .863/.0971 = 8.888$

conclusion:
  Reject H_o; there is sufficient evidence to reject the claim that $\rho = 0$ and to conclude that $\rho \neq 0$ (in fact, that $\rho > 0$).

Yes; based on this result there is a significant positive linear correlation, suggesting that researchers might be able to reduce their laboratory expenses by measuring only one of these two variables.

C. Tar is the better choice, since it is more highly correlated with nicotine.

21. A. Actual-fiveday relationship?

a. n = 31
Σx = 1069
Σy = 1082
Σx² = 37381
Σy² = 38352
Σxy = 37628

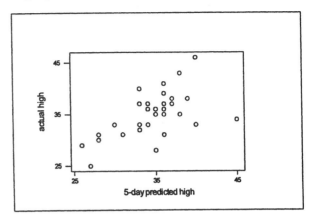

b. n(Σxy) - (Σx)(Σy) = 31(37628) - (1069)(1082) = 9810
n(Σx²) - (Σx)² = 31(37381) - (1069)² = 16050
n(Σy²) - (Σy)² = 31(38352) - (1082)² = 18188
r = [n(Σxy) - (Σx)(Σy)]/[√n(Σx²) - (Σx)² · √n(Σy²) - (Σy)²]
= 9810/[√16050 · √18188] = .574

c. $H_0: \rho = 0$
$H_1: \rho \neq 0$
$\alpha = .05$
C.R. r < -.361   OR   C.R. t < $-t_{29,.025}$ = -2.045
r > .361   t > $t_{29,.025}$ = 2.045

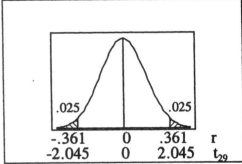

calculations: calculations:
r = .574 $t_r = (r - \mu_r)/s_r$
                         = (.574 - 0)/√(1-(.574)²)/29
conclusion: = .574/.1520 = 3.777

Reject H_0; there is sufficient evidence to reject the claim that $\rho = 0$ and to conclude that $\rho \neq 0$ (in fact, that $\rho > 0$).

Yes; there is a linear correlation between the 5-day forecast and the actual temperature. No; a correlation means only that there is a relationship between the values, not that they agree with each other. If the 5-day forecasts were always exactly 20 degrees too cold, for example, there would be a perfected correlation but not agreement.

B. Actual-oneday relationship?

a. n = 31
Σx = 1033
Σy = 1082
Σx² = 35107
Σy² = 38352
Σxy = 36489

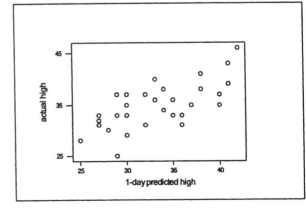

b. $n(\Sigma xy) - (\Sigma x)(\Sigma y) = 31(36489) - (1033)(1082) = 13453$
$n(\Sigma x^2) - (\Sigma x)^2 = 31(35107) - (1033)^2 = 21228$
$n(\Sigma y^2) - (\Sigma y)^2 = 31(38352) - (1082)^2 = 18188$
$r = [n(\Sigma xy) - (\Sigma x)(\Sigma y)]/[\sqrt{n(\Sigma x^2) - (\Sigma x)^2} \cdot \sqrt{n(\Sigma y^2) - (\Sigma y)^2}]$
$= 13453/[\sqrt{21228} \cdot \sqrt{18188}] = .685$

c. $H_o: \rho = 0$
$H_1: \rho \neq 0$
$\alpha = .05$
C.R. $r < -.361$ OR   C.R. $t < -t_{29,.025} = -2.045$
$r > .361$ $t > t_{29,.025} = 2.045$

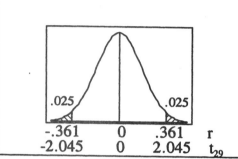

calculations: calculations:
$r = .685$ $t_r = (r - \mu_r)/s_r$
                        $= (.685 - 0)/\sqrt{(1-(.685)^2)/29}$
                        $= .685/.1353 = 5.059$

conclusion:
Reject H_o; there is sufficient evidence to reject the claim that $\rho = 0$ and to conclude that $\rho \neq 0$ (in fact, that $\rho > 0$).

Yes; there is a linear correlation between the 1-day forecast and the actual temperature. No; a correlation means only that there is a relationship between the values, not that they agree with each other. If the 1-day forecasts were always exactly 20 degrees too cold, for example, there would be a perfected correlation but not agreement.

C. One would expect the 1-day forecasts to have a higher correlation with the actual temperatures than the 5-day forecasts would, and the results support this expectation. But even a very high correlation means only that there is a relationship between the predicted and the actual temperatures, not that they agree with each other. If the predictions were always exactly 20 degrees too cold, for example, there would be a perfected correlation but not agreement.

22. A. Conductivity-temperature relationship?
a. $n = 61$
$\Sigma x = 1853.0$
$\Sigma y = 2874.5$
$\Sigma x^2 = 56457.63$
$\Sigma y^2 = 139786.58$
$\Sigma xy = 87422.02$

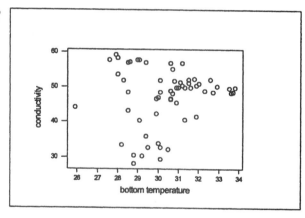

b. $n(\Sigma xy) - (\Sigma x)(\Sigma y) = 61(87422.02) - (1853.0)(2874.5) = 6294.72$
$n(\Sigma x^2) - (\Sigma x)^2 = 61(56457.63) - (1853.0)^2 = 10306.43$
$n(\Sigma y^2) - (\Sigma y)^2 = 61(139786.58) - (2874.5)^2 = 264231.13$
$r = [n(\Sigma xy) - (\Sigma x)(\Sigma y)]/[\sqrt{n(\Sigma x^2) - (\Sigma x)^2} \cdot \sqrt{n(\Sigma y^2) - (\Sigma y)^2}]$
$= 6294.72/[\sqrt{10306.43} \cdot \sqrt{264231.13}] = .121$

c. H_o: $\rho = 0$
   H_1: $\rho \neq 0$
   $\alpha = .05$
   C.R. $r < -.254$ OR   C.R. $t < -t_{59,.025} = -2.000$
        $r > .254$ $t > t_{59,.025} = 2.000$

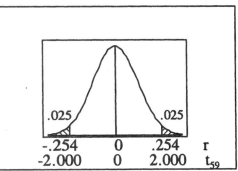

calculations: calculations:
   r = .121 $t_r = (r - \mu_r)/s_r$
                                $= (.121 - 0)/\sqrt{(1-(.121)^2)/59}$
                                $= .121/.1292 = .933$
conclusion:
   Do not reject H_o; there is not sufficient evidence to reject the claim that $\rho = 0$.

B. Conductivity-rainfall relationship?
a. n = 61
   Σx = 16.51
   Σy = 2874.5
   Σx^2 = 22.6397
   Σy^2 = 139786.58
   Σxy = 744.641

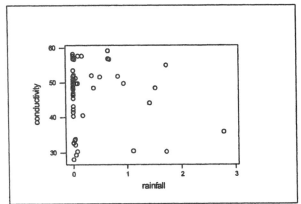

b. $n(\Sigma xy) - (\Sigma x)(\Sigma y) = 61(744.641) - (16.51)(2874.5) = -2034.894$
   $n(\Sigma x^2) - (\Sigma x)^2 = 61(22.6397) - (16.51)^2 = 1108.4416$
   $n(\Sigma y^2) - (\Sigma y)^2 = 61(139786.58) - (2874.5)^2 = 264231.13$
   $r = [n(\Sigma xy) - (\Sigma x)(\Sigma y)]/[\sqrt{n(\Sigma x^2) - (\Sigma x)^2} \cdot \sqrt{n(\Sigma y^2) - (\Sigma y)^2}\ ]$
     $= -2034.894/[\sqrt{1108.4416} \cdot \sqrt{264231.13}\ ] = -.119$

c. H_o: $\rho = 0$
   H_1: $\rho \neq 0$
   $\alpha = .05$
   C.R. $r < -.254$ OR   C.R. $t < -t_{59,.025} = -2.000$
        $r > .254$ $t > t_{59,.025} = 2.000$

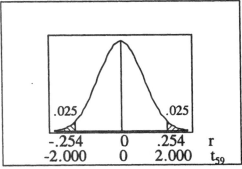

calculations: calculations:
   r = -.119 $t_r = (r - \mu_r)/s_r$
                                $= (-.119 - 0)/\sqrt{(1-(-.119)^2)/59}$
                                $= -.119/.1293 = -.920$
conclusion:
   Do not reject H_o; there is not sufficient evidence to reject the claim that $\rho = 0$.

C. Since the correlation between conductivity and salinity is nearly 1.00, the above conclusions regarding conductivity apply to salinity – i.e., neither bottom temperature nor rainfall amount has a significant linear correlation with salinity. Notice that the negative sign for the calculated sample correlation in part (b) agrees with the notion that high rainfall means low salinity – which, if there is close to a perfect positive correlation between salinity and conductivity, means low conductivity.

23. A. Interval-duration relationship?

a. n = 50
Σx = 10832
Σy = 4033
Σx² = 2513280
Σy² = 332331
Σxy = 903488

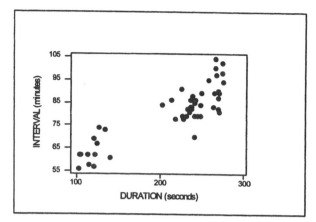

b. $n(\Sigma xy) - (\Sigma x)(\Sigma y) = 50(903488) - (10832)(4033) = 1488944$
$n(\Sigma x^2) - (\Sigma x)^2 = 50(2513280) - (10832)^2 = 8331776$
$n(\Sigma y^2) - (\Sigma y)^2 = 50(332331) - (4033)^2 = 351461$
$r = [n(\Sigma xy) - (\Sigma x)(\Sigma y)]/[\sqrt{n(\Sigma x^2) - (\Sigma x)^2} \cdot \sqrt{n(\Sigma y^2) - (\Sigma y)^2}\,]$
$= 1488944/[\sqrt{8331776} \cdot \sqrt{351461}\,] = .870$

c. $H_o: \rho = 0$
$H_1: \rho \neq 0$
$\alpha = .05$
C.R. $r < -.279$ OR C.R. $t < -t_{48,.025} = -1.960$
    $r > .279$ $t > t_{48,.025} = 1.960$

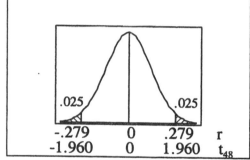

calculations: calculations:
r = .870 $t_r = (r - \mu_r)/s_r$
                       $= (.870 - 0)/\sqrt{(1-(.870)^2)/48}$
conclusion: $= .870/.0711 = 12.231$

Reject H_o; there is sufficient evidence to reject the claim that $\rho = 0$ and to conclude that $\rho \neq 0$ (in fact, that $\rho > 0$).
Yes; there is a significant positive linear correlation, suggesting that the interval after an eruption is related to the duration of the eruption. NOTE: The longer the duration of an eruption, the more pressure has been released and the longer it will take the geyser to build back up for another eruption. In fact, the park rangers use the duration of one eruption to predict the time of the next eruption.

B. Interval-height relationship?

a. n = 50
Σx = 6904
Σy = 4033
Σx² = 961150
Σy² = 332331
Σxy = 556804

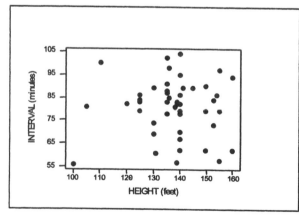

b. $n(\Sigma xy) - (\Sigma x)(\Sigma y) = 50(556804) - (6904)(4033) = -3632$
$n(\Sigma x^2) - (\Sigma x)^2 = 50(961150) - (6904)^2 = 392284$
$n(\Sigma y^2) - (\Sigma y)^2 = 50(332331) - (4033)^2 = 351461$

$r = [n(\Sigma xy) - (\Sigma x)(\Sigma y)]/[\sqrt{n(\Sigma x^2) - (\Sigma x)^2} \cdot \sqrt{n(\Sigma y^2) - (\Sigma y)^2}]$
  $= -3632/[\sqrt{392284} \cdot \sqrt{351461}] = -.00978$

c. $H_o: \rho = 0$
   $H_1: \rho \neq 0$
   $\alpha = .05$
   C.R. $r < -.279$ OR C.R. $t < -t_{48,.025} = -1.960$
        $r > .279$ $t > t_{48,.025} = 1.960$

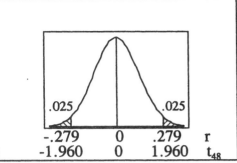

   calculations: calculations:
      $r = -.00978$ $t_r = (r - \mu_r)/s_r$
                               $= (-.00978 - 0)/\sqrt{(1-(-.00978)^2)/48}$
                               $= -.00978/.144 = -.068$
   conclusion:
      Do not reject H_o; there is not sufficient evidence to reject the claim that $\rho = 0$.
      No; there is not a significant linear correlation, suggesting that the interval after an eruption is not so related to the height of the eruption.

C. Duration is the more relevant predictor of the interval until the next eruption -- because it has a significant correlation with interval, while height does not.

24. A. Price-weight relationship? [NOTE: For convenience, we measure the price y in $1000's.]

   a. n = 30
      Σx = 50.65
      Σy = 433.23
      Σx² = 109.74
      Σy² = 12378.86
      Σxy = 1027.03

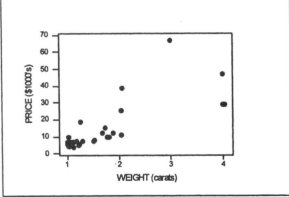

   b. $n(\Sigma xy) - (\Sigma x)(\Sigma y) = 30(1027.03) - (50.65)(433.23) = 8867.80$
      $n(\Sigma x^2) - (\Sigma x)^2 = 30(109.74) - (50.65)^2 = 726.78$
      $n(\Sigma y^2) - (\Sigma y)^2 = 30(12378.86) - (433.23)^2 = 183677.57$
      $r = [n(\Sigma xy) - (\Sigma x)(\Sigma y)]/[\sqrt{n(\Sigma x^2) - (\Sigma x)^2} \cdot \sqrt{n(\Sigma y^2) - (\Sigma y)^2}]$
        $= 8867.80/[\sqrt{726.78} \cdot \sqrt{183677.57}] = .767$

   c. $H_o: \rho = 0$
      $H_1: \rho \neq 0$
      $\alpha = .05$
      C.R. $r < -.361$ OR C.R. $t < -t_{28,.025} = -2.048$
           $r > .361$ $t > t_{28,.025} = 2.048$

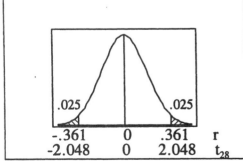

      calculations: calculations:
         $r = .767$ $t_r = (r - \mu_r)/s_r$
                                  $= (.767 - 0)/\sqrt{(1-(.767)^2)/28}$
      conclusion: $= .767/.121 = 6.325$
         Reject H_o; there is sufficient evidence to reject the claim that $\rho = 0$ and to conclude that $\rho \neq 0$ (in fact, that $\rho > 0$).

Yes; based on this result there is a significant positive linear correlation between the price of a diamond and its weight in carats.

B. Price-color relationship? [NOTE: For convenience, we measure the price y in $1000's.]

a. n = 30
   Σx = 140
   Σy = 433.23
   Σx² = 782
   Σy² = 12378.86
   Σxy = 1630.03

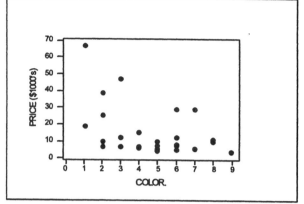

b. $n(\Sigma xy) - (\Sigma x)(\Sigma y) = 30(1630.03) - (140)(433.23) = -11751.30$
   $n(\Sigma x^2) - (\Sigma x)^2 = 30(782) - (140)^2 = 3860$
   $n(\Sigma y^2) - (\Sigma y)^2 = 30(12378.86) - (433.23)^2 = 183677.57$
   $r = [n(\Sigma xy) - (\Sigma x)(\Sigma y)]/[\sqrt{n(\Sigma x^2) - (\Sigma x)^2} \cdot \sqrt{n(\Sigma y^2) - (\Sigma y)^2}]$
   $= -11751.30/[\sqrt{3860} \cdot \sqrt{183677.57}\,] = -.441$

c. $H_o: \rho = 0$
   $H_1: \rho \neq 0$
   $\alpha = .05$
   C.R. r < -.361   OR   C.R. $t < -t_{28,.025} = -2.048$
       r > .361 $t > t_{28,.025} = 2.048$

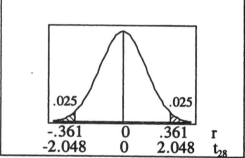

calculations: calculations:
  r = -.441 $t_r = (r - \mu_r)/s_r$
                        $= (-.441 - 0)/\sqrt{(1-(-.441)^2)/28}$
conclusion: $= -.441/.170 = -2.600$

Reject H_o; there is sufficient evidence to reject the claim that $\rho = 0$ and to conclude that $\rho \neq 0$ (in fact, that $\rho < 0$).

Yes; based on this result there is a significant negative linear correlation between price and color – i.e., the lower the color number, the higher the price.

C. When determining the value of a diamond, weight is more important than color. The weight of a diamond is more highly correlated (i.e., has an r value farther from 0) with the selling price than is the color.

25. A linear correlation coefficient very close to zero indicates <u>no</u> significant linear correlation and no tendencies can be inferred.

26. A significant linear correlation indicates that the factors are associated, not that there is a cause and effect relationship. Even if there is a cause and effect relationship, correlation analysis cannot identify which factor is the cause and which factor is the effect.

27. A linear correlation coefficient very close to zero indicates no significant <u>linear</u> correlation, but there may some other type of relationship between the variables.

28. A significant linear correlation between group averages indicates nothing about the relationship between the individual scores -- which may be uncorrelated, correlated in the opposite direction, or have different correlations within each of the groups.

29. There are n=13 data points.
    From the scatterplot, it appears that the correlation is about +.75.
    $H_o: \rho = 0$
    $H_1: \rho \neq 0$
    $\alpha = .05$
    C.R. $r < -.553$ OR   C.R. $t < -t_{11,.025} = -2.201$
          $r > .553$ $t > t_{11,.025} = 2.201$

    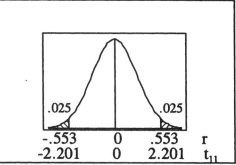

    calculations: calculations:
       r = .75 $t_r = (r - \mu_r)/s_r$
                              $= (.75 - 0)/\sqrt{(1-(.75)^2)/11}$
                              $= .75/.199 = 3.761$
    conclusion:
       Reject H_o; there is sufficient evidence to reject the claim that $\rho = 0$ and to conclude that $\rho \neq 0$ (in fact, that $\rho > 0$).

    Yes; based on this result there is a significant positive linear correlation between the mortality rate and the rate of infections acquired in intensive care units.

30. The following table gives the values for y, x, x^2, log x, $\sqrt{x}$ and 1/x. The rows at the bottom of the table give the sum of the values (i.e., Σv), the sum of squares of the values (i.e., Σv^2), the sum of each value times the corresponding y value (i.e., Σvy), and the quantity $n\Sigma v^2 - (\Sigma v)^2$ needed in subsequent calculations.

	y	x	x^2	log x	$\sqrt{x}$	1/x
	.11	1.3	1.69	.1139	1.1402	.7692
	.38	2.4	5.76	.3802	1.5492	.4167
	.41	2.6	6.76	.4150	1.6125	.3846
	.45	2.8	7.84	.4472	1.6733	.3571
	.39	2.4	5.76	.3802	1.5492	.4167
	.48	3.0	9.00	.4771	1.7321	.3333
	.61	4.1	16.81	.6128	2.0248	.2439
Σv	2.83	18.6	53.62	2.8264	11.2814	2.9216
Σv^2	1.2817	53.62	539.95	1.2774	18.6000	1.3850
Σvy		8.258	25.495	1.2795	4.7989	1.0326
$n\Sigma v^2-(\Sigma v)^2$	.9630	29.38	904.55	.9533	2.9300	1.1593

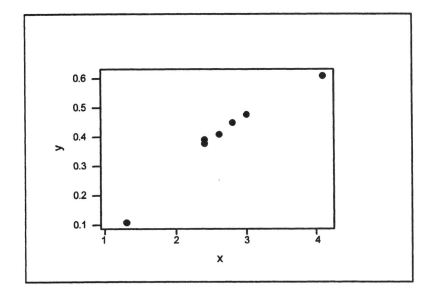

In general, $r = [n(\Sigma vy) - (\Sigma v)(\Sigma y)]/[\sqrt{n(\Sigma v^2) - (\Sigma v)^2} \cdot \sqrt{n(\Sigma y^2) - (\Sigma y)^2}]$

a. For $v = x$,
$r = [7(8.258) - (18.6)(2.83)]/[\sqrt{29.38} \cdot \sqrt{.9630}] = .9716$

b. For $v = x^2$,
$r = [7(25.495) - (53.62)(2.83)]/[\sqrt{904.55} \cdot \sqrt{.9630}] = .9053$

c. For $v = \log x$,
$r = [7(1.2795) - (2.8264)(2.83)]/[\sqrt{.9533} \cdot \sqrt{.9630}] = .9996$

d. For $v = \sqrt{x}$,
$r = [7(4.7989) - (11.2814)(2.83)]/[\sqrt{2.9300} \cdot \sqrt{.9630}] = .9918$

e. For $v = 1/x$,
$r = [7(1.0326) - (2.9216)(2.83)]/[\sqrt{29.38} \cdot \sqrt{1.1593}] = -.9842$

In each case, the critical values from Table A-6 for testing significance at the .05 level are $\pm.754$. While all the correlations are significant, the largest value for r occurs in part (c).

31. a. For $\pm t_{48,.025} = \pm 1.960$,
the critical values are $r = \pm 1.960/\sqrt{(\pm 1.960)^2 + 48} = \pm.272$.

b. For $\pm t_{73,.05} = \pm 1.645$,
the critical values are $r = \pm 1.645/\sqrt{(\pm 1.645)^2 + 73} = \pm.189$.

c. For $-t_{18,.05} = -1.734$,
the critical value is $r = -1.734/\sqrt{(-1.734)^2 + 18} = -.378$.

d. For $t_{8,.05} = 1.860$,
the critical value is $r = 1.860/\sqrt{(1.860)^2 + 8} = .549$.

e. For $t_{10,.01} = 2.764$,
the critical value is $r = 2.764/\sqrt{(2.764)^2 + 10} = .658$.

32. The following summary statistics apply to all parts of the exercise.

both genders	females only	males only
n = 16	n = 8	n = 8
$\Sigma x = 2408$	$\Sigma x = 1042$	$\Sigma x = 1366$
$\Sigma y = 195$	$\Sigma y = 34$	$\Sigma y = 161$
$\Sigma x^2 = 369474$	$\Sigma x^2 = 135958$	$\Sigma x^2 = 233516$
$\Sigma y^2 = 3463$	$\Sigma y^2 = 164$	$\Sigma y^2 = 3299$
$\Sigma xy = 31884$	$\Sigma xy = 4414$	$\Sigma xy = 27470$

a. There appears to be a strong positive linear correlation between weight and remote use.

b. $n(\Sigma xy) - (\Sigma x)(\Sigma y) = 16(31884) - (2408)(195) = 40584$
$n(\Sigma x^2) - (\Sigma x)^2 = 16(369474) - (2408)^2 = 113120$
$n(\Sigma y^2) - (\Sigma y)^2 = 16(3463) - (195)^2 = 17383$
$r = [n(\Sigma xy) - (\Sigma x)(\Sigma y)]/[\sqrt{n(\Sigma x^2) - (\Sigma x)^2} \cdot \sqrt{n(\Sigma y^2) - (\Sigma y)^2}]$
$= 40584/[\sqrt{113120} \cdot \sqrt{17383}] = .915$

Since CV $= \pm.497$ at the .05 level of significance, conclude that there is a positive correlation between weight and remote use.

c. $n(\Sigma xy) - (\Sigma x)(\Sigma y) = 8(4414) - (1042)(34) = -116$
$n(\Sigma x^2) - (\Sigma x)^2 = 8(135958) - (1042)^2 = 1900$
$n(\Sigma y^2) - (\Sigma y)^2 = 8(164) - (34)^2 = 156$
$r = [n(\Sigma xy) - (\Sigma x)(\Sigma y)]/[\sqrt{n(\Sigma x^2) - (\Sigma x)^2} \cdot \sqrt{n(\Sigma y^2) - (\Sigma y)^2}]$
$= -116/[\sqrt{1900} \cdot \sqrt{156}] = -.213$

Since CV $= \pm.707$ at the .05 level of significance, there is not evidence of a linear correlation between weight and remote use for the females.

d. $n(\Sigma xy) - (\Sigma x)(\Sigma y) = 8(27470) - (1366)(161) = -166$
$n(\Sigma x^2) - (\Sigma x)^2 = 8(233516) - (1366)^2 = 2172$
$n(\Sigma y^2) - (\Sigma y)^2 = 8(3299) - (161)^2 = 471$
$r = [n(\Sigma xy) - (\Sigma x)(\Sigma y)]/[\sqrt{n(\Sigma x^2) - (\Sigma x)^2} \cdot \sqrt{n(\Sigma y^2) - (\Sigma y)^2}]$
$= -166/[\sqrt{2172} \cdot \sqrt{471}] = -.164$

Since CV = ±.707 at the .05 level of significance, there is not evidence of a linear correlation between weight and remote use for the males.

e. Combining different groups can give a misleading impression of the true correlation within the separate groups.

33. For r = .600, (1+r)/(1-r) = (1.6)/(.4) = 4. Following the procedure outlined in the text,

Step a. $z_{.025} = 1.960$

Step b. $w_L = \frac{1}{2} \cdot \ln(4) - 1.960/\sqrt{47} = .407$
$w_R = \frac{1}{2} \cdot \ln(4) + 1.960/\sqrt{47} = .979$

Step c. $(e^{.814} - 1)/(e^{.814} + 1) < \rho < (e^{1.958} - 1)/(e^{1.958} + 1)$
$(1.258)/(3.258) < \rho < (6.086)/(8.086)$
$.386 < \rho < .753$

NOTE: While the distribution of r is not normal, $\tanh^{-1}(r)$ [i.e., the inverse hyperbolic tangent of r] follows a normal distribution with $\mu_r = \tanh^{-1}(\rho)$ and $\sigma_r = 1/\sqrt{n-3}$. The steps above are equivalent to finding $\tanh^{-1}(r)$, constructing a confidence interval for $\tanh^{-1}(\rho)$, and then applying the hyperbolic tangent function to the endpoints to produce a confidence interval for ρ.

9-3 Regression

1. a. For n=20, CV=±.444; since r=.987>.444, use the regression line for predictions.
    $\hat{y} = 6.00 + 4.00x$
    $\hat{y}_{3.00} = 6.00 + 4.00(3.00) = 18.00$
    b. For n=20, CV=±.444; since r=.052<.444, use the mean for predictions.
    $\hat{y} = \bar{y}$
    $\hat{y}_{3.00} = \bar{y} = 5.00$

2. a. For n=30, CV=±.361; since r=-.123>-.361, use the mean for predictions.
    $\hat{y} = \bar{y}$
    $\hat{y}_{2.00} = \bar{y} = 8.00$
    b. For n=30, CV=±.361; since r=-.567<-.361, use the regression line for predictions.
    $\hat{y} = 7.00 - 2.00x$
    $\hat{y}_{2.00} = 7.00 - 2.00(2.00) = 3.00$

3. For n=8, CV=±.707; since r=.993>.707, use the regression line for predictions.
    $\hat{y} = -187 + 11.3x$
    $\hat{y}_{52} = -187 + 11.3(52) = 401$

4. For n=21, CV=±.444; since r=-.133>-.444, use the mean for predictions.
    $\hat{y} = \bar{y}$
    $\hat{y}_{1200} = \bar{y} = 51.4$

NOTE: For exercises 5-24, the exact summary statistics (i.e., without any rounding) are given on the right. While the intermediate calculations on the left are presented rounded to various degrees of accuracy, the entire unrounded values were preserved in the calculator until the end.

5. $\bar{x} = 2.00$
    $\bar{y} = 2.00$
    $b_1 = [n(\Sigma xy) - (\Sigma x)(\Sigma y)]/[n(\Sigma x^2) - (\Sigma x)^2]$
    $= [5(20) - (10)(10)]/[5(30) - (10)^2]$
    $= 0/50$
    $= 0.00$

    $n = 5$
    $\Sigma x = 10$
    $\Sigma y = 10$
    $\Sigma x^2 = 30$
    $\Sigma y^2 = 344$
    $\Sigma xy = 20$

$b_o = \bar{y} - b_1\bar{x}$
$= 2.00 - (0.00)(2.00) = 2.00$
$\hat{y} = b_o + b_1x = 2.00 + 0.00x$

6. $\bar{x} = 3.20$ $n = 5$
$\bar{y} = 8.20$ $\Sigma x = 16$
$b_1 = [n(\Sigma xy) - (\Sigma x)(\Sigma y)]/[n(\Sigma x^2) - (\Sigma x)^2]$ $\Sigma y = 41$
$= [5(185) - (16)(41)]/[5(70) - (16)^2]$ $\Sigma x^2 = 70$
$= 269/94$ $\Sigma y^2 = 495$
$= 2.862$ $\Sigma xy = 185$
$b_o = \bar{y} - b_1\bar{x}$
$= 8.20 - (2.862)(3.20) = -.597$
$\hat{y} = b_o + b_1x = -.597 + 2.862x$

7. a. $\bar{x} = 2.80$ $n = 10$
$\bar{y} = 2.80$ $\Sigma x = 28$
$b_1 = [n(\Sigma xy) - (\Sigma x)(\Sigma y)]/[n(\Sigma x^2) - (\Sigma x)^2]$ $\Sigma y = 28$
$= 576/636$ $\Sigma x^2 = 142$
$= .906$ $\Sigma y^2 = 142$
$b_o = \bar{y} - b_1\bar{x}$ $\Sigma xy = 136$
$= 2.80 - .906(2.80) = .264$
$\hat{y} = b_o + b_1x = .264 + .906x$

b. $\bar{x} = 2.00$ $n = 9$
$\bar{y} = 2.00$ $\Sigma x = 18$
$b_1 = [n(\Sigma xy) - (\Sigma x)(\Sigma y)]/[n(\Sigma x^2) - (\Sigma x)^2]$ $\Sigma y = 18$
$= 0/54$ $\Sigma x^2 = 42$
$= .000$ $\Sigma y^2 = 42$
$b_o = \bar{y} - b_1\bar{x}$ $\Sigma xy = 136$
$= 2.00 - .000(2.00) = 2.00$
$\hat{y} = b_o + b_1x = 2.00 + 0.00x$

c. The two regression lines are very different, illustrating that one point can affect the regression dramatically.

NOTE: In the exercises that follow, this manual uses the full accuracy of b_o and b_1 when calculating $\hat{y}$. Using only the rounded values as stated in the equation produces slightly different answers. More detail for calculating the numerator and denominator of b_1 is given in the corresponding exercise in the previous section.

8. $\bar{x} = 61.55$ $n = 11$
$\bar{y} = 3.04$ $\Sigma x = 677$
$b_1 = [n(\Sigma xy) - (\Sigma x)(\Sigma y)]/[n(\Sigma x^2) - (\Sigma x)^2]$ $\Sigma y = 33.4$
$= 1071.2/15826 = .0677$ $\Sigma x^2 = 43105$
$b_o = \bar{y} - b_1\bar{x}$ $\Sigma y^2 = 126.04$
$= 3.04 - (.0677)(61.55) = -1.13$ $\Sigma xy = 2153.0$
$\hat{y} = b_o + b_1x = -1.13 + .0677x$
$\hat{y}_{80} = \bar{y} = 3.04$ [no significant correlation]

9. $\bar{x} = 23.525$ $n = 8$
$\bar{y} = 6.500$ $\Sigma x = 188.2$
$b_1 = [n(\Sigma xy) - (\Sigma x)(\Sigma y)]/[n(\Sigma x^2) - (\Sigma x)^2]$ $\Sigma y = 52.0$
$= -656.64/59381.08 = -.0111$ $\Sigma x^2 = 11850.04$
$b_o = \bar{y} - b_1\bar{x}$ $\Sigma y^2 = 402.90$
$= 6.5000 - (-.0111)(23.525) = 6.76$ $\Sigma xy = 1141.22$
$\hat{y} = b_o + b_1x = 6.76 - .0111x$
$\hat{y}_{16} = \bar{y} = 6.5$ [no significant correlation]
In this instance, the predicted value is far from actual value.

10. $\bar{x} = 70.22$
    $\bar{y} = 121.00$
    $b_1 = [n(\Sigma xy) - (\Sigma x)(\Sigma y)]/[n(\Sigma x^2) - (\Sigma x)^2]$
       $= 666/171.5 = 3.88$
    $b_0 = \bar{y} - b_1\bar{x}$
       $= 121.00 - (3.88)(70.22) = -151.7$
    $\hat{y} = b_0 + b_1 x = -151.7 + 3.88x$
    $\hat{y}_{69} = -151.7 + 3.88(69) = 116$

    $n = 9$
    $\Sigma x = 632.0$
    $\Sigma y = 1089$
    $\Sigma x^2 = 44399.50$
    $\Sigma y^2 = 132223$
    $\Sigma xy = 76546.0$

11. $\bar{x} = 133.93$
    $\bar{y} = 88.64$
    $b_1 = [n(\Sigma xy) - (\Sigma x)(\Sigma y)]/[n(\Sigma x^2) - (\Sigma x)^2]$
       $= 11447/14881 = .769$
    $b_0 = \bar{y} - b_1\bar{x}$
       $= 88.64 - (.769)(133.93) = -14.4$
    $\hat{y} = b_0 + b_1 x = -14.4 + .769x$
    $\hat{y}_{122} = -14.4 + .769(112) = 79$

    $n = 14$
    $\Sigma x = 1875$
    $\Sigma y = 1241$
    $\Sigma x^2 = 252179$
    $\Sigma y^2 = 111459$
    $\Sigma xy = 167023$

12. $\bar{x} = 59.750$
    $\bar{y} = 147.077$
    $b_1 = [n(\Sigma xy) - (\Sigma x)(\Sigma y)]/[n(\Sigma x^2) - (\Sigma x)^2]$
       $= 129.010/4076 = .0317$
    $b_0 = \bar{y} - b_1\bar{x}$
       $= 147.077 - (.0317)(59.750) = 145.2$
    $\hat{y} = b_0 + b_1 x = 145.2 + .0317x$
    $\hat{y}_{73} = \bar{y} = 147.077$ [no significant correlation]
    The predicted value is over 3 minutes off; considering the variability in times, this seems like a large error.

    $n = 8$
    $\Sigma x = 478$
    $\Sigma y = 1176.617$
    $\Sigma x^2 = 29070$
    $\Sigma y^2 = 173068.664832$
    $\Sigma xy = 70318.992$

13. $\bar{x} = 14.583$
    $\bar{y} = 175.205$
    $b_1 = [n(\Sigma xy) - (\Sigma x)(\Sigma y)]/[n(\Sigma x^2) - (\Sigma x)^2]$
       $= 77407.62/31235 = 2.478$
    $b_0 = \bar{y} - b_1\bar{x}$
       $= 175.205 - (2.478)(14.583) = 139.1$
    $\hat{y} = b_0 + b_1 x = 139.1 + 2.478x$
    $\hat{y}_{40} = \bar{y} = 175.2$ [no significant correlation]

    $n = 12$
    $\Sigma x = 175$
    $\Sigma y = 2102.46$
    $\Sigma x^2 = 5155$
    $\Sigma y^2 = 601709.8240$
    $\Sigma xy = 37111.51$

14. $\bar{x} = 4.743$
    $\bar{y} = 47.793$
    $b_1 = [n(\Sigma xy) - (\Sigma x)(\Sigma y)]/[n(\Sigma x^2) - (\Sigma x)^2]$
       $= 9691.14/1814.60 = 5.34$
    $b_0 = \bar{y} - b_1\bar{x}$
       $= 47.793 - (5.34)(4.743) = 22.5$
    $\hat{y} = b_0 + b_1 x = 22.5 + 5.34x$
    $\hat{y}_{4.0} = 22.5 + 5.34(4.0) = 43.8$
    Circumferences are more easily and more accurately measured than heights.

    $n = 14$
    $\Sigma x \doteq 66.4$
    $\Sigma y = 669.1$
    $\Sigma x^2 = 444.54$
    $\Sigma y^2 = 37365.71$
    $\Sigma xy = 3865.67$

15. $\bar{x} = .0219$
    $\bar{y} = 3.7625$
    $b_1 = [n(\Sigma xy) - (\Sigma x)(\Sigma y)]/[n(\Sigma x^2) - (\Sigma x)^2]$
       $= .402/.1063 = 3.78$
    $b_0 = \bar{y} - b_1\bar{x}$
       $= 3.7625 - (3.78)(.0219) = 3.68$
    $\hat{y} = b_0 + b_1 x = 3.68 + 3.78x$
    $\hat{y}_{.05} = \bar{y} = 3.76$ [no significant correlation]

    $n = 16$
    $\Sigma x = .35$
    $\Sigma y = 60.2$
    $\Sigma x^2 = .0143$
    $\Sigma y^2 = 227.24$
    $\Sigma xy = 1.342$

S-294 INSTRUCTOR'S SOLUTIONS Chapter 9

16. $\bar{x} = 57.44$ $\qquad\qquad\qquad\qquad\qquad$ $n = 50$
$\bar{y} = 32.46$ $\qquad\qquad\qquad\qquad\qquad\qquad$ $\Sigma x = 2872$
$b_1 = [n(\Sigma xy) - (\Sigma x)(\Sigma y)]/[n(\Sigma x^2) - (\Sigma x)^2]$ $\qquad$ $\Sigma y = 1623$
$\quad = 4027294/26497516 = .152$ $\qquad\qquad$ $\Sigma x^2 = 694918$
$b_o = \bar{y} - b_1\bar{x}$ $\qquad\qquad\qquad\qquad\qquad\qquad$ $\Sigma y^2 = 268331$
$\quad = 32.46 - (.152)(57.44) = 23.7$ $\qquad\qquad$ $\Sigma xy = 173771$
$\hat{y} = b_o + b_1x = 23.7 + .152x$
$\hat{y}_0 = \bar{y} = 32$ [no significant correlation]

17. $\bar{x} = 240.875$ $\qquad\qquad\qquad\qquad\qquad$ $n = 40$
$\bar{y} = 25.740$ $\qquad\qquad\qquad\qquad\qquad\qquad$ $\Sigma x = 9635$
$b_1 = [n(\Sigma xy) - (\Sigma x)(\Sigma y)]/[n(\Sigma x^2) - (\Sigma x)^2]$ $\qquad$ $\Sigma y = 1029.6$
$\quad = 861868.8/53959535 = .0160$ $\qquad\qquad$ $\Sigma x^2 = 3669819$
$b_o = \bar{y} - b_1\bar{x}$ $\qquad\qquad\qquad\qquad\qquad\qquad$ $\Sigma y^2 = 27984.46$
$\quad = 25.740 - (.0160)(240.875) = 21.9$ $\qquad$ $\Sigma xy = 269551.6$
$\hat{y} = b_o + b_1x = 21.9 + .0160x$
$\hat{y}_{500} = 21.9 + .0160(500) = 29.9$

18. $\bar{x} = 70.73$ $\qquad\qquad\qquad\qquad\qquad\qquad$ $n = 12$
$\bar{y} = 6.50$ $\qquad\qquad\qquad\qquad\qquad\qquad$ $\Sigma x = 848.8$
$b_1 = [n(\Sigma xy) - (\Sigma x)(\Sigma y)]/[n(\Sigma x^2) - (\Sigma x)^2]$ $\qquad$ $\Sigma y = 78.0$
$\quad = -3342.48/16933.04 = -.197$ $\qquad\qquad$ $\Sigma x^2 = 61449.54$
$b_o = \bar{y} - b_1\bar{x}$ $\qquad\qquad\qquad\qquad\qquad\qquad$ $\Sigma y^2 = 572.94$
$\quad = 6.50 - (-.197)(70.73) = 20.5$ $\qquad\qquad$ $\Sigma xy = 5238.66$
$\hat{y} = b_o + b_1x = 20.5 - .197x$
$\hat{y}_{50} = 20.5 - .197(50) = 10.6$

19. a. selling price/asking price relationship
$\bar{x} = 178.34$ $\qquad\qquad\qquad\qquad\qquad\qquad$ $n = 50$
$\bar{y} = 170.34$ $\qquad\qquad\qquad\qquad\qquad\qquad$ $\Sigma x = 8917$
$b_1 = [n(\Sigma xy) - (\Sigma x)(\Sigma y)]/[n(\Sigma x^2) - (\Sigma x)^2]$ $\qquad$ $\Sigma y = 8517.0$
$\quad = 14117561/15438961 = .914$ $\qquad\qquad$ $\Sigma x^2 = 1899037$
$b_o = \bar{y} - b_1\bar{x}$ $\qquad\qquad\qquad\qquad\qquad\qquad$ $\Sigma y^2 = 1710311.50$
$\quad = 170.34 - (.914)(178.34) = 7.26$ $\qquad\qquad$ $\Sigma xy = 1801273$
$\hat{y} = b_o + b_1x = 7.26 + .914x$
$\hat{y}_{200} = 7.26 + .914(200) = 190.1$ [$190,100]

b. NOTE: These x-y values are reversed from what they were in the corresponding exercise in section 9.2.
selling price/taxes relationship
$\bar{x} = 3706.10$ $\qquad\qquad\qquad\qquad\qquad\qquad$ $n = 50$
$\bar{y} = 170.34$ $\qquad\qquad\qquad\qquad\qquad\qquad$ $\Sigma x = 8517.0$
$b_1 = [n(\Sigma xy) - (\Sigma x)(\Sigma y)]/[n(\Sigma x^2) - (\Sigma x)^2]$ $\qquad$ $\Sigma y = 185305$
$\quad = 253353140/12976286 = 19.5$ $\qquad\qquad$ $\Sigma x^2 = 1710311.50$
$b_o = \bar{y} - b_1\bar{x}$ $\qquad\qquad\qquad\qquad\qquad\qquad$ $\Sigma y^2 = 809069344$
$\quad = 3706.10 - (19.5)(170.34) = 380.3$ $\qquad$ $\Sigma xy = 36631916.5$
$\hat{y} = b_o + b_1x = 380.3 + 19.5x$
$\hat{y}_{400} = 380.3 + 19.5(400) = 8190$

20. a. nicotine/tar relationship
$\bar{x} = 12.1$ $\qquad\qquad\qquad\qquad\qquad\qquad$ $n = 29$
$\bar{y} = .941$ $\qquad\qquad\qquad\qquad\qquad\qquad$ $\Sigma x = 351$
$b_1 = [n(\Sigma xy) - (\Sigma x)(\Sigma y)]/[n(\Sigma x^2) - (\Sigma x)^2]$ $\qquad$ $\Sigma y = 27.3$
$\quad = 1133.2/17420 = .0651$ $\qquad\qquad\qquad$ $\Sigma x^2 = 4849$
$b_o = \bar{y} - b_1\bar{x}$ $\qquad\qquad\qquad\qquad\qquad\qquad$ $\Sigma y^2 = 28.45$
$\quad = .941 - (.0651)(12.1) = .154$ $\qquad\qquad$ $\Sigma xy = 369.5$
$\hat{y} = b_o + b_1x = .154 + .0651x$

$\hat{y}_{15} = .154 + .0651(15) = 1.1$

b. nicotine/CO relationship
$\bar{x} = 12.4$ $n = 29$
$\bar{y} = .941$ $\Sigma x = 359$
$b_1 = [n(\Sigma xy) - (\Sigma x)(\Sigma y)]/[n(\Sigma x^2) - (\Sigma x)^2]$ $\Sigma y = 27.3$
     $= 981.5/16206 = .0606$ $\Sigma x^2 = 5003$
$b_0 = \bar{y} - b_1\bar{x}$ $\Sigma y^2 = 28.45$
     $= .941 - (.0606)(12.4) = .192$ $\Sigma xy = 371.8$
$\hat{y} = b_0 + b_1 x = .192 + .0606x$
$\hat{y}_{15} = .192 + .0606(15) = 1.1$

21. a. actual/five-day relationship
$\bar{x} = 34.48$ $n = 31$
$\bar{y} = 34.90$ $\Sigma x = 1069$
$b_1 = [n(\Sigma xy) - (\Sigma x)(\Sigma y)]/[n(\Sigma x^2) - (\Sigma x)^2]$ $\Sigma y = 1082$
     $= 9810/16050 = .611$ $\Sigma x^2 = 37381$
$b_0 = \bar{y} - b_1\bar{x}$ $\Sigma y^2 = 38352$
     $= 34.90 - (.611)(34.48) = 13.8$ $\Sigma xy = 37628$
$\hat{y} = b_0 + b_1 x = 13.8 + .611x$
$\hat{y}_{28} = 13.8 + .611(28) = 31$

b. actual/one-day relationship
$\bar{x} = 33.32$ $n = 31$
$\bar{y} = 34.90$ $\Sigma x = 1033$
$b_1 = [n(\Sigma xy) - (\Sigma x)(\Sigma y)]/[n(\Sigma x^2) - (\Sigma x)^2]$ $\Sigma y = 1082$
     $= 13453/21228 = .634$ $\Sigma x^2 = 35107$
$b_0 = \bar{y} - b_1\bar{x}$ $\Sigma y^2 = 38352$
     $= 34.90 - (.634)(33.32) = 13.8$ $\Sigma xy = 36489$
$\hat{y} = b_0 + b_1 x = 13.8 + .634x$
$\hat{y}_{28} = 13.8 + .634(28) = 32$

c. The predicted value in part (b) is better, because the correlation was higher.

22. a. conductivity/temperature relationship
$\bar{x} = 30.38$ $n = 61$
$\bar{y} = 47.12$ $\Sigma x = 1853.0$
$b_1 = [n(\Sigma xy) - (\Sigma x)(\Sigma y)]/[n(\Sigma x^2) - (\Sigma x)^2]$ $\Sigma y = 2874.5$
     $= 6294.72/10306.43 = .611$ $\Sigma x^2 = 56457.63$
$b_0 = \bar{y} - b_1\bar{x}$ $\Sigma y^2 = 139786.58$
     $= 47.12 - (.611)(30.38) = 28.6$ $\Sigma xy = 87422.02$
$\hat{y} = b_0 + b_1 x = 28.6 + .611x$
$\hat{y}_{30} = \bar{y} = 47.12$ [no significant correlation]

b. conductivity/rainfall relationship
$\bar{x} = .271$ $n = 61$
$\bar{y} = 47.12$ $\Sigma x = 16.51$
$b_1 = [n(\Sigma xy) - (\Sigma x)(\Sigma y)]/[n(\Sigma x^2) - (\Sigma x)^2]$ $\Sigma y = 2874.5$
     $= -2034.894/1108.4416 = -1.84$ $\Sigma x^2 = 22.6397$
$b_0 = \bar{y} - b_1\bar{x}$ $\Sigma y^2 = 139786.58$
     $= 47.12 - (-1.84)(.271) = 47.6$ $\Sigma xy = 744.641$
$\hat{y} = b_0 + b_1 x = 47.6 - 1.84x$
$\hat{y}_0 = \bar{y} = 47.12$ [no significant correlation]

c. No; since neither relationship involves a significant correlation, the predicted values are not likely to be very accurate.

23. a. interval/duration relationship
    $\bar{x} = 216.64$ $n = 50$
    $\bar{y} = 80.66$ $\Sigma x = 10832$
    $b_1 = [n(\Sigma xy) - (\Sigma x)(\Sigma y)]/[n(\Sigma x^2) - (\Sigma x)^2]$ $\Sigma y = 4033$
    $\quad = 1488944/8331776 = .179$ $\Sigma x^2 = 2513280$
    $b_0 = \bar{y} - b_1\bar{x}$ $\Sigma y^2 = 332331$
    $\quad = 80.66 - (.179)(216.64) = 41.9$ $\Sigma xy = 903488$
    $\hat{y} = b_0 + b_1 x = 41.9 + .179x$
    $\hat{y}_{210} = 41.9 + .179(210) = 79.5$ minutes

    b. interval/height relationship
    $\bar{x} = 138.08$ $n = 50$
    $\bar{y} = 80.66$ $\Sigma x = 6904$
    $b_1 = [n(\Sigma xy) - (\Sigma x)(\Sigma y)]/[n(\Sigma x^2) - (\Sigma x)^2]$ $\Sigma y = 4033$
    $\quad = -3632/392284 = -.00926$ $\Sigma x^2 = 961150$
    $b_0 = \bar{y} - b_1\bar{x}$ $\Sigma y^2 = 332331$
    $\quad = 80.66 - (-.00926)(138.08) = 81.9$ $\Sigma xy = 556804$
    $\hat{y} = b_0 + b_1 x = 81.9 - .00926x$
    $\hat{y}_{275} = \bar{y} = 80.7$ minutes [no significant correlation]

    c. The predicted time in part (a) is better, since interval and duration are significantly correlated. Since interval and height are not significantly correlated, the predicted time in part (b) did not even use the height data.

24. a. price/weight relationship [NOTE: For convenience, we measure the price y in $1000's.]
    $\bar{x} = 1.688$ $n = 30$
    $\bar{y} = 14.441$ $\Sigma x = 50.65$
    $b_1 = [n(\Sigma xy) - (\Sigma x)(\Sigma y)]/[n(\Sigma x^2) - (\Sigma x)^2]$ $\Sigma y = 433.23$
    $\quad = 8867.80/726.78 = 12.202$ $\Sigma x^2 = 109.74$
    $b_0 = \bar{y} - b_1\bar{x}$ $\Sigma y^2 = 12378.86$
    $\quad = 14.441 - (12.202)(1.688) = -6.159$ $\Sigma xy = 1027.03$
    $\hat{y} = b_0 + b_1 x = -6.159 + 12.202x$
    $\hat{y}_{1.5} = -6.159 + 12.202(1.5) = 12.143$ [$12,143]

    b. price/color relationship [NOTE: For convenience, we measure the price y in $1000's.]
    $\bar{x} = 4.667$ $n = 50$
    $\bar{y} = 14.441$ $\Sigma x = 140$
    $b_1 = [n(\Sigma xy) - (\Sigma x)(\Sigma y)]/[n(\Sigma x^2) - (\Sigma x)^2]$ $\Sigma y = 433.23$
    $\quad = -11751.3/3860 = -3.044$ $\Sigma x^2 = 782$
    $b_0 = \bar{y} - b_1\bar{x}$ $\Sigma y^2 = 12378.86$
    $\quad = 14.441 - (-3.044)(4.667) = 28.648$ $\Sigma xy = 1630.03$
    $\hat{y} = b_0 + b_1 x = 28.648 + (-3.044)x$
    $\hat{y}_3 = 28.648 - 3.044(3) = 19.515$ [$19,515]

    c. The predicted time in part (a) is better, since there is a stronger correlation between price and weight than between price and color.

25. Yes; the point is an outlier, since it is far from the other data points – 120 is far from the other boat values, and 160 is far from the other manatee death values.
    No; the point is not an influential one, since it will not change the regression line by very much – the original regression line predicts $\hat{y} = -113 + 2.27(120) = 159.4$, and so the new point is consistent with the others.

26. Yes; the point is an outlier, since it is far from the other data points – 120 is far from the other boat values, and 10 is far from the other manatee death values.
    Yes; the point is an influential one, since it will dramatically alter the regression line – the original regression line predicts $\hat{y} = -113 + 2.27(120) = 159.4$, and so the new regression line will have to change considerably to come close to (120,10).

INSTRUCTOR'S SOLUTIONS Chapter 9   S-297

27. <u>original data</u>
    $n = 5$
    $\Sigma x = 4,234,178$
    $\Sigma y = 576$
    $\Sigma x^2 = 3,595,324,583,102$
    $\Sigma y^2 = 67552$
    $\Sigma xy = 491,173,342$

   <u>original data divided by 1000</u>
    $n = 5$
    $\Sigma x = 4,234.178$
    $\Sigma y = 576$
    $\Sigma x^2 = 3,595,324.583102$
    $\Sigma y^2 = 67552$
    $\Sigma xy = 491,173.342$

   <u>original data</u>
    $\bar{x} = 846835.6$
    $\bar{y} = 115.2$
    $n\Sigma xy - (\Sigma x)(\Sigma y) = 16,980,182$
    $n\Sigma x^2 - (\Sigma x)^2 = 48,459,579,826$
    $b_1 = 16,980,182/48,459,579,826$
        $= .0003504$
    $b_0 = \bar{y} - b_1\bar{x}$
        $= 115.2 - .0003504(846835.6)$
        $= -181.53$
    $\hat{y} = b_0 + b_1 x$
        $= -181.53 + .0003504x$

   <u>original data divided by 1000</u>
    $\bar{x} = 846.8356$
    $\bar{y} = 115.2$
    $n\Sigma xy - (\Sigma x)(\Sigma y) = 16,980.182$
    $n\Sigma x^2 - (\Sigma x)^2 = 48,459.579826$
    $b_1 = 16,980.182/48,459.579826$
        $= .3504$
    $b_0 = \bar{y} - b_1\bar{x}$
        $= 115.2 - .3504(846.8356)$
        $= -181.53$
    $\hat{y} = b_0 + b_1 x$
        $= -181.53 + .3504x$

   Dividing each x by 1000 multiplies b_1, the coefficient of x in the regression equation, by 1000; multiplying the x coefficient by 1000 and dividing x by 1000 will "cancel out" and all predictions remain the same.
   Dividing each y by 1000 divides both b_1 and b_0 by 1000; consistent with the new "units" for y, all predictions will also turn out divided by 1000.

28. Using $\hat{y} = 8 + 3x$ produces the table at the right. As expected $\Sigma(y-\hat{y})^2$ is less [364 < 374] using the proper regression equation $\hat{y} = 5 + 4x$ as given in the text.

x	y	$\hat{y}$	$y-\hat{y}$	$(y-\hat{y})^2$
1	4	11	-7	49
2	24	14	10	100
4	8	20	-12	144
5	32	23	9	81
12	68	68	0	374

29. •original data

x	y
2.0	12.0
2.5	18.7
4.2	53.0
10.0	225.0

    $n = 4$
    $\Sigma x = 18.7$
    $\Sigma y = 308.7$
    $\Sigma x^2 = 127.89$
    $\Sigma y^2 = 53927.69$
    $\Sigma xy = 2543.35$

    $b_1 = [n(\Sigma xy) - (\Sigma x)(\Sigma y)]/[n(\Sigma x^2) - (\Sigma x)^2] = 4400.71/161.87 = 27.2$
    $b_0 = \bar{y} - b_1\bar{x} = (308.7/4) - (27.2)(18.7/4) = -49.9$
    $\hat{y} = b_0 + b_1 x = -49.9 + 27.2x$
    $r = [n(\Sigma xy) - (\Sigma x)(\Sigma y)]/[\sqrt{n(\Sigma x^2) - (\Sigma x)^2} \cdot \sqrt{n(\Sigma y^2) - (\Sigma y)^2}]$
      $= 4400.71/[\sqrt{161.87} \cdot \sqrt{120415.07}] = .9968$

    •using ln(x) for x

x	y
.693	12.0
.916	18.7
1.435	53.0
2.303	225.0

    $n = 4$
    $\Sigma x = 5.347$
    $\Sigma y = 308.7$
    $\Sigma x^2 = 8.681$
    $\Sigma y^2 = 53927.69$
    $\Sigma xy = 619.594$

    $b_1 = [n(\Sigma xy) - (\Sigma x)(\Sigma y)]/[n(\Sigma x^2) - (\Sigma x)^2] = 827.7220/6.134071 = 134.9$
    $b_0 = \bar{y} - b_1\bar{x} = (308.7/4) - (134.9)(5.347/4) = 103.2$
    $\hat{y} = b_0 + b_1 \cdot \ln(x) = -103.2 + 134.9 \cdot \ln(x)$ [since the "x" is really ln(x)]

S-298 INSTRUCTOR'S SOLUTIONS Chapter 9

$r = [n(\Sigma xy) - (\Sigma x)(\Sigma y)]/[\sqrt{n(\Sigma x^2) - (\Sigma x)^2} \cdot \sqrt{n(\Sigma y^2) - (\Sigma y)^2}]$
$= 827.7220/[\sqrt{6.134071} \cdot \sqrt{120415.07}] = .9631$

Based on the value of the associated correlations (.9968 > .9631), the equation using the original data seems to fit the data better than the equation using ln(x) instead of x.

NOTE: Both x and y (perhaps, especially y) seem to grow exponentially. A wiser choice for a transformation might be to use both ln(x) for x and ln(y) for y [or, perhaps, only ln(y) for y].

30. If $H_o: \rho = 0$ is true, there is no linear correlation between x and y and $\hat{y} = \bar{y}$ is the appropriate prediction for y for any x.
If $H_o: \beta_1 = 0$ is true then the true regression line is $y = \beta_o + 0x = \beta_o$, and the best estimate for β_o is $b_o = \bar{y} - 0\bar{x} = \bar{y}$, producing the line $\hat{y} = \bar{y}$.
Since both null hypotheses imply precisely the same result, they are equivalent.

31. Refer to the table below at the left, where $\hat{y} = -112.71 + 2.2741x$ and the residual is $y - \hat{y}$. The residual plot is given below at the right, with x on the horizontal axis and the residuals on the vertical axis.

year	x	y	$\hat{y}$	residual
1991	68	53	41.9	11.1
1992	68	38	41.9	-3.9
1993	67	35	39.7	-4.7
1994	70	49	46.5	2.5
1995	71	42	48.8	-6.8
1996	73	60	53.3	6.7
1997	76	54	60.1	-6.1
1998	81	67	71.5	-4.5
1999	83	82	76.0	6.0
2000	84	78	78.3	-0.3
	741	558	558.0	0.0

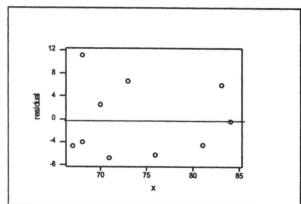

9-4 Variation and Prediction Intervals

1. The coefficient of determination is $r^2 = (.8)^2 = .64$.
The portion of the total variation explained by the regression line is $r^2 = .64 = 64\%$

2. The coefficient of determination is $r^2 = (-.6)^2 = .36$
The portion of the total variation explained by the regression line is $r^2 = .36 = 36\%$.

3. The coefficient of determination is $r^2 = (-.503)^2 = .253$.
The portion of the total variation explained by the regression line is $r^2 = .253 = 25.3\%$.

4. The coefficient of determination is $r^2 = (.636)^2 = .404$
The portion of the total variation explained by the regression line is $r^2 = .404 = 40.4\%$.

5. Since $r^2 = .924$, $r = +.961$ (positive because $b_1 = 12.5$ is positive).
From Table A-6, the critical values necessary for significance are $\pm.361$.
Since $.961 > .361$, we conclude there is a significant positive linear correlation between the tar and nicotine contents of cigarettes.

6. $R^2 = 92.4\%$

INSTRUCTOR'S SOLUTIONS Chapter 9   S-299

7. Use 1.2599, from the "fit" value on the Minitab display. This also agrees with the regression equation: $\hat{y} = .154 + .0651x$
$$\hat{y}_{17} = .154 + .0651(17) = 1.26$$

8. $1.07 < y_{17} < 1.45$
We are 95% confident that the interval from 1.07 mg to 1.45 mg contains the true amount of nicotine for an individual cigarette with a tar level of 17 mg.

NOTE: The following summary statistics apply to exercises 9-12 and 13-16. They may be used in the chapter formulas to calculate any values needed in the process of working the problems.

exercise 9	exercise 10	exercise 11	exercise 12
n = 9	n = 14	n = 14	n = 12
$\Sigma x = 632.0$	$\Sigma x = 1875$	$\Sigma x = 66.4$	$\Sigma x = 51.4$
$\Sigma y = 1089$	$\Sigma y = 1241$	$\Sigma y = 669.1$	$\Sigma y = 969.0$
$\Sigma x^2 = 44399.50$	$\Sigma x^2 = 252179$	$\Sigma x^2 = 444.54$	$\Sigma x^2 = 220.42$
$\Sigma y^2 = 132223$	$\Sigma y^2 = 111459$	$\Sigma y^2 = 37365.71$	$\Sigma y^2 = 78487.82$
$\Sigma xy = 76546.0$	$\Sigma xy = 167023$	$\Sigma xy = 3865.67$	$\Sigma xy = 4144.61$
see also 9.2-3, #10	see also 9.2-3, #11	see also 9.2-3, #14	not used previously

9. The predicted values were calculated using the regression line $\hat{y} = -151.6977 + 3.883382x$.

x	y	$\hat{y}$	$\bar{y}$	$\hat{y}-\bar{y}$	$(\hat{y}-\bar{y})^2$	$y-\hat{y}$	$(y-\hat{y})^2$	$y-\bar{y}$	$(y-\bar{y})^2$
71.0	125	124.02	121	3.0204	9.123	0.97958	0.960	4	16
70.5	119	122.08	121	1.0787	1.164	-3.07874	9.479	-2	4
71.0	128	124.02	121	3.0204	9.123	3.97958	15.837	7	49
72.0	128	127.90	121	6.9038	47.663	0.09619	0.009	7	49
70.0	119	120.14	121	-0.8630	0.745	-1.13702	1.293	-2	4
70.0	127	120.14	121	-0.8630	0.745	6.86298	47.100	6	36
66.5	105	106.55	121	-14.4548	208.941	-1.54520	2.388	-16	256
70.0	123	120.14	121	-0.8630	0.745	2.86298	8.197	2	4
71.0	115	124.02	121	3.0204	9.123	-9.02042	81.368	-6	36
632.0	1089	1089.00	1089	0.0001	287.371	-0.00006	166.630	0	454

a. The explained variation is $\Sigma(\hat{y}-\bar{y})^2 = 287.37$
b. The unexplained variation is $\Sigma(y-\hat{y})^2 = 166.63$
c. The total variation is $\Sigma(y-\bar{y})^2 = 454$
d. $r^2 = \Sigma(\hat{y}-\bar{y})^2/\Sigma(y-\bar{y})^2 = 287.37/454 = .6330$
e. $s_e^2 = \Sigma(y-\hat{y})^2/(n-2) = 166.63/7 = 23.8043$
   $s_e = 4.8790$

NOTE: A table such as the one in the preceding problem organizes the work and provides all the values needed to discuss variation. In such a table, the following must always be true (except for minor discrepancies due to rounding) and can be used as a check before proceeding.
* $\Sigma y = \Sigma \hat{y} = \Sigma \bar{y}$ * $\Sigma(\hat{y}-\bar{y}) = \Sigma(y-\hat{y}) = \Sigma(y-\bar{y}) = 0$ * $\Sigma(y-\bar{y})^2 + \Sigma(y-\hat{y})^2 = \Sigma(y-\bar{y})^2$

10. The predicted values were calculated using the regression line $\hat{y} = -14.37981 + .7692359x$.

x	y	$\hat{y}$	$\bar{y}$	$\hat{y}-\bar{y}$	$(\hat{y}-\bar{y})^2$	$y-\hat{y}$	$(y-\hat{y})^2$	$y-\bar{y}$	$(y-\bar{y})^2$
138	82	91.77	88.64	3.1319	9.808	-9.7747	95.546	-6.64	44.13
130	91	85.62	88.64	-3.0220	9.133	5.3791	28.935	2.36	5.56
135	100	89.47	88.64	0.8241	0.679	10.5330	110.943	11.36	128.98
140	100	93.31	88.64	4.6703	21.812	6.6868	44.713	11.36	128.98
120	80	77.93	88.64	-10.7144	114.798	2.0715	4.291	-8.64	74.70
125	90	81.77	88.64	-6.8682	47.173	8.2253	67.656	1.36	1.84
120	80	77.93	88.64	-10.7144	114.798	2.0715	4.291	-8.64	74.70
130	80	85.62	88.64	-3.0220	9.133	-5.6209	31.594	-8.64	74.70
130	80	85.62	88.64	-3.0220	9.133	-5.6209	31.594	-8.64	74.70
144	98	96.39	88.64	7.7473	60.020	1.6098	2.592	9.36	87.56
143	105	95.62	88.64	6.9780	48.693	9.3791	87.967	16.36	267.55
140	85	93.31	88.64	4.6703	21.812	-8.3132	69.110	-3.64	13.27
130	70	85.62	88.64	-3.0220	9.133	-15.6209	244.011	-18.64	347.56
150	100	101.01	88.64	12.3627	152.836	-1.0056	1.011	11.36	128.98
1875	1241	1241.00	1240.96	-0.0006	628.960	-0.0000	824.254	0.04	1453.21

S-300 INSTRUCTOR'S SOLUTIONS Chapter 9

a. The explained variation is $\Sigma(\hat{y}-\bar{y})^2 = 628.96$
b. The unexplained variation is $\Sigma(y-\hat{y})^2 = 824.25$
c. The total variation is $\Sigma(y-\bar{y})^2 = 1453.21$
d. $r^2 = \Sigma(\hat{y}-\bar{y})^2/\Sigma(y-\bar{y})^2 = 628.96/1453.21 = .4328$
e. $s_e^2 = \Sigma(y-\hat{y})^2/(n-2) = 824.254/12 = 68.6878$
   $s_e = 8.2878$

11. The predicted values were calculated using the regression line $\hat{y} = 22.46293 + 5.340648x$.

x	y	$\hat{y}$	$\bar{y}$	$\hat{y}-\bar{y}$	$(\hat{y}-\bar{y})^2$	$y-\hat{y}$	$(y-\hat{y})^2$	$y-\bar{y}$	$(y-\bar{y})^2$
1.8	21.0	32.076	47.793	-15.7169	247.02	-11.0761	122.68	-26.793	717.86
1.9	33.5	32.610	47.793	-15.1828	230.52	0.8898	0.79	-14.293	204.29
1.8	24.6	32.076	47.793	-15.7169	247.02	-7.4761	55.89	-23.193	537.92
2.4	40.7	35.280	47.793	-12.5125	156.56	5.4195	29.37	-7.093	50.31
5.1	73.2	49.700	47.793	1.9072	3.64	23.4998	552.24	25.407	645.52
3.1	24.9	39.019	47.793	-8.7741	76.98	-14.1189	199.34	-22.893	524.09
5.5	40.4	51.836	47.793	4.0435	16.35	-11.4365	130.79	-7.393	54.66
5.1	45.3	49.700	47.793	1.9072	3.64	-4.4002	19.36	-2.493	6.22
8.3	53.5	66.790	47.793	18.9973	360.90	-13.2903	176.63	5.707	32.57
13.7	93.8	95.630	47.793	47.8368	2288.36	-1.8298	3.35	46.007	2116.64
5.3	64.0	50.768	47.793	2.9754	8.85	13.2316	175.08	16.207	262.67
4.9	62.7	48.632	47.793	0.8391	0.70	14.0679	197.91	14.907	222.22
3.7	47.2	42.223	47.793	-5.5697	31.02	4.9767	24.77	-0.593	0.35
3.8	44.3	42.757	47.793	-5.0356	25.36	1.5426	2.38	-3.493	12.20
66.4	669.1	669.100	669.102	-0.0019	3696.93	-0.0001	1690.58	-0.002	5387.51

a. The explained variation is $\Sigma(\hat{y}-\bar{y})^2 = 3696.93$
b. The unexplained variation is $\Sigma(y-\hat{y})^2 = 1690.58$
c. The total variation is $\Sigma(y-\bar{y})^2 = 5387.51$
d. $r^2 = \Sigma(\hat{y}-\bar{y})^2/\Sigma(y-\bar{y})^2 = 3696.93/5387.51 = .6862$
e. $s_e^2 = \Sigma(y-\hat{y})^2/(n-2) = 1690.58/12 = 140.8817$
   $s_e = 11.8694$

12. The predicted values were calculated using the regression line $\hat{y} = 179.88 - 23.143x$.

x	y	$\hat{y}$	$\bar{y}$	$\hat{y}-\bar{y}$	$(\hat{y}-\bar{y})^2$	$y-\hat{y}$	$(y-\hat{y})^2$	$y-\bar{y}$	$(y-\bar{y})^2$
4.1	85.3	84.994	80.75	4.2437	18.009	0.30630	0.094	4.55	20.703
4.2	84.3	82.679	80.75	1.9294	3.723	1.62059	2.626	3.55	12.603
4.2	79.5	82.679	80.75	1.9294	3.723	-3.17941	10.109	-1.25	1.563
4.4	82.5	78.051	80.75	-2.6992	7.286	4.44920	19.795	1.75	3.063
4.3	80.2	80.365	80.75	-0.3849	0.148	-0.16511	0.027	-0.55	0.303
4.2	84.6	82.679	80.75	1.9294	3.723	1.92059	3.689	3.85	14.822
4.5	79.2	75.737	80.75	-5.0135	25.135	3.46349	11.996	-1.55	2.403
4.5	70.9	75.737	80.75	-5.0135	25.135	-4.83650	23.392	-9.85	97.022
4.3	78.6	80.365	80.75	-0.3849	0.148	-1.76511	3.116	-2.15	4.623
4.0	86.2	87.308	80.75	6.5580	43.007	-1.10801	1.228	5.45	29.702
4.4	74.0	78.051	80.75	-2.6992	7.286	-4.05080	16.409	-6.75	45.563
4.3	83.7	80.365	80.75	-0.3849	0.148	3.33489	11.122	2.95	8.702
51.4	969.0	969.010	969.00	0.0098	137.470	-0.00989	103.601	0.00	241.070

a. The explained variation is $\Sigma(\hat{y}-\bar{y})^2 = 137.47$
b. The unexplained variation is $\Sigma(y-\hat{y})^2 = 103.60$
c. The total variation is $\Sigma(y-\bar{y})^2 = 241.07$
d. $r^2 = \Sigma(\hat{y}-\bar{y})^2/\Sigma(y-\bar{y})^2 = 137.47/241.07 = .5702$
e. $s_e^2 = \Sigma(y-\hat{y})^2/(n-2) = 103.601/10 = 10.3601$
   $s_e = 3.2187$

13. a. $\hat{y} = -151.6977 + 3.883382x$.
    $\hat{y}_{69} = -151.6977 + 3.883382(69) = 116.29$

b. preliminary calculations
   n = 9
   $\Sigma x = 632.0$ $\bar{x} = (\Sigma x)/n = 632.0/9 = 70.222$
   $\Sigma x^2 = 44399.50$ $n\Sigma x^2 - (\Sigma x)^2 = 9(44399.50) - (632.0)^2 = 171.50$
   $\hat{y} \pm t_{n-2,\alpha/2} s_e \sqrt{1 + 1/n + n(x_o - \bar{x})^2/[n\Sigma x^2 - (\Sigma x)^2]}$
   $\hat{y}_{69} \pm t_{7,.025}(4.8790)\sqrt{1 + 1/9 + 9(69-70.222)^2/[171.50]}$
   $116.29 \pm (2.365)(4.8790)\sqrt{1.18948}$
   116.29 ± 12.58
   $103.7 < y_{69} < 128.9$

14. a. $\hat{y} = -14.37981 + .7692359x$
       $\hat{y}_{120} = -14.37981 + .7692359(120) = 77.92$
    b. preliminary calculations
       n = 14
       $\Sigma x = 1875$ $\bar{x} = (\Sigma x)/n = 1875/14 = 133.929$
       $\Sigma x^2 = 252179$ $n\Sigma x^2 - (\Sigma x)^2 = 14(252179) - (1875)^2 = 14881$
       $\hat{y} \pm t_{n-2,\alpha/2} s_e \sqrt{1 + 1/n + n(x_o - \bar{x})^2/[n\Sigma x^2 - (\Sigma x)^2]}$
       $\hat{y}_{120} \pm t_{12,.025}(8.2878)\sqrt{1 + 1/14 + 14(120-133.929)^2/[14881]}$
       $77.92 \pm (2.179)(8.2878)\sqrt{1.25395}$
       77.92 ± 20.22
       $57.7 < y_{120} < 98.2$

15. a. $\hat{y} = 22.46293 + 5.340648x$
       $\hat{y}_{4.0} = 22.46293 + 5.340648(4.0) = 43.83$
    b. preliminary calculations
       n = 14
       $\Sigma x = 66.4$ $\bar{x} = 66.4/14 = 4.743$
       $\Sigma x^2 = 444.54$ $n\Sigma x^2 - (\Sigma x)^2 = 14(444.54) - (66.4)^2 = 1814.60$
       $\hat{y} \pm t_{n-2,\alpha/2} s_e \sqrt{1 + 1/n + n(x_o - \bar{x})^2/[n\Sigma x^2 - (\Sigma x)^2]}$
       $\hat{y}_{4.0} \pm t_{12,.005}(11.8694)\sqrt{1 + 1/14 + 14(4.0-4.743)^2/[1814.60]}$
       $43.83 \pm (3.055)(11.8694)\sqrt{1.07569}$
       43.83 ± 37.61
       $6.2 < y_{4.0} < 81.4$

16. a. $\hat{y} = 179.88 - 23.143x$
       $\hat{y}_{4.0} = 179.88 - 23.143(4.0) = 87.31$
    b. preliminary calculations
       n = 12
       $\Sigma x = 51.4$ $\bar{x} = (\Sigma x)/n = 51.4/12 = 4.283$
       $\Sigma x^2 = 220.42$ $n\Sigma x^2 - (\Sigma x)^2 = 12(220.42) - (51.4)^2 = 3.08$
       $\hat{y} \pm t_{n-2,\alpha/2} s_e \sqrt{1 + 1/n + n(x_o - \bar{x})^2/[n\Sigma x^2 - (\Sigma x)^2]}$
       $\hat{y}_{4.0} \pm t_{10,.005}(3.2187)\sqrt{1 + 1/12 + 12(4.0-4.283)^2/[3.08]}$
       $87.31 \pm (3.169)(3.2187)\sqrt{1.39610}$
       87.31 ± 12.05
       $75.3 < y_{4.0} < 99.4$

Exercises 17-20 refer to the chapter problem of Table 9-1. They use the following, which are calculated and/or discussed at various places in the text,
   n = 10
   $\Sigma x = 741$ $\hat{y} = -112.7098976 + 2.274087687x$
   $\Sigma x^2 = 55289$ $s_e = 6.6123487$
and the values obtained below.

S-302   INSTRUCTOR'S SOLUTIONS Chapter 9

$\bar{x} = (\Sigma x)/n = 741/10 = 74.1$
$n\Sigma x^2 - (\Sigma x)^2 = 10(55289) - (741)^2 = 3809$

17. $\hat{y}_{85} = -112.7098976 + 2.274087687(85) = 80.58$
$\hat{y} \pm t_{n-2,\alpha/2} s_e \sqrt{1 + 1/n + n(x_o-\bar{x})^2/[n\Sigma x^2-(\Sigma x)^2]}$
$\hat{y}_{85} \pm t_{8,.005}(6.6123487)\sqrt{1 + 1/10 + 10(85-74.1)^2/[3809]}$
$80.58 \pm (3.355)(6.6123487)\sqrt{1.41192}$
80.58 ± 26.36
$54.2 < y_{85} < 106.9$

18. $\hat{y}_{85} = -112.7098976 + 2.274087687(85) = 80.58$
$\hat{y} \pm t_{n-2,\alpha/2} s_e \sqrt{1 + 1/n + n(x_o-\bar{x})^2/[n\Sigma x^2-(\Sigma x)^2]}$
$\hat{y}_{85} \pm t_{8,.05}(6.6123487)\sqrt{1 + 1/10 + 10(85-74.1)^2/[3809]}$
$80.58 \pm (1.860)(6.6123487)\sqrt{1.41192}$
80.58 ± 14.61
$66.0 < y_{85} < 95.2$

19. $\hat{y}_{90} = -112.7098976 + 2.274087687(90) = 91.96$
$\hat{y} \pm t_{n-2,\alpha/2} s_e \sqrt{1 + 1/n + n(x_o-\bar{x})^2/[n\Sigma x^2-(\Sigma x)^2]}$
$\hat{y}_{90} \pm t_{8,.025}(6.6123487)\sqrt{1 + 1/10 + 10(90-74.1)^2/[3809]}$
$91.96 \pm (2.306)(6.6123487)\sqrt{1.76372}$
91.96 ± 20.25
$71.7 < y_{90} < 112.2$

20. $\hat{y}_{90} = -112.7098976 + 2.274087687(90) = 91.96$
$\hat{y} \pm t_{n-2,\alpha/2} s_e \sqrt{1 + 1/n + n(x_o-\bar{x})^2/[n\Sigma x^2-(\Sigma x)^2]}$
$\hat{y}_{90} \pm t_{8,.005}(6.6123487)\sqrt{1 + 1/10 + 10(90-74.1)^2/[3809]}$
$91.96 \pm (3.355)(6.6123487)\sqrt{1.76372}$
91.96 ± 29.46
$62.5 < y_{90} < 121.4$

21. This exercise uses the following values from the chapter problem of Table 9-1, which are calculated and/or discussed at various places in the text,
    $n = 10$ $\Sigma x = 741$ $b_o = -112.7098976$ $s_e = 6.6123487$
              $\Sigma x^2 = 55289$ $b_1 = 2.274087687$
    and the values obtained below.
    $\bar{x} = (\Sigma x)/n = 741/10 = 74.1$
    $\Sigma x^2 - (\Sigma x)^2/n = 55289 - (741)^2/10 = 380.9$

    a. $b_o \pm t_{n-2,\alpha/2} s_e \sqrt{1/n + \bar{x}^2/[\Sigma x^2-(\Sigma x)^2/n]}$
    $-112.7098976 \pm t_{8,.025}(6.6123487)\sqrt{1/10 + (74.1)^2/[380.9]}$
    $-112.7098976 \pm (2.306)(6.6123487)\sqrt{14.5153}$
    $-112.7098976 \pm 58.093686$
    $-170.8 < \beta_o < 54.6$

    b. $b_1 \pm t_{n-2,\alpha/2} s_e / \sqrt{\Sigma x^2-(\Sigma x)^2/n}$
    $2.274087687 \pm t_{8,.025}(6.6123487)/\sqrt{380.9}$
    $2.274087687 \pm (2.306)(6.6123487)/\sqrt{380.9}$
    $2.274087687 \pm .781285$
    $1.5 < \beta_1 < 3.1$

22. a. If $s_e=0$, then $\Sigma(y-\hat{y})^2=0$ – which means that all the observed data lie on the regression line and $r=1.00$ or $r=-1.00$.

b. If $\Sigma(\hat{y}-\bar{y})^2=0$, then $\hat{y}=\bar{y}$ for all x – which means the regression line is parallel to the x axis and has slope 0.

23. a. Since $s_e^2 = \Sigma(y-\hat{y})^2/(n-2)$, $(n-2)\cdot s_e^2 = \Sigma(y-\hat{y})^2$.
And so (explained variation) = $(n-2)\cdot s_e^2$.
b. Let EXPV = explained variation = $\Sigma(\hat{y}-\bar{y})^2$.
Let UNXV = unexplained variation = $\Sigma(y-\hat{y})^2$.
Let TOTV = total variation = $\Sigma(y-\bar{y})^2$.
$$r^2 = (EXPV)/(TOTV)$$
$$(TOTV)\cdot r^2 = (EXPV)$$
$$(EXPV + UNXV)\cdot r^2 = (EXPV)$$
$$(EXPV)\cdot r^2 + (UNXV)\cdot r^2 = (EXPV)$$
$$(UNXV)\cdot r^2 = (EXPV) - (EXPV)\cdot r^2$$
$$(UNXV)\cdot r^2 = (EXPV)\cdot(1-r^2)$$
$$(UNXV)\cdot r^2/(1-r^2) = (EXPV)$$
And so (explained variation) = $[r^2/(1-r^2)]\cdot$(unexplained variation)
c. If $r^2=.900$, than $r=\pm.949$.
Since the regression line has a negative slope (i.e., $b_1=-2$), choose the negative root.
And so the linear correlation coefficient is $-.949$.

24. $\hat{y}_{85} = -112.7098976 + 2.274087687(85) = 80.58$
$\hat{y} \pm t_{n-2,\alpha/2}s_e\sqrt{1/n + n(x_o-\bar{x})^2/[n\Sigma x^2-(\Sigma x)^2]}$
$\hat{y}_{85} \pm t_{8,.025}(6.6123487)\sqrt{1/10 + 10(85-74.1)^2/[3809]}$
$80.58 \pm (2.306)(6.6123487)\sqrt{0.41192}$
80.58 ± 9.79
$70.8 < \mu_{y|x=85} < 90.4$

9-5 Multiple Regression

1. WEIGHT = $-271.71 - .870$(HEADLENGTH)$+ .554$(LENGTH) $+12.153$(CHEST)
$\hat{y} = -271.71 - .870x_1 + .554x_2 + 12.153x_3$

2. a. P-value = 0.000 (accurate to 3 decimals)
b. $R^2 = 92.8\% = .928$
c. adjusted $R^2 = 92.4\% = .924$

3. Yes, because the adjusted R^2 is .924. Although the P-value is less than .05, that alone in multiple regression problems does not necessarily indicate the regression is of practical significance. Increasing the number of x variables, like increasing the sample size in the previous chapters, can create statistical significance that is not of practical value. The adjusted R^2 takes into account the number of x variables.

4. a. $\hat{y} = -271.71 - .870x_1 + .554x_2 + 12.153x_3$
$\hat{y}_{14.0,70.0,50.0} = -271.71 - .870(14.0) + .554(70.0) + 12.153(50.0) = 362.54$ lbs
b. The predicted weight of 362.5 is off by 42.5 lbs. For this particular bear the prediction is reasonable, but not extremely close.
NOTE: In general, s measures the typical spread around some reference point. In a single set of data, s measures the typical spread around the mean. In regression problems, s_e measures the spread around the regression line. Minitab's $s_e=33.66$ suggests that the typical predicted weight is off by about 34 pounds one way or the other. That the predicted weight for this bear was off by about 42.5 pounds is about the accuracy that should be expected.

5. Use HWY, the highway fuel consumption, because it has the highest adjusted R^2 [.853] among all the regressions with a single predictor variable.

6. Use HWY and WT, highway fuel consumption and weight, because it had the highest adjusted R^2 [.861] among all the regressions with two predictor variables.

7. Use HWY and WT, highway fuel consumption and weight, because it had the highest adjusted R^2 [.861] among all the regressions given. A reasonable argument could be made for using HWY alone, since its adjusted R^2 [.853] is only slightly less – i.e., the addition of a second predictor variable does not substantially improve the regression.

8. $\hat{y} = 4.6 + .794x_1 - .00209x_2$
   $\hat{y}_{35,2675} = 4.6 + .794(35) - .00209(2675) = 26.8$ mpg
   Since adjusted $R^2 = .861$ is fairly high, we expect this to be a "good" estimate.
   Whether of not we expect this to be an accurate estimate depends on s_e, the spread around the regression line. The note below shows that $s_e^2 = [1 - $ adjusted $R^2]s_y^2$. In this case, $s_e^2 = [1-.861](4.234)^2 = 2.492$ and $s_e = 1.58$. Loosely speaking, the typical value is about 1.58 mpg above or below the regression line – and about 95% of the values will be within $2 \cdot (1.58) \approx 3.2$ mpg of the regression line. Since the prediction could be off either way by about 3.2 mpg without being considered an unusual result, the prediction is not extremely accurate.
   NOTE: For k predictor variables, $s_e^2 = \Sigma(y-\hat{y})^2/(n-1-k)$; always, $s_y^2 = \Sigma(y-\bar{y})^2/(n-1)$. Start with (unexplained variation)/(total variation) = $\Sigma(y-\hat{y})^2/\Sigma(y-\bar{y})^2 = 1 - R^2$ and solve for s_e^2 as follows.
   $$\Sigma(y-\hat{y})^2/\Sigma(y-\bar{y})^2 = 1 - R^2$$
   $$\Sigma(y-\hat{y})^2 = (1 - R^2) \cdot \Sigma(y-\bar{y})^2$$
   $$\Sigma(y-\hat{y})^2/(n-1) = (1 - R^2) \cdot \Sigma(y-\bar{y})^2/(n-1)$$
   $$[(n-1-k)/(n-1)] \cdot \Sigma(y-\hat{y})^2/(n-1-k) = (1 - R^2) \cdot \Sigma(y-\bar{y})^2/(n-1)$$
   $$[(n-1-k)/(n-1)] \cdot s_e^2 = (1 - R^2) \cdot s_y^2$$
   $$s_e^2 = [(n-1)/(n-1-k)] \cdot (1 - R^2) \cdot s_y^2$$
   $$s_e^2 = [1 - \text{adjusted } R^2] \cdot s_y^2$$

9. a. HEIGHT = 21.56 + .6899(MOTHER)
      $\hat{y} = 21.56 + .6899x_1$ adjusted $R^2 = .347$
   b. HEIGHT = 45.72 + .2926(FATHER)
      $\hat{y} = 45.72 + .2926x_2$ adjusted $R^2 = .050$
   c. HEIGHT = 9.80 + .6580(MOTHER) + .2004(FATHER)
      $\hat{y} = 9.80 + .6580x_1 - .2004x_2$ adjusted $R^2 = .366$
   d. The regression in part (c) is the best, because it has the highest adjusted R^2. The regression in part (a) is almost as good, however, as adding a second variable improves the regression very little.
   e. No, since the adjusted R^2 of the best equation is only .366.

10. a. FREAD = 91.83 - .8428(WORDS)
       $\hat{y} = 91.83 - .8428x_1$ adjusted $R^2 = .256$
    b. FREAD = 179.88 - 23.143(CHARACTERS)
       $\hat{y} = 179.88 - 23.143x_2$ adjusted $R^2 = .527$
    c. FREAD = 171.92 - .547(WORDS) - 19.606(CHARACTERS)
       $\hat{y} = 171.92 - .547x_1 - 19.606x_2$ adjusted $R^2 = .625$
    d. The regression in part (c) is the best, because it has the highest adjusted R^2.
    e. Yes, since the adjusted R^2 of the best equation is .625.

11. a. CALORIES = 3.680 + 3.782(FAT)
       $\hat{y} = 3.680 + 3.782x_1$ adjusted $R^2 = .067$
    b. CALORIES = 3.464 + 1.012(SUGAR)
       $\hat{y} = 3.464 + 1.012x_2$ adjusted $R^2 = .556$

c. CALORIES = 3.403 + 3.198(FAT) + .982(SUGAR)
   $\hat{y} = 3.403 + 3.198x_1 - .982x_2$ adjusted R^2 = .628
d. The regression in part (c) is the best, because it has the highest adjusted R^2.
e. Yes, since the adjusted R^2 of the best equation .628.

12. a. HHSIZE = 2.841 + .1803(FOOD)
   $\hat{y} = 2.841 + .1803x_1$ adjusted R^2 = .078
b. HHSIZE = 1.080 + 1.3761(PLASTIC)
   $\hat{y} = 1.090 + 1.3761x_2$ adjusted R^2 = .556
c. HHSIZE = 1.143 + .6580(FOOD) + .2004(PLASTIC)
   $\hat{y} = 1.143 - .0298x_1 + 1.4184x_2$ adjusted R^2 = .551
d. The regression in part (b) is the best, because it has the highest adjusted R^2.
e. No; while the adjusted R^2 of the best equation is a moderately high .556, the s_e = 1.302 is large relative to the household sizes – it is likely that confidence intervals for y will include negative values and have a truncated practical lower limit of 0 (which is really non-informative).

13. To find the best variables for a regression to predict NICOTINE, consider the correlations. Arranged in one table, the correlations are as follows.

	TAR	NICOTINE	CO
TAR	1.000		
NICOTINE	.961	1.000	
CO	.934	.863	1.000

The variable having the highest correlation [.961] with NICOTINE is TAR. The positive correlation indicates cigarettes with more nicotine tend to contain more tar. The variable having the second highest correlation [.863] with NICOTINE is CO. But since the correlation [.934] between TAR and CO is so high, those two variables can be accurately predicted from each other and give duplicate information. They should not both be included. The best multiple regression equation to predict NICOTINE, therefore, is probably the simple linear regression
   NICOTINE = .1540 + .0651(TAR)
   R^2 = 92.4% = .924   adjusted R^2 = 92.1% = .921
   overall P-value = .000

14. a. To find the best variables for a regression to predict PRICE, consider the correlations. Arranged in one table, the correlations are as follows.

	PRICE	CARAT	COLOR	CLARITY
PRICE	1.000			
CARAT	.767	1.000		
COLOR	-.441	.013	1.000	
CLARITY	-.200	.046	.179	1.000

The variable having the highest correlation [.767] with PRICE is CARAT. The variable having the second highest correlation [-.441] with PRICE is COLOR. Since the correlation between CARAT and COLOR is essentially 0 [.013], these variable are not related and contain different information. Both should be included in the model. Since CLARITY has little correlation with PRICE, it probably will add little – although its even smaller correlations with the other variables could mean that it explains significant price subtleties that the other variables fail to address. The following results confirm this analysis.
   PRICE = 8221 + 12297(CARAT) - 3116(COLOR)
   R^2 = 79.3% = .793   adjusted R^2 = 77.8% = .778
   overall P-value = .000
   PRICE = 15469 + 12409(CARAT) - 2918(COLOR) - 2322(CLARITY)
   R^2 = 81.8% = .818   adjusted R^2 = 79.7% = .797
   overall P-value = .000

The best multiple regression is the one with all three predictor variables – but since the addition of CLARITY does not make any practical improvement, an argument could be made for using the regression with two predictor variables.

b. Consider the relevant correlations as given below.

	DEPTH	TABLE	COLOR
DEPTH	1.000		
TABLE	.035	1.000	
COLOR	-.028	.084	1.000

For n=30, Table A-6 gives r=±.361 as the critical value for significance at the .05 level. DEPTH and TABLE have no significant linear relationship with COLOR, but one or both of them could have a significant non-linear effect.

15. To find the best variables for a regression to predict the selling price SELL, consider the correlations. Arranged in one table, the correlations are as follows.

	SELL	LIST	AREA	ROOM	BEDR	BATH	AGEH	ACRE	TAXH
SELL	1.000								
LIST	.997	1.000							
AREA	.879	.892	1.000						
ROOM	.560	.571	.751	1.000					
BEDR	.335	.320	.476	.657	1.000				
BATH	.640	.640	.668	.555	.458	1.000			
AGEH	-.147	-.130	.125	.371	.141	-.023	1.000		
ACRE	.169	.167	.177	.282	.037	.301	.304	1.000	
TAXH	.899	.907	.810	.513	.314	.583	-.189	.060	1.000

The variable having the highest correlation [.997] with SELL is LIST. This positive correlation indicates that the houses with the higher selling prices tend to be those with the high listing prices. This would be the best single variable for predicting selling price. The best additional predictor variable to add would be one having a low correlation with LIST (so that it is not merely duplicating the information provided by LIST). This suggests that AGEH or ACRE would be appropriate variables to consider next. But since the seller and/or realtor presumably considered each of the other variables when determining LIST, there is a sense in which the information of the other variables is already contained in LIST. The best multiple regression equation with SELL as the dependent variable, therefore, is probably
    SELL = 7.26 + .914(LIST)
    R^2 = 99.5% = .995   adjusted R^2 = 99.5% = .995
    overall P-value = .000

The adjusted R^2 and overall P-value indicate that this is a suitable regression equation for predicting selling price. Since the adjusted R^2 is so close to 1.000, these is little room for improvement and no additional variable could be of any value.

16. Letting $x_1 = x$ and $x_2 = x^2$, the multiple regression equation is
    $\hat{y} = 1.143 - .0298x_1 + 1.4184x_2$
    R^2 = 100.0% = 1.000   adjusted R^2 = 99.9% = .999
    overall P-value = .000

Accurate to three decimal places, the coefficient of multiple determination is R^2 = 1.000. Since this value is extremely close to 100% (i.e., 1.000 exactly), these is almost a perfect fit between the equation and the data.

9-6 Modeling

1. The graph appears to be that of a quadratic function.
    Try a regression of the form $y = ax^2 + bx + c$.
    This produces $y = 2x^2 - 12x + 18$ with adjusted R^2 = 100.0%, a perfect fit.

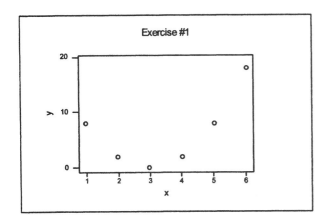
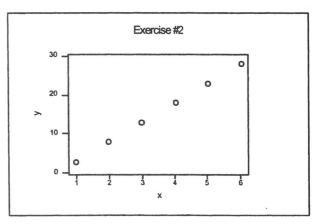

2. The graph appears to be that of a straight line function.
   Try a regression of the form y = ax + b.
   This produces y = 5x -2 with adjusted R^2 = 100.0%, a perfect fit.

3. The graph appears to be that of an exponential function.
   The equation has the form $y = a \cdot b^x$
   $$\ln y = \ln (a \cdot b^x) \text{ [taking the natural log of both sides]}$$
   $$= \ln a + x \cdot (\ln b)$$
   Try a regression of the form ln y = c + dx
   This produces ln y = -.00009 + 1.1x with adjusted $R^2 \approx$ 100.0%, a "perfect" fit.
   solving for the original parameters: ln a = -.0009   ln b = 1.1
   $$a = e^{-.00009} \qquad b = e^{1.1}$$
   $$= 1.00 \qquad\qquad = 3.00$$
   which yields the exponential equation: $y = (1.00) \cdot (3.00)^x$

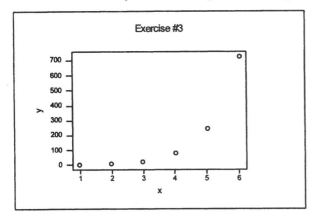
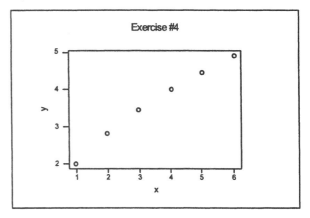

4. The graph appears to be that of a straight line function.
   Try a regression of the form y = ax + b.
   This produces y = .570x + 1.61 with adjusted R^2 = 98.1%, an excellent fit.
   The graph could also be a power function with an exponent less than 1.00.
   The equation has the form $y = ax^b$
   $$\ln y = \ln (a \cdot x^b) \text{ [taking the natural log of both sides]}$$
   $$= \ln a + b \cdot (\ln x)$$
   This produces ln y = .6931 + .5000·(ln x) with adjusted $R^2 \approx$ 100.0%, a "perfect" fit.
   solving for the original parameter: ln a = .6931
   $$a = e^{.6931}$$
   $$= 2.00$$
   which yields the power equation: $y = 2.00 \cdot x^{.500}$
   The power function gives a slightly better fit than the linear function.

5. Code the years as 1,2,3,... for 1980,1981,1982,...
The graph appears to be either that of a straight line or that of a quadratic function.
Try a regression of the form $y = ax + b$.
   This produces $y = 2.6455x + 14.329$ with adjusted $R^2 = 79.2\%$.
Try a regression of the form $y = ax^2 + bx + c$.
   This produces $y = .05167x^2 + 1.509x + 18.686$ with adjusted $R^2 = 79.0\%$.
The linear model appears to be the better model.
   $\hat{y} = 2.6455x + 14.329$
   $\hat{y}_{22} = 2.6455(22) + 14.329 = 73$, which is considerably lower than the actual value 82.

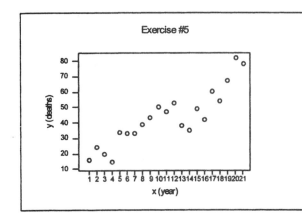
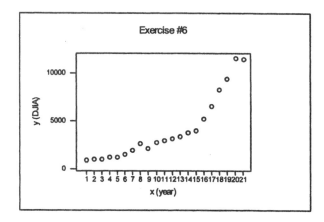

6. Code the years as 1,2,3,... for 1980,1981,1982,...
The graph appears to be either that of a quadratic function or an exponential function.
Try a regression of the form $y = ax^2 + bx + c$.
   This produces $y = 38.085x^2 - 346.7x + 1932.7$ with adjusted $R^2 = 96.0\%$.
The exponential function has the form $y = a \cdot b^x$
   $\ln y = \ln(a \cdot b^x)$ [taking the natural log of both sides]
   $= \ln a + x \cdot (\ln b)$
Try a regression of the form $\ln y = c + dx$
This produces $\ln y = 6.63687 + .126249x$ with adjusted $R^2 \approx 97.4\%$.
   solving for the original parameters: $\ln a = 6.63687$ $\ln b = .126249$
                                         $a = e^{6.63687}$ $b = e^{.126249}$
                                         $= 762.704$ $= 1.13456$
   which yields the exponential equation: $y = (762.704) \cdot (1.13456)^x$
The exponential model appears to be the better model.
   $\hat{y} = (762.704) \cdot (1.13456)^x$
   $\hat{y}_{22} = (762.704) \cdot (1.13456)^{22} = 12,262$, which is higher than the actual 11,350.

7. Code the years as 1,2,3,... for 1990,1991,1992,...
The graph appears to be either that of a straight line or that of a quadratic function.
Try a regression of the form $y = ax + b$.
   This produces $y = 56.982x + 340.382$ with adjusted $R^2 = 99.5\%$.
Try a regression of the form $y = ax^2 + bx + c$.
   This produces $y = 1.2145x^2 + 42.408x + 371.958$ with adjusted $R^2 = 99.9\%$.
The quadratic model is the better model.
   $\hat{y} = 1.2145x^2 + 42.408x + 371.958$
   $\hat{y}_{16} = 1.2145(16)^2 + 42.408(16) + 371.958 = 1361$

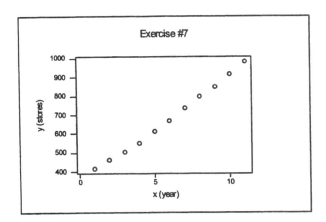
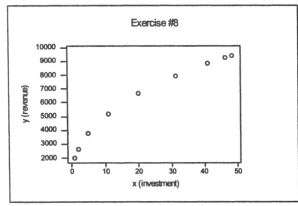

8. The graph appears to be that of a power function with an exponent less than 1.00.
   The equation has the form $y = ax^b$
   $$\ln y = \ln(a \cdot x^b) \text{ [taking the natural log of both sides]}$$
   $$= \ln a + b \cdot (\ln x)$$
   This produces $\ln y = 7.60112 + .3999937 \cdot (\ln x)$ with adjusted $R^2 \approx 100.0\%$.
   solving for the original parameter: $\ln a = 7.60112$
   $$a = e^{7.60112}$$
   $$= 2000.435$$
   which yields the power equation: $y = 2000.435 \cdot x^{.3999937}$

9. a. Moore's law is modeled by an exponential function.
      If y has value "a" at year x=1 and doubles every 1 year, $y = a \cdot 2^{x-1} = a \cdot 2^{(x-1)/1}$.
      If y has value "a" at year x=1 and doubles every 1.5 years, $y = a \cdot 2^{(x-1)/1.5}$.
      NOTE: $y = a \cdot 2^{(x-1)/1.5} = a \cdot 2^{(2/3)(x-1)} = a \cdot 2^{(2/3)x} \cdot 2^{-2/3} = [a/2^{2/3}] \cdot [(2^{2/3})^x] = (a/2^{2/3}) \cdot (1.587)^x$
   b. The exponential function has the form $y = a \cdot b^x$
      $$\ln y = \ln(a \cdot b^x) \text{ [taking the natural log of both sides]}$$
      $$= \ln a + x \cdot (\ln b)$$
      Try a regression of the form $\ln y = c + dx$
      This produces $\ln y = .8417 + .311794x$ with adjusted $R^2 \approx 99.3\%$.
      solving for the original parameters: $\ln a = .8417 \quad \ln b = .311794$
      $$a = e^{.8417} \quad b = e^{.311794}$$
      $$= 2.3203 \quad = 1.36587$$
      which yields the exponential equation: $y = (2.3203) \cdot (1.36587)^x$
      With adjusted $R^2 = .993$, the exponential model appears to be the best model.
   c. No; considering the base of x in the exponential model, Moore's law appears to be off.
      Part (a) gives the base for a quantity that doubles every 18 months as 1.587; part (b) gives an actual base of $1.366 = 2^{.45}$, which corresponds to doubling every 1/.45 years (or about every 27 months).

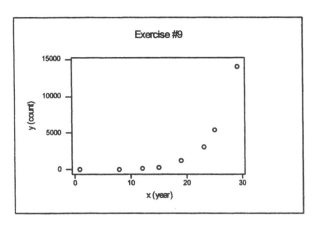

10. Linear predicted values were calculated using $\hat{y} = -61.93 + 27.200x$.
Quadratic predicted values were calculated using $\hat{q} = 2.76690x^2 - 6.00288x + 10.01212$.

x	y	$\hat{y}$	$\hat{q}$	$(y-\hat{y})^2$	$(q-\hat{q})^2$
1	5	-34.73	6.78	1578.47	3.1549
2	10	-7.53	9.07	307.30	0.8573
3	17	19.67	16.91	7.13	0.0089
4	31	46.87	30.27	251.86	0.5310
5	50	74.07	49.17	579.36	0.6879
6	76	101.27	73.60	638.57	5.7423
7	106	128.47	103.57	504.90	5.9019
8	132	155.67	139.07	560.27	50.0033
9	179	182.87	180.11	14.98	1.2228
10	227	210.07	226.67	286.62	0.1062
11	281	237.27	278.78	1912.31	4.9451
66	1114	1113.97	1114.00	6641.78	73.1616

a. The sum of the squares of the residuals (i.e., the unexplained variation) resulting from the linear model is $\Sigma(y-\hat{y})^2 = 6441.78$.
b. The sum of the squares of the residuals (i.e., the unexplained variation) resulting from the quadratic model is $\Sigma(q-\hat{q})^2 = 73.16$.
c. According to the sum of squares criterion, the quadratic model for these data is far superior to the linear one.
NOTE: $\Sigma(y-\hat{y}) = \Sigma(q-\hat{q}) = 0$. This demonstrates there is no unique model for which the sum of the residuals is zero. In general, the goal is to find the model for which the sum of the squares of the residuals is a minimum.

11. Using the formula provided for $\hat{y}$ produces the following table.

x	y	$\bar{y}$	$\hat{y}$	$(y-\bar{y})^2$	$(y-\hat{y})^2$
1	5	101.27	9.63	9268.4	21.420
2	10	101.27	14.57	8330.7	20.927
3	17	101.27	21.94	7101.9	24.403
4	31	101.27	32.76	4938.3	3.088
5	50	101.27	48.33	2628.9	2.797
6	76	101.27	70.10	638.7	34.799
7	106	101.27	99.35	22.3	44.217
8	132	101.27	136.59	944.2	21.092
9	179	101.27	180.91	6041.5	3.633
10	227	101.27	229.57	15807.3	6.626
11	281	101.27	278.54	32301.9	6.051
66	1114			88024.2	189.052

a. $\Sigma(y-\hat{y})^2 = 189.1$
b. $R^2 = 1 - \Sigma(y-\hat{y})^2/\Sigma(y-\bar{y})^2 = 1 - 189.1/88024.2 = .9979$
c. For the quadratic model, $R^2 = .9991688$ is given in th text.
Solving $R^2 = 1 - \Sigma(y-\hat{y})^2/\Sigma(y-\bar{y})^2$ gives $\Sigma(y-\hat{y})^2 = (1 - R^2)\cdot\Sigma(y-\bar{y})^2$
$= (.0008312)\cdot(88024.2)$
$= 73.2$
The quadratic model is better because its $\Sigma(y-\hat{y})^2$ is lower and its R^2 is higher.

Review Exercises

1. $n = 8$
$\Sigma x = 276.6$
$\Sigma y = 1.66$
$\Sigma x^2 = 10680.48$
$\Sigma y^2 = .35220$
$\Sigma xy = 57.191$

$n(\Sigma xy) - (\Sigma x)(\Sigma y) = 8(57.191) - (276.6)(1.66) = -1.628$
$n(\Sigma x^2) - (\Sigma x)^2 = 8(10680.48) - (276.6)^2 = 8936.28$
$n(\Sigma y^2) - (\Sigma y)^2 = 8(.3522) - (1.66)^2 = .0620$
$r = [n(\Sigma xy) - (\Sigma x)(\Sigma y)]/[\sqrt{n(\Sigma x^2) - (\Sigma x)^2} \cdot \sqrt{n(\Sigma y^2) - (\Sigma y)^2}]$
$= -1.628/[\sqrt{8936.28} \cdot \sqrt{.0620}] = -.069$

H_o: $\rho = 0$
H_1: $\rho \neq 0$
$\alpha = .05$
C.R. $r < -.707$ OR C.R. $t < -t_{6,.025} = -2.447$
    $r > .707$ $t > t_{6,.025} = 2.447$

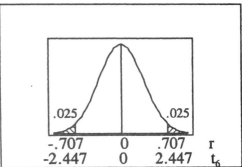

calculations: calculations:
  $r = -.069$ $t_r = (r - \mu_r)/s_r$
                        $= (-.069 - 0)/\sqrt{(1-(-.069)^2)/6}$
                        $= -.069/.4073 = -.170$
conclusion:
  Do not reject H_o; there is not sufficient evidence to reject the claim that $\rho = 0$.
  No; there is no significant linear correlation between age and BAC.

2. a. $n = 6$ $n(\Sigma xy) - (\Sigma x)(\Sigma y) = 6(5308.7436)-(441.84)(63.58) = 3760.2744$
   $\Sigma x = 441.84$ $n(\Sigma x^2) - (\Sigma x)^2 = 6(36754.1408) - (441.84)^2 = 25302.2592$
   $\Sigma y = 63.58$ $n(\Sigma y^2) - (\Sigma y)^2 = 6(809.5364) - (63.58)^2 = 414.8020$
   $\Sigma x^2 = 36754.1408$
   $\Sigma y^2 = 809.5364$ $r = [n(\Sigma xy) - (\Sigma x)(\Sigma y)]/[\sqrt{n(\Sigma x^2) - (\Sigma x)^2} \cdot \sqrt{n(\Sigma y^2) - (\Sigma y)^2}]$
   $\Sigma xy = 5308.7436$ $= 3760.2744/[\sqrt{25302.2592} \cdot \sqrt{414.8020}] = .828$

   H_o: $\rho = 0$
   H_1: $\rho \neq 0$
   $\alpha = .05$
   C.R. $r < -.811$ OR C.R. $t < -t_{4,.025} = -2.776$
       $r > .811$ $t > t_{4,.025} = 2.776$

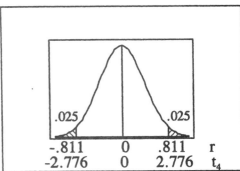

   calculations: calculations:
     $r = .828$ $t_r = (r - \mu_r)/s_r$
                           $= (.828 - 0)/\sqrt{(1-(.828)^2)/4}$
                           $= .828/.2802 = 2.955$
   conclusion:
     Reject H_o; there is sufficient evidence to reject the claim that $\rho = 0$ and conclude that
     $\rho \neq 0$ (in fact, that $\rho > 0$).
  b. $\bar{x} = 441.84/6 = 73.64$
     $\bar{y} = 63.58/6 = 10.597$
     $b_1 = [n(\Sigma xy) - (\Sigma x)(\Sigma y)]/[n(\Sigma x^2) - (\Sigma x)^2]$
         $= 3760.2744/25302.2592 = .1486$
     $b_o = \bar{y} - b_1\bar{x}$
         $= 10.597 - (.1486)(73.64) = -.3473$
     $\hat{y} = b_o + b_1 x = -.347 + .149x$
     The tips left are slightly less than 15% of the bill – to follow the established pattern,
     leave 35¢ less than 14.9% of the bill.

3. Let x be the price and y be the consumption.
   $n = 10$ $n(\Sigma xy) - (\Sigma x)(\Sigma y) = 10(4.74330) - (13.66)(3.468)$
   $\Sigma x = 13.66$ $= .06012$
   $\Sigma y = 3.468$ $n(\Sigma x^2) - (\Sigma x)^2 = 10(18.6694) - (13.66)^2$
   $\Sigma x^2 = 18.6694$ $= .0984$
   $\Sigma y^2 = 1.234888$ $n(\Sigma y^2) - (\Sigma y)^2 = 10(1.234888) - (3.468)^2$
   $\Sigma xy = 4.74330$ $= .321856$

$r = [n(\Sigma xy) - (\Sigma x)(\Sigma y)]/[\sqrt{n(\Sigma x^2) - (\Sigma x)^2} \cdot \sqrt{n(\Sigma y^2) - (\Sigma y)^2}]$
$= [.06012]/[\sqrt{.0984} \cdot \sqrt{.321856}] = .338$

a. $H_o: \rho = 0$
$H_1: \rho \neq 0$
$\alpha = .05$
C.R. $r < -.632$ OR C.R. $t < -t_{8,.025} = -2.306$
  $r > .632$ $t > t_{8,.025} = 2.306$

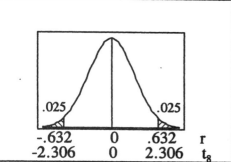

calculations: calculations:
  $r = .338$ $t_r = (r - \mu_r)/s_r$
                         $= (.338 - 0)/\sqrt{(1-(.338)^2)/8}$
                         $= .338/.333 = 1.015$

conclusion:
Do not reject H_o; there is not sufficient evidence to reject the claim that $\rho = 0$.

b. $r^2 = (.338)^2 = .114 = 11.4\%$
c. $b_1 = [n(\Sigma xy) - (\Sigma x)(\Sigma y)]/[n(\Sigma x^2) - (\Sigma x)^2] = .06012/.0984 = .611$
   $b_o = \bar{y} - b_1\bar{x} = (3.468/10) - (.611)(13.66/10) = -.488$
   $\hat{y} = b_o + b_1 x$
   $= -.488 + .611x$

d. $\hat{y} = -.488 + .611x$
   $\hat{y}_{1.38} = \bar{y} = .3468$ pints per capita per week [no significant correlation]

4. Let x be the income and y be the consumption.
   $n = 10$ $n(\Sigma xy) - (\Sigma x)(\Sigma y) = 10(1230.996) - (3548)(3.468)$
   $\Sigma x = 3548$ $= 5.496$
   $\Sigma y = 3.468$ $n(\Sigma x^2) - (\Sigma x)^2 = 10(1259524) - (3548)^2$
   $\Sigma x^2 = 1259524$ $= 6936$
   $\Sigma y^2 = 1.234888$ $n(\Sigma y^2) - (\Sigma y)^2 = 10(1.234888) - (3.468)^2$
   $\Sigma xy = 1230.996$ $= .321856$

$r = [n(\Sigma xy) - (\Sigma x)(\Sigma y)]/[\sqrt{n(\Sigma x^2) - (\Sigma x)^2} \cdot \sqrt{n(\Sigma y^2) - (\Sigma y)^2}]$
$= [5.496]/[\sqrt{6936} \cdot \sqrt{.321856}] = .116$

a. $H_o: \rho = 0$
$H_1: \rho \neq 0$
$\alpha = .05$
C.R. $r < -.632$ OR C.R. $t < -t_{8,.025} = -2.306$
  $r > .632$ $t > t_{8,.025} = 2.306$

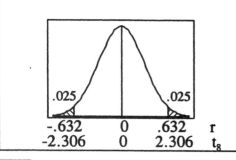

calculations: calculations:
  $r = .116$ $t_r = (r - \mu_r)/s_r$
                         $= (.116 - 0)/\sqrt{(1-(.116)^2)/8}$
                         $= .116/.351 = .331$

conclusion:
Do not reject H_o; there is not sufficient evidence to reject the claim that $\rho = 0$.

b. $r^2 = (.116)^2 = .013 = 1.3\%$
c. $b_1 = [n(\Sigma xy) - (\Sigma x)(\Sigma y)]/[n(\Sigma x^2) - (\Sigma x)^2] = 5.496/6936 = .000792$
   $b_o = \bar{y} - b_1\bar{x} = (3.468/10) - (.000792)(3548/10) = .0657$
   $\hat{y} = b_o + b_1 x$
   $= .0657 + .000792x$

d. $\hat{y} = .0657 + .000792x$
   $\hat{y}_{365} = \bar{y} = .3468$ pints per capita per week [no significant correlation]

5. Let x be the temperature and y be the consumption.
$n = 10$ $n(\Sigma xy) - (\Sigma x)(\Sigma y) = 10(189.038) - (526)(3.468)$
$\Sigma x = 526$ $= 66.212$
$\Sigma y = 3.468$ $n(\Sigma x^2) - (\Sigma x)^2 = 10(29926) - (526)^2$
$\Sigma x^2 = 29926$ $= 22584$
$\Sigma y^2 = 1.234888$ $n(\Sigma y^2) - (\Sigma y)^2 = 10(1.234888) - (3.468)^2$
$\Sigma xy = 189.038$ $= .321856$
$r = [n(\Sigma xy) - (\Sigma x)(\Sigma y)]/[\sqrt{n(\Sigma x^2) - (\Sigma x)^2} \cdot \sqrt{n(\Sigma y^2) - (\Sigma y)^2}]$
  $= [66.212]/[\sqrt{22584} \cdot \sqrt{.321856}] = .777$

a. $H_0: \rho = 0$
   $H_1: \rho \neq 0$
   $\alpha = .05$
   C.R. $r < -.632$ OR C.R. $t < -t_{8,.025} = -2.306$
        $r > .632$ $t > t_{8,.025} = 2.306$

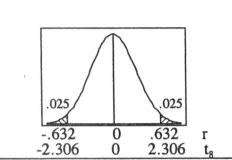

   calculations: calculations:
   $r = .777$ $t_r = (r - \mu_r)/s_r$
                     $= (.777 - 0)/\sqrt{(1-(.777)^2)/8}$
                     $= .777/.223 = 3.486$
   conclusion:
   Reject H_0; there is sufficient evidence to reject the claim that $\rho = 0$ and to conclude that $\rho \neq 0$ (in fact, that $\rho > 0$).

b. $r^2 = (.777)^2 = .604 = 60.4\%$
c. $b_1 = [n(\Sigma xy) - (\Sigma x)(\Sigma y)]/[n(\Sigma x^2) - (\Sigma x)^2] = 66.212/22584 = .00293$
   $b_0 = \bar{y} - b_1\bar{x} = (3.468/10) - (.00293)(526/10) = .193$
   $\hat{y} = b_0 + b_1 x$
       $= .193 + .00293x$
d. $\hat{y} = .193 + .00293x$
   $\hat{y}_{32} = .193 + .00293(32) = .2864$ pints per capita per week

6. Using Minitab and the given notation,
$\hat{y} = -.053 + .747x_1 - .00220x_2 + .00303x_3$
$R^2 = 72.6\% = .726$
adjusted $R^2 = 58.9\% = .589$
overall P-value $= .040$
Yes; since the overall P-value of .040 is less than .05, the regression equation can be used to predict ice cream consumption. Individually, x_1 and x_2 were not significantly related to consumption; only x_3 could be used to predict consumption. This suggests that variables x_1 and x_2 might not be making a worthwhile contribution to the regression, and that the equation from exercise #5 (using x_3 alone) might actually be a better and more efficient predictive equation.
To check this, calculate the adjusted R^2 for exercise #5.
   adjusted $R^2 = 1 - \{(n-1)/[n-(k+1)]\} \cdot (1-R^2)$
             $= 1 - \{9/[10-(2)]\} \cdot (1-R^2) = 1 - (9/8) \cdot (1-.777^2) = .554$
Since the adjusted R^2 of .589 is higher than the adjusted R^2 of .554 from exercise #3, however, there is a well-defined statistical sense in which this multiple regression equation is the best of the equations considered in these exercises.

Cumulative Review

1. Let x be the F-K Grade Level score and y be the Flesch Reading Ease score.
   n = 12
   $\Sigma x = 101.2$
   $\Sigma y = 793.8$
   $\Sigma x^2 = 897.82$
   $\Sigma y^2 = 53189.10$
   $\Sigma xy = 6540.86$

   $n(\Sigma xy) - (\Sigma x)(\Sigma y) = 12(6540.86) - (101.2)(793.8)$
   $= -1842.24$
   $n(\Sigma x^2) - (\Sigma x)^2 = 12(897.82) - (101.2)^2$
   $= 532.40$
   $n(\Sigma y^2) - (\Sigma y)^2 = 12(53189.10) - (793.8)^2$
   $= 8150.76$

   $r = [n(\Sigma xy) - (\Sigma x)(\Sigma y)]/[\sqrt{n(\Sigma x^2) - (\Sigma x)^2} \cdot \sqrt{n(\Sigma y^2) - (\Sigma y)^2}]$
   $= [-1842.24]/[\sqrt{532.40} \cdot \sqrt{8150.76}] = -.884$

   a. $H_o: \rho = 0$
      $H_1: \rho \neq 0$
      $\alpha = .05$
      C.R. $r < -.576$ OR   C.R. $t < -t_{10,.025} = -2.228$
           $r > .576$ $t > t_{10,.025} = 2.228$

   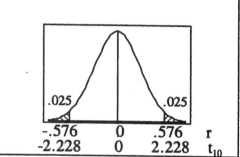

   calculations: calculations:
   $r = -.884$ $t_r = (r - \mu_r)/s_r$
                           $= (-.884 - 0)/\sqrt{(1-(-.884)^2)/10}$
   conclusion: $= -.884/.1476 = -5.991$
      Reject H_o; there is sufficient evidence to reject the claim that $\rho = 0$ and to conclude that $\rho \neq 0$ (in fact, that $\rho < 0$).

   b. $b_1 = [n(\Sigma xy) - (\Sigma x)(\Sigma y)]/[n(\Sigma x^2) - (\Sigma x)^2] = -1842.24/532.40 = -3.460$
      $b_o = \bar{y} - b_1\bar{x} = (793.80/12) - (-3.460)(101.20/12) = 95.33$
      $\hat{y} = b_o + b_1 x$
      $= 95.33 - 3.460x$

   c. Yes; in general it's always possible to test wether two parameter values are equal. No; since the two measures do not use the same approach or the same units to the problem, it would not make sense to test wether the values are numerically equal.

   d. First find the mean and standard deviation of y, the Flesch Reading Ease scores.
      $\bar{y} = (\Sigma y)/n = 793.8/12 = 66.15$
      $s^2 = [n(\Sigma y^2) - (\Sigma y)^2]/[n(n-1)] = [12(53189.10) - (793.8)^2]/[12(11)]$
      $= (8150.76)/132 = 61.748$
      $s = 7.858$
      $\bar{x} \pm t_{11,.025} \cdot s/\sqrt{n}$
      $66.15 \pm 2.201 \cdot (7.858)/\sqrt{12}$
      66.15 ± 4.99
      $61.16 < \mu < 71.14$

2. Let x and y be as given.
   n = 12
   $\Sigma x = 1189$
   $\Sigma y = 1234$
   $\Sigma x^2 = 118599$
   $\Sigma y^2 = 127724$
   $\Sigma xy = 122836$

   $n(\Sigma xy) - (\Sigma x)(\Sigma y) = 12(122836) - (1189)(1234)$
   $= 6806$
   $n(\Sigma x^2) - (\Sigma x)^2 = 12(118599) - (1189)^2$
   $= 9467$
   $n(\Sigma y^2) - (\Sigma y)^2 = 12(127724) - (1234)^2$
   $= 9932$

   $r = [n(\Sigma xy) - (\Sigma x)(\Sigma y)]/[\sqrt{n(\Sigma x^2) - (\Sigma x)^2} \cdot \sqrt{n(\Sigma y^2) - (\Sigma y)^2}]$
   $= [6806]/[\sqrt{9467} \sqrt{9932}] = .702$

   a. $\bar{x} = 1189/12 = 99.1$ $s_x = \sqrt{9467/(12 \cdot 11)} = 8.47$
   b. $\bar{y} = 1234/12 = 102.8$ $s_y = \sqrt{9932/(12 \cdot 11)} = 8.67$

c. No; there does not appear to be a difference between the means of the two populations. In exploring the relationship between the IQ's of twins, such a two sample approach is not appropriate because it completely ignores the pairings of the scores.

d. Combing the scores produces the following summary statistics.
summary statistics: $n = 24$, $\Sigma x = 2423$, $\Sigma x^2 = 246,323$, $\bar{x} = 100.958$, $s = 8.5997$
original claim: $\mu \neq 100$
H_0: $\mu = 100$
H_1: $\mu \neq 100$
$\alpha = .05$
C.R. $t < -t_{23,.025} = -2.069$
$t > t_{23,.025} = 2.069$
calculations:
$t_{\bar{x}} = (\bar{x} - \mu)/s_{\bar{x}}$
$= (100.958 - 100)/(8.5997/\sqrt{24})$
$= .958/1.755 = .546$
P-value $= 2 \cdot P(t_{23} > .546) > .20$

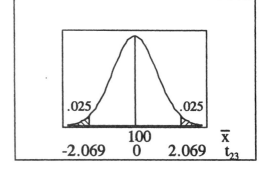

conclusion:
Do not reject H_0; there is not sufficient evidence to conclude that $\mu \neq 120$.

e. Correlation would be appropriate for answering the question "is there a (linear) relationship?"
H_0: $\rho = 0$
H_1: $\rho \neq 0$
$\alpha = .05$ [assumed]
C.R. $r < -.576$ OR C.R. $t < -t_{10,.025} = -2.228$
$r > .576$ $t > t_{10,.025} = 2.228$

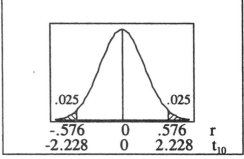

calculations: calculations:
$r = .702$ $t_r = (r - \mu_r)/s_r$
                   $= (.702 - 0)/\sqrt{(1-(.702)^2)/10}$
conclusion: $= .702/.225 = 3.116$

Reject H_0; there is sufficient evidence to reject the claim that $\rho = 0$ and to conclude that $\rho \neq 0$ (in fact, $\rho > 0$).
Ordinarily, the conclusion would be that about $R^2 = (.702)^2 = .493 = 49.3\%$ of the variation in x can be explained in terms of y (and vice-versa). In this context that means that about 49.3% of the variation among the IQ's in one group can be explained in terms of the IQ's of their twins (i.e, in terms of heredity). In simplest terms, it seems intelligence is about ½ due to heredity and ½ due to environment.

NOTE: Exercise #2 contains at least two interesting subtleties. (1) Correlation addresses only whether there is a relationship between the IQ's, and not whether the IQ's are close to each other. If each older twin had an IQ 20 points higher than the corresponding younger twin, there would be a perfect correlation – but the twins would not be similar in IQ at all. Beware of the misconception that correlation implies similarity. (2) Within each pair, the older twin was designated x. That was an arbitrary decision to produce an objective rule. The x-y designation within pairs could just as properly have been made alphabetically, randomly, or by another rule. But the rule affects the results. If it happens to designate all the twins with the higher IQ as x, the correlation rises to $r = .810$. If it happens to designate the twins with the higher IQ as x in the first six pairs and as y in the last six, the correlation falls to $r = .633$.

One technique which addresses both of the above issues (i.e., it tests for similarity and does not depend upon an x-y designation at all) involves comparing the variation within pairs to the overall variation in IQ's. If there is significantly less variability between the twins than there is variability in the general population (or between non-identical twins raised apart), then there is a significant relationship (i.e., in the sense of similarity) between the IQ's of identical twins.

Chapter 10

Multinomial Experiments and Contingency Tables

10-2 Multinomial Experiments: Goodness-of-Fit

1. a. $H_o: p_1 = p_2 = p_3 = p_4 = .25$
    b. Since $\Sigma O = (5+6+8+13) = 32$, the hypothesis of equally likely categories indicates
       $E_i = .25(32) = 8$ for $i=1,2,3,4$.
    c. $\chi^2 = \Sigma[(O-E)^2/E] = (5-8)^2/8 + (6-8)^2/8 + (8-8)^2/8 + (13-8)^2/8$
       $= 9/8 + 4/8 + 0/8 + 25/8$
       $= 38/8 = 4.750$
    d. $\chi^2_{3,.05} = 7.815$
    e. There is not enough evidence to reject the claim that the four categories are equally likely.

2. a. $H_o: p_1 = p_2 = p_3 = .20, p_4 = .30, p_5 = .10$
    b. Since $\Sigma O = (9+8+13+14+6) = 50$, the hypothesis indicates
       $E_i = .20(50) = 10$ for $i=1,2,3$
       $E_4 = .30(50) = 15$
       $E_5 = .10(50) = 5$
    c. $\chi^2 = \Sigma[(O-E)^2/E] = (9-10)^2/10 + (8-10)^2/10 + (13-10)^2/10 + (14-15)^2/15 + (6-5)^2/5$
       $= 1/10 + 4/10 + 9/10 + 1/15 + 1/5$
       $= .100 + .400 + .900 + .067 + .200 = 1.667$
    d. $\chi^2_{4,.05} = 9.488$
    e. There is not enough evidence to reject the claim that the five categories occur as specified.

3. a. $\chi^2_{37,.10} = 51.085$
    b. Since $\chi^2_{37,.10} = 51.085 > 38.232 > 29.051 = \chi^2_{37,.90}$, $.10 < $ P-value $ < .90$.
    c. There is not enough evidence to reject the claim that the 38 results are equally likely.

4. a. $\chi^2_{9,.05} = 16.919$
    b. Since $\chi^2_{9,.10} = 14.684 > 8.815 > 4.168 = \chi^2_{9,.90}$, $.10 < $ P-value $ < .90$.
    c. There is not enough evidence to reject the claim that the observed outcomes agree with the expected frequencies. Yes, the machine appears to be working correctly.

NOTES FOR THE REMAINING EXERCISES:
(1) In multinomial problems, always verify that $\Sigma E = \Sigma O$ before proceeding. If these sums are not equal, then an error has been made and further calculations have no meaning.
(2) As in the previous uses of the chi-squared distribution, the accompanying illustrations follow the "usual" shape – even though that shape is not correct for df=1 or df=2.

5. H_o: $p_1 = p_2 = p_3 = \ldots = p_6 = 1/6$
   H_1: at least one of the proportions is different from 1/6
   $\alpha = .05$
   C.R. $\chi^2 > \chi^2_{5,.05} = 11.071$
   calculations:

outcome	O	E	$(O-E)^2/E$
1	27	33.33	1.2033
2	31	33.33	.1633
3	42	33.33	2.2533
4	40	33.33	1.3333
5	28	33.33	.8533
6	32	33.33	.0533
	200	200.00	5.8600

   $\chi^2 = \Sigma[(O-E)^2/E] = 5.860$

   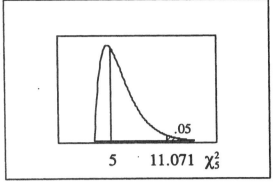

   conclusion:
   Do not reject H_o; there is not sufficient evidence to conclude that at least one of the proportions is different from 1/6.
   No, this particular loaded die did not behave noticeably differently from a fair die.

6. H_o: $p_{LF} = p_{RF} = p_{LR} = p_{RR} = .25$
   H_1: at least one of the proportions is different from .25
   $\alpha = .05$
   C.R. $\chi^2 > \chi^2_{3,.05} = 7.815$
   calculations:

tire	O	E	$(O-E)^2/E$
LF	11	10	.100
RF	15	10	2.500
LR	8	10	.400
RR	6	10	1.600
	40	40	4.600

   $\chi^2 = \Sigma[(O-E)^2/E] = 4.600$

   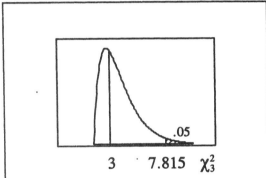

   conclusion:
   Do not reject H_o; there is not sufficient evidence to conclude that at least one of the proportions is different from .25.
   While we cannot be 95% certain that there is a tendency to pick one tire more than any other, it might be worthwhile to take a larger sample to see if the trend toward picking front tires becomes significant.

7. H_o: $p_{Sun} = p_{Mon} = p_{Tue} = \ldots = p_{Sat} = 1/7$
   H_1: at least one of the proportions is different from 1/7
   $\alpha = .05$
   C.R. $\chi^2 > \chi^2_{6,.05} = 12.592$
   calculations:

day	O	E	$(O-E)^2/E$
Sun	31	25.714	1.0865
Mon	20	25.714	1.2698
Tue	20	25.714	1.2698
Wed	22	25.714	.5365
Thu	22	25.714	.5365
Fri	29	25.714	.4198
Sat	36	25.714	4.1143
	180	180.000	9.2332

   $\chi^2 = \Sigma[(O-E)^2/E] = 9.233$

   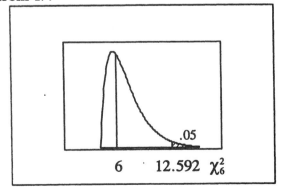

   conclusion:
   Do not reject H_o; there is not sufficient evidence to conclude that at least one of the proportions is different from 1/7.

S-318   INSTRUCTOR'S SOLUTIONS Chapter 10

8. H_o: $p_{Sun} = p_{Mon} = p_{Tue} = \ldots = p_{Sat} = 1/7$
    H_1: at least one of the proportions is different from 1/7
    $\alpha = .05$
    C.R. $\chi^2 > \chi^2_{6,.05} = 12.592$
    calculations:

day	O	E	$(O-E)^2/E$
Sun	40	30.857	2.709
Mon	24	30.857	1.524
Tue	25	30.857	1.112
Wed	28	30.857	.265
Thu	29	30.857	.112
Fri	32	30.857	.042
Sat	38	30.857	1.653
	216	216.000	7.417

    $\chi^2 = \Sigma[(O-E)^2/E] = 7.417$

    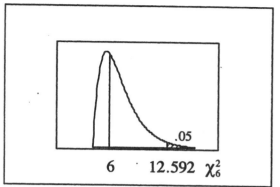

    conclusion:
    Do not reject H_o; there is not sufficient evidence to conclude that at least one of the proportions is different from 1/7.
    There is not enough evidence to reject with 95% confidence the theory that fatal DWI car crashes are caused by those who drink daily.

9. H_o: $p_{Mon} = p_{Tue} = p_{Wed} = p_{Thu} = p_{Fri} = 1/5$
    H_1: at least one of the proportions is different from 1/5
    $\alpha = .05$ [assumed]
    C.R. $\chi^2 > \chi^2_{4,.05} = 9.488$
    calculations:

day	O	E	$(O-E)^2/E$
Mon	31	29.4	.087
Tue	42	29.4	5.400
Wed	18	29.4	4.420
Thu	25	29.4	.659
Fri	31	29.4	.087
	147	147.0	10.653

    $\chi^2 = \Sigma[(O-E)^2/E] = 10.653$

    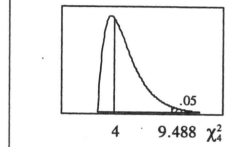

    conclusion:
    Reject H_o; there is sufficient evidence to conclude that at least one of the proportions is different from 1/5.
    The accidents seem to occur less frequently on Wednesday, and with increasing frequency toward the start and end of the week. This could be due to Monday workers still thinking about the past weekend and Friday workers thinking about the coming weekend. The only exception to this pattern is Tuesday.

10. H_o: $p_F = p_M = p_B = 1/3$
    H_1: at least one of the proportions is different from 1/3
    $\alpha = .05$ [assumed]
    C.R. $\chi^2 > \chi^2_{2,.05} = 5.991$
    calculations:

place	O	E	$(O-E)^2/E$
Front	17	10.33	4.301
Middle	9	10.33	.172
Back	5	10.33	2.753
	31	31.00	7.226

    $\chi^2 = \Sigma[(O-E)^2/E] = 7.226$

    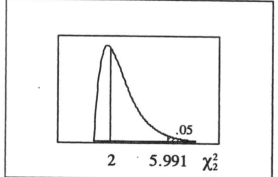

    conclusion:
    Reject H_o; there is sufficient evidence to support the claim that A students are not evenly distributed throughout the class room.

Not necessarily; since it is probably the type of student that determines the seat position and not vice versa. Reasons supporting the possibility that seat position determines the type of student, however, include the following. (1) Sitting in the front and in the midst of good students makes it more likely that a student will pay attention and learn more. (2) Sitting in the front with good students builds relationships and makes it more likely that a student will study with those same people outside class. (3) Sitting in the front with good students could affect the professor's perception and (especially in classes where grading is more subjective) and make it more likely that a student will get a better grade – or at least the benefit of the doubt in borderline cases.

11. H_o: $p_1 = p_2 = p_3 = \ldots = p_8 = 1/8$
    H_1: at least one of the proportions is different from 1/8
    $\alpha = .05$ [assumed]
    C.R. $\chi^2 > \chi^2_{7,.05} = 14.067$
    calculations:

start	O	E	(O-E)²/E
1	29	18	6.722
2	19	18	.056
3	18	18	.000
4	25	18	2.722
5	17	18	.056
6	10	18	3.556
7	15	18	.500
8	11	18	2.722
	144	144	16.333

    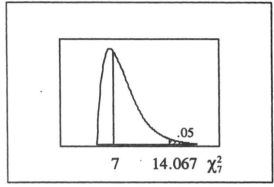

    $\chi^2 = \Sigma[(O-E)^2/E] = 16.333$
    conclusion:
    Reject H_o; there is sufficient evidence to conclude that at least one of the proportions is different from 1/8 – i.e., that the probabilities of winning in the different starting positions are not all the same.

12. H_o: $p_0 = p_1 = p_2 = \ldots = p_9 = 1/10$
    H_1: at least one of the proportions is different from 1/10
    $\alpha = .05$ [assumed]
    C.R. $\chi^2 > \chi^2_{9,.05} = 16.919$
    calculations:

digit	O	E	(O-E)²/E
0	16	8	8.000
1	0	8	8.000
2	14	8	4.500
3	0	8	8.000
4	17	8	10.125
5	0	8	8.000
6	16	8	8.000
7	0	8	8.000
8	17	8	10.125
9	0	8	8.000
	80	80	80.750

    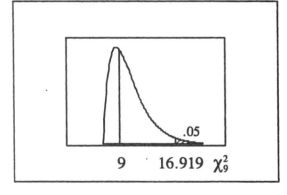

    $\chi^2 = \Sigma[(O-E)^2/E] = 80.750$
    conclusion:
    Reject H_o; there is sufficient evidence to conclude that at least one of the proportions is different from 1/10 – i.e., that last digits are not uniformly distributed.
    All of the observed last digits are even numbers. Further inspection reveals that all 80 of the pulse rates per minute are multiples of 4 – because the standard technique for taking pulses manually is to count the beats for 15 seconds and then multiply by 4. One can infer that these pulse rates were obtained manually (i.e., not by a machine) by the standard technique.

13. H_o: $p_0 = p_1 = p_2 = \ldots = p_9 = 1/10$
    H_1: at least one of the proportions is different from 1/10
    $\alpha = .05$
    C.R. $\chi^2 > \chi^2_{9,.05} = 16.919$
    calculations:

digit	O	E	(O-E)²/E
0	18	16	0.2500
1	12	16	1.0000
2	14	16	0.2500
3	9	16	3.0625
4	17	16	0.0625
5	20	16	1.0000
6	21	16	1.5625
7	26	16	6.2500
8	7	16	5.0625
9	16	16	0.0000
	160	160	18.5000

    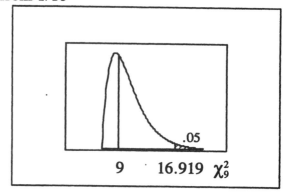

    $\chi^2 = \Sigma[(O-E)^2/E] = 18.500$

    conclusion:
    Reject H_o; there is sufficient evidence to conclude that at least one of the proportions is different from 1/10 – i.e., that the digits are not equally likely.
    If the .01 level of significance is used, the critical region is $\chi^2 > \chi^2_{9,.01} = 21.666$ and the decision changes. While we can be 95% confident that the digits are not being equally selected, we cannot be 99% confident that such is the case. If the digits are not being equally selected, (1) something is wrong with the device/method used to make the selections and (2) a person could use that knowledge to increase his chances of winning.

14. **NOTE:** The goodness of fit tests works with actual frequencies, and not with percents or proportions. In this case, there is not enough accuracy in the reported observed proportions to determine the exact observed frequencies – e.g., any $108958 \leq x \leq 110382$ yields the $x/1424287 = 7.7\%$ reported for January. Furthermore, the reported observed percents sum to 100.1% – guaranteeing that the calculations will be flawed due to rounding in the given numbers. One approach is as follows.
    For sample size n, $O_i = n\hat{p}_i$ and $E_i = np_i$ mean that the calculated test statistic can be rewritten as
    $$\chi^2 = \Sigma[(O-E)^2/E] = \Sigma[(n\hat{p}-np)^2/np] = \Sigma[n^2(\hat{p}-p)^2/np] = \Sigma n[(\hat{p}-p)^2/p] = n\Sigma[(\hat{p}-p)^2/p]$$
    In this particular problem, calculate $\Sigma[(\hat{p}-p)^2/p]$ and use $\chi^2 = n\Sigma[(\hat{p}-p)^2/p]$. This allows one to work with more convenient numbers. Furthermore, since $\Sigma\hat{p} = 1.001$ use $p = 1.001/12 = .0834167$ to test the claim that the reported occurrences are equally spread among the 12 categories.

    H_o: $p_{Jan} = p_{Feb} = p_{Mar} = \ldots = p_{Dec} = 1/12$
    H_1: at least one of the proportions is different from 1/12
    $\alpha = .01$
    C.R. $\chi^2 > \chi^2_{11,.05} = 24.725$
    calculations:

month	$\hat{p}$	p	$(\hat{p}-p)^2/p$
Jan	.077	.0834	.00049360
Feb	.074	.0834	.00106303
Mar	.084	.0834	.00000408
Apr	.083	.0834	.00000208
May	.092	.0834	.00088319
Jun	.086	.0834	.00008000
Jul	.090	.0834	.00051956
Aug	.089	.0834	.00037370
Sep	.086	.0834	.00008000
Oct	.087	.0834	.00015393
Nov	.076	.0834	.00065943
Dec	.077	.0834	.00049360
	1.001	1.0010	.00480620

    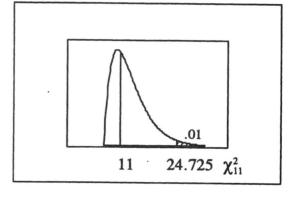

    $\chi^2 = n\Sigma[(\hat{p}-p)^2/p] = (1424287)(.00480620) = 6845.408$

conclusion:
   Reject H_o; there is sufficient evidence to conclude that at least one of the proportions is different from 1/12 – i.e., that the crimes are not equally distributed among the 12 months.
The evidence so overwhelmingly calls for rejection, even though the listed percentages do not differ all that dramatically, because the sample size is so large. The 12 months are certainly not <u>exactly</u> equal, and a large enough sample will detect even the smallest differences – even though they may be of no practical importance. In this case there do appear to be some practical differences, as it appears that there may be less violent crime from November to February – i.e., during the winter months.

15. H_o: $p_{Bro}=.3$, $p_{Yel}=.2$, $p_{Red}=.2$, $p_{Ora}=.1$, $p_{Gre}=.1$, $p_{Blu}=.1$
    H_1: at least one of the proportions is not as claimed
    $\alpha = .05$
    C.R. $\chi^2 > \chi^2_{5,.05} = 11.071$
    calculations:

color	O	E	$(O-E)^2/E$
Bro	33	30	.300
Yel	26	20	1.800
Red	21	20	.050
Ora	8	10	.400
Gre	7	10	.900
Blu	5	10	2.500
	100	100	5.950

    $\chi^2 = \Sigma[(O-E)^2/E] = 5.950$

    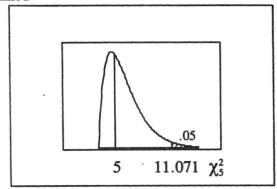

    conclusion:
        Do not reject H_o; there is not sufficient evidence to conclude that at least one of the proportions is not as claimed.

16. H_o: $p_1=.16$, $p_2=.44$, $p_3=.27$, $p_4=.13$
    H_1: at least one of the proportions is different from the license proportions
    $\alpha = .05$
    C.R. $\chi^2 > \chi^2_{3,.05} = 7.815$
    calculations:

group	O	E	$(O-E)^2/E$
1: <25	36	14.08	34.125
2: 25-44	21	38.72	8.109
3: 45-64	12	23.76	5.821
4: >64	19	11.44	4.996
	88	88.00	53.051

    $\chi^2 = \Sigma[(O-E)^2/E] = 53.051$

    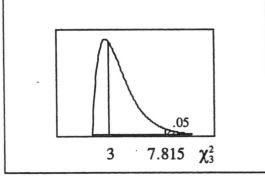

    conclusion:
        Reject H_o; there is sufficient evidence to conclude that at least one of the proportions is different from the license proportions.
    Yes; the "under 25" group appears to have a disproportionately large number of crashes – but it would be fairer, although much more difficult, to base the E values on the proportion of miles driven and not on the proportion of licenses possessed.

17. H_o: $p_0 = p_1 = p_2 = ... = p_9 = 1/10$
    H_1: at least one of the proportions is different from 1/10
    $\alpha = .05$
    C.R. $\chi^2 > \chi^2_{9,.05} = 16.919$

S-322 INSTRUCTOR'S SOLUTIONS Chapter 10

calculations:

digit	O	E	$(O-E)^2/E$
0	8	10	.400
1	8	10	.400
2	12	10	.400
3	11	10	.100
4	10	10	.000
5	8	10	.400
6	9	10	.100
7	8	10	.400
8	12	10	.400
9	14	10	1.600
	100	100	4.200

$\chi^2 = \Sigma[(O-E)^2/E] = 18.500$

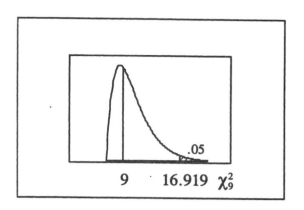

conclusion:
    Do not reject H_o; there is not sufficient evidence to conclude that at least one of the proportions is different from 1/10.

18. H_o: $p_0 = p_1 = p_2 = \ldots = p_9 = 1/10$
    H_1: at least one of the proportions is different from 1/10
    $\alpha = .05$
    C.R. $\chi^2 > \chi^2_{9,.05} = 16.919$
    calculations:

digit	O	E	$(O-E)^2/E$
0	0	10	10.000
1	17	10	4.900
2	17	10	4.900
3	1	10	8.100
4	17	10	4.900
5	16	10	3.600
6	0	10	10.000
7	16	10	3.600
8	16	10	3.600
9	0	10	10.000
	100	100	63.600

$\chi^2 = \Sigma[(O-E)^2/E] = 63.600$

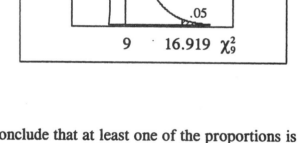

conclusion:
    Reject H_o; there is sufficient evidence to conclude that at least one of the proportions is different from 1/10 – i.e., that the digits do not appear with equal frequencies.
The result differs dramatically from that of exercise #17. Since 22/7 is a repeating decimal, once the repetitions begin only those digits will appear and other digits will not appear at all.

19. H_o: the leading digit proportions conform to Benford's law.
    H_1: at least one of the proportions is different from those specified by Benford's law.
    $\alpha = .05$
    C.R. $\chi^2 > \chi^2_{8,.05} = 15.507$
    calculations:

digit	O	p	E	$(O-E)^2/E$
1	72	.301	60.2	2.3130
2	23	.176	35.2	4.2284
3	26	.125	25.0	.0400
4	20	.097	19.4	.0186
5	21	.079	15.8	1.7114
6	18	.067	13.4	1.5791
7	8	.058	11.6	1.1172
8	8	.051	10.2	.4745
9	4	.046	9.2	2.9391
	200	1.000	200.0	14.4213

$\chi^2 = \Sigma[(O-E)^2/E] = 14.421$

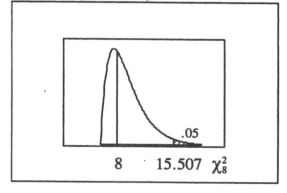

conclusion:
> Do not reject H_o; there is not sufficient evidence to conclude that at least one of the proportions is different those specified by Benford's law.

20. H_o: the proportions conform to those specified by the Poisson distribution
H_1: at least one of the proportions is different from those specified by the Poisson
$\alpha = .05$
C.R. $\chi^2 > \chi^2_{4,.05} = 9.488$
calculations:

hits	O	E	$(O-E)^2/E$
0	229	227.5	.00989
1	211	211.4	.00076
2	93	97.9	.24525
3	35	30.5	.66393
4+	8	8.7	.05632
	576	576.0	.97615

$\chi^2 = \Sigma[(O-E)^2/E] = .976$

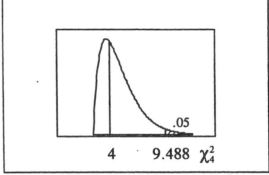

conclusion:
> Do not reject H_o; there is not sufficient evidence to conclude that at least one of the proportions is different from those specified by the Poisson distribution.

21. H_o: $p_{LF} = p_{RF} = p_{LR} = p_{RR} = .25$
H_1: at least one of the proportions is different from .25
$\alpha = .05$
C.R. $\chi^2 > \chi^2_{3,.05} = 7.815$
calculations:

tire	O	E	$(O-E)^2/E$
LF	11	23.5	6.649
RF	15	23.5	3.074
LR	8	23.5	10.223
RR	60	23.5	56.691
	94	94.0	76.638

$\chi^2 = \Sigma[(O-E)^2/E] = 76.638$

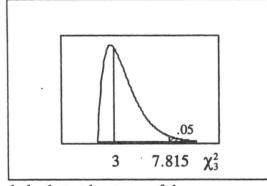

conclusion:
> Reject H_o; there is sufficient evidence to conclude that at least one of the proportions is different from .25.

In general, an "outlier" is a point that is different from the normal pattern. Since that (i.e., "difference from the normal pattern") is essentially what the goodness-of-fit test is designed to detect, an outlier will tend to increase the probability of rejecting H_o.

22. Use the lower tail of the χ^2 distribution to determine the critical region. If the null hypothesis is true and only random fluctuation exists, the expected value of $\chi^2 = \Sigma[(O-E)^2/E]$ is the degrees of freedom df. If $\Sigma[(O-E)^2/E]$ is significantly greater than df, then there is more variation than can be expected by chance and the data do not fit the null hypothesis. If $\Sigma[(O-E)^2/E]$ is significantly less than df, then there is less variation than can be expected by chance and the data fit the null hypothesis too well.

23. NOTE: Both outcomes having the same expected frequency is equivalent to $p_1 = p_2 = .5$.
  a. H_o: $p_1 = p_2 = .5$
  H_1: at least one of the p_i is different from .5
  $\alpha = .05$
  C.R. $\chi^2 > \chi^2_{1,.05} = 3.841$

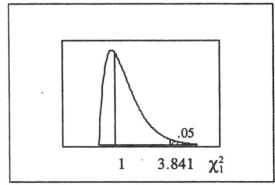

calculations

type	O	E	O-E	$(O-E)^2$	$(O-E)^2/E$
A	f_1	$(f_1+f_2)/2$	$(f_1-f_2)/2$	$(f_1-f_2)^2/4$	$[(f_1-f_2)^2/4]/[(f_1+f_2)/2]$
B	f_2	$(f_1+f_2)/2$	$(f_2-f_1)/2$	*$(f_1-f_2)^2/4$	$[(f_1-f_2)^2/4]/[(f_1+f_2)/2]$
	f_1+f_2	f_1+f_2			$[(f_1-f_2)^2/2]/[(f_1+f_2)/2]$

*NOTE: $(f_2-f_1)^2 = (f_1-f_2)^2$

$\chi^2 = \Sigma[(O-E)^2/E] = [(f_1-f_2)^2/2]/[(f_1+f_2)/2] = (f_1-f_2)^2/(f_1+f_2)$

b. H_o: p = .5
H_1: p ≠ .5
α = .05
C.R. z < $-z_{.025}$ = -1.960
    z > $z_{.025}$ = 1.960
calculations:
$z_{\hat{p}} = (\hat{p} - \mu_{\hat{p}})/\sigma_{\hat{p}}$
$= [f_1/(f_1+f_2) - .5]/\sqrt{(.5)(.5)/(f_1+f_2)}$
$= [.5(f_1-f_2)/(f_1+f_2)]/\sqrt{(.5)(.5)/(f_1+f_2)}$
$= [(f_1-f_2)/(f_1+f_2)]/\sqrt{1/(f_1+f_2)}$
$= (f_1-f_2)/\sqrt{f_1+f_2}$

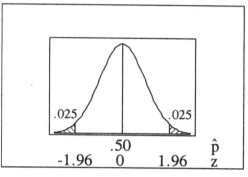

Since $[(f_1-f_2)/\sqrt{f_1+f_2}]^2 = (f_1-f_2)^2/(f_1+f_2)$ and $(\pm 1.960)^2 = 3.841$, $z^2 = \chi^2$.

24. a. Use the binomial formula $P(x) = [n!/x!(n-x)!] \cdot p^x \cdot (1-p)^{n-x}$ with n=3 and p=1/3.
   $P(x=0) = 1(1/3)^0(2/3)^3 = 8/27 = .296$
   $P(x=1) = 3(1/3)^1(2/3)^2 = 12/27 = .444$
   $P(x=2) = 3(1/3)^2(2/3)^1 = 6/27 = .222$
   $P(x=3) = 1(1/3)^3(2/3)^0 = 1/27 = .037$

b.
x	O	E	$(O-E)^2/E$
0	89	88.89	.000
1	133	133.33	.001
2	52	66.67	3.227
3	26	11.11	19.951
	300	300.00	23.179

c. H_o: there is goodness of fit to the binomial
   distribution with n=3 and p=1/3
   H_1: there is not goodness of fit
   α = .05
   C.R. $\chi^2 > \chi^2_{3,.05}$ = 7.815
   calculations:
   $\chi^2 = \Sigma[(O-E)^2/E]$
       = 23.179

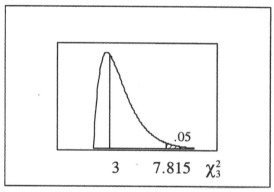

conclusion:
   Reject H_o; there is sufficient evidence to conclude that the observed frequencies do
   not fit a binomial distribution with n=3 and p=1/3.
   NOTE: While the phrase "goodness of fit" is the accepted statistical terminology in this
   context, the test could also be expressed as
      H_o: p_0 = 8/27, p_1 = 12/27, p_2 = 6/27, p_3 = 1/27
      H_1: at least p_i is not as claimed

25. a. Refer to the illustration below.
   $P(x<79.5) = P(z<-1.37) = .0853$
   $P(79.5<x<95.5) = P(z<-.30) - P(z<-1.37) = .3821 - .0853 = .2968$
   $P(95.5<x<110.5) = P(z<.70) - P(z<-.30) = .7580 - .3821 = .3759$
   $P(110.5<x<120.5) = P(z<1.37) - P(z<.70) = .9147 - .7850 = .1567$
   $P(x>120.5) = 1 - P(z<1.37) = 1 - .9147 = .0853$

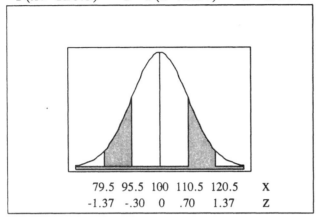

b.
score	O	E	$(O-E)^2/E$
< 80	20	17.06	.507
80- 95	20	59.36	26.099
96-110	80	75.18	.309
111-120	40	31.34	2.393
>120	40	17.06	30.847
	200	200.00	60.154

c. H_o: there is goodness of fit to the normal distribution with $\mu=100$ and $\sigma=15$
H_1: there is not goodness of fit
$\alpha = .01$
C.R. $\chi^2 > \chi^2_{4,.01} = 13.277$
calculations:
$\chi^2 = \Sigma[(O-E)^2/E]$
$= 60.154$

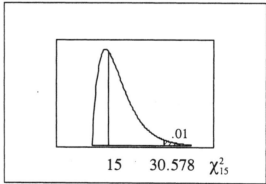

conclusion:
Reject H_o; there is sufficient evidence to conclude that the observed frequencies do not fit a normal distribution with $\mu=100$ and $\sigma=15$.

10-3 Contingency Tables: Independence and Homogeneity

NOTE: For each row and each column it must be true that $\Sigma O = \Sigma E$. After the marginal row and column totals are calculated, both the row totals and the column totals must sum to produce the same grand total. If either of the preceding is not true, then an error has been made and further calculations have no meaning. In addition, the following are true for all χ^2 contingency table analyses in this manual.
* The E values for each cell are given in parentheses below the O values.
* The addends used to calculate the χ^2 test statistic follow the physical arrangement of the cells in the original contingency table. This practice makes it easier to monitor the large number of intermediate steps involved and helps to prevent errors caused by missing or double-counting cells.
* The accompanying chi-square illustrations follow the "usual" shape, even though that shape is not correct for df=1 or df=2.

1. H_o: ethnicity and being stopped are independent*
   H_1: ethnicity and being stopped are related*
   $\alpha = .05$
   C.R. $\chi^2 > \chi^2_{1,.05} = 3.841$
   calculations:

   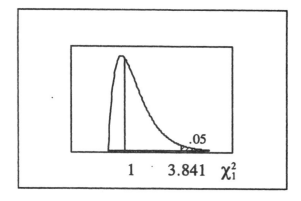

	ETHNICITY		
	B	W	
STOPPED? Y	24 (21.375)	147 (149.625)	171
N	176 (178.625)	1253 (1250.375)	1429
	200	1400	1600

   $\chi^2 = \Sigma[(O-E)^2/E]$
   $= .3224 + .0461$
   $\quad .0386 + .0055 = .413$

   conclusion:
   Do not reject H_o; there is not sufficient evidence to reject the claim that ethnicity and being stopped are independent.
   NOTE: The P-value from Minitab can also be used to make the decision. Since .521 > .05, do not reject H_o.
   No, we cannot conclude that racial profiling is being used.
   *NOTE: While the table claims to summarize "results for randomly selected drivers stopped by police," it appears that fixed numbers (200 black, 1400 white) of the ethnic groups were asked whether or not they had been stopped by the police. If so, the test should use "H_o: the proportion of drivers stopped is the same for both ethnic groups" to test homogeneity and not independence.

2. H_o: wearing a helmet and receiving facial injuries are independent
   H_1: wearing a helmet and receiving facial injuries are related
   $\alpha = .05$
   C.R. $\chi^2 > \chi^2_{1,.05} = 3.841$
   calculations:

   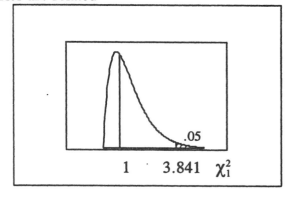

	HELMET		
	Y	N	
INJURY Y	30 (45.11)	182 (166.89)	212
N	83 (67.89)	236 (251.11)	319
	113	418	531

   $\chi^2 = \Sigma[(O-E)^2/E]$
   $= 5.0640 + 1.3690$
   $\quad 3.3654 + .9098 = 10.708$

   conclusion:
   Reject H_o; there is sufficient evidence to conclude that wearing a helmet and receiving facial injuries are related.
   NOTE: The P-value from TI-83 Plus can also be used to make the decision. Since .00107 < .05, reject H_o.
   NOTE: There is a statistical relationship, but not necessarily a cause-and-effect relationship. This test does not really address the question "Does a helmet seem to be effective in helping to prevent facial injuries in a crash?" The fact that people who wear helmets receive statistically fewer facial injuries could be due to other factors than the helmet preventing the injury – e.g., people who wear helmets might be safer people who go slower and whose accidents are less serious.

INSTRUCTOR'S SOLUTIONS Chapter 10   S-327

3. H_o: response is independent of company status
   H_1: response is related to company status
   $\alpha = .05$
   C.R. $\chi^2 > \chi^2_{1,.05} = 3.841$
   calculations:

   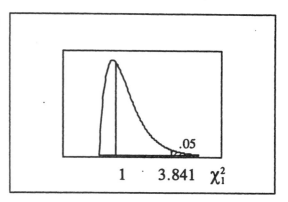

   ```
                OPINION
                Y         N
           W │  192    │   244   │ 436
   STATUS    │(181.60) │(254.40) │
           W │   40    │    81   │ 121
             │ (50.40) │ (70.60) │
               232        325      557
   ```

   $\chi^2 = \Sigma[(O-E)^2/E]$
   $= .5954 + .4250$
   $\, 2.1454 + 1.5316 = 4.697$

   conclusion:
   Reject H_o; there is sufficient evidence to reject the claim that response is independent of company status and to conclude that the two variables are related.
   Yes, the decision changes if the .01 level is used – since $4.697 > \chi^2_{1,.01} = 6.635$.
   No, workers and bosses no not appear to agree on this issue.

4. H_o: subject's truthfulness and polygraph reading are independent
   H_1: subject's truthfulness and polygraph reading are related
   $\alpha = .05$
   C.R. $\chi^2 > \chi^2_{1,.05} = 3.841$
   calculations:

   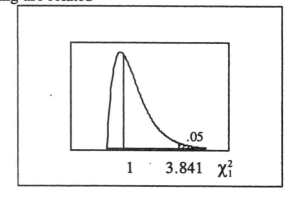

   ```
                POLYGRAPH
                T         L
           T │   65    │   15    │ 80
   SUBJECT   │ (54.40) │ (25.60) │
           L │    3    │   17    │ 20
             │ (13.60) │  (6.40) │
                68        32       100
   ```

   $\chi^2 = \Sigma[(O-E)^2/E]$
   $= 2.0654 + 4.3891$
   $\, 8.2618 + 17.5562 = 32.273$

   conclusion:
   Reject H_o; there is sufficient evidence to reject the claim that a subject's truthfulness and the polygraph reading are independent and to conclude that those variables are related.
   The results suggest that the two variables are related – i.e., that one can be predicted from the other. The results do not suggest how accurate that prediction might be.

5. H_o: there is homogeneity of proportions across gender of interviewer
   H_1: there is not homogeneity of proportions across gender of interviewer
   $\alpha = .01$
   C.R. $\chi^2 > \chi^2_{1,.01} = 6.635$
   calculations:

   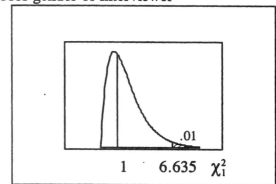

   ```
                INTERVIEWER
                M         F
           A │  512    │   336   │ 848
   RESPONSE  │(565.3)  │ (282.7) │
           D │  288    │    64   │ 352
             │(234.7)  │ (117.3) │
               800        400      1200
   ```

   $\chi^2 = \Sigma[(O-E)^2/E]$
   $= 5.031 + 10.063$
   $\, 12.121 + 24.242 = 51.458$

conclusion:
  Reject H_o; there is sufficient evidence to reject the claim that the proportion of agree/disagree responses are homogeneous across the gender of the interviewer.

6. H_o: the proportion who pass is homogeneous across the groups
H_1: the proportion who pass is not homogeneous across the groups
$\alpha = .05$
C.R. $\chi^2 > \chi^2_{1,.05} = 3.841$
calculations:

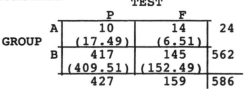

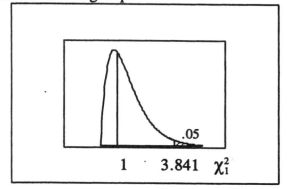

$\chi^2 = \Sigma[(O-E)^2/E]$
$= 3.2062 + 8.6105$
$ 0.1369 + 0.3677 = 12.321$

conclusion:
  Reject H_o; there is sufficient evidence to reject the claim that the proportion who pass is homogeneous across the groups.
  Yes, the test appears to "discriminate" in the sense that it distinguished between the two groups. This does bot necessarily indicate anything unfair (e.g., that the test is culturally biased) or illegal (e.g., that the minority group's tests were graded unfairly).

7. The lack of precision in the reported results prevents this exercise from having a well-defined unique solution. Any $482 \leq x \leq 491$, for example, produces the $x/1014 = 48\%$ males as stated. The exercise is worked using the closest integers that yield the stated percents, and the possible extreme solutions are given in a NOTE at the end.
H_o: gender and fear are flying are independent
H_1: gender and fear of flying are related
$\alpha = .05$
C.R. $\chi^2 > \chi^2_{1,.05} = 3.841$
calculations:

		GENDER		
		M	F	
FEAR?	Y	58 (111.42)	174 (120.58)	232
	N	429 (375.58)	353 (406.42)	782
		487	527	1014

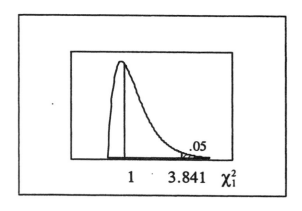

$\chi^2 = \Sigma[(O-E)^2/E]$
$= 25.615 + 23.671$
$ 7.599 + 7.023 = 63.908$

conclusion:
  Reject H_o; there is sufficient evidence to reject the claim that gender and fear of flying are independent and to conclude that the two variables are related.
NOTE: The following spread over more than 10 in the calculated χ^2 is possible from the following frequencies which also meet all the criteria in the statement of the exercise.

		GENDER		
		M	F	
FEAR?	Y	60 (110.76)	173 (122.24)	233
	N	422 (371.24)	359 (409.76)	781
		482	532	1014

$\chi^2 = \Sigma[(O-E)^2/E] = 57.559$

		GENDER		
		M	F	
FEAR?	Y	57 (112.34)	175 (119.66)	232
	N	434 (378.66)	348 (403.34)	782
		491	523	1014

$\chi^2 = \Sigma[(O-E)^2/E] = 68.533$

8. H_o: success in stopping to smoke is independent of the method used
H_1: success in stopping to smoke is related to the method used
$\alpha = .05$
C.R. $\chi^2 > \chi^2_{1,.05} = 3.841$
calculations:

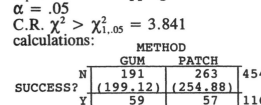

	METHOD		
	GUM	PATCH	
SUCCESS? N	191 (199.12)	263 (254.88)	454
Y	59 (50.88)	57 (65.12)	116
	250	320	570

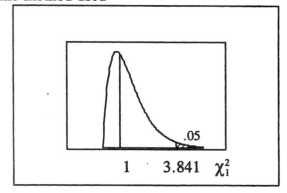

$\chi^2 = \Sigma[(O-E)^2/E]$
$= 0.331 + 0.259$
$\quad 1.297 + 1.013 = 2.900$

conclusion:
Do not reject H_o; there is not sufficient evidence to reject the claim that success in stopping to smoke is independent of the method used.
If someone wants to stop smoking, it appears that the method used does not make a difference – and that a person's motivation is probably the most important factor.
NOTE: While the text asked for a test of independence, it appears from the data that the sample sizes of 250 and 320 may have been fixed and selected from the populations of gum and patch users. If that were the case, this should be a test of homogeneity of proportions – and not of independence of variables.

9. H_o: success in stopping to smoke is independent of the method used
H_1: success in stopping to smoke is related to the method used.
$\alpha = .05$
C.R. $\chi^2 > \chi^2_{2,.05} = 5.991$
calculations:

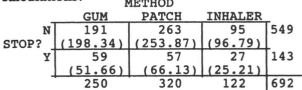

	METHOD			
	GUM	PATCH	INHALER	
STOP? N	191 (198.34)	263 (253.87)	95 (96.79)	549
Y	59 (51.66)	57 (66.13)	27 (25.21)	143
	250	320	122	692

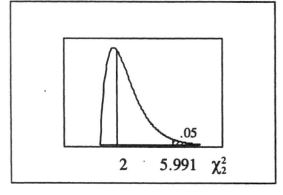

$\chi^2 = \Sigma[(O-E)^2/E]$
$= 0.271 + 0.328 + 0.033$
$\quad 1.042 + 1.260 + 0.127 = 3.062$

conclusion:
Do not reject H_o; there is not sufficient evidence to reject the claim that success in stopping to smoke is independent of the method used.

10. H_o: smoking and level of education are independent
H_1: smoking and level of education are related
$\alpha = .05$
C.R. $\chi^2 > \chi^2_{2,.05} = 5.991$
calculations:

	EDUCATION			
	PRIM	MIDD	COLL	
SMOKE Y	606 (564.53)	1234 (1210.5)	100 (164.97)	1940
N	205 (246.47)	505 (528.5)	137 (72.03)	847
	811	1739	237	2787

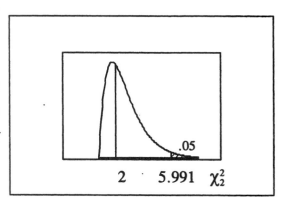

$\chi^2 = \Sigma[(O-E)^2/E]$
$= 3.047 + 0.456 + 25.589$
$\quad 6.978 + 1.045 + 58.610 = 95.725$

conclusion:
> Reject H_o; there is sufficient evidence to reject the claim that smoking and level of education are independent and to conclude that the two variables are related.

It appears (in China) that those with more education are more likely to have never smoked – as most dramatically illustrated by those who have a college level education.

11. H_o: death by homicide and occupation are independent
    H_1: death by homicide and occupation are related
    $\alpha = .05$ [assumed]
    C.R. $\chi^2 > \chi^2_{3,.05} = 7.815$
    calculations:

    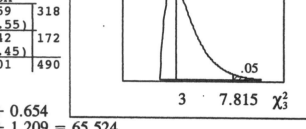

	OCCUPATION				
	POL	CAS	TAX	GUA	
Y	82	107	70	59	318
H	(112.92)	(75.28)	(64.25)	(65.55)	
N	92	9	29	42	172
	(61.08)	(40.72)	(34.75)	(35.45)	
	174	116	99	101	490

    $\chi^2 = \Sigma[(O-E)^2/E]$
    $= 8.468 + 13.364 + 0.515 + 0.654$
    $15.655 + 24.708 + 0.952 + 1.209 = 65.524$

    conclusion:
    > Reject H_o; there is sufficient evidence to reject the claim that death by homicide and occupation are independent.

    Yes, the occupation cashier appears to be significantly more prone to death by homicide.

12. H_o: scanner accuracy and type of price are independent
    H_1: scanner accuracy and type of price are related
    $\alpha = .05$ [assumed]
    C.R. $\chi^2 > \chi^2_{2,.05} = 5.991$
    calculations:

    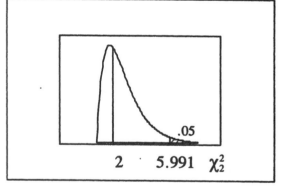

		PRICE		
		regular	special	
SCANNER	under	20	7	27
		(13.81)	(13.19)	
ACCURACY	over	15	29	44
		(22.51)	(21.49)	
	exact	384	364	748
		(382.68)	(365.32)	
		419	400	819

    $\chi^2 = \Sigma[(O-E)^2/E]$
    $= 2.7710 + 2.9016$
    $2.5058 + 2.2648$
    $0.0046 + 0.0048 = 10.814$

    conclusion:
    > Reject H_o; there is sufficient evidence conclude that scanner accuracy is related to the type of price being scanned.

    Since it appears that disproportionately more overcharges occur with advertised special prices, consumers should be especially diligent to monitor the price flashed when items on special are being scanned.

13. Refer to the following table.

		AGE						
		18-21	22-29	30-39	40-49	50-59	60+	
CO-OP	Y	73	255	245	136	138	202	1049
		(73.13)	(239.4)	(242.0)	(132.3)	(143.6)	(218.5)	
	N	11	20	33	16	27	49	156
		(10.87)	(35.6)	(36.0)	(19.7)	(21.4)	(33.5)	
		84	275	278	152	165	251	1205

H_o: age and cooperation are independent
H_1: age and cooperation are related
$\alpha = .01$
C.R. $\chi^2 > \chi^2_{5,.01} = 15.086$

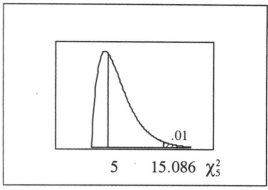

calculations:
[refer to the table at the start of the problem]
$\chi^2 = \Sigma[(O-E)^2/E]$
$= .0002 + 1.0168 + .0369 + .1022 + .2214 + 1.2468$
$.0014 + 6.8371 + .2484 + .6875 + 1.4886 + 8.3838 = 20.271$

conclusion:
Reject H_o; there is sufficient evidence to conclude that a person's age and level of cooperation are related.
Yes, those aged 60+ appear to be particularly more uncooperative than the others.

14. H_o: gun storage method and firearm training are independent
H_1: gun storage method and firearm training are related
$\alpha = .05$
C.R. $\chi^2 > \chi^2_{1,.05} = 3.841$
calculations:

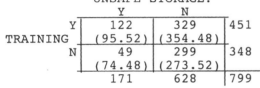

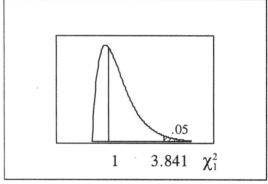

$\chi^2 = \Sigma[(O-E)^2/E]$
$= 6.7252 + 1.8312$
$8.7158 + 2.3732 = 19.645$

conclusion:
Reject H_o; there is sufficient evidence to conclude that gun storage method is related to whether the owner had formal firearm training.
No, the relationship appears to be in the negative direction – i.e., there was more unsafe storage than expected by chance alone among those who had formal firearm training. Since the test identifies only relationship and not cause and effect, it is not necessarily true that the training course had a negative effect. One possible likely explanation is that those who consider guns important and use them regularly are more likely to (a) keep their firearms handy [i.e., loaded and unlocked] and (b) support courses in firearm training.

15. H_o: type of crime and criminal/victim connection are independent
H_1: type of crime and criminal/victim connection are related
$\alpha = .05$
C.R. $\chi^2 > \chi^2_{2,.05} = 5.991$
calculations:

	CRIME			
	H	R	A	
S	12	379	727	1118
C/V	(29.93)	(284.64)	(803.43)	
A	39	106	642	787
	(21.07)	(200.36)	(565.57)	
	51	485	1369	1905

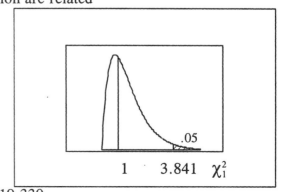

$\chi^2 = \Sigma[(O-E)^2/E]$
$= 10.7418 + 31.2847 + 7.2715$
$15.2600 + 44.4425 + 10.3298 = 119.330$

conclusion:
Reject H_o; there is sufficient evidence to reject the claim that the type of crime and the criminal/victim connection are related and to conclude that the two variables are related.

16. H_o: amount of smoking and seat belt use are independent
H_1: amount of smoking and seat belt use are related
$\alpha = .05$ [assumed]
C.R. $\chi^2 > \chi^2_{3,.05} = 7.815$
calculations:

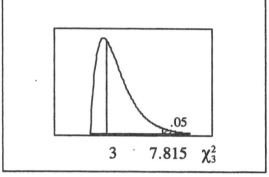

		NUMBER OF CIGARETTES				
		0	1-14	15-34	35+	
U S E	Y	175 (171.5)	20 (19.59)	42 (43.94)	6 (7.941)	243
	N	149 (152.5)	17 (17.41)	41 (39.06)	9 (7.059)	216
		324	37	83	15	459

$\chi^2 = \Sigma[(O-E)^2/E]$
$= .0702 + .0087 + .0858 + .4745$
$.0790 + .0097 + .0965 + .5338 = 1.358$

conclusion:
Do not reject H_o; there is not sufficient evidence to conclude that amount of smoking and seat belt use are related.
No; while the direction of the data agrees with that theory, we cannot be 95% certain that the patterns in these data are anything other than chance occurrences.

17. H_o: sentence and plea are independent
H_1: sentence and plea are related
$\alpha = .05$
C.R. $\chi^2 > \chi^2_{1,.05} = 3.841$
calculations:

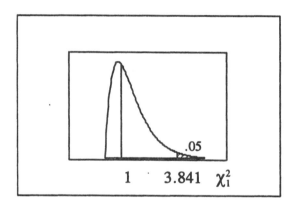

		PLEA		
		G	NG	
SENTENCE	P	392 (418.48)	58 (31.52)	450
	NP	564 (537.52)	14 (40.48)	578
		956	72	1028

$\chi^2 = \Sigma[(O-E)^2/E]$
$= 1.6759 + 22.2518$
$1.3047 + 17.3241 = 42.557$

conclusion:
Reject H_o; there is sufficient evidence to reject the claim that sentence and plea are independent and to conclude that a person's sentence and his original plea are related.
Yes; assuming that those who are really guilty will indeed be convicted with a trial, these results suggest that a guilty plea should be encouraged. But the study reported only those who plead not guilty and were convicted in trials. Suppose there were also guilty 50 persons who plead not guilty and were acquitted. Including them in the no prison category changes the conclusion entirely and yields the following.

		PLEA		
		G	NG	
SENTENCE	P	392 (399.07)	58 (50.93)	450
	NP	564 (556.93)	64 (71.07)	628
		956	112	1078

$\chi^2 = \Sigma[(O-E)^2/E]$
$= .125 + .982$
$.090 + .704 = 1.901$

18. H_0: size of home advantage and sport are independent
H_1: size of home advantage and sport are related
$\alpha = .10$
C.R. $\chi^2 > \chi^2_{3,.10} = 6.251$
calculations:

```
              SPORT
W       bask    base   hock   foot
I  H    127     53     50     57   | 287
N       (115.97)(58.57)(54.47)(57.99)
N  V    71      47     43     42   | 203
E       (82.03) (41.43)(38.53)(41.01)
R       198     100    93     99   | 490
```

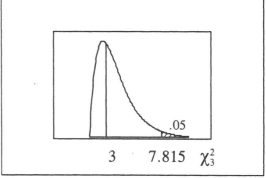

$\chi^2 = \Sigma[(O-E)^2/E]$
$= 1.049 + .530 + .367 + .017$
$ 1.483 + .749 + .519 + .024 = 4.737$

conclusion:
Do not reject H_0; there is not sufficient evidence to conclude that size of home advantage and the sport are related.

19. H_0: getting a headache and the amount pf drug used are independent
H_1: getting a headache and the amount of drug used are related
$\alpha = .05$
C.R. $\chi^2 > \chi^2_{3,.05} = 7.815$
calculations:

```
           DOSE (mg)
A          0       10      20/40    80
C  Y       19      47      8        6     | 80
H          (16.10) (51.45) (6.86)   (5.60)
E  N       251     816     107      88    | 203
?          (253.90)(811.55)(108.14) (88.40)
           270     863     115      94    | 1342
```

$\chi^2 = \Sigma[(O-E)^2/E]$
$= .524 + .384 + .191 + .028$
$ .033 + .024 + .012 + .002 = 1.199$

conclusion:
Do not reject H_0; there is not sufficient evidence to conclude that size of home advantage and the sport are related.

20. Refer to the following table and calculations.

```
                        EXERCISE LEVEL
                <200    200-599   600-1499   ≥1500
         0      4997    5205      5784       4155     | 20141
                (5584.18)(5197.50)(5481.33)  (3877.99)
SMOKING  <15    604     484       447        359      | 1894
LEVEL           (525.12)(488.76) (515.45)   (364.67)
         ≥15    1403    830       644        350      | 3227
                (894.70)(832.75) (878.22)   (621.33)
                7004    6519      6875       4864     | 25262
```

$\chi^2 = \Sigma[(O-E)^2/E]$
$= 61.742 + 0.011 + 16.713 + 19.787$
$ 11.849 + 0.046 + 9.089 + 0.088$
$ 288.777 + 0.009 + 62.467 + 118.490$
$= 589.069$

H_o: smoking and exercise levels are independent
H_1: smoking and exercise levels are related
$\alpha = .05$
C.R. $\chi^2 > \chi^2_{6,.05} = 12.592$
calculations:

[refer to the table and calculations at the start of the problem]

$\chi^2 = \Sigma[(O-E)^2/E]$
$= 589.069$

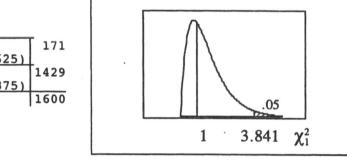

conclusion:
Reject H_o; there is sufficient evidence to reject the claim that level of smoking and level of exercise are independent and to conclude that the two variables are related.

21. H_o: ethnicity and being stopped are independent
H_1: ethnicity and being stopped are related
$\alpha = .05$
C.R. $\chi^2 > \chi^2_{1,.05} = 3.841$
calculations:

	ETHNICITY		
STOPPED?	B	W	
Y	24 (21.375)	147 (149.625)	171
N	176 (178.625)	1253 (1250.375)	1429
	200	1400	1600

$\chi^2 = \Sigma[(|O-E|-.5)^2/E]$
$= .2113 + .0302$
$.0253 + .0036 = .270$

conclusion:
Do not reject H_o; there is not sufficient evidence to reject the claim that ethnicity and being stopped are independent.
No, we cannot conclude that racial profiling is being used.
Without the correction for continuity [see the solution for exercise #1 for the details] the calculated test statistic is .413. Since $(|O-E|-.5)^2 < (O-E)^2$ whenever $|O-E| > .25$, Yates' correction generally lowers the calculated test statistic.

22. a. In the format of this section, the 2x2 contingency table is

a	b	R1
c	d	R2
C1	C2	N

where R1 = a+b, R2 = c+d
C1 = a+c, C2 = b+d
N = C1+C2 = R1+R2 = a+b+c+d

Since the $\Sigma O = \Sigma E$ for each row and column, the (O-E) "misses" must add out to zero in each row and column. Let $a - E_a = M$, then $b - E_b = -M$
$c - E_c = -M$
$d - E_d = M$

NOTE: This is a complex exercise. We proceed by presenting fact 1 and fact 2, then verifying the given equation.

fact 1: $N^2 = N \cdot N$
$= (C1+C2) \cdot (R1+R2)$
$= C1 \cdot R1 + C1 \cdot R2 + C2 \cdot R1 + C2 \cdot R2$

fact 2: $M = a - E_a$
$= a - (a+c)(a+b)/(a+b+c+d)$
$= [a(a+b+c+d) - (a+c)(a+b)]/(a+b+c+d)$
$= [a^2 + ab + ac + ad - a^2 - ab - ac - bc]/N$
$= (ad - bc)/N$

Together, these facts yield
$M^2 = (ad - bc)^2/N^2$
$= (ad - bc)^2/(C1 \cdot R1 + C1 \cdot R2 + C2 \cdot R1 + C2 \cdot R2)$ from fact 1

$\chi^2 = \Sigma[(O-E)^2/E]$
$= M^2/(C1 \cdot R1/N) + M^2/(C2 \cdot R1/N) + M^2/(C1 \cdot R2/N) + M^2/(C2 \cdot R2/N)$
$= (N \cdot M^2)/(C1 \cdot R1) + (N \cdot M^2)/(C2 \cdot R1) + (N \cdot M^2)/(C1 \cdot R2) + (N \cdot M^2)/(C2 \cdot R2)$
$= (N \cdot M^2 \cdot C2 \cdot R2)/(C1 \cdot R1 \cdot C2 \cdot R2) + (N \cdot M^2 \cdot C1 \cdot R2)/(C1 \cdot R1 \cdot C2 \cdot R2) +$
$\qquad (N \cdot M^2 \cdot C2 \cdot R1)/(C1 \cdot R1 \cdot C2 \cdot R2) + (N \cdot M^2 \cdot C1 \cdot R1)/(C1 \cdot R1 \cdot C2 \cdot R2)$
$= N \cdot M^2 \cdot (C2 \cdot R2 + C1 \cdot R2 + C2 \cdot R1 + C1 \cdot R1)/(C1 \cdot R1 \cdot C2 \cdot R2)$
$= N \cdot (ad - bc)^2/(C1 \cdot R1 \cdot C2 \cdot R2)$ from fact 2
$= N \cdot (ad - bc)^2/(R1 \cdot R2 \cdot C2 \cdot C1)$
$= (a+b+c+d) \cdot (ad - bc)^2/[(a+b)(c+d)(b+d)(a+c)]$

b. In the format of a two-population proportion problem, the 2x2 contingency table is

```
              GROUP
               1   2
          Y |  a | b |  x    where x = a+b,  n-x = c+d
SUCCESS     |----|---|           n₁ = a+c,  n₂ = b+d
          N |  c | d | n-x       n = n₁+n₂ = n+(n-x) = a+b+c+d
              n₁  n₂   n         p̄ = x/n = (a+b)/(a+b+c+d)
```

When testing $H_0: p_1 - p_2 = 0$, $z_{\hat{p}_1 - \hat{p}_2} = ((\hat{p}_1 - \hat{p}_2) - 0)/\sqrt{\bar{p}\bar{q}/n_1 + \bar{p}\bar{q}/n_2}$.
We consider the numerator NUM and the denominator DEN separately.

$\text{NUM} = \hat{p}_1 - \hat{p}_2$ \qquad\qquad\qquad $(\text{DEN})^2 = \bar{p}\bar{q}/n_1 + \bar{p}\bar{q}/n_2$
$\quad = a/n_1 - b/n_2$ \qquad\qquad\qquad\qquad $= \bar{p}\bar{q}[1/n_1 + 1/n_2]$
$\quad = [an_2 - bn_1]/n_1 n_2$ \qquad\qquad\qquad $= \bar{p}\bar{q}[(n_2 + n_1)/n_1 n_2]$
$\quad = [a(b+d) - b(a+c)]/n_1 n_2$ \qquad\qquad $= \bar{p}\bar{q}[n/n_1 n_2]$
$\quad = [ad - ac]/n_1 n_2$ \qquad\qquad\qquad\qquad $= [x/n] \cdot [(n-x)/n] \cdot [n/n_1 n_2]$
$(\text{NUM})^2 = (ad - bc)^2/n_1 n_2 n_1 n_2$ \qquad\qquad $= [x(n-x)]/[nn_1 n_2]$

$z^2 = (\text{NUM})^2 \cdot [1/(\text{DEN})^2]$
$\quad = [(ad - bc)^2/n_1 n_2 n_1 n_2] \cdot [nn_1 n_2]/[x(n-x)]$
$\quad = n \cdot (ad - bc)^2/[x(n-x) n_2 n_1]$
$\quad = (a+b+c+d) \cdot (ad - bc)^2/[(a+b)(c+d)(b+d)(a+c)]$
$\quad = \chi^2$ as calculated in part (a) above

Review Exercises

1. $H_0: p_{Mon} = p_{Tue} = p_{Wed} = p_{Thu} = p_{Fri} = 1/5$
   $H_1:$ at least one of the proportions is different from 1/5
   $\alpha = .05$
   C.R. $\chi^2 > \chi^2_{4,.05} = 9.488$
   calculations:

day	O	E	$(O-E)^2/E$
Mon	98	75	7.053
Tue	68	75	.653
Wed	89	75	2.613
Thu	64	75	1.613
Fri	56	75	4.813
	375	375	16.747

   $\chi^2 = \Sigma[(O-E)^2/E] = 16.747$

   conclusion:
   Reject H_0; there is sufficient evidence to conclude that at least one of the proportions is

different from 1/5.
Since the calls are not uniformly distributed over the days of the business week, it would not make sense to have the staffing level the same each day – unless the company staffed to handle the higher numbers of calls and had other work available to do during slack times.

2. H_o: $p_{Mon} = p_{Tue} = p_{Wed} = \ldots = p_{Sun} = 1/7$
H_1: at least one of the proportions is different from 1/7
$\alpha = .05$
C.R. $\chi^2 > \chi^2_{6,.05} = 12.592$
calculations:

day	O	E	$(O-E)^2/E$
Mon	74	66.29	.8978
Tue	60	66.29	.5961
Wed	66	66.29	.0012
Thu	71	66.29	.3353
Fri	51	66.29	3.5249
Sat	66	66.29	.0012
Sun	76	66.29	1.4236
	464	464.00	6.7801

$\chi^2 = \Sigma[(O-E)^2/E] = 6.780$

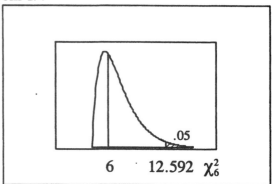

conclusion:
Do not reject H_o; there is not sufficient evidence to reject the claim that the gunfire death rates are the same for the different days of the week.
No, there is not support for the theory theat more gunfire deaths occur on weekends.

3. Refer to the following table and calculations.

			CRIME				
		arson	rape	violence	stealing	coining	fraud	
USE	Y	50	88	155	379	18	63	753
		(49.11)	(79.21)	(139.93)	(358.55)	(16.90)	(109.31)	
	N	43	62	110	300	14	144	673
		(43.89)	(70.79)	(125.07)	(320.45)	(15.10)	(97.69)	
		93	150	265	679	32	207	1426

$\chi^2 = \Sigma[(O-E)^2/E]$
$= .016 + 0.976 + 1.622 + 1.167 + .072 + 19.617$
$.018 + 1.092 + 1.815 + 1.306 + .080 + 21.949 = 49.731$

H_o: alcohol use and type of crime are independent
H_1: alcohol use and type of crime are related
$\alpha = .05$ [assumed]
C.R. $\chi^2 > \chi^2_{5,.05} = 11.071$
calculations:

[refer to the table and calculations at the start of the problem]

$\chi^2 = \Sigma[(O-E)^2/E]$
$= 49.731$

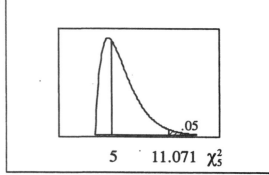

conclusion:
Reject H_o; there is sufficient evidence to conclude that alcohol use and the type of crime committed are related.
Fraud seems to be different from the other crimes in that it is more likely to be committed by someone who abstains from alcohol.

4. H_o: success in stopping to smoke is independent of the method used
H_1: success in stopping to smoke is related to the method used
$\alpha = .05$
C.R. $\chi^2 > \chi^2_{1,.05} = 3.841$

calculations: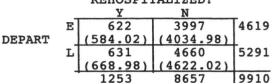

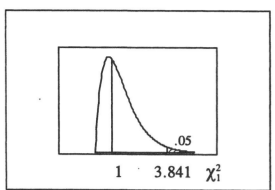

$\chi^2 = \Sigma[(O-E)^2/E]$
= 2.470 + 0.358
  2.157 + 0.312 = 5.297

conclusion:
Reject H_o; there is sufficient evidence to reject the claim that whether a newborn is rehospitalized is independent of how soon it was discharged and to conclude that the two variables are related.
Yes, the conclusion changes if the level of significance becomes .01 [CV = 6.635]. This means we can be 95% confidence such a relationship exists, but not 99% confident.

Cumulative Review Exercises

1. scores in order: 66 75 77 80 82 84 89 94; n = 8 $\Sigma x = 647$ $\Sigma x^2 = 52847$
   $\bar{x} = 647/8 = 80.9$
   $\tilde{x} = (80 + 82)/2 = 81.0$
   R = 94 - 66 = 28
   $s^2 = [8(52847) - (647)^2]/[8(7)] = 74.41$
   s = 8.6
   five-number summary:
   $x_1 = 66$
   $P_{25} = (75 + 77)/2 = 76.0$ [(.25)8 = 2, a whole number, use $x_{2.5}$]
   $P_{50} = \tilde{x} = 81.0$
   $P_{75} = (84 + 89)/2 = 86.5$ [(.75)8 = 6, a whole number, use $x_{6.5}$]
   $x_8 = 94$

2. Consider the table at the right.

	A	B	C	D	
M	66	80	82	75	303
F	77	89	94	84	344
	143	169	176	159	647

   a. P(C) = 176/647 = .272
   b. P(M) = 303/647 = .468
   c. P(M or C) = P(M) + P(C) - P(M and C)
         = (303/647) + (176/647) - (82/647) = 397/647 = .614
   d. $P(F_1 \text{ and } F_2) = P(F_1) \cdot P(F_2|F_1) = (344/647) \cdot (343/646) = .282$

3. H_o: gender and selection are independent
   H_1: gender and selection are related
   α = .05 [assumed]
   C.R. $\chi^2 > \chi^2_{3,.05} = 7.815$
   calculations:

   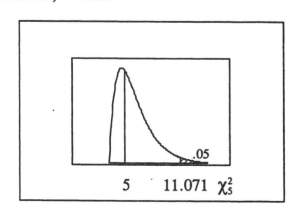

		SELECTION			
G		A	B	C	D	
E	M	66	80	82	75	303
N		(66.97)	(79.15)	(82.42)	(74.46)	
D	F	77	89	94	84	344
E		(76.03)	(89.85)	(93.58)	(84.54)	
R		143	169	176	159	647

   $\chi^2 = \Sigma[(O-E)^2/E]$
   = .014 + .009 + .002 + .004
     .012 + .008 + .002 + .003 = .055

conclusion:
  Do not reject H_o; there is not sufficient evidence to reject the claim that gender and selection are independent.

4. $n = 4$ $\Sigma xy = 26210$ $n(\Sigma xy) - (\Sigma x)(\Sigma y) = 4(26210) - (303)(344) = 608$
   $\Sigma x = 303$ $\Sigma x^2 = 23105$ $n(\Sigma x^2) - (\Sigma x)^2 = 4(23105) - (303)^2 = 611$
   $\Sigma y = 344$ $\Sigma y^2 = 29742$ $n(\Sigma y^2) - (\Sigma y)^2 = 4(29742) - (344)^2 = 632$
   $r = [n(\Sigma xy) - (\Sigma x)(\Sigma y)]/[\sqrt{n(\Sigma x^2) - (\Sigma x)^2} \cdot \sqrt{n(\Sigma y^2) - (\Sigma y)^2}\,]$
     $= 608/[\sqrt{611} \cdot \sqrt{632}] = .978$
   original claim: $\rho \neq 0$
   $H_o: \rho = 0$
   $H_1: \rho \neq 0$
   $\alpha = .05$ [assumed]
   C.R. $r < -.950$ OR C.R. $t < -t_{2,.025} = -4.303$
       $r > .950$ $t > t_{2,.025} = 4.303$

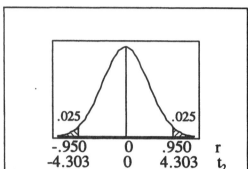

   calculations: calculations:
     $r = .978$ $t_r = (r - \mu_r)/s_r$
                             $= (.978 - 0)/\sqrt{(1-(.978)^2)/2}$
                             $= .978/.146 = 6.696$
   conclusion:
     Reject H_o; there is sufficient evidence to conclude that $\rho \neq 0$ (in fact, $\rho > 0$).

5. $d = y - x$: 11 9 12 9;
   $n = 4$ $\Sigma d = 41$ $\Sigma d^2 = 427$ $\bar{d} = 10.25$ $s_d = 1.50$
   original claim: $\mu_d > 0$
   $H_o: \mu_d = 0$
   $H_1: \mu_d > 0$
   $\alpha = .05$ [assumed]
   C.R. $t > t_{3,.05} = 2.353$

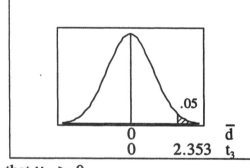

   calculations:
     $t_{\bar{d}} = (\bar{d} - \mu_{\bar{d}})/s_{\bar{d}}$
         $= (10.25 - 0)/(1.50/\sqrt{4}\,)$
         $= 10.25/.750 = 13.667$
   conclusion:
     Reject H_o; there is sufficient evidence to conclude that $\mu_d > 0$.

6. summary statistics
     males: $n_1 = 4$ $\Sigma x_1 = 303$ $\Sigma x_1^2 = 23105$ $\bar{x}_1 = 75.75$ $s_1^2 = 50.917$
     females: $n_2 = 4$ $\Sigma x_2 = 344$ $\Sigma x_2^2 = 29742$ $\bar{x}_2 = 86.00$ $s_2^2 = 52.667$
   $\bar{x}_1 - \bar{x}_2 = 75.75 - 86.00 = -10.25$
   $H_o: \mu_1 - \mu_2 = 0$
   $H_1: \mu_1 - \mu_2 \neq 0$
   $\alpha = .05$ [assumed]
   C.R. $t < -t_{3,.025} = -3.182$
       $t > t_{3,.025} = 3.182$
   calculations:
     $t_{\bar{x}_1 - \bar{x}_2} = (\bar{x}_1 - \bar{x}_2 - \mu_{\bar{x}_1 - \bar{x}_2})/s_{\bar{x}_1 - \bar{x}_2}$
       $= (-10.25 - 0)/\sqrt{50.917/4 + 52.667/4}$
       $= -10.25/5.089 = -2.014$

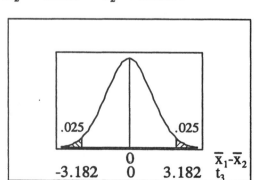

   conclusion:
     Do not reject H_o; there is not sufficient evidence to reject the claim that $\mu_1 - \mu_2 = 0$.

Chapter 11

Analysis of Variance

11-2 One-Way ANOVA

1. a. H_o: $\mu_1 = \mu_2 = \mu_3$
    b. H_1: at least one mean is different
    c. $F^2_{33} = 8.98$
    d. $F^2_{33,.05} = 3.3158$ [closest entry in Table A-5]
    e. P-value = .001
    f. Since $8.98 > 3.3158$, reject H_o and conclude the three mean F-K Grade Level scores are not all the same.

2. a. H_o: $\mu_1 = \mu_2 = \mu_3 = \mu_4 = \mu_5$
    b. H_1: at least one mean is different
    c. $F^4_{43} = 2.9493$
    d. $F^4_{43,.05} = 2.5888$ [from Excel]
    e. P-value = .0307
    f. Yes; since $2.9493 > 2.5888$, there is sufficient evidence at the .05 level to support the claim that the means for the five different laboratories are not all the same.

3. a. H_o: $\mu_1 = \mu_2 = \mu_3$
    b. H_1: at least one mean is different
    c. $F^2_{108} = .1887$
    d. $F^2_{108,.05} = 3.0804$ [from Excel]
    e. P-value = .8283
    f. No; since $.1887 < 3.0804$, there is not sufficient evidence to support the claim that the means for three different age groups are not all the same.

4. a. H_o: $\mu_1 = \mu_2 = \mu_3$
    b. H_1: at least one mean is different
    c. $F^2_{37} = 1.65$
    d. P-value = .205
    e. No; since $.205 > .05$, there is not sufficient evidence to support the claim that the means for the three different age groups are not all the same.

NOTE: This section is calculation-oriented. Do not get so involved with the formulas that you miss concepts. This manual arranges the calculations to promote both computational efficiency and understanding of the underlying principles. The following notation is used in this section.

k = the number of groups
n_i = the number of scores in group i (where i = 1,2,...,k)
$\bar{x}_i$ = the mean of group i
s_i^2 = the variance of group i
$\bar{\bar{x}}$ = the overall mean of all the scores in all the groups
  = $\Sigma n_i \bar{x}_i / \Sigma n_i$ = the (weighted) mean of the group means
  = $\Sigma \bar{x}_i / k$ = simplified form when each group has equal size n
s_B^2 = the variance between the groups
  = $\Sigma n_i (\bar{x}_i - \bar{\bar{x}})^2 / (k-1)$
  = $n \Sigma (\bar{x}_i - \bar{\bar{x}})^2 / (k-1) = n s_{\bar{x}}^2$ = simplified form when each group has equal size n

S-340 INSTRUCTOR'S SOLUTIONS Chapter 11

s_p^2 = the variance within the groups
   = $\Sigma df_i s_i^2 / \Sigma df_i$ = the (weighted) mean of the group variances
   = the two-sample formula for s_p^2 generalized to k samples
   = $\Sigma s_i^2/k$ = simplified form when each group has equal size n
numerator df = k-1
denominator df = Σdf_i
   = k(n-1) = simplified form when each group has equal size n
$F = s_B^2/s_p^2$ = (variance between groups)/(variance within groups)

5. Since each group has equal size n, the simplified forms can be used.
   The following preliminary values are identified and/or calculated.

	subcompact	compact	midsize	full size
n	5	5	5	5
Σx	3344	2779	2434	2689
Σx^2	2470638	1577659	1297312	1541765
$\bar{x}$	668.8	555.8	486.8	537.8
s^2	58542.7	8272.8	28110.2	23905.2

k = 4 $\bar{\bar{x}} = \Sigma \bar{x}_i/k$
n = 5   = 562.3
$s_{\bar{x}}^2 = \Sigma(\bar{x}_i-\bar{\bar{x}})^2/(k-1)$ $s_p^2 = \Sigma s_i^2/k$
   = 5895   = 29707.725

H_o: $\mu_1 = \mu_2 = \mu_3 = \mu_4$
H_1: at least one mean is different
$\alpha = .05$
C.R. $F > F^3_{16,.05} = 3.2389$
calculations:
   $F = ns_{\bar{x}}^2/s_p^2$
      = 5(5895)/29797.725 = .9922

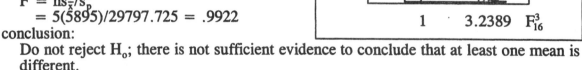

conclusion:
   Do not reject H_o; there is not sufficient evidence to conclude that at least one mean is different.
   No, these data do not suggest that larger cars are safer.

6. Since each group has equal size n, the simplified forms can be used.
   The following preliminary values are identified and/or calculated.

	subcompact	compact	midsize	full size
n	5	5	5	5
Σx	252	265	244	230
Σx^2	12880	14131	11952	10782
$\bar{x}$	50.4	53.0	48.8	46.0
s^2	44.8	21.5	11.2	50.5

k = 4 $\bar{\bar{x}} = \Sigma \bar{x}_i/k$
n = 5   = 49.55
$s_{\bar{x}}^2 = \Sigma(\bar{x}_i-\bar{\bar{x}})^2/(k-1)$ $s_p^2 = \Sigma s_i^2/k$
   = 8.597   = 32.000

H_o: $\mu_1 = \mu_2 = \mu_3 = \mu_4$
H_1: at least one mean is different
$\alpha = .05$
C.R. $F > F^3_{16,.05} = 3.2389$
calculations:
   $F = ns_{\bar{x}}^2/s_p^2$
      = 5(8.597)/32.000 = 1.3432

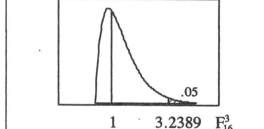

conclusion:
   Do not reject H_o; there is not sufficient evidence to conclude that at least one mean is different.
   No, these data do not suggest that larger cars are safer.

7. Since each group has equal size n, the simplified forms can be used.
The following preliminary values are identified and/or calculated.

	4000 BC	1850 BC	150 AD
n	9	9	9
Σx	1194	1210	1243
Σx^2	158544	162768	171853
$\bar{x}$	132.67	134.44	138.11
s^2	17.500	11.278	22.611

$k = 3$ $\qquad\qquad\bar{\bar{x}} = \Sigma\bar{x}_i/k$
$n = 9$ $\qquad\qquad\quad = 135.07$
$s_{\bar{x}}^2 = \Sigma(\bar{x}_i-\bar{\bar{x}})^2/(k-1)$ $\quad s_p^2 = \Sigma s_i^2/k$
$\quad\; = 15.416/2$ $\qquad\qquad = 51.389/3$
$\quad\; = 7.708$ $\qquad\qquad\quad = 17.130$

$H_o: \mu_1 = \mu_2 = \mu_3$
$H_1:$ at least one mean is different
$\alpha = .05$
C.R. $F > F_{2,24,.05} = 3.4028$
calculations:
$\quad F = ns_{\bar{x}}^2/s_p^2$
$\qquad = 9(7.708)/17.130 = 4.0498$
conclusion:
 Reject H_o; there is sufficient evidence to conclude that at least one mean is different.

8. Since each group has equal size n, the simplified forms can be used.
The following preliminary values are identified and/or calculated.

	sunny	cloudy	rainy
n	6	6	6
Σx	81.4	76.5	72.3
Σx^2	1105.14	975.59	871.55
$\bar{x}$	13.567	12.750	12.050
s^2	.1627	.0430	.0670

$k = 3$ $\qquad\qquad\bar{\bar{x}} = \Sigma\bar{x}_i/k$
$n = 6$ $\qquad\qquad\quad = 49.55$
$s_{\bar{x}}^2 = \Sigma(\bar{x}_i-\bar{\bar{x}})^2/(k-1)$ $\quad s_p^2 = \Sigma s_i^2/k$
$\quad\; = .5762$ $\qquad\qquad\; = .0909$

$H_o: \mu_1 = \mu_2 = \mu_3$
$H_1:$ at least one mean is different
$\alpha = .05$
C.R. $F > F_{2,15,.05} = 3.6823$
calculations:
$\quad F = ns_{\bar{x}}^2/s_p^2$
$\qquad = 6(.5762)/.0909 = 38.0379$
conclusion:
 Reject H_o; there is sufficient evidence to conclude that at least one mean is different. While it appears that sunny days provide more solar energy that either cloudy or rainy days, all that can be concluded from this test is that there is at least one difference. Determining exactly what that difference is requires more advanced techniques. It could be, for example, that the difference between sunny and rainy days is significant but the difference between sunny and cloudy days is not.

9. The following preliminary values are identified and/or calculated.

	R	O	Y	Br	Bl	G	total
n	21	8	26	33	5	7	100
Σx	19.104	7.401	23.849	30.123	4.507	6.846	91.470
Σx^2	17.934278	6.862429	21.904543	27.546793	4.075729	6.018734	83.802506
$\bar{x}$	.90971	.92513	.91272	.91282	.90140	.92657	.91470
s^2	.000755	.002226	.001144	.001562	.003280	.001499	

$\bar{\bar{x}} = \Sigma n_i \bar{x}_i / \Sigma n_i$
  $= [21(.90971) + 8(.92513) + 26(.91272) + 33(.91282) + 5(.90140) + 7(.92657)]/100$
  $= 91.470/100$
  $= .91470$ [NOTE: $\bar{\bar{x}}$ must always agree with the $\bar{x}$ in the "total" column.]

$\Sigma n_i(\bar{x}_i - \bar{\bar{x}})^2 = 21(.90971-.91470)^2 + 8(.92513-.91470)^2 + 26(.91272-.91470)^2 +$
  $33(.91282-.91470)^2 + 5(.90140-.91470)^2 + 7(.92657-.91470)^2$
  $= .00355$

$\Sigma df_i s_i^2 = 20(.000755) + 7(.002226) + 25(.001144) +$
  $32(.001562) + 4(.003280) + 6(.001499)$
  $= .13135$

$s_B^2 = \Sigma n_i(\bar{x}_i - \bar{\bar{x}})^2/(k-1)$
  $= .00355/5 = .00071$

$s_p^2 = \Sigma df_i s_i^2 / \Sigma df_i$
  $= .13135/94 = .001398$

$H_o: \mu_R = \mu_O = \mu_Y = \mu_{Br} = \mu_{Bl} = \mu_G$
H_1: at least one mean is different
$\alpha = .05$
C.R. $F > F^5_{94,.05} = 2.2899$
calculations:
  $F = s_B^2/s_p^2$
  $= .00071/.001398 = .5081$
conclusion:

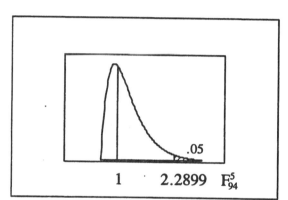

Do not reject H_o; there is not sufficient evidence to conclude that at least one mean is different.

No. If the intent is to make the different colors have the same mean, there is no evidence that this is not being accomplished. Corrective action is not required.

NOTE: An ANOVA table may be completed as follows:

source	SS	df	MS	F
Trt	.00355	5	.00071	.51
Error	.13135	94	.00140	
Total	.13490	99		

$F = MS_{Trt}/MS_{Err}$
  $= .00071/.00140$
  $= .51$

(1) Enter $SS_{Trt} = \Sigma n_i(\bar{x}_i - \bar{\bar{x}})^2$ and $SS_{Err} = \Sigma df_i s_i^2$ values from the preliminary calculations.
(2) Enter $df_{Trt} = k-1$ and $df_{Err} = \Sigma df_i = \Sigma(n_i-1) = \Sigma n_i - k$.
(3) Add the SS and df columns to find SS_{Tot} and df_{Tot}. [The df_{Tot} must equal $\Sigma n_i - 1$.]
(4) Calculate $MS_{Trt} = SS_{Trt}/df_{Trt}$ and $MS_{Err} = SS_{Err}/df_{Err}$.
(5) Calculate $F = MS_{Trt}/MS_{Err}$.

As a final check, calculate s^2 (i.e., the variance of all the scores in one large group) two different ways as indicated below. If these answers agree, the problem is probably correct.

* from the "total" column in the table for the preliminary calculations:
  $s^2 = [n\Sigma x^2 - (\Sigma x)^2]/[n(n-1)]$
    $= [100(83.802506) - (91.470)^2]/[100(99)]$
    $= 13.4897/9900 = .001363$
* from the "total" row of the ANOVA table
  $s^2 = SS_{Tot}/df_{Tot}$
    $= .13490/99 = .001363$

10. The following preliminary values are identified and/or calculated.

	Bonds	McGwire	Sosa	total
n	73	70	66	209
Σx	29468	29296	26720	85484
Σx^2	11962970	12403606	10900378	35266954
$\bar{x}$	403.671	418.514	404.848	409.014
s^2	938.696	2069.732	1274.254	

$\bar{\bar{x}} = \Sigma n_i \bar{x}_i / \Sigma n_i$
$= [73(403.671) + 70(418.514) + 66(404.848)]/209 = 409.014$

$\Sigma n_i(\bar{x}_i - \bar{\bar{x}})^2 = 73(403.671 - 409.014)^2 + 70(418.514 - 409.104)^2 + 66(404.848 - 409.014)^2$
$= 9546.945$

$\Sigma df_i s_i^2 = 72(938.696) + 69(2069.732) + 65(1274.254)$
$= 293224.13$

$s_B^2 = \Sigma n_i(\bar{x}_i - \bar{\bar{x}})^2/(k-1)$
$= 9546.945/2 = 4773.47$

$s_p^2 = \Sigma df_i s_i^2 / \Sigma df_i$
$= 293224.13/206 = 1423.42$

$H_o: \mu_B = \mu_M = \mu_S$
H_1: at least one mean is different
$\alpha = .05$
C.R. $F > F^2_{206,.05} = 3.0718$
calculations:
$F = s_B^2/s_p^2$
$= 4773.47/1423.42 = 3.3535$

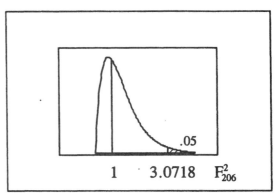

conclusion:
Reject H_o; there is sufficient evidence to conclude that at least one mean is different.

No. Apparently greater distance of home runs is not associated with greater number of home runs, for Bonds has the lowest distance but the greatest numbers. One can sometimes say, for example: "He doesn't hit home runs very often – but when he hits one, it really goes!"

11. The following preliminary values are identified and/or calculated. ["shelf 3" = shelf 3/4]

	shelf 1	shelf 2	shelf 3	total
n	6	6	4	16
Σx	1.19	2.73	.80	4.72
Σx^2	.3853	1.2445	.1846	1.8144
$\bar{x}$	.1983	.4550	.2000	.2950
s^2	.029857	.000470	.008200	

$\bar{\bar{x}} = \Sigma n_i \bar{x}_i / \Sigma n_i$
$= [6(.1983) + 6(.4550) + 4(.2000)]/16 = .2950$

$\Sigma n_i(\bar{x}_i - \bar{\bar{x}})^2 = 6(.1983 - .2950)^2 + 6(.4550 - .2950)^2 + 4(.2000 - .295)^2 = .24577$

$\Sigma df_i s_i^2 = 5(.029857) + 5(.000470) + 3(.008200) = .17624$

$s_B^2 = \Sigma n_i(\bar{x}_i - \bar{\bar{x}})^2/(k-1) = .24577/2 = .12288$

$s_p^2 = \Sigma df_i s_i^2 / \Sigma df_i = .17624/13 = .01356$

$H_o: \mu_1 = \mu_2 = \mu_3$
H_1: at least one mean is different
$\alpha = .05$
C.R. $F > F^2_{13,.05} = 3.8056$
calculations:
$F = s_B^2/s_p^2$
$= .12288/.01356 = 9.0645$

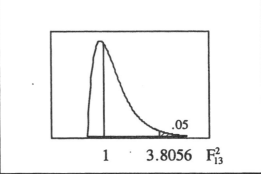

conclusion:
Reject H_o; there is sufficient evidence to conclude that at least one mean is different.
It appears that the cereals are not placed on the shelves at random, but that those with higher sugar amounts are on shelf 2 – at children's eye-level.

12. Since each group has equal size n, the simplified forms can be used.
The following preliminary values are identified and/or calculated.

	smokers	exposed ns	nonexposed
n	40	40	40
Σx	6899	2423	654
Σx^2	1746819	890393	163200
$\overline{x}$	172.475	60.575	16.350
s^2	14279.846	19067.174	3910.438

$k = 3$ $\overline{\overline{x}} = \Sigma \overline{x}_i / k$
$n = 40$ $= 83.133$
$s_{\overline{x}}^2 = \Sigma(\overline{x}_i - \overline{\overline{x}})^2/(k-1)$ $s_p^2 = \Sigma s_i^2/k$
     $= 12950.825/2$ $= 37257.458/3$
     $= 6475.41$ $= 12419.15$

$H_o: \mu_1 = \mu_2 = \mu_3$
H_1: at least one mean is different
$\alpha = .05$
C.R. $F > F_{117,.05}^2 = 3.0718$

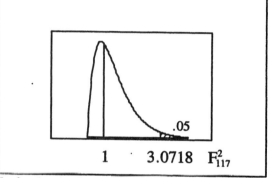

calculations:
   $F = ns_{\overline{x}}^2/s_p^2$
     $= 40(6475.41)/12419.15 = 20.8562$
conclusion:
Reject H_o; there is sufficient evidence to conclude that at least one mean is different.
The results seem to support the claim that secondhand smoke is harmful, but such a statement cannot be from the ANOVA. The conclusion is only that at least one is different. Determining exactly what that difference is requires more advanced techniques. It could be, for example, that the exposed and non-exposed non-smokers have significantly less cotinine levels than the smokers – but that their levels are not significantly different from each other.

13. a. $_5C_2 = 5!/2!3! = 10$
 b. P(no type I error in one t test) = .95
    assuming independence, P(no Type I error in ten t tests) = $(.95)^{10} = .599$
 c. P(no type I error in one F test) = .95
 d. the analysis of variance test

14. The following preliminary values are identified and/or calculated.

	Clancy	Rowling
n	12	12
Σx	848.8	969.0
Σx^2	61449.54	78487.82
$\overline{x}$	70.733	80.750
s^2	128.2806	21.9155

a. original claim: $\mu_1 - \mu_2 = 0$
   $\overline{x}_1 - \overline{x}_2 = 70.733 - 80.750 = -10.017$
   $s_p^2 = [(n_1-1) \cdot s_1^2 + (n_2-1) \cdot s_2^2]/(n_1 + n_2 - 2)$
     $= [(11)(128.2806) + (11)(21.9155)]/(22)$
     $= 75.098$ [with df = $df_1 + df_2 = 22$]
$H_o: \mu_1 - \mu_2 = 0$
$H_1: \mu_1 - \mu_2 \neq 0$
$\alpha = .05$
C.R. $t < -t_{22,.025} = -2.074$
     $t > t_{22,.025} = 2.074$

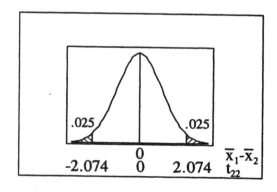

calculations:
   $t_{\overline{x}_1 - \overline{x}_2} = (\overline{x}_1 - \overline{x}_2 - \mu_{\overline{x}_1 - \overline{x}_2})/s_{\overline{x}_1 - \overline{x}_2}$
     $= (-10.017 - 0)/\sqrt{(75.098)/12 + (75.098)/12}$
     $= -10.017/3.5378 = -2.83129$

conclusion:
Reject H_o; there is sufficient evidence to reject the claim that $\mu_1-\mu_2 = 0$ and to conclude that $\mu_1-\mu_2 \neq 0$ (in fact, that $\mu_1-\mu_2 < 0$).

b. Since each group has equal size n, the simplified forms can be used. In addition to the summary statistics above, the following preliminary values are noted.

$k = 2$ $\quad\quad \bar{\bar{x}} = \Sigma\bar{x}_i/k$
$n = 12$ $\quad\quad\quad\quad = 75.7417$
$s_{\bar{x}}^2 = \Sigma(\bar{x}_i-\bar{\bar{x}})^2/(k-1) \quad s_p^2 = \Sigma s_i^2/k$
$\quad = 50.1668/1 \quad\quad\quad = 150.1961/2$
$\quad = 50.1668 \quad\quad\quad\quad = 75.098$

$H_o: \mu_1 = \mu_2 = \mu_3$
$H_1:$ at least one mean is different
$\alpha = .05$
C.R. $F > F^1_{22,.05} = 4.3009$

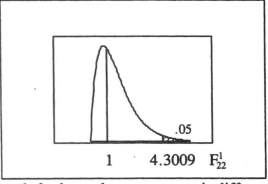

calculations:
$F = ns_{\bar{x}}^2/s_p^2$
$\quad = 12(50.1668)/75.098 = 8.016$
conclusion:
Reject H_o; there is sufficient evidence to conclude that at least one mean is different.

c. critical value: $t^2 = (\pm 2.074)^2 = 4.301 = F$
calculated statistic: $t^2 = (-2.83129)^2 = 8.016 = F$ [In general $t^2_{df,\alpha/2} = F^1_{df,\alpha}$.]

11-3 Two-Way ANOVA

NOTE: The formulas and principles in this section are logical extensions of the previous ones.
$SS_{Row} = \Sigma n_i(\bar{x}_i-\bar{\bar{x}})^2$ for i=1,2,3... [for each row]
$SS_{Col} = \Sigma n_j(\bar{x}_j-\bar{\bar{x}})^2$ for j=1,2,3... [for each column]
$SS_{Tot} = \Sigma(x-\bar{\bar{x}})^2$ [for all the x's]
When there is only one observation per cell the unexplained variation is
$\quad SS_{Err} = SS_{Tot} - SS_{Row} - SS_{Col}$
and there is not enough data to measure interaction.
When there is more than one observation per cell the unexplained variation (i.e., failure of items in the same cell to respond the same) is
$\quad SS_{Err} = \Sigma(x-\bar{x}_{ij})^2 = \Sigma df_{ij}s_{ij}^2$ [for each cell – i,e., for each i,j (row,col) combination]
and the interaction sum of squares is
$\quad SS_{Int} = SS_{Tot} - SS_{Row} - SS_{Col} - SS_{Err}$.
Since the data will be analyzed from statistical software packages, however, the above formulas need not be used by hand.

1. The term "two-way" refers to the fact that the data can be analyzed from two different perspectives. There are two factors acting simultaneously that may or may not affect the response, and this technique allows for simultaneous analyses of both factors.
The term "analysis of variance" refers to the fact that we are making judgments about means by analyzing variances. If the variance between the means is larger than the unexplained natural variation in the problem, then the means are declared to be different.

2. Two separate tests cannot address the possibility of interaction between the factors. In general, the two-way test is more powerful since the unexplained variation (the denominator portion of the F statistic) is smaller (because more of the variation can be explained by the other factor), thus increasing the likelihood of identifying population differences by having a large F ratio.

3. If there is a significant interaction, no general statements can be made about either of the factors – differences in the levels of factor A may be significant at one level of factor B, but not at another level of factor B.

4. The two-way tables in chapter 10 are frequency counts of categorical data and not numerical values of quantitative data. Tables in chapter 10 have total df=(r-1)(c-1) because the entries in the table are not independent of each other – the frequency in one cell limits the possible values for the frequencies in adjacent cells. Tables in chapter 11 have total df=n-1 because the entries in the table are independent of each other – the only restriction being that the nth value must be whatever is necessary to maintain the value of the overall $\bar{\bar{x}}$.

5. H_o: there is no site-age interaction effect
   H_1: there is a site-age interaction effect
   $\alpha = .05$
   C.R. $F > F^4_{18,.05} = 2.9277$
   calculations:
   $F = MS_{INT}/MS_E$
   $= 4.43/3.44$
   $= 1.28$
   P-value $= .313$ [from Minitab]

   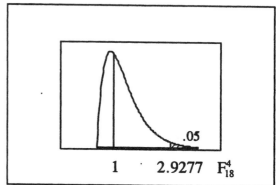

   conclusion:
   Do not reject H_o; there is not sufficient evidence to conclude that there is any interaction between a falcon's site and its age (in determining its DDT level).

6. H_o: $\mu_1 = \mu_2 = \mu_3$ [there is no site effect]
   H_1: at least one μ_i is different
   $\alpha = .05$
   C.R. $F > F^2_{18,.05} = 3.5546$
   calculations:
   $F = MS_{SITE}/MS_E$
   $= 8892.70/3.44$
   $= 2581.75$
   P-value $= .000$ [from Minitab]

   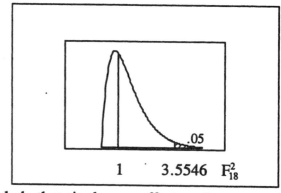

   conclusion:
   Reject H_o; there is sufficient evidence to conclude that site has an effect on the amount of DDT..

7. H_o: $\mu_1 = \mu_2 = \mu_3$ [there is no age effect]
   H_1: at least one μ_i is different
   $\alpha = .05$
   C.R. $F > F^2_{18,.05} = 3.5546$
   calculations:
   $F = MS_{AGE}/MS_E$
   $= 860.59/3.44$
   $= 249.85$
   P-value $= .000$ [from Minitab]

   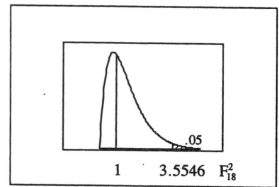

   conclusion:
   Reject H_o; there is sufficient evidence to conclude that age has an effect on the amount of DDT.

8. H_o: there is no gender-type interaction effect
   H_1: there is a gender-type interaction effect
   $\alpha = .05$
   C.R. $F > F^1_{36,.05} = 4.0847$
   calculations:
   $F = MS_{INT}/MS_E$
   $= 31528/10465$
   $= 3.01$
   P-value $= .091$ [from Minitab]

   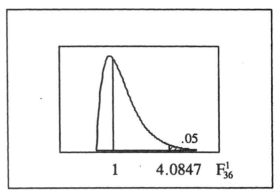

   conclusion:
   Do not reject H_o; there is not sufficient evidence to conclude that there is any interaction between a person's gender and the type of SAT test.

9. H_o: $\mu_M = \mu_F$ [there is no gender effect]
   H_1: the gender means are different
   $\alpha = .05$
   C.R. $F > F^1_{36,.05} = 4.0847$
   calculations:
   $F = MS_{GENDER}/MS_E$
   $= 52635/10465$
   $= 5.03$
   P-value $= .031$ [from Minitab]

   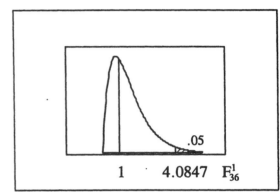

   conclusion:
   Reject H_o; there is sufficient evidence to conclude that gender affects SAT score.

10. H_o: $\mu_{Verbal} = \mu_{Math}$ [there is no type effect]
    H_1: the type means are different
    $\alpha = .05$
    C.R. $F > F^1_{36,.05} = 4.0847$
    calculations:
    $F = MS_{TEST}/MS_E$
    $= 6027/10465$
    $= .58$
    P-value $= .453$ [from Minitab]

    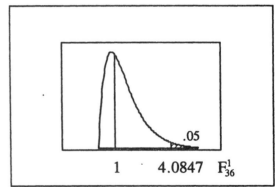

    conclusion:
    Do not reject H_o; there is not sufficient evidence to conclude that the type of test has an effect on SAT scores.

11. H_o: $\mu_1 = \mu_2 = ... = \mu_{24}$ [there is no subject effect]
    H_1: at least one μ_i is different
    $\alpha = .05$
    C.R. $F > F^{23}_{69,.05} = 1.7001$
    calculations:
    $F = MS_{SUBJ}/MS_E$
    $= 140.5/36.3$
    $= 3.87$
    P-value $= .000$ [from Minitab]

    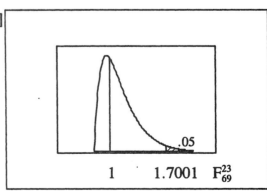

    conclusion:
    Reject H_o; there is sufficient evidence to conclude that the choice of subject has an effect on the hearing test score.
    This makes practical sense, because all people do not have the same level of hearing.

12. H_o: $\mu_1=\mu_2=\mu_3=\mu_4$ [there is no list effect]
H_1: at least one μ_i is different
$\alpha = .05$
C.R. $F > F^3_{69,.05} = 2.7581$
calculations:
$F = MS_{LIST}/MS_E$
$= 306.8/36.3$
$= 8.45$
P-value = .000 [from Minitab]

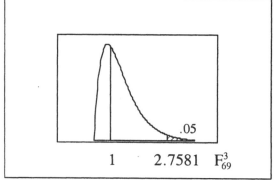

conclusion:
Reject H_o; there is sufficient evidence to conclude that choice of word list has an effect on the hearing test score.

13. While it is assumed the ANOVA table will be obtained using a statistical software package, the actual calculations necessary to construct the table by hand are given below. There are n=4 observations in each of the indicated age-gender cells.

```
                        AGE
              under 20      20-40          over 40
              Σx  = 288     Σx  = 288      Σx  = 288
G             Σx² = 21536   Σx² = 21120    Σx² = 21152
E    male     x̄ = 72        x̄ = 72         x̄ = 72          x̄ = 72
N             s² = 266.667  s² = 128       s² = 138.667
D
E             Σx  = 284     Σx  = 300      Σx  = 264
R             Σx² = 20272   Σx² = 22736    Σx² = 17504
     female   x̄ = 71        x̄ = 75         x̄ = 66          x̄ = 70.667
              s² = 36       s² = 78.667    s² = 26.667

              x̄ = 71.5      x̄ = 73.5       x̄ = 69         Σx  = 1712      x̄̄ = 71.333
                                                           Σx² = 124320   s² = 95.536
```

$SS_{Gender} = \Sigma n_i(\bar{x}_i-\bar{\bar{x}})^2 = 12(72-71.333)^2 + 12(70.667-71.333)^2 = 10.667$
$SS_{Age} = \Sigma n_j(\bar{x}_j-\bar{\bar{x}})^2 = 8(71.5-71.333)^2 + 8(73.5-71.333)^2 + 8(69-71.333)^2 = 81.333$
$SS_{Tot} = \Sigma(x-\bar{\bar{x}})^2 = df \cdot s^2 = 23(95.536) = 2197.333$
$SS_{Err} = \Sigma(x-\bar{x}_{ij})^2 = \Sigma df_{ij} s^2_{ij} = 3(266.667) + 3(128) + 3(138.667) +$
$\qquad\qquad\qquad\qquad\qquad\qquad 3(36) + 3(78.667) + 3(26.667) = 2024$
$SS_{Int} = SS_{Tot} - SS_{Gender} - SS_{Age} - SS_{Err}$
$\quad = 2197.333 - 10.667 - 81.333 - 2024 = 81.333$

The resulting ANOVA table is used for the test of hypothesis in this exercise.

Source	df	SS	MS	F
Gender	1	10.667	10.667	.0949
Age	2	81.333	40.667	.3617
Interaction	2	81.333	40.667	.3617
Error	18	2024.000	112.444	
Total	23	2197.333		

H_o: there is no gender-age interaction effect
H_1: there is a gender-age interaction effect
$\alpha = .05$
C.R. $F > F^2_{18,.05} = 3.5546$
calculations:
$F = MS_{INT}/MS_E$
$= 40.667/112.444$
$= .3617$

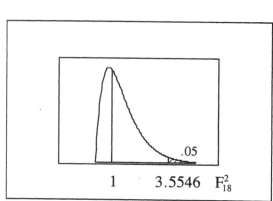

conclusion:
Do not reject H_o; there is not sufficient evidence to conclude that there is any interaction between a person's gender and age (in determining pulse rate).

H_o: $\mu_M = \mu_F$ [there is no gender effect]
H_1: at least one μ_i is different
$\alpha = .05$
C.R. $F > F^1_{18,.05} = 4.4139$
calculations:
$F = MS_{GENDER}/MS_E$
$= 10.667/112.444$
$= .0949$

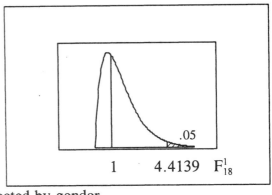

conclusion:
Do not reject H_o; there is not sufficient evidence to conclude that pulse rates are affected by gender.

H_o: $\mu_1 = \mu_2 = \mu_3$ [there is no age effect]
H_1: at least one μ_i is different
$\alpha = .05$
C.R. $F > F^2_{18,.05} = 3.5546$
calculations:
$F = MS_{AGE}/MS_E$
$= 40.667/112.444$
$= .3617$

conclusion:
Do not reject H_o; there is not sufficient evidence to conclude that pulse rates are affected by age.

14. While it is assumed the ANOVA table will be obtained using a statistical software package, the actual calculations necessary to construct the table by hand are given below. Since there is only 1 observation per cell, there is no direct estimate for the natural variability among like engines – in such cases all variability not attributed to the main factors is used to determine the error variance, and no test can be made for interaction.

```
                CYLINDERS
                4    6    8
         M     33   30   28  | 30.333        For the entire data set, Σx = 173
TRANS    A     31   27   24  | 27.333                                  Σx² = 5039
               32  28.5  26  | 28.833                                   s² = 10.167
```

$SS_{Trans} = \Sigma n_i(\bar{x}_i - \bar{\bar{x}})^2 = 3(30.333-28.833)^2 + 3(27.333-28.833)^2 = 13.5$
$SS_{Cyl} = \Sigma n_j(\bar{x}_j - \bar{\bar{x}})^2 = 2(32-28.833)^2 + 2(28.5-28.833)^2 + 2(26-28.833)^2 = 36.333$
$SS_{Tot} = \Sigma(x-\bar{\bar{x}})^2 = df \cdot s^2 = 5(10.167) = 50.833$
$SS_{Err} = SS_{Tot} - SS_{Trans} - SS_{Cyl} = 50.833 - 13.500 - 36.333 = 1.000$

The resulting ANOVA table is used for the test of hypothesis in this exercise.

Source	df	SS	MS	F
Transmission	1	13.500	13.500	27.000
Cylinders	2	36.333	18.167	36.333
Error	2	1.000	0.500	
Total	5	50.833		

H_o: $\mu_M = \mu_A$ [there is no transmission effect]
H_1: at least one μ_i is different
$\alpha = .05$
C.R. $F > F^1_{2,.05} = 18.513$
calculations:
$F = MS_{TRANS}/MS_E$
$= 13.500/.500$
$= 27.000$

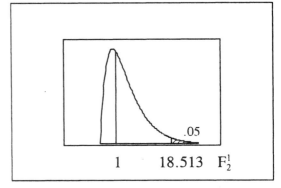

conclusion:
  Reject H_o; there is sufficient evidence to conclude that fuel consumption amounts are effected by the type of transmission.

H_o: $\mu_4 = \mu_6 = \mu_8$ [there is no cylinder effect]
H_1: at least one μ_i is different
$\alpha = .05$
C.R. $F > F^2_{2,.05} = 19.000$
calculations:
  $F = MS_{CYL}/MS_E$
  $= 18.167/.500$
  $= 36.333$

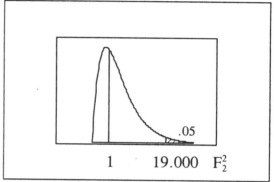

conclusion:
  Reject H_o; there is sufficient evidence to conclude that fuel consumption is affected by the number of cylinders.

15. a. No change. The ANOVA calculated statistics are ratios of variances. Since adding the same value to each score does not affect the variances, the ANOVA table will not change.
    b. No change. The ANOVA calculated statistics are ratios of variances. Since multiplying each score by the same nonzero constant will multiply the all the variances by the square of that constant, the numerators and denominators of each ANOVA ratio will be multiplied by the same constant and the ratios will not change.
    c. No change. The same values will appear in different positions, but referring to the same factors as before.
    d. Depends. The change will affect the mean for row 1, the mean for column 1, and the variability within the first cell. In general, all calculated ANOVA statistics will change.

Review Exercises

1. H_o: $\mu_A = \mu_B = \mu_C$
   H_1: at least one μ_i is different
   $\alpha = .05$
   C.R. $F > F^2_{14,.05} = 3.7389$
   calculations:
     $F = s_B^2/s_p^2$
     $= .0038286/.0000816$
     $= 46.90$
     P-value $= .000$ [from Minitab]

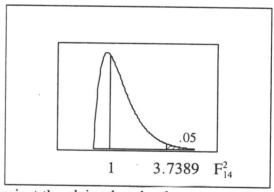

   conclusion:
     Reject H_o; there is not sufficient evidence to reject the claim that the three groups have the same mean and to conclude that at least one mean is different.

2. Since each group has equal size n, the simplified forms can be used.
   The following preliminary values are identified and/or calculated.

	oceanside	oceanfront	bayside	bayfront
n	6	6	6	6
Σx	2294	2956	1555	3210
Σx^2	944,422	1,494,108	435,465	1,772,790
$\bar{x}$	382.333	492.667	259.167	535.000
s^2	13469.867	7557.067	6492.167	11088.000

$k = 4$ $\bar{\bar{x}} = \Sigma\bar{x}_i/k$
$n = 6$ $= 417.292$
$s_{\bar{x}}^2 = \Sigma(\bar{x}_i-\bar{\bar{x}})^2/(k-1)$ $s_p^2 = \Sigma s_i^2/k$
    $= 45762.243/3$ $= 38607.100/4$
    $= 15254.081$ $= 9651.775$

H_o: $\mu_{OS} = \mu_{OF} = \mu_{BS} = \mu_{BF}$
H_1: at least one mean is different
$\alpha = .05$
C.R. $F > F_{20,.05}^3 = 3.0984$
calculations:
   $F = ns_{\bar{x}}^2/s_p^2$
     $= 6(15254.081)/9651.775 = 9.4827$

conclusion:
   Reject H_o; there is sufficient evidence to conclude that at least one mean is different.

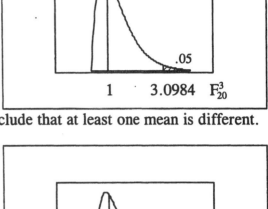

3. H_o: there is no gender-major interaction effect
    H_1: there is a gender-major interaction effect
    $\alpha = .05$
    C.R. $F > F_{12,.05}^2 = 3.8853$
    calculations:
      $F = MS_{INT}/MS_E$
        $= 7.1/37.8$
        $= .19$
      P-value $= .832$ [from Minitab]

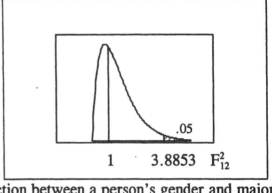

    conclusion:
      Do not reject H_o; there is not sufficient
      evidence to conclude that there is any interaction between a person's gender and major (in estimating length).

4. H_o: $\mu_M = \mu_F$ [there is no gender effect]
    H_1: the gender means are different
    $\alpha = .05$
    C.R. $F > F_{12,.05}^1 = 4.7472$
    calculations:
      $F = MS_{GENDER}/MS_E$
        $= 29.4/37.8$
        $= .78$
      P-value $= .395$ [from Minitab]

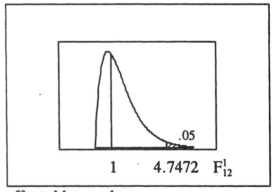

    conclusion:
      Do not Reject H_o; there is not sufficient
      evidence to conclude that estimated length is affected by gender.

5. H_o: $\mu_M = \mu_B = \mu_L$ [there is no major effect]
    H_1: the major means are different
    $\alpha = .05$
    C.R. $F > F_{12,.05}^2 = 3.8853$
    calculations:
      $F = MS_{MAJOR}/MS_E$
        $= 5.1/37.8$
        $= .13$
      P-value $= .876$ [from Minitab]

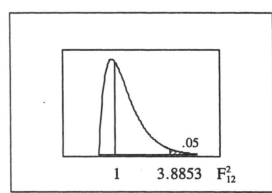

    conclusion:
      Do not reject H_o; there is not sufficient
      evidence to conclude that estimated length is affected by major.

6. a. H_o: $\mu_M = \mu_A$ [there is no transmission effect]
    H_1: at least one μ_i is different
    $\alpha = .05$
    C.R. $F > F^1_{2,.05} = 18.513$
    calculations:
    $F = MS_{TRANS}/MS_E$
    $= .667/.667$
    $= 1.00$
    P-value = .423 [from Minitab]

    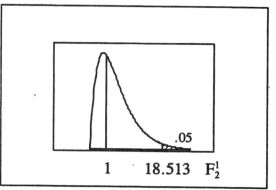

    conclusion:
    Do not reject H_o; there is not sufficient evidence to conclude that emitted greenhouse gases are affected by the type of transmission.

    b. H_o: $\mu_4 = \mu_6 = \mu_8$ [there is no cylinder effect]
    H_1: at least one μ_i is different
    $\alpha = .05$
    C.R. $F > F^2_{2,.05} = 19.000$
    calculations:
    $F = MS_{CYL}/MS_E$
    $= 4.667/.667$
    $= 7.000$
    P-value = .125 [from Minitab]

    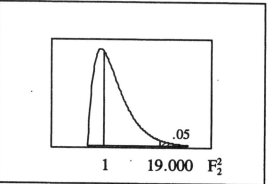

    conclusion:
    Do not reject H_o; there is not sufficient evidence to conclude that emitted greenhouse gases are affected by the number of cylinders.

    c. No; we cannot conclude there in no effect, but only that there is not enough evidence to be 95% certain that there is an effect. The very small sample sizes make it unlikely that the data would provide enough evidence to support any claim, no matter how valid it might be.

Cumulative Review Exercises

1. summary statistics: $n = 52$ $\Sigma x = 5.22$ $\Sigma x^2 = 4.0416$
    a. $\bar{x} = (\Sigma x)/n = 5.22/52 = .100$
    b. $s^2 = [n(\Sigma x^2) - (\Sigma x)^2]/[n(n-1)] = [52(4.0416) - (5.22)^2]/[52(51)] = .06897$
    $s = .263$
    c. from the ordered list:
    $x_1 = 0$
    $Q_1 = P_{25} = (x_{13} + x_{14})/2 = (0+0)/2 = 0$
    $\tilde{x} = P_{50} = (x_{26} + x_{27})/2 = (0+0)/2 = 0$
    $Q_3 = P_{75} = (x_{39} + x_{40})/2 = (.01 + .01)/2$
    $= .01$
    $x_{52} = 1.41$
    d. $\bar{x} + 2s = .100 + 2(.263) = .626$
    The outliers are the two points above .626: .92 and 1.41
    e. One possible histogram is given at the right, with a class width of .25 inches.
    f. No; the distributions are not approximately normal.

    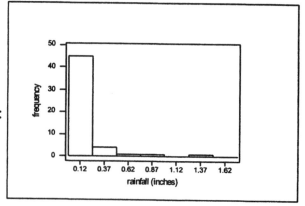

INSTRUCTOR'S SOLUTIONS Chapter 11   S-353

g. let R = rain falls on a selected Monday in Boston
   Since there was rain on 19 of the 52 observed Mondays, estimate P(R) = 19/52 = .365.

2. For each color, the scores are listed in order at the right.
   The following preliminary values are determined.

	Red	Green	Blue
n	20	20	20
Σx	19210	19600	20920
Σx²	19030374	20298724	22859414
x̄	960.5	980.0	1046.0
s²	30482.579	57406.526	51426.000

#	Red	Green	Blue
01	621	499	706
02	699	583	793
03	743	630	821
04	813	780	848
05	855	793	866
06	858	828	892
07	896	907	915
08	896	916	939
09	898	993	996
10	908	996	1004
11	921	1025	1013
12	996	1071	1039
13	1030	1111	1068
14	1092	1121	1097
15	1095	1147	1131
16	1130	1153	1159
17	1133	1180	1244
18	1179	1188	1370
19	1190	1229	1408
20	1257	1450	1611

   a. $\bar{x}_R = 960.5$, $\bar{x}_G = 980.5$, $\bar{x}_B = 1046.0$
      The means are fairly close, but the one for the blue M&M's may be significantly higher – a statistical test is needed to decide for sure.

   b. $\tilde{x}_R = (908+921)/2 = 914.5$
      $\tilde{x}_G = (996+1025)/2 = 1010.5$
      $\tilde{x}_B = (1004+1013)/2 = 1008.5$
      The medians are fairly close, but the one for the red M&M's may be significantly lower – a statistical test is needed to decide for sure.

   c. $s_R = 174.6$, $s_G = 239.6$, $s_B = 226.8$
      The standard deviations are fairly close, but the one for the red M&M's may be significantly lower – a statistical test is needed to decide for sure.

   d. Let the scores from the red M&M's be group 1.
      original claim: $\mu_1 - \mu_2 = 0$
      $\bar{x}_1 - \bar{x}_2 = 960.5 - 980.0 = -19.5$
      H_o: $\mu_1 - \mu_2 = 0$
      H_1: $\mu_1 - \mu_2 \neq 0$
      $\alpha = .05$ [assumed]
      C.R. $t < -t_{19,.025} = -2.093$
           $t > t_{19,.025} = 2.093$
      calculations:
      $t_{\bar{x}_1-\bar{x}_2} = (\bar{x}_1-\bar{x}_2 - \mu_{\bar{x}_1-\bar{x}_2})/s_{\bar{x}_1-\bar{x}_2}$
      $= (-19.5 - 0)/\sqrt{(30483)/20 + (57407)/20}$
      $= -19.6/66.291 = -.294$
      P-value = $2 \cdot P(t_{19} < -.29) > .20$

      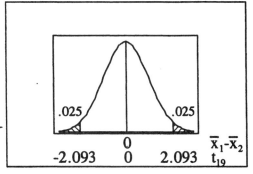

      conclusion:
         Do not reject H_o; there is not sufficient evidence to reject the claim that $\mu_1 - \mu_2 = 0$.

   e. $\bar{x} \pm t_{19,.025} \cdot s/\sqrt{n}$
      $960.5 \pm 2.093 \cdot (174.593)/\sqrt{20}$
      960.5 ± 81.7
      $878.8 < \mu < 1042.2$

   f. Since each group has equal size n, the simplified forms can be used. In addition to the summary statistics above, the following preliminary values are noted.
      $k = 3$ $\bar{\bar{x}} = \Sigma\bar{x}_i/k$
      $n = 20$ $= 995.5$
      $s_{\bar{x}}^2 = \Sigma(\bar{x}_i - \bar{\bar{x}})^2/(k-1)$ $s_p^2 = \Sigma s_i^2/k$
      $= 4015.5/2$ $= 139315.105/3$
      $= 2007.75$ $= 46438.368$

H_o: $\mu_R = \mu_G = \mu_B$
H_1: at least one mean is different
$\alpha = .05$
C.R. F > $F^2_{57,.05}$ = 3.1504
calculations:
$F = ns^2_{\bar{x}}/s^2_p$
$= 20(2007.75)/46438.368$
$= .8647$

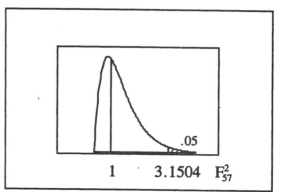

conclusion:
Do not reject H_o; there is not sufficient evidence to conclude that at least one mean is different.

3. a. Let x = the weight of a newborn baby.
normal distribution
$\mu = 7.54$
$\sigma = 1.09$
P(x > 8.00)
$= P(z > .42)$
$= 1 - P(z < .42)$
$= 1 - .6628$
$= .3372$

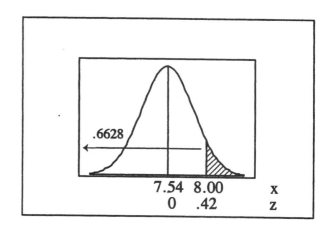

b. normal distribution,
since the original distribution is so
$\mu_{\bar{x}} = \mu = 7.54$
$\sigma_{\bar{x}} = \sigma/\sqrt{n} = 1.09/\sqrt{16} = .2725$
$P(\bar{x} > 8.00)$
$= P(z > 1.69)$
$= 1 - P(z < 1.69)$
$= 1 - .9545$
$= .0455$

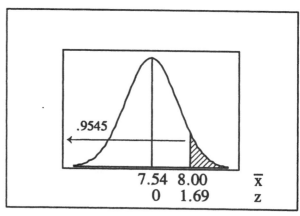

c. Let B = a baby weighs more than 7.54 lbs.
P(B) = .5000, for each birth
$P(B_1 \text{ and } B_2 \text{ and } B_3) = P(B_1) \cdot P(B_2) \cdot P(B_3)$
$= (.5000) \cdot (.5000) \cdot (.5000)$
$= .125$

Chapter 12

Nonparametric Statistics

12-2 Sign Test

1. 10 +'s and 5 –'s
   n = 15 ≤ 25; use C.V. = 3, from Table A-7
   Since x=5>3, do not reject the null hypothesis of no difference.

2. 6 +'s and 16 –'s
   n = 21 ≤ 25; use C.V. = 5, from Table A-7
   Since x=6>5, do not reject the null hypothesis of no difference.

3. 50 +'s and 40 –'s
   n = 90 > 25; use C.V. = -1.96, from the z table
   z = [(x+.5) - (n/2)]/[$\sqrt{n}$/2] = [40.5 - 45]/[$\sqrt{90}$/2] = -4.5/4.743 = -.95
   Since z=-.95>1.96, do not reject the null hypothesis of no difference.

4. 10 +'s and 30 –'s
   n = 40 > 25; use C.V. = -1.96, from the z table
   z = [(x+.5) - (n/2)]/[$\sqrt{n}$/2] = [10.5 - 20]/[$\sqrt{40}$/2] = -9.5/3.162 = -3.00
   Since z=-3.00<-1.96, reject the null hypothesis of no difference.
   Since there were more –'s, conclude that the second variable has the larger scores.

NOTE for n ≤ 25: Table A-7 gives only x_L, the <u>lower</u> critical value for the sign test. Accordingly, the text lets x be the <u>smaller</u> of the number of +'s or the number of –'s and warns the user to use common sense to avoid concluding the reverse of what the data indicates. But the problem's symmetry means that the upper critical value is x_U = n-x_L and that μ_x = n/2, the natural expected value for x when H_o is true. For completeness, this manual indicates those values whenever using the sign test.
    Letting x <u>always</u> be the number of +'s is an alternative approach that maintains the natural agreement between the alternative hypothesis and the critical region and is consistent with the logic and notation of parametric tests.

NOTE for n > 25: The correction for continuity is a conservative adjustment intending to make less likely a false rejection of H_o by shifting the x value .5 units toward the middle. When x is the smaller of the number of +'s or the number of –'s, this always involves replacing x with x+.5. In the alternative approach suggested above, x is replaced with either x+.5 or x-.5 according to which one shifts the value toward the middle
    – i.e., with x+.5 when x < μ_x = (n/2),
       and with x-.5 when x > μ_x = (n/2).
    The formula given is the usual one for converting a score into its standard score, using the correction for continuity, where the mean and standard deviation of the x's are (n/2) and $\sqrt{n}$/2 respectively.

$$z = \frac{x - \mu_x}{\sigma_x} = \frac{(x \pm .5) - (n/2)}{\sqrt{n}/2}$$

NOTE: The manual follows the lower tail method of the text, with the addition of upper tail references as indicated above. Exercises that are naturally upper-tailed (see #9, #10, #11 in this section) are worked both ways.

5. Let the reported heights be group 1.
   claim: median difference ≠ 0

male	1	2	3	4	5	6	7	8	9	10	11	12
R-M	+	+	-	+	+	-	+	-	-	-	+	+

   n = 12: 7+'s and 5-'s
   H_o: median difference = 0
   H_1: median difference ≠ 0
   $\alpha = .05$
   C.R. $x \leq x_{L,12,.025} = 2$
   $x \geq x_{U,12,.025} = 12-2 = 10$
   calculations:
   x = 5 (using less frequent count)
   x = 7 (using + count)

   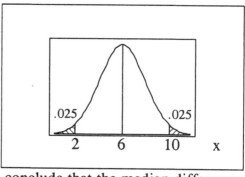

   conclusion:
   Do not reject H_o; there is not sufficient evidence to conclude that the median difference is different from 0.

NOTE: Compare the above results to those obtained in exercise #6 of section 8-4, where a parametric test was used on the same data. Although the text does not so indicate, several exercises in this chapter are re-tests of data that was previously analyzed parametrically; comparisons of the two results are usually informative and provide statistical insights.

6. Let the reported heights be group 1.
   claim: median difference = 0

male	1	2	3	4	5	6	7	8	9	10	11	12
R-M	+	+	+	+	+	+	+	+	-	+	+	

   n = 12: 11+'s and 1-'s
   H_o: median difference = 0
   H_1: median difference ≠ 0
   $\alpha = .05$
   C.R. $x \leq x_{L,12,.025} = 2$
   $x \geq x_{U,12,.025} = 12-2 = 10$
   calculations:
   x = 1 (using less frequent count)
   x = 11 (using + count)

   conclusion:
   Reject H_o; there is not sufficient evidence reject the claim that the median difference is 0 and to conclude that the median difference is different from 0 (in fact, that the median difference is greater than 0).

7. claim: median < 98.6

temp	1	2	3	4	5	6	7	8	9	10	11	12
t-98.6	-	-	0	-	-	+	-	-	-	-	-	-

   n = 11: 1+'s and 10-'s
   H_o: median = 98.6
   H_1: median < 98.6
   $\alpha = .05$
   C.R. $x \leq x_{L,11,.05} = 2$
   calculations:
   x = 1

   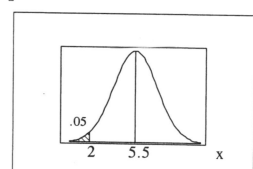

   conclusion:
   Reject H_o; there is sufficient evidence to conclude that the median is less than 98.6.

8. claim: median < 12

   ```
   wgt  | 1  2  3  4  5  6  7  8  9  10 11 12 13 14 15
   w-12 | -  -  -  -  -  -  -  0  +  -  -  -  -  -
   ```
   n = 14: 1 +'s and 13 -'s
   H_0: median = 12
   H_1: median < 12
   $\alpha = .05$
   C.R. $x \leq x_{L,14,.05} = 3$
   calculations:
   $x = 1$

   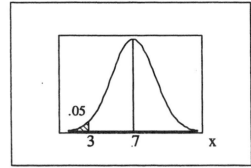

   conclusion:
   Reject H_0; there is sufficient evidence to conclude that the median is less than 12.

9. Let x = the number who did not vote.
   claim: $p < .5$
   701 +'s (voted)   301 -'s (did not vote)
   n = 1002 +'s or -'s
   Since n > 25, use z with
   $\mu_x = n/2 = 1002/2 = 501$
   $\sigma_x = \sqrt{n}/2 = \sqrt{1002}/2 = 15.827$
   H_0: p = .5
   H_1: p < .5
   $\alpha = .05$ [assumed]
   C.R. $z < -z_{.05} = -1.645$
   calculations:
   $x = 301$
   $z_x = [(x+.5)-\mu_x]/\sigma_x$
   $= [301.5 - 501]/15.827$
   $= -199.5/15.827 = -12.605$

   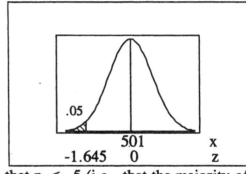

   conclusion:
   Reject H_0; there is sufficient evidence to conclude that p < .5 (i.e., that the majority of the people say they did vote in the election).

NOTE: Exercises #9 is a one-tailed test naturally worded toward the upper tail. For such an exercise, working with only the lower critical region necessitates notation contrary to the its natural wording. The exercise may be worked more naturally in the manner in which it is stated as follows. See the NOTES preceding exercise #5 of this section.

Let x = the number who voted.
claim: $p > .5$
   701 +'s (voted)   301 -'s (did not vote)
   n = 1002 +'s or -'s
Since n > 25, use z with
   $\mu_x = n/2 = 1002/2 = 501$
   $\sigma_x = \sqrt{n}/2 = \sqrt{1002}/2 = 15.827$
H_0: p = .5
H_1: p > .5
$\alpha = .05$ [assumed]
C.R. $z > z_{.05} = 1.645$
calculations:
   $x = 701$
   $z_x = [(x-.5)-\mu_x]/\sigma_x$
   $= [700.5 - 501]/15.827$
   $= 199.5/15.827 = 12.605$

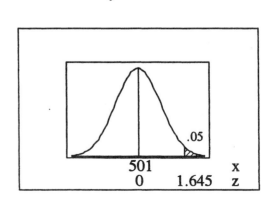

conclusion:
Reject H_o; there is sufficient evidence to conclude that $p > .5$ (i.e., that the majority of the people say they voted in the election).

10. Let x = the number who were <u>not</u> smoking one year later.
claim: $p < .5$
```
    41 +'s (smoking)      30 -'s (not smoking)
    n = 71 +'s or -'s
```
Since n > 25, use z with
$\mu_x = n/2 = 71/2 = 35.5$
$\sigma_x = \sqrt{n}/2 = \sqrt{71}/2 = 4.213$
H_o: $p = .5$
H_1: $p < .5$
$\alpha = .05$
C.R. $z < -z_{.05} = -1.645$
calculations:
x = 30
$z_x = [(x+.5)-\mu_x]/\sigma_x$
$= [30.5 - 35.5]/4.213$
$= -5/4.213 = -1.187$

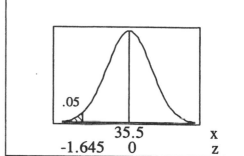

conclusion:
Do not reject H_o; there is not sufficient evidence to conclude that $p < .5$ (i.e., that the majority of the people <u>are</u> smoking one year later).

NOTE: Exercises #10 is a one-tailed test naturally worded toward the upper tail. For such an exercise, working with only the lower critical region necessitates notation contrary to the its natural wording. The exercise may be worked more naturally in the manner in which it is stated as follows. See the NOTES preceding exercise #5 of this section.

Let x = the number who were smoking one year later
claim: $p > .5$
```
    41 +'s (smoking)      31 -'s (not smoking)
    n = 71 +'s or -'s
```
Since n > 25, use z with
$\mu_x = n/2 = 71/2 = 35.5$
$\sigma_x = \sqrt{n}/2 = \sqrt{71}/2 = 4.213$
H_o: $p = .5$
H_1: $p > .5$
$\alpha = .05$
C.R. $z > z_{.05} = 1.645$
calculations:
x = 41
$z_x = [(x-.5)-\mu_x]/\sigma_x$
$= [40.5 - 35.5]/4.213$
$= 5/4.213 = 1.187$

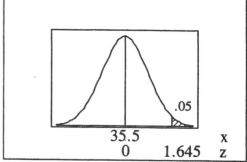

conclusion:
Do not reject H_o; there is not sufficient evidence to conclude that $p > .5$ (i.e., that the majority of the people are smoking one year later).

11. claim: median > 12
```
    33 +'s      1 -        2 0's
    n = 34 +'s or -'s
```
Since n > 25, use z with
$\mu_x = n/2 = 34/2 = 17$
$\sigma_x = \sqrt{n}/2 = \sqrt{34}/2 = 2.915$

H_o: median = 12
H_1: median > 12
α = .05 [assumed]
C.R. z < $-z_{.05}$ = -1.645   OR   C.R. z > $z_{.05}$ = 1.645
calculations: calculations:
  x = 1 (less frequent count)   x = 33 (+ count)
  z_x = [(x+.5)-μ_x]/σ_x z_x = [(x-.5)-μ_x]/σ_x
     = [1.5 - 17]/2.915   = [32.5 - 17]/2.915
     = -15.5/2.915 = -5.316   = 15.5/2.915 = 5.316
conclusion:
  Reject H_o; there is sufficient evidence to conclude that median > 12.
Yes; assuming that the intent of the company is to produce a product with a median weight slightly larger than the stated weight of 12 ounces, it appears that the cans are being filled correctly.

12. claim: median = 3.5
      60 +'s   10 -'s
    n = 70 +'s or -'s
    Since n > 25, use z with
      μ_x = n/2 = 70/2 = 35
      σ_x = $\sqrt{n}$/2 = $\sqrt{70}$/2 = 4.183
    H_o: median = 3.5
    H_1: median ≠ 3.5
    α = .05 [assumed]
    C.R. z < $-z_{.025}$ = -1.960
        z > $z_{.025}$ = 1.960

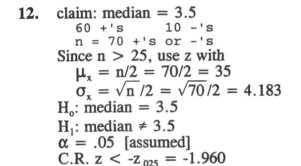

    calculations: OR   calculations:
      x = 10 (less frequent count)   x = 60 (+ count)
      z_x = [(x+.5)-μ_x]/σ_x z_x = [(x-.5)-μ_x]/σ_x
         = [10.5 - 35]/4.183   = [59.5 - 35]/4.183
         = -24.5/4.183 = -5.857   = 24.5/4.183 = 5.857
    conclusion:
      Reject H_o; there is sufficient evidence to reject the claim that median = 3.5 and conclude that median ≠ 3.5 (in fact, that median > 3.5).
    Yes; assuming that the intent of the company is to produce a product with a median weight slightly larger than the stated weight of 3.5 ounces, it appears that the packets are being filled correctly.

13. claim: median > 77
      37 +'s   13 -'s   no 0's
    n = 50 +'s or -'s
    Since n > 25, use z with
      μ_x = n/2 = 50/2 = 25
      σ_x = $\sqrt{n}$/2 = $\sqrt{25}$/2 = 3.536
    H_o: median = 77
    H_1: median > 77
    α = .05 [assumed]
    C.R. z < $-z_{.05}$ = -1.645   OR   C.R. z > $z_{.05}$ = 1.645
    calculations: calculations:
      x = 13 (less frequent count)   x = 37 (+ count)
      z_x = [(x+.5)-μ_x]/σ_x z_x = [(x-.5)-μ_x]/σ_x
         = [13.5 - 25]/3.536   = [36.5 - 25]/3.536
         = -11.5/3.536 = -3.253   = 11.5/3.536 = 3.253
    conclusion:
      Reject H_o; there is sufficient evidence to conclude that median > 77.

14. Let the actual temperatures be group 1 – i.e., difference = actual - predicted.
    claim: median difference ≠ 0
    ```
        13 +'s      15 -'s       3 0's
        n = 28 +'s or -'s
    ```
    Since n > 25, use z with
    $\mu_x = n/2 = 28/2 = 14$
    $\sigma_x = \sqrt{n}/2 = \sqrt{28}/2 = 2.646$
    H_0: median difference = 0
    H_1: median difference ≠ 0
    $\alpha = .05$ [assumed]
    C.R. $z < -z_{.025} = -1.960$
    $z > z_{.025} = 1.960$

    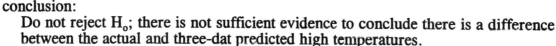

    calculations:
    x = 13
    $z_x = [(x+.5)-\mu_x]/\sigma_x$
    $= [13.5 - 14]/2.646$
    $= -.5/2.646 = -.189$

    conclusion:
    Do not reject H_0; there is not sufficient evidence to conclude there is a difference between the actual and three-dat predicted high temperatures.

15. claim: median < 100
    ```
        40 +'s      60 -'s      21 0's
    ```
    a. Using the usual method: discard the 0's.
    ```
        n = 100 +'s or -'s
    ```
    Since n > 25, use z with
    $\mu_x = n/2 = 100/2 = 50$
    $\sigma_x = \sqrt{n}/2 = \sqrt{100}/2 = 5.000$
    H_0: median = 100
    H_1: median < 100
    $\alpha = .05$
    C.R. $z < -z_{.05} = -1.645$
    calculations:
    x = 40
    $z_x = [(x+.5)-\mu_x]/\sigma_x$
    $= [40.5 - 50]/5.000$
    $= -9.5/5.000 = -1.900$

    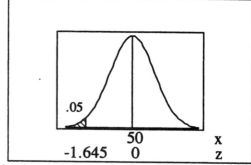

    conclusion:
    Reject H_0; there is sufficient evidence to conclude that median < 100.

    b. Using the second method: count half the zeros each way (and discarding the odd zero).
    ```
        50 +'s      70 -'s
        n = 120 +'s or -'s
    ```
    Since n > 25, use z with
    $\mu_x = n/2 = 120/2 = 60$
    $\sigma_x = \sqrt{n}/2 = \sqrt{120}/2 = 5.477$
    H_0: median = 0
    H_1: median < 0
    $\alpha = .05$
    C.R. $z < -z_{.05} = -1.645$
    calculations:
    x = 50
    $z_x = [(x+.5)-\mu_x]/\sigma_x$
    $= [50.5 - 60]/5.477$
    $= -9.5/5.477 = -1.734$

    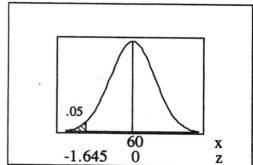

    conclusion:
    Reject H_0; there is sufficient evidence to conclude that median < 100.

c. Using the third method: count the zeros in a one-tailed test to favor H_o.
   61 +'s   60 -'s
   n = 161 +'s or -'s
   Since n > 25, use z with
   μ_x = n/2 = 121/2 = 60.5
   σ_x = $\sqrt{n}$ /2 = $\sqrt{121}$ /2 = 5.500
   H_o: median = 0
   H_1: median < 0
   α = .05
   C.R. z < -$z_{.05}$ = -1.645
   calculations:
   x = 60
   z_x = [(x+.5)-μ_x]/σ_x
       = [60.5 - 60.5]/5.500
       = 0/5.500 = 0
   conclusion:
   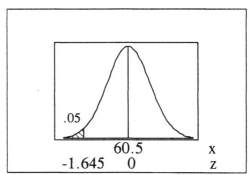
   Do not reject H_o; there is not sufficient evidence to conclude that median < 100.

16. Use the binomial table for p=.5.
    Reject for lower-tailed x values summing to less than .03.
    For 1 ≤ n ≤ 5, P(x=0) > .03
    For 6 ≤ n ≤ 8, P(x≤0) ≤ .03 but P(x≤1) > .03
    For 9 ≤ n ≤ 11, P(x≤1) ≤ .03 but P(x≤2) > .03
    For 12 ≤ n ≤ 13, P(x≤2) ≤ .03 but P(x≤3) > .03
    For 14 ≤ n ≤ 15, P(x≤3) ≤ .03 but P(x≤4) > .03
    In summary, the desired critical values are as follows.
    n: 1 2 3 4 5 6 7 8 9 10 11 12 13 14 15
    x: * * * * * 0 0 0 1 1 1 2 2 3 3

17. Let p = the proportion of women hired.
    claim: p < .5
    a. using the sign test
       18 +'s   36 -'s
       n = 54 +'s or -'s
       Since n > 25, use z with
       μ_x = n/2 = 54/2 = 27
       σ_x = $\sqrt{n}$ /2 = $\sqrt{54}$ /2 = 3.674
       H_o: p = .5
       H_1: p < .5
       α = .01
       C.R. z < -$z_{.01}$ = -2.326
       calculations:
       x = 18
       z_x = [(x+.5)-μ_x]/σ_x
           = [18.5 - 27]/3.674
           = -8.5/3.674 = -2.313
       conclusion:
       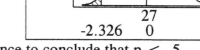
       Do not reject H_o; there is not sufficient evidence to conclude that p < .5.
    b. Using the exact binomial distribution, P-value = P(x≤18|n=54 and p=.5).
       The test in part (a) uses the normal approximation
       P(x≤18) = P_c(x<18.5) = P(z<-2.313) ≈ .0104 > .01 [do not reject H_o].
       Using the exact binomial requires using the binomial formula 19 times.
       P(x) = [n!/x!(n-x)!]·p^x·$(1-p)^{n-x}$
       P(x) = [54!/x!54!]·$(.5)^x$·$(.5)^{54-x}$ for x = 0,1,2,...,18
       The statDisk program indicates
       P(x≤18) = P(x=0) + P(x=1) +...+ P(x=18) = .00992 < .01 [reject H_o].

12-3 Wilcoxon Signed-Ranks Test for Matched Pairs

NOTE: Table A-8 gives only T_L, the <u>lower</u> critical value for the signed-ranks test. Accordingly, the text lets T be the <u>smaller</u> of the sum of positive ranks or the sum of the negative ranks and warns the user to use common sense to avoid concluding the reverse of what the data indicates. But the problem's symmetry means that the upper critical value is $T_U = \Sigma R - T_L$ and that $\mu_T = \Sigma R/2$, the natural expected value for T when H_o is true.

Letting T <u>always</u> be the sum of the positive ranks is an alternative approach that maintains the natural agreement between the alternative hypothesis and the critical region and is consistent with the logic and notation of parametric tests. The manual follows the lower tail method of the text, with the addition of upper tail references as indicated above.

NOTE 2: This manual follows the text and the directions to the exercises of this section by using "the populations have the same distribution" as the null hypothesis. To be more precise, the signed-rank test doesn't test "distributions" but tests the "location" (i.e., central tendency – as opposed to variation) of distributions. The test discerns whether one group taken as a whole tends to have higher or lower scores than another group taken as a whole. The test does not discern whether one group is more variable than another. This distinction is reflected in the wording of the conclusion when rejecting H_o. Notice also that the signed-rank test measures overall differences between the groups and <u>not</u> whether the two groups give the same results for individuals subjects. If the group 1 scores, for example, were higher for half the subjects and lower for the other half of the subjects by the same amounts, then $\Sigma R-$ would equal $\Sigma R+$ (so we could not reject H_o), but the distributions would be very different.

NOTE 3: This manual uses a minus sign preceding ranks associated with negative differences. The ranks themselves are not negative, but the use of the minus sign helps to organize the information.

1. claim: the populations have the same distribution

   ```
   x-y  |  0   -1    2    5    8    9   10    10
   R    |  -   -1    2    3    4    5  6.5   6.5
   ΣR- =  1
   ΣR+ = 27              n = 7 non-zero ranks
   ΣR  = 28
   ```
   check: $\Sigma R = n(n+1)/2 = 7(8)/2 = 28$
   H_o: the populations have the same distribution
   H_1: the populations have different distributions
   $\alpha = .05$
   C.R. $T \leq T_{L,7,.025} = 2$
   $T \geq T_{U,7,.025} = 28-2 = 26$
   calculations
   T = 1 (using the smaller ranks)
   T = 27 (using the positive ranks)
   conclusion:
   Reject H_o; there is sufficient evidence to reject the claim that the populations have the same distribution and to conclude that they have different distributions (in fact, that the x scores are greater).

2. claim: the populations have the same distribution

   ```
   x-y  |  0   -2   -3   -5   -7   -8   10   -12   -12
   R    |  -   -1   -2   -3   -4   -5    6  -7.5  -7.5
   ΣR- = 30
   ΣR+ =  6              n = 8 non-zero ranks
   ΣR  = 36
   ```
   check: $\Sigma R = n(n+1)/2 = 8(9)/2 = 36$

H_o: the populations have the same distribution
H_1: the populations have different distributions
$\alpha = .05$
C.R. $T \leq T_{L,8,.025} = 4$
$T \geq T_{U,8,.025} = 36-4 = 32$
calculations
$T = 6$

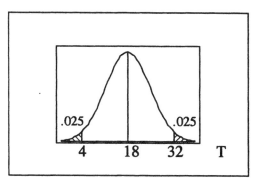

conclusion:
Do not reject H_o; there is not sufficient evidence to reject the claim that the populations have the same distribution.

3. Let the reported heights be group 1.
claim: the populations have the same distribution
```
R-M|  .1   1.1  -1.9   1.7   .7  -.6   .5  -3.0  -1.6  -11.2   1.0   1.2
R  |  1     6   -10    9    4   -3    2   -11    -8    -12     5     7
ΣR- = 44
ΣR+ = 34                n = 12 non-zero ranks
ΣR  = 78
```
check: $\Sigma R = n(n+1)/2 = 12(13)/6 = 78$
H_o: the populations have the same distribution
H_1: the populations have different distributions
$\alpha = .05$
C.R. $T \leq T_{L,12,.025} = 14$
$T \geq T_{U,12,.025} = 78-14 = 64$
calculations
$T = 34$

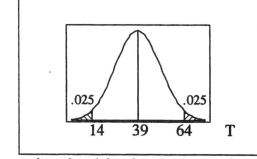

conclusion:
Do not reject H_o; there is not sufficient evidence to reject the claim that the populations have the same distribution.

4. Let the reported heights be group 1.
claim: the populations have the same distribution
```
R-M|  1.2    .1   7.95    .4   1.8   .1   1.6   .1   1.4   -.9    .4   1.2
R  |  7.5    2    12     4.5   11    2    10    2    9     -6    4.5   7.5
ΣR- =  6
ΣR+ = 72                n = 12 non-zero ranks
ΣR  = 78
```
check: $\Sigma R = n(n+1)/2 = 12(13)/6 = 78$
H_o: the populations have the same distribution
H_1: the populations have different distributions
$\alpha = .05$
C.R. $T \leq T_{L,12,.025} = 14$
$T \geq T_{U,12,.025} = 78-14 = 64$
calculations
$T = 6$ (using the smaller ranks)
$T = 72$ (using the positive ranks

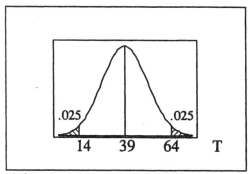

conclusion:
Reject H_o; there is sufficient evidence to reject the claim that the populations have the same distribution and to conclude that they have different distributions (in fact, that the reported heights are greater).

5. Let Sitting be group 1.
   claim: the populations have the same distribution

pair	1	2	3	4	5	6	7	8	9	10
Si-Su	.99	1.60	.98	.82	1.01	1.54	.21	.70	1.67	1.32
R	5	9	4	3	6	8	1	2	10	7

   $\Sigma R- = 0$
   $\Sigma R+ = 55$
   $\Sigma R = 55$ $n = 10$ non-zero ranks
   check: $\Sigma R = n(n+1)/2 = 10(11)/2 = 55$
   H_o: the populations have the same distribution
   H_1: the populations have different distributions
   $\alpha = .05$
   C.R. $T \le T_{L,10,.025} = 8$
   $T \ge T_{U,10,.025} = 55-8 = 47$
   calculations:
      $T = 0$ (using the smaller ranks)
      $T = 55$ (using the positive ranks)

   conclusion:
      Reject H_o; there is sufficient evidence to reject the claim that the populations have the same distribution and to conclude that they have different distributions (in fact, the sitting positions have larger capacity scores).

6. Let Before be group 1.
   claim: the populations have different distributions (i.e., the drug has an effect)

pair	A	B	C	D	E	F	G	H	I	J	K	L
B-A	9	4	21	3	20	31	17	26	19	10	23	33
R	3	2	8	1	7	11	5	10	6	4	9	12

   $\Sigma R- = 0$
   $\Sigma R+ = 78$
   $\Sigma R = 78$ $n = 12$ non-zero ranks
   check: $\Sigma R = n(n+1)/2 = 12(13)/2 = 78$
   H_o: the populations have the same distribution
   H_1: the populations have different distributions
   $\alpha = .05$ [assumed]
   C.R. $T \le T_{L,12,.025} = 14$
   $T \ge T_{U,12,.025} = 78-14 = 64$
   calculations:
      $T = 0$ (using the smaller ranks)
      $T = 78$ (using the positive ranks)
   conclusion:
      Reject H_o; there is sufficient evidence to reject the claim that the populations have the same distribution and to conclude that they have different distributions (in fact, the Before group has higher scores).

7. The 31 "actual minus predicted" differences and their signed ranks are as follows.

   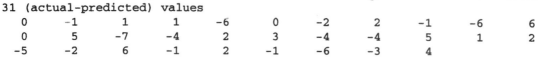

   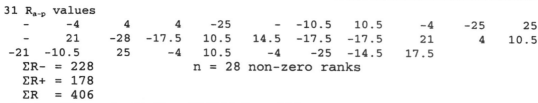

   $\Sigma R- = 228$
   $\Sigma R+ = 178$ $n = 28$ non-zero ranks
   $\Sigma R = 406$
   check: $\Sigma R = n(n+1)/2 = 28(29)/2 = 406$

H_o: the populations have the same distribution
H_1: the populations have different distributions
$\alpha = .05$ [assumed]
C.R. $T \leq T_{L,28,.025} = 117$
$T \geq T_{U,28,.025} = 406-117 = 289$
calculations:
$T = 178$

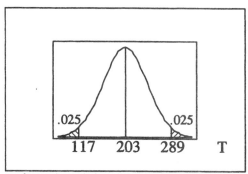

conclusion:
   Do not reject H_o; there is not sufficient evidence to reject the claim that the populations have the same distribution.

8. Ignoring the 16 movies with times of 0 for both alcohol and tobacco use leaves 34 movies to be analyzed. The 34 "tobacco minus alcohol" differences and the signed ranks of the differences are as follows.

```
34 (tobacco-alcohol) values
 143    88   540    -4    37   158    18  -113   248    17  -131
 -39   -34  -249    74     9   -76  -123    -1    -1   -23   168
  -4   132    90     1     1   117     5    91    81    -4    55
 -22
```

```
34 R_{t-a} values
  29    21    34    -6    15    30    11   -24    32    10   -27
 -16   -14   -33    18     9   -19   -26  -2.5  -2.5   -13    31
  -6    28    22   2.5   2.5    25     8    23    20    -6    17
 -12
```
$\Sigma R- = 207$ n = 34 non-zero ranks
$\Sigma R+ = 388$
$\Sigma R = 595$
check: $\Sigma R = n(n+1)/2 = 34(35)/2 = 595$
For n = 34 ranks, use the z approximation with
$\mu_T = n(n+1)/4$
$ = 34(35)/4 = 297.5$
$\sigma_T = \sqrt{[n(n+1)(2n+1)/24]}$
$ = \sqrt{[34(35)(69)/24]} = 58.491$

H_o: the populations have the same distribution
H_1: the populations have different distributions
$\alpha = .05$
C.R. $z < -z_{.025} = -1.960$
$z > z_{.025} = 1.960$
calculations:
$T = 207$
$z = (T - \mu_T)/\sigma_T$
$ = (207 - 297.5)/58.491$
$ = -90.5/58.491 = -1.547$

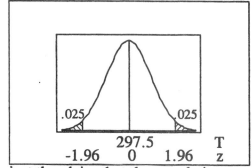

conclusion:
   Do not reject H_o, there is not sufficient evidence reject the claim that the populations have the same distribution.
For such films that show any tobacco or alcohol use, there appears to be no difference in the times those products are used on screen.

9. The 106 differences obtained by subtracting 98.6 from each temperature, and the signed ranks of the differences are as follows.

```
106 (x-98.6) values
  0      0    -.6   -.6    .4   -.2   -.2   -.2   -.2    0     0
 .2      0   -1.6  -1.6   0.2  -1.0  -.9    .2   -.6   -.6   -.3
-.1    -1.3    .1  -1.2    .3    0    .9  -1.1  -1.3  -1.0   -.4
1.0     .1    .8   -.4   -.6    0    0   -1.4   -.2    0    -.4
-.6    -.8   -.6   -.2    0     0   -.8    .4  -2.1  -1.0   -.6
-1.7  -1.0  -1.5   -.7   -.2  -1.3  -.6  -1.1  -1.0   -.4   -.1
 .2     .1   -.8   -.6  -1.5  -1.2   .8   -.2    0    -.2   -.1
 0     -.3    .1    .2    .5    0   -.7    .2   -.6    .1   -.1
 .3    -.2    0   -1.5   -.7    .2   .1   -1.0   -.4    .6   -.8
-.6    -.2   -.8   -.2  -1.2   -.6 -1.6
```

```
106 R_{x-98.6} values
  -      -   -48.5 -48.5  37.0 -16.5 -16.5 -16.5 -16.5    -      -
 26.0    -   -88.0 -88.0  26.0 -71.0 -66.5  26.0 -48.5 -48.5 -30.5
 -5.5 -81.0   5.5 -78.0  32.5    -   66.5 -75.5 -81.0 -71.0 -37.0
 71.0   5.5  64.5 -37.0 -48.5    -     -   -83.0 -16.5    -  -37.0
-48.5 -61.0 -48.5 -16.5    -     -  -61.0  37.0 -91.0 -71.0 -48.5
-90.0 -71.0 -85.0 -57.0 -16.5 -81.0 -48.5 -75.5 -71.0 -37.0  -5.5
 26.0   5.5 -61.0 -48.5 -85.0 -78.0  64.5 -16.5    -  -16.5  -5.5
  -   -30.5   5.5  26.0  41.0    -  -57.0  26.0 -48.5   5.5  -5.5
 32.5 -16.5    -  -85.0 -57.0  26.0   5.5 -71.0 -37.0  48.5 -61.0
-48.5 -16.5 -61.0 -16.5 -78.0 -48.5 -88.0
```
ΣR- = 3476   n = 91 non-zero's to be ranked
ΣR+ = 710
ΣR = 4186
check: ΣR = n(n+1)/2 = (91)(92)/2 = 4186
For n = 91 ranks, use the z approximation with
  $\mu_T = n(n+1)/4$
       $= 91(92)/4 = 2093$
  $\sigma_T = \sqrt{[n(n+1)(2n+1)/24]}$
       $= \sqrt{[91(92)(183)/24]} = 252.66$

H_o: median = 98.6
H_1: median ≠ 98.6
$\alpha = .05$
C.R. $z < -z_{.025} = -1.960$
     $z > z_{.025} = 1.960$
calculations:
  T = 710
  $z = (T - \mu_T)/\sigma_T$
    $= (710 - 2093)/252.66$
    $= -1383/252.66 = -5.473$

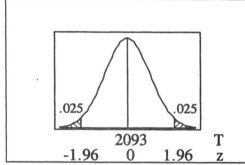

conclusion:
Reject H_o, there is sufficient evidence reject the claim that median = 98.6 and to conclude that median ≠ 98.6 (in fact, median < 98.6).

12-4 Wilcoxon Rank-Sum Test for Two Independent Samples

NOTE: As in the previous section, the manual follows the wording in the text and tests the hypothesis that "the populations have the same distribution" with the understanding that the test detects only differences in location and not differences in variability. In addition, always letting R = ΣR_1 guarantees agreement and consistency between the directions of H_1 and the C.R. as in the previous chapters.

1. Below are the ordered scores for each group.
   claim: the populations have the same distribution

grp 1	R	grp 2	R
1	1	2	2
3	3	5	5
4	4	7	7
6	6	9	9
8	8	11	10
12	11	13	12
15	14	14	13
16	15	18	17
17	16	19	18
22	20	20	19
26	22.5	25	21
	120.5	26	22.5
			155.5

   $n_1 = 11 \qquad \Sigma R_1 = 120.5$
   $n_2 = 12 \qquad \Sigma R_2 = 155.5$

   $n = \Sigma n = 23 \qquad \Sigma R = 276$

   check: $\Sigma R = n(n+1)/2$
   $= 23(24)/2$
   $= 276$

   $R = \Sigma R_1 = 120.5$
   $\mu_R = n_1(n+1)/2$
   $= 11(24)/2$
   $= 132$
   $\sigma_R^2 = n_1 n_2(n+1)/12$
   $= (11)(12)(24)/12$
   $= 264$

   H_o: the populations have the same distribution
   H_1: the populations have different distributions
   $\alpha = .05$
   C.R. $z < -z_{.025} = -1.96$
   $z > z_{.025} = 1.96$
   calculations:
   $z_R = (R - \mu_R)/\sigma_R$
   $= (120.5 - 132)/\sqrt{264}$
   $= -11.5/16.248 = -.708$

   conclusion:
   Do not reject H_o; there is not sufficient evidence to reject the claim that the populations have the same distribution.

   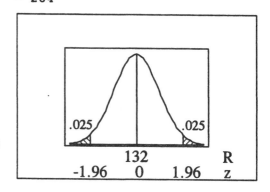

2. Below are the ordered scores for each group.
   claim: the populations have the same distribution

grp 1	R	grp 2	R
1	1	22	10.5
3	2	25	12
4	3	28	14
6	4	33	15
8	5	34	16
12	6	35	17
15	7	37	18
16	8	39	19
17	9	41	20
22	10.5	43	21
26	13	45	22
	68.5		184.5

   $n_1 = 11 \qquad \Sigma R_1 = 68.5$
   $n_2 = 11 \qquad \Sigma R_2 = 184.5$

   $n = \Sigma n = 22 \qquad \Sigma R = 253$

   check: $\Sigma R = n(n+1)/2$
   $= 22(23)/2$
   $= 253$

   $R = \Sigma R_1 = 68.5$
   $\mu_R = n_1(n+1)/2$
   $= 11(23)/2$
   $= 126.5$
   $\sigma_R^2 = n_1 n_2(n+1)/12$
   $= (11)(11)(23)/12$
   $= 231.917$

   H_o: the populations have the same distribution
   H_1: the populations have different distributions
   $\alpha = .05$
   C.R. $z < -z_{.025} = -1.96$
   $z > z_{.025} = 1.96$
   calculations:
   $z_R = (R - \mu_R)/\sigma_R$
   $= (68.5 - 126.5)/\sqrt{231.917}$
   $= -58/15.229 = -3.809$

   conclusion:
   Reject H_o; there is sufficient evidence to reject the claim that the populations have the same distribution and to conclude that the population distributions are different (in fact, that population 1 has the smaller values).

   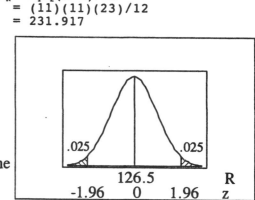

S-368 INSTRUCTOR'S SOLUTIONS Chapter 12

3. Below are the ordered scores for each group. The group listed first is considered group 1.
claim: the populations have the same distribution

O-C	R	Con	R
.210	1	.334	7.5
.287	2	.349	11
.288	3	.402	12
.304	4	.413	14
.305	5	.429	15
.308	6	.445	16
.334	7.5	.460	18.5
.340	9	.476	20.5
.344	10	.483	21
.407	13	.501	22
.455	17	.519	23
.463	19	.594	24
	96.5		203.5

$n_1 = 12$ $\Sigma R_1 = 96.5$
$n_2 = 12$ $\Sigma R_2 = 203.5$

$n = \Sigma n = 24$ $\Sigma R = 300.0$

check: $\Sigma R = n(n+1)/2$
$= 24(25)/2$
$= 300$

$R = \Sigma R_1 = 96.5$
$\mu_R = n_1(n+1)/2$
$= 12(25)/2$
$= 150$
$\sigma_R^2 = n_1 n_2 (n+1)/12$
$= (12)(12)(25)/12$
$= 300$

H_0: the populations have the same distribution
H_1: the populations have different distributions
$\alpha = .01$
C.R. $z < -z_{.005} = -2.575$
    $z > z_{.005} = 2.575$
calculations:
$z_R = (R - \mu_R)/\sigma_R$
$= (96.5 - 150)/\sqrt{300}$
$= -53.5/17.321 = -3.089$

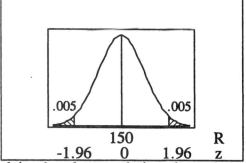

conclusion:
Reject H_0; there is sufficient evidence to reject the claim that the populations have the same distribution and to conclude that they have different distributions (in fact, that population 1 has smaller volumes).
Based on this result, we can be 99% confident that there are biological factors related to obsessive-compulsive disorders.

4. Below are the ordered scores for each group. The group listed first is considered group 1.
claim: the populations have different distributions

L to H	R	H to L	R
25	1	225	5
169	2	300	6
175	3	354	7
200	4	372	8
500	11.5	400	9
560	13	428	10
800	16	500	11.5
842	17	640	14
856	18	750	15
1110	19	1200	20
1252	21	1500	22
1560	23	1876	24
2040	26	2000	25
4000	29.5	2050	27
5000	31	3600	28
5635	32	4000	29.5
10000	33	23410	34
40320	35	42000	36
42200	37	49000	38
49654	39	52836	40
	411.0	64582	41
		100000	42
			492.0

$n_1 = 20$ $\Sigma R_1 = 411$
$n_2 = 22$ $\Sigma R_2 = 492$

$n = \Sigma n = 42$ $\Sigma R = 903$

check: $\Sigma R = n(n+1)/2$
$= 42(43)/2$
$= 903$

$R = \Sigma R_1 = 411$

$\mu_R = n_1(n+1)/2$
$= 20(43)/2$
$= 430$

$\sigma_R^2 = n_1 n_2 (n+1)/12$
$= (20)(22)(43)/12$
$= 1576.67$

H_o: the populations have the same distribution
H_1: the populations have different distributions
$\alpha = .05$ [assumed]
C.R. $z < -z_{.025} = -1.960$
$z > z_{.025} = 1.960$
calculations:
$z_R = (R - \mu_R)/\sigma_R$
$= (411 - 430)/\sqrt{1576.67}$
$= -19/39.707 = -.479$

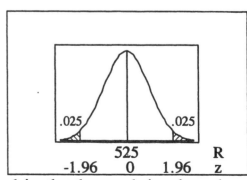

conclusion:
Do not reject H_o; there is not sufficient evidence to conclude that the populations have different distributions.

5. Below are the ordered scores for each group. The group listed first is considered group 1.

E to D	R	D to E	R
7.10	1	26.63	11
16.32	2	26.68	12
20.60	3	27.24	14
21.06	4	27.62	15
21.13	5	29.34	21
21.96	6	29.49	22
24.23	7	30.20	24
24.64	8	30.26	25
25.49	9	32.34	29
26.43	10	32.54	30
26.69	13	33.53	34
27.85	16	33.62	35
28.02	17	34.02	36
28.71	18	35.32	37
28.89	19	35.91	38
28.90	20	42.91	41
30.02	23		424
30.29	26		
30.72	27		
31.73	28		
32.83	31		
32.86	32		
33.31	33		
38.81	39		
39.29	40		
	437		

$n_1 = 25$ $\Sigma R_1 = 437$
$n_2 = 16$ $\Sigma R_2 = 424$

$n = \Sigma n = 41$ $\Sigma R = 861$

check: $\Sigma R = n(n+1)/2$
$= 41(42)/2$
$= 861$

$R = \Sigma R_1 = 437$

$\mu_R = n_1(n+1)/2$
$= 25(42)/2$
$= 525$

$\sigma_R^2 = n_1 n_2 (n+1)/12$
$= (25)(16)(42)/12$
$= 1400$

H_o: the populations have the same distribution
H_1: the populations have different distributions
$\alpha = .05$
C.R. $z < -z_{.025} = -1.96$
$z > z_{.025} = 1.96$
calculations:
$z_R = (R - \mu_R)/\sigma_R$
$= (437 - 525)/\sqrt{1400}$
$= -88/37.417 = -2.352$

conclusion:
Reject H_o; there is sufficient evidence to reject the claim that the populations have the same distribution and to conclude that they have different distributions (in fact, that population 1 has lower scores).

6. Below are the ordered scores for each group. The group listed first is considered group 1.
claim: the populations have the same distribution

Red	R	Bro	R
.870	6	.856	1
.872	8	.858	2
.874	9	.860	3
.882	12	.866	4
.888	13	.867	5
.891	15	.871	7
.897	16.5	.875	10
.898	18.5	.876	11
.908	25	.889	14
.908	25	.897	16.5
.908	25	.898	18.5
.911	29	.900	20
.912	30	.902	21.5
.913	31	.902	21.5
.920	35.5	.904	23
.924	39.5	.909	27.5
.924	39.5	.909	27.5
.933	45	.914	32.5
.936	46.5	.914	32.5
.952	48	.919	34
.983	52	.920	35.5
	569.0	.921	37
		.923	38
		.928	41
		.930	42.5
		.930	42.5
		.932	44
		.936	46.5
		.955	49
		.965	50
		.976	51
		.988	53
		1.033	54
			916.0

$n_1 = 21$ $\Sigma R_1 = 569$
$n_2 = 33$ $\Sigma R_2 = 916$
$n = \Sigma n = 54$ $\Sigma R = 1485$

check: $\Sigma R = n(n+1)/2$
$= 54(55)/2$
$= 1485$

$R = \Sigma R_1 = 569$

$\mu_R = n_1(n+1)/2$
$= 21(55)/2$
$= 577.5$

$\sigma_R^2 = n_1 n_2 (n+1)/12$
$= (21)(33)(55)/12$
$= 3176.25$

H_0: the populations have the same distribution
H_1: the populations have different distributions
$\alpha = .05$
C.R. $z < -z_{.025} = -1.96$
$z > z_{.025} = 1.96$
calculations:
$z_R = (R - \mu_R)/\sigma_R$
$= (569 - 577.5)/\sqrt{3176.25}$
$= -8.5/56.358 = -.151$
conclusion:
Do not reject H_0; there is not sufficient evidence
to reject the claim that the populations have the same distribution.

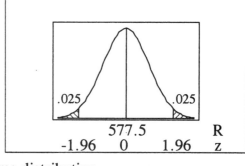

7. See the data summary on the data page.
claim: the populations have different distributions
preliminary calculations:

$n_1 = 12$ $\Sigma R_1 = 86.5$
$n_2 = 12$ $\Sigma R_2 = 213.5$
$n = \Sigma n = 24$ $\Sigma R = 300.0$

$R = \Sigma R_1 = 86.5$
$\mu_R = n_1(n+1)/2 = 12(25)/2 = 150$
$\sigma_R^2 = n_1 n_2 (n+1)/12 = (12)(12)(25)/12 = 300$

check: $\Sigma R = n(n+1)/2$
$= 24(25)/2$
$= 300$

H_0: the populations have the same distribution
H_1: the populations have different distributions
$\alpha = .05$
C.R. $z < -z_{.025} = -1.96$
$z > z_{.025} = 1.96$
calculations:
$z_R = (R - \mu_R)/\sigma_R$
$= (86.5 - 150)/\sqrt{300}$
$= -63.5/17.321 = -3.666$
conclusion:
Reject H_0; there is sufficient evidence to conclude
the populations have different distributions (in fact, that the Tolstoy scores are higher).

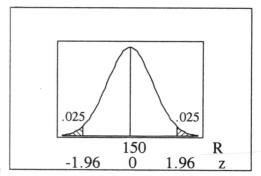

INSTRUCTOR'S SOLUTIONS Chapter 12 S-371

Below are the ordered scores for exercises #7, #8, #9 and #10. The group listed first is considered group 1.

	exercise #7				exercise #8				exercise #9				exercise #10			
#	Rowl	R	Tols	R	McGw	R	Bond	R	w	R	e	R	male	R	fema	R
01	3.2	1.0	5.9	10.5	340	3.0	320	1.5	15	4.5	12	1.0	19.6	6.0	17.7	1.0
02	3.7	2.0	5.9	10.5	341	4.0	320	1.5	16	9.5	15	4.5	19.9	9.0	18.3	2.0
03	4.1	3.0	6.1	12.5	350	7.0	347	5.0	16	9.5	15	4.5	20.7	12.5	19.2	3.0
04	4.4	4.5	7.2	16.0	350	7.0	350	7.0	16	9.5	15	4.5	21.5	16.0	19.3	4.0
05	4.4	4.5	7.7	17.0	360	11.5	360	11.5	16	9.5	15	4.5	21.6	17.0	19.6	6.0
06	4.9	6.0	8.2	18.0	360	11.5	360	11.5	17	13.5	15	4.5	22.7	21.0	19.6	6.0
07	5.2	7.0	8.4	19.0	360	11.5	360	11.5	17	13.5	17	13.5	23.2	24.0	19.8	8.0
08	5.6	8.0	8.6	20.0	369	17.0	361	15.0	18	21.0	17	13.5	23.3	25.0	20.5	10.0
09	5.7	9.0	9.8	21.0	370	20.5	365	16.0	18	21.0	18	21.0	23.4	26.0	20.6	11.0
10	6.1	12.5	10.9	22.0	370	20.5	370	20.5	18	21.0	18	21.0	23.5	27.5	20.7	12.5
11	6.7	14.0	11.0	23.0	370	20.5	370	20.5	18	21.0	18	21.0	23.8	30.5	21.2	14.0
12	6.9	15.0	11.5	24.0	370	20.5	375	25.5	18	21.0	18	21.0	23.8	30.5	21.4	15.0
13		86.5		213.5	377	28.0	375	25.5	18	21.0	18	21.0	24.2	34.0	21.9	18.5
14					380	33.5	375	25.5	19	33.0	19	33.0	24.5	35.0	21.9	18.5
15					380	33.5	375	25.5	19	33.0	19	33.0	24.6	36.5	22.0	20.0
16					380	33.5	380	33.5	19	33.0	19	33.0	24.6	36.5	22.8	22.5
17					380	33.5	380	33.5	19	33.0	19	33.0	25.2	39.5	22.8	22.5
18					380	33.5	380	33.5	20	45.0	19	33.0	25.5	41.0	23.5	27.5
19					385	40.0	380	33.5	21	53.0	19	33.0	25.6	42.0	23.8	30.5
20					385	40.0	380	33.5	21	53.0	19	33.0	26.2	44.5	23.8	30.5
21					388	42.0	385	40.0	21	53.0	19	33.0	26.2	44.5	24.0	33.0
22					390	45.5	390	45.5	22	59.5	19	33.0	26.3	46.0	25.1	38.0
23					390	45.5	390	45.5	22	59.5	20	45.0	26.4	47.5	25.2	39.5
24					390	45.5	391	49.0	22	59.5	20	45.0	26.4	47.5	26.0	43.0
25					390	45.5	394	50.0	23	66.5	20	45.0	26.6	50.0	26.5	49.0
26					398	52.0	396	51.0	23	66.5	20	45.0	26.7	51.0	27.5	56.0
27					400	55.5	400	55.5	24	73.5	20	45.0	26.9	52.0	28.5	60.0
28					400	55.5	400	55.5	24	73.5	20	45.0	27.0	53.0	28.7	61.5
29					409	61.0	400	55.5	24	73.5	20	45.0	27.1	54.0	28.9	63.0
30					410	69.0	400	55.5	24	73.5	20	45.0	27.4	55.0	29.1	64.0
31					410	69.0	404	59.0	24	73.5	20	45.0	27.8	57.0	29.7	65.0
32					410	69.0	405	60.0	25	80.0	20	45.0	28.1	58.0	29.8	66.0
33					410	69.0	410	69.0	25	80.0	21	53.0	28.3	59.0	29.9	67.0
34					410	69.0	410	69.0	26	85.0	21	53.0	28.7	61.5	30.9	68.5
35					420	89.0	410	69.0	26	85.0	22	59.5	30.9	68.5	31.0	70.0
36					420	89.0	410	69.0	27	88.5	22	59.5	31.4	71.0	31.7	72.0
37					420	89.0	410	69.0	29	92.5	22	59.5	31.9	73.0	33.5	77.0
38					420	89.0	410	69.0	29	92.5	22	59.5	32.1	74.0	37.7	78.0
39					420	89.0	410	69.0	30	95.5	22	59.5	33.1	75.0	40.6	79.0
40					423	96.0	410	69.0	30	95.5	23	66.5	33.2	76.0	44.9	80.0
41					425	97.0	410	69.0	31	100.0	23	66.5		1727.5		1512.5
42					430	104.5	410	69.0	31	100.0	23	66.5				
43					430	104.5	411	77.0	32	104.0	23	66.5				
44					430	104.5	415	78.5	33	106.5	24	73.5				
45					430	104.5	415	78.5	33	106.5	24	73.5				
46					430	104.5	416	80.0	34	109.0	24	73.5				
47					430	104.5	417	81.5	38	116.5	25	80.0				
48					430	104.5	417	81.5	39	118.0	25	80.0				
49					440	117.0	420	89.0	40	119.0	25	80.0				
50	(McGw/Bond continued)				440	117.0	420	89.0	41	121.0	26	85.0	(east continued)			
51	470	134.0	430	104.5	440	117.0	420	89.0	41	121.0	26	85.0	32	104.0		
52	470	134.0	430	104.5	450	124.0	420	89.0	42	123.5	26	85.0	32	104.0		
53	470	134.0	435	111.5	450	124.0	420	89.0	43	125.0	27	88.5	34	109.0		
54	478	136.0	435	111.5	450	124.0	420	89.0	45	126.0	28	90.5	34	109.0		
55	480	137.0	436	113.0	450	124.0	420	89.0	48	129.0	28	90.5	35	111.5		
56	500	139.0	440	117.0	452	127.0	420	89.0	66	131.0	30	95.5	35	111.5		
57	510	140.5	440	117.0	458	129.0	429	98.0		3861.0	30	95.5	36	113.0		
58	510	140.5	440	117.0	460	130.5	430	104.5			31	100.0	37	114.5		
59	527	142.0	440	117.0	460	130.5	430	104.5			31	100.0	37	114.5		
60	550	143.0	442	121.0	461	132.0	430	104.5			31	100.0	38	116.5		
		5468.5	450	124.0		←		←			→		41	121.0		
			454	128.0									42	123.5		
			488	138.0									47	127.0		
				4827.5									48	129.0		
													48	129.0		
														4785.0		

8. See the data summary on the data page.
claim: the populations have different distributions
preliminary calculations:

$n_1 = 70$ $\Sigma R_1 = 5468.5$
$n_2 = 73$ $\Sigma R_2 = 4827.5$
$n = \Sigma n = 143$ $\Sigma R = 10296.0$

$R = \Sigma R_1 = 5468.5$
$\mu_R = n_1(n+1)/2 = 70(144)/2 = 5040$
$\sigma_R^2 = n_1 n_2(n+1)/12 = (70)(73)(144)/12 = 61320$

check: $\Sigma R = n(n+1)/2$
$= 143(144)/2$
$= 10296$

H_o: the populations have the same distribution
H_1: the populations have different distributions
$\alpha = .05$
C.R. $z < -z_{.025} = -1.96$
$z > z_{.025} = 1.96$
calculations:
$z_R = (R - \mu_R)/\sigma_R$
$= (5468.5 - 5040)/\sqrt{61320}$
$= 428.5/247.629 = 1.730$

conclusion:
Do not reject H_o; there is not sufficient evidence to reject the claim that the samples come from populations with the same distribution.

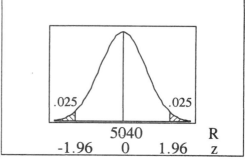

9. See the data summary on the data page.
claim: the populations have different distributions
preliminary calculations:

$n_1 = 56$ $\Sigma R_1 = 3861$
$n_2 = 75$ $\Sigma R_2 = 4785$
$n = \Sigma n = 131$ $\Sigma R = 8646$

$R = \Sigma R_1 = 3861$
$\mu_R = n_1(n+1)/2 = 56(132)/2 = 3696$
$\sigma_R^2 = n_1 n_2(n+1)/12 = (56)(75)(132)/12 = 46200$

check: $\Sigma R = n(n+1)/2$
$= 131(132)/2$
$= 8646$

H_o: the populations have the same distribution
H_1: the populations have different distributions
$\alpha = .05$
C.R. $z < -z_{.025} = -1.96$
$z > z_{.025} = 1.96$
calculations:
$z_R = (R - \mu_R)/\sigma_R$
$= (3861 - 3696)/\sqrt{46200}$
$= 165/214.942 = .768$

conclusion:
Do not reject H_o; there is not sufficient evidence to reject the claim that the ages come from populations with the same distribution.

10. See the data summary on the data page.
claim: the populations have different distributions
preliminary calculations:

$n_1 = 40$ $\Sigma R_1 = 1727.5$
$n_2 = 40$ $\Sigma R_2 = 1512.5$
$n = \Sigma n = 80$ $\Sigma R = 3240.0$

$R = \Sigma R_1 = 1727.5$
$\mu_R = n_1(n+1)/2 = 40(81)/2 = 1620$
$\sigma_R^2 = n_1 n_2(n+1)/12 = (40)(40)(81)/12 = 10800$

check: $\Sigma R = n(n+1)/2$
$= 80(81)/2$
$= 3240$

H_o: the populations have the same distribution
H_1: the populations have different distributions
$\alpha = .05$
C.R. $z < -z_{.025} = -1.96$
$z > z_{.025} = 1.96$
calculations:
$z_R = (R - \mu_R)/\sigma_R$
$= (1727.5 - 1620)/\sqrt{10800}$
$= 107.5/103.923 = 1.034$

conclusion:
Do not reject H_o; there is not sufficient evidence to reject the claim that the BMI values come from populations with the same distribution.

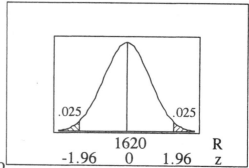

INSTRUCTOR'S SOLUTIONS Chapter 12   S-373

11. The denominators of the Mann-Whitney and Wilcoxon statistics are the same.
    The numerator of the Mann-Whitney statistic, assuming $R = \Sigma R_1$, is
    $$U - n_1n_2/2 = [n_1n_2 + n_1(n_1+1)/2 - R] - n_1n_2/2$$
    $$= n_1n_2/2 + n_1(n_1+1)/2 - R$$
    $$= (n_1/2)(n_2+n_1+1) - R$$
    $$= -[R - n_1(n_1+n_2+1)/2] = -[\text{numerator of Wilcoxon statistic}]$$
    For the readability example in this section,
    $$U = n_1n_2 + n_1(n_1+1)/2 - R$$
    $$= (13)(12) + (13)(14)/2 - 236.5$$
    $$= 10.5$$
    $\mu_U = n_1n_2/2 = (13)(12)/2 = 78$
    $\sigma_U^2 = \sigma_R^2 = n_1n_2(n+1)/12 = (13)(12)(26)/12 = 338$
    $z_U = (U - \mu_U)/\sigma_U$
    $\quad = (10.5 - 78)/\sqrt{338}$
    $\quad = -67.5/18.385 = -3.67$
    As predicted, z_U has the same absolute value but the opposite sign of z_R.

12. With $n_1 = 2$ and $n_2 = 2$ (so that $n = n_1 + n_2 = 4$), the test of hypothesis is the same as in the previous exercises except that the samples are not large enough to use the normal approximation and the critical region must be based on $R = \Sigma R_1$ instead of z. Let the A's be considered group 1.
    H_o: the populations have the same distribution
    H_1: the populations have different distributions
    $\alpha = .05$
    C.R. R < ?
    $\quad\quad$ R > ?

    a. There are 4!/2!2! = 6 possible arrangements of 2 scores from A and 2 scores from B. Those 6 arrangements and their associated $R = \Sigma R_1$ values are given at the right.

rank 1 2 3 4	R
A A B B	3
A B A B	4
A B B A	5
B A A B	5
B A B A	6
B B A A	7

    b. If H_o is true and populations A and B are identical, then there is no reason to expect any of the arrangements over another and each of the 6 arrangements is equally likely. The possible values of R and the probability associated with value is as given below.

R	P(R)	NOTE:	R·P(R)	R²·P(R)
3	1/6		3/6	9/6
4	1/6		4/6	16/6
5	2/6		10/6	50/6
6	1/6		6/6	36/6
7	1/6		7/6	49/6
	1		30/6	160/6

    $\mu_R = \Sigma R \cdot P(R)$
    $\quad = 30/6 = 5$
    $\sigma_R^2 = \Sigma R^2 \cdot P(R) - (\mu_R)^2$
    $\quad = 160/6 - 5^2$
    $\quad = 10/6$

    NOTE: The fact that each of the arrangements is equally likely can be proven using conditional probability formulas for selecting A's and B's from a finite population of 2 A's and 2 B's as follows.
    $P(A_1A_2B_3B_4) = P(A_1) \cdot P(A_2) \cdot P(B_3) \cdot P(B_4) = (2/4) \cdot (1/3) \cdot (2/2) \cdot (1/1) = 4/24 = 1/6$
    $P(A_1B_2A_3B_4) = P(A_1) \cdot P(B_2) \cdot P(A_3) \cdot P(B_4) = (2/4) \cdot (2/3) \cdot (1/2) \cdot (1/1) = 4/24 = 1/6$
    $P(A_1B_2B_3A_4) = P(A_1) \cdot P(B_2) \cdot P(B_3) \cdot P(A_4) = (2/4) \cdot (2/3) \cdot (1/2) \cdot (1/1) = 4/24 = 1/6$
    $P(B_1A_2A_3B_4) = P(B_1) \cdot P(A_2) \cdot P(A_3) \cdot P(B_4) = (2/4) \cdot (2/3) \cdot (1/2) \cdot (1/1) = 4/24 = 1/6$
    $P(B_1A_2B_3A_4) = P(B_1) \cdot P(A_2) \cdot P(B_3) \cdot P(A_4) = (2/4) \cdot (2/3) \cdot (1/2) \cdot (1/1) = 4/24 = 1/6$
    $P(B_1B_2A_3A_4) = P(B_1) \cdot P(B_2) \cdot P(A_3) \cdot P(A_4) = (2/4) \cdot (1/3) \cdot (2/2) \cdot (1/1) = 4/24 = 1/6$
    In addition, the μ_R and σ_R^2 values calculated above using the probability formulas agree with the formulas
    $\mu_R = n_1(n+1)/2 = 2(5)/2 = 5$
    $\sigma_R^2 = n_1n_2(n+1)/12 = (2)(2)(5)/12 = 20/12 = 10/6$

given in this section. But since the distribution – as evidenced by the P(R) values – is not normal, we cannot use these values to convert to z scores for use with Table A-2.

c. The P(R) values indicate that the smallest α at which the test can be performed is the 2/6 = .333 level with
C.R. R ≤ 3
R ≥ 7
as illustrated at the right -- using discrete bars instead of a continuous normal distribution. Even with the smallest or largest possible R values (i.e., R=3 or R=7) it would not be possible to reject H_o using α = .05.

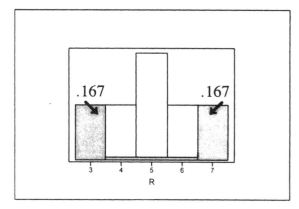

12-5 Kruskal-Wallis Test

NOTE: As in the previous sections, the manual follows the wording in the text and tests the hypothesis that "the populations have the same distribution" with the understanding that the test detects only differences in location and not differences in variability.

1. H_o: the populations have the same distribution
   H_1: the populations have different distributions
   α = .05
   C.R. H > $\chi^2_{2,.05}$ = 5.991
   calculations:
   H = $[12/n(n+1)] \cdot [\Sigma(R_i^2/n_i)] - 3(n+1)$
       = .58 [from Minitab]
   P-value = $P(\chi^2_2 > .58)$
           = .747 [from Minitab]
   conclusion:
   Do not reject H_o; there is not sufficient evidence to reject the claim that the different ages have times with identical population distributions.
   NOTE: Compare these results to those of the parametric test in section 11-2, exercise #3.

   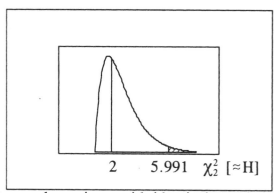

2. H_o: the populations have the same distribution
   H_1: the populations have different distributions
   α = .05
   C.R. H > $\chi^2_{2,.05}$ = 5.991
   calculations:
   H = $[12/n(n+1)] \cdot [\Sigma(R_i^2/n_i)] - 3(n+1)$
       = 11.93 [from Minitab]
   P-value = $P(\chi^2_2 > 11.93)$
           = .003 [from Minitab]
   conclusion:
   Reject H_o; there is sufficient evidence to reject the claim that the different years have time intervals with identical population distributions and to conclude that the populations have different distributions.
   Yes, it appears that the eruption behavior of Old Faithful is changing over time.

3. Below are the ordered scores for each group. The group listed first is group 1, etc.

subco	R	compa	R	midsi	R	fulls	R
420	4	442	6	259	1	360	2
428	5	514	9	454	7	384	3
681	16	525	10.5	469	8	602	12
898	19	643	13	525	10.5	656	15
917	20	655	14	727	18	687	17
	64		52.5		44.5		49

$n_1 = 5$ $R_1 = 64$
$n_2 = 5$ $R_2 = 52.5$
$n_3 = 5$ $R_3 = 44.5$
$n_4 = 5$ $R_4 = 49$

$n = \Sigma n = 20$ $\Sigma R = 210$

check:
$\Sigma R = n(n+1)/2$
$= 20(21)/2$
$= 210$

H_o: the populations have the same distribution
H_1: the populations have different distributions
$\alpha = .05$ [assumed]
C.R. $H > \chi^2_{3,.05} = 7.815$
calculations:
$H = [12/n(n+1)] \cdot [\Sigma(R_i^2/n_i)] - 3(n+1)$
$= [12/20(21)] \cdot [(64)^2/5 + (52.5)^2/5$
$+ (44.5)^2/5 + (49)^2/5] - 3(21)$
$= [.0286] \cdot [2246.7] - 63$
$= 1.191$

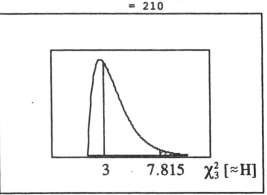

conclusion:
Do not reject H_o; there is not sufficient evidence to support the claim that the head injury measurements have different population distributions.

No; for this particular measurement, the data do not show that the heavier cars are safer in a crash. While the subcompact head injury values might seem higher, they are not statistically significant and could have occurred by chance when sampling from identical populations.

4. Below are the scores for each group. The group listed first is group 1, etc.

sunny	R	cloud	R	rainy	R
13.5	15	12.7	9.5	12.1	3
13.0	12.0	12.5	7	12.2	4.5
13.2	14	12.6	8	12.3	6
13.9	17	12.7	9.5	11.9	2
13.8	16	13.0	12.0	11.6	1
14.0	18	13.0	12.0	12.2	4.5
	92.0		58.0		21.0

$n_1 = 6$ $R_1 = 92.0$
$n_2 = 6$ $R_2 = 58.0$
$n_3 = 6$ $R_3 = 21.0$

$n = \Sigma n = 18$ $\Sigma R = 171.0$

check:
$\Sigma R = n(n+1)/2$
$= 18(19)/2$
$= 171$

H_o: the populations have the same distribution
H_1: the populations have different distributions
$\alpha = .05$
C.R. $H > \chi^2_{2,.05} = 5.991$
calculations:
$H = [12/n(n+1)] \cdot [\Sigma(R_i^2/n_i)] - 3(n+1)$
$= [12/18(19)] \cdot [(92)^2/6 + (58)^2/6$
$+ (21)^2/6] - 3(19)$
$= [.0351] \cdot [2044.8] - 57$
$= 14.749$

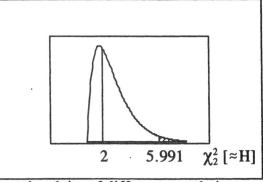

conclusion:
Reject H_o; there is sufficient evidence to support the claim of different population distributions.

No; the identification of any pairwise significant differences among the populations requires more advanced techniques. We may not conclude from this test, for example, that the difference between sunny days and cloudy days is significant.

NOTE: Compare these results to those of the parametric test in section 11-2, exercise #8.

5. Below are the ordered scores for each group. The group listed first is group 1, etc.

4000	R	1850	R	150	R
125	1	129	4	128	2
129	4	129	4	136	14
131	6	134	10	137	17.5
132	7.5	134	10	137	17.5
132	7.5	136	14	138	21.5
134	10	136	14	139	24
135	12	137	17.5	141	25
138	21.5	137	17.5	142	26
138	21.5	138	21.5	145	27
	91.0		112.5		174.5

$n_1 = 9 \quad R_1 = 91$
$n_2 = 9 \quad R_2 = 112.5$
$n_3 = 9 \quad R_3 = 174.5$

$n = \Sigma n = 27 \quad \Sigma R = 378.0$

check:
$\Sigma R = n(n+1)/2$
$= 27(28)/2$
$= 378$

H_o: the populations have the same distribution
H_1: the populations have different distributions
$\alpha = .05$
C.R. $H > \chi^2_{2,.05} = 5.991$
calculations:
$H = [12/n(n+1)] \cdot [\Sigma(R_i^2/n_i)] - 3(n+1)$
$= [12/27(28)] \cdot [(91)^2/9 + (112.5)^2/9$
$+ (174.5)^2/9] - 3(28)$
$= [.0159] \cdot [5709.7] - 84$
$= 6.631$

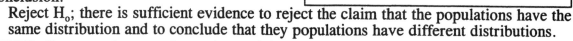

conclusion:
Reject H_o; there is sufficient evidence to reject the claim that the populations have the same distribution and to conclude that they populations have different distributions.

NOTE: Compare these results to those of the parametric test in section 11-2, exercise #7.

6. Below are the scores (in order) for each group. The group listed first is group 1, etc.

1	R	2	R	3	R	4	R	5	R
2.9	9.5	2.7	2	2.8	5.5	2.7	2	2.8	5.5
2.9	9.5	3.2	19.5	2.8	5.5	2.7	2	3.1	15
3.0	12	3.3	24	2.8	5.5	2.9	9.5	3.5	31.5
3.1	15	3.4	28	3.2	19.5	2.9	9.5	3.7	36
3.1	15	3.4	28	3.3	24	3.2	19.5	4.1	44
3.1	15	3.6	34	3.3	24	3.2	19.5	4.1	44
3.1	15	3.8	39	3.5	31.5	3.3	24	4.2	46.5
3.7	36	3.8	39	3.5	31.5	3.3	24		222.5
4.7	36	4.0	42	3.5	31.5	3.4	28		
3.9	41	4.1	44	3.8	39		138.0		
4.2	46.5	4.3	48		217.5				
	250.5		347.5						

$n_1 = 11 \quad R_1 = 250.5$
$n_2 = 11 \quad R_2 = 347.5$
$n_3 = 10 \quad R_3 = 217.5$
$n_4 = 9 \quad R_4 = 138.0$
$n_5 = 7 \quad R_5 = 222.5$

$n = \Sigma n = 48 \quad \Sigma R = 1176.0$

check:
$\Sigma R = n(n+1)/2$
$= 48(49)/2$
$= 1176$

H_o: the populations have the same distribution
H_1: the populations have different distributions
$\alpha = .05$ [assumed]
C.R. $H > \chi^2_{4,.05} = 9.488$
calculations:
$H = [12/n(n+1)] \cdot [\Sigma(R_i^2/n_i)] - 3(n+1)$
$= [12/48(49)] \cdot [(250.5)^2/11 + (347.5)^2/11$
$+ (217.5)^2/10 + (138)^2/9$
$+ (222.5)^2/7] - 3(49)$
$= [.00510] \cdot [30601.356] - 147 = 9.129$

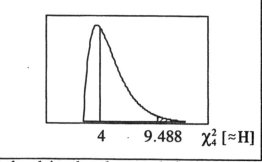

conclusion:
Reject H_o; there is sufficient evidence to reject the claim that the populations have the same distribution and to conclude that they have different distributions.

7. Below are the ordered weights (in thousandths of a gram) for each group. The group listed first is group 1, etc.

red	R	ora	R	yel	R	bro	R	blu	R	gre	R
870	9.5	861	5	868	8	856	2	838	1	890	26
872	12	897	31	876	16.5	858	3	870	9.5	902	39.5
874	13	898	34	877	18	860	4	875	14.5	902	39.5
882	21	903	42	879	19.5	866	6	956	87	911	52
888	24	920	61	879	19.5	867	7	968	90.5	930	72
891	27	942	82	886	22.5	871	11		202.5	949	83.5
897	31	971	92	886	22.5	875	14.5			1002	98
898	34	1009	99	892	28	876	16.5				410.5
908	46		446	893	29	889	25				
908	46			900	36.5	897	31				
908	46			906	44	898	34				
911	52			910	50	900	36.5				
912	54			911	52	902	39.5				
913	55			917	58	902	39.5				
920	61			921	63.5	904	43				
924	67			924	67	909	48.5	n_1 = 21	R_1 = 998.0		
924	67			926	69	909	48.5	n_2 = 8	R_2 = 446.0		
933	75			934	76	914	56.5	n_3 = 26	R_3 = 1392.5		
936	77.5			939	79	914	56.5	n_4 = 33	R_4 = 1600.5		
952	85			940	80	919	59	n_5 = 5	R_5 = 202.5		
983	95			941	81	920	61	n_6 = 7	R_6 = 410.5		
	998.0			949	83.5	921	63.5				
				960	88	923	65	n = Σn =100	ΣR = 5050.0		
				968	90.5	928	70				
				978	94	930	72	check:			
				989	97	930	72	ΣR = n(n+1)/2			
					1392.5	932	74	= 100(101)/2			
						936	77.5	= 5050			
						955	86				
						965	89				
						976	93				
						988	96				
						1033	100				
							1600.5				

H_o: the populations have the same distribution
H_1: the populations have different distributions
α = .05 [assumed]
C.R. H > $\chi^2_{5,.05}$ = 11.071
calculations:
$H = [12/n(n+1)] \cdot [\Sigma(R_i^2/n_i)] - 3(n+1)$
$= [12/100(101)] \cdot [(998)^2/21 + (446)^2/8 + (1392.5)^2/26 + (1600.5)^2/33 + (202.5)^2/5 + (410.5)^2/7] - 3(101)$
$= [.00119] \cdot [256770.7] - 303 = 2.074$

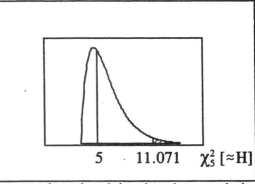

conclusion:
Do not reject H_o; there is not sufficient evidence to reject the claim that the populations have the same distribution.

No; the results do not indicate a problem that requires corrective action.

NOTE: Compare these results to those of the parametric test in section 11-2, exercise #9.

8. Below are the ordered distances. The one listed first is considered player 1, etc. Since the data are grouped, ΣR for each player is found by by adding his (count)·R products.

McGwire			Sosa			Bonds		
dist	count	R	dist	count	R	dist	count	R
340	1	3.5	340	1	3.5	320	2	1.5
341	1	5	344	1	6	347	1	7
350	2	10.5	350	3	10.5	350	1	10.5
360	3	17	360	1	17	360	3	17
369	1	28	364	2	22.5	361	1	21
370	4	34	365	1	24.5	365	1	24.5
377	1	45	366	1	26	370	2	34
380	5	53.5	368	1	27	375	4	42.5
385	2	63	370	5	34	380	5	53.5
388	1	65.5	371	1	40	385	1	63
390	4	70.5	380	6	53.5	390	2	70.5
398	1	78	388	1	65.5	391	1	75
400	2	84	390	2	70.5	394	1	76
409	1	93	400	5	84	396	1	77
410	5	103.5	405	1	91.5	400	4	84
420	5	132	410	5	103.5	404	1	90
423	1	143	414	1	115	405	1	91.5
425	1	144	415	1	117	410	10	103.5
430	7	154.5	420	8	132	411	1	114
440	3	175.5	430	6	154.5	415	2	117
450	4	184.5	433	2	164.5	416	1	119
452	1	188	434	2	166.5	417	2	120.5
458	1	190	440	3	175.5	420	8	132
460	2	192	450	1	184.5	429	1	145
461	1	194	460	1	192	430	5	154.5
470	3	196	480	2	200	435	2	168.5
478	1	198	482	1	202	436	1	170
480	1	200	500	1	204.5	440	4	175.5
500	1	204.5		66	6533.5	442	1	181
510	2	206.5				450	1	184.5
527	1	208				454	1	189
550	1	209				488	1	203
	70	8156.5					73	7255.0

$n_1 = 70$ $R_1 = 8156.5$
$n_2 = 66$ $R_2 = 6533.5$
$n_3 = 73$ $R_3 = 7255.0$
$n = \Sigma n = 209$ $\Sigma R = 21945.0$

check:
$\Sigma R = n(n+1)/2$
$= 209(210)/2$
$= 21945$

H_o: the populations have the same distribution
H_1: the populations have different distributions
$\alpha = .05$
C.R. $H > \chi^2_{2,.05} = 5.991$
calculations:
$H = [12/n(n+1)] \cdot [\Sigma(R_i^2/n_i)] - 3(n+1)$
$= [12/209(210)] \cdot [(8156.5)^2/70$
$+ (6533.5)^2/66 + (7255)^2/73] - 3(210)$
$= [.0002734] \cdot [2318201.776] - 630$
$= 3.821$

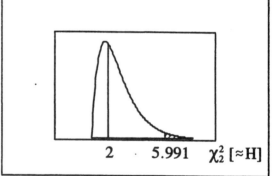

conclusion:
Do not reject H_o; there is not sufficient evidence to reject the claim that the populations of distances for the 3 players have the same distribution.

NOTE: Compare these results to those of the parametric test in section 11-2, exercise #10.

9. $H = [12/n(n+1)] \cdot [\Sigma(R_i^2/n_i)] - 3(n+1)$, which depends only on the rank of each score and the number of scores within each group.

a. Adding or subtracting a constant to each score does not change the number of scores within each group. Since adding or subtracting a constant to each score does not affect the order of the scores, their ranks and the calculated H statistic are not affected.

b. Multiplying or dividing each score by a positive constant does not change the number of scores within each group. Since multiplying or dividing each score by a positive constant does not affect the order of the scores, their ranks and the calculated H statistic are not affected.

c. Changing a single sample value so that it becomes an outlier will not change the overall values of the ranks (i.e., they will still vary from 1 to n), but it may change individual ranks by ±1 and affect how the ranks are distributed among the groups. In general, the effect on the value of the test statistic H will be minimal unless the change places the largest (or smallest) rank in the group which previously had the smallest (or largest) ranks.

10. For 15 total scores in 3 groups of 5 each, the sum of the ranks is 15(16)/2 = 120.
The largest H occurs when there is maximum separation of the groups.
$R_1 = 1 + 2 + 3 + 4 + 5 = 15$
$R_2 = 6 + 7 + 8 + 9 + 10 = 40$
$R_3 = 11 + 12 + 13 + 14 + 15 = 65$
$H = [12/n(n+1)] \cdot [\Sigma(R_i^2/n_i)] - 3(n+1)$
$= [12/(15)(16)] \cdot [15^2/5 + 40^2/5 + 65^2/5] - 3(16) = 12.5$
The smallest H results when $R_1 = R_2 = R_3 = 40$.
$H = [12/n(n+1)] \cdot [\Sigma(R_i^2/n_i)] - 3(n+1)$
$= [12/(15)(16)] \cdot [40^2/5 + 40^2/5 + 40^2/5] - 3(16) = 0$

11.
rank	t	T = t³- t
4.5	2	6
9.5	2	6
12	3	24
		36

correction factor:
$1 - \Sigma T/(n^3-n) = 1 - 36/(18^3-18)$
$= 1 - 36/5814$
$= .99380805$

The original calculated test statistic is H = 14.74853801
The corrected calculated test statistic is H = 14.74853801/.99380805 = 14.8404
No; the corrected value of H does not differ substantially from the original one.
NOTE: Be careful when counting the number of tied ranks; in addition to the easily recognized ".5's" that result from an even number of ties, there may be multiple whole number ranks resulting from an odd number of ties.

12. The algebra is tedious and the outline of the proof is sketched below.
The key useful fact relating $n_1 + n_2 = n$ to R_1 and R_2 is
$R_1 + R_2 = n(n+1)/2$ which implies $(n+1) = 2(R_1+R_2)/n$ and $(n+1)^2 = 4(R_1+R_2)^2/n^2$.
Also, let $D = n_1n_2(n_1+n_2+1)/12 = n_1n_2(n+1)/12$.
(1) $H = [12/n(n+1)][R_1^2/n_1 + R_2^2/n_2] - 3(n+1)$
$= [12/n(n+1)][(n_2R_1^2 + n_1R_2^2)/n_1n_2 - n(n+1)^2/4]$
$= [12/n_1n_2(n+1)][(n_2R_1^2 + n_1R_2^2)/n - n_1n_2(n+1)^2/4]$
$= [1/D][(n_2R_1^2 + n_1R_2^2)/n - n_1n_2(R_1+R_2)^2/n^2]$
$= [1/D][A]$
(2) $z = [1/\sqrt{D}][R_1 - n_1(n+1)/2]$
$= [1/\sqrt{D}][R_1 - n_1(R_1+R_2)/n]$
$= [1/\sqrt{D}][(n_2R_1 - n_1R_2)/n]$
$z^2 = [1/D][(n_2R_1 - n_1R_2)^2/n^2]$
$= [1/D][B]$
For calculated values, the following algebra shows that A = B and, therefore, that $z^2 = H$.
$A = (n_2R_1^2 + n_1R_2^2)/n - n_1n_2(R_1+R_2)^2/n^2$
$= [(n_1+n_2)/n] \cdot (n_2R_1^2 + n_1R_2^2)/n - n_1n_2(R_1+R_2)^2/n^2]$
$= (n_1n_2R_1^2 + n_1^2R_2^2 + n_2^2R_1^2 + n_1n_2R_2^2 - n_1n_2R_1^2 - 2n_1n_2R_1R_2 - n_1n_2R_2^2)/n^2$
$= (n_1^2R_2^2 + n_2^2R_1^2 + -2n_1n_2R_1R_2)/n^2$
$= (n_2R_1 - n_1R_2)^2/n^2$
$= B$
For critical values, tables show that $(z_{\alpha/2})^2 = \chi^2_{1,\alpha}$ for any α.
For example, $(z_{.025})^2 = (1.96)^2 = 3.841 = \chi^2_{1,.05}$

12-6 Rank Correlation

NOTE: This manual calculates $d = R_x - R_y$, thus preserving the sign of d. This convention means Σd must equal 0 and provides a check for the assigning and differencing of the ranks. In addition, it must always be true that $\Sigma R_x = \Sigma R_y = n(n+1)/2$.

1. In each case the n=5 pairs are pairs of ranks, called R_x and R_y below to stress that fact.

   a.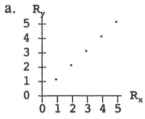

R_x	R_y	d	d^2
1	1	0	0
3	3	0	0
5	5	0	0
4	4	0	0
2	2	0	0
15	15	0	0

   $r_s = 1 - [6(\Sigma d^2)]/[n(n^2-1)]$
   $= 1 - [6(0)]/[5(24)]$
   $= 1 - 0$
   $= 1$

   Yes; there appears to be a perfect positive correlation between R_x and R_y.

   b.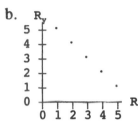

R_x	R_y	d	d^2
1	5	-4	16
2	4	-2	4
3	3	0	0
4	2	2	4
5	1	4	16
15	15	0	40

   $r_s = 1 - [6(\Sigma d^2)]/[n(n^2-1)]$
   $= 1 - [6(40)]/[5(24)]$
   $= 1 - 2$
   $= -1$

   Yes; there appears to be a perfect negative correlation between R_x and R_y.

   c.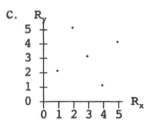

R_x	R_y	d	d^2
1	2	-1	1
2	5	-3	9
3	3	0	0
4	1	3	9
5	4	1	1
15	15	0	20

   $r_s = 1 - [6(\Sigma d^2)]/[n(n^2-1)]$
   $= 1 - [6(20)]/[5(24)]$
   $= 1 - 1$
   $= 0$

   No; there does not appear to be any correlation between R_x and R_y.

2. a. Since $n \leq 30$, use Table A-9.
   CV: $r_s = \pm.450$
   b. Since $n > 30$, use Formula 12-1.
   CV: $r_s = \pm 1.960/\sqrt{49} = \pm.280$
   c. Since $n > 30$, use Formula 12-1.
   CV: $r_s = \pm 2.327/\sqrt{39} = \pm.373$
   d. Since $n \leq 30$, use Table A-9.
   CV: $r_s = \pm.689$
   e. Since $n > 30$, use Formula 12.1.
   CV: $r_s = \pm 2.05/\sqrt{81} = \pm.228$

3. The following table summarizes the calculations.

R_x	R_y	d	d^2
2	2	0	0
6	7	-1	1
3	7	-3	9
5	4	1	1
7	5	2	4
10	8	2	4
9	9	0	0
8	10	-2	4
4	3	1	1
1	1	0	0
55	55	0	24

   $r_s = 1 - [6(\Sigma d^2)]/[n(n^2-1)]$
   $= 1 - [6(24)]/[10(99)]$
   $= 1 - .145$
   $= .855$

   $H_0: \rho_s = 0$
   $H_1: \rho_s \neq 0$
   $\alpha = .05$
   C.R. $r_s < -.648$
   $r_s > .648$
   calculations:
   $r_s = .855$

   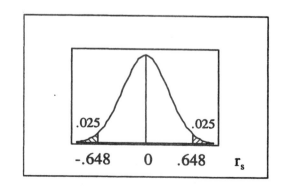

conclusion:
Reject H_o; there is sufficient evidence to reject the claim that $\rho_s = 0$ and to conclude that $\rho_s \neq 0$ (in fact, that $\rho > 0$).
Yes; it does appear that salary increases as stress increases.

IMPORTANT NOTE: The rank correlation is correctly calculated using the ranks in the Pearson product moment correlation formula 9-1 of chapter 9 to produce
$$r_s = [\Sigma R_x R_y - (\Sigma R_x)(\Sigma R_y)]/[\sqrt{\Sigma R_x^2 - (\Sigma R_x)^2} \cdot \sqrt{\Sigma R_y^2 - (\Sigma R_y)^2}]$$
Since $\Sigma R_x = \Sigma R_y = 1+2+\ldots+n = n(n+1)/2$ [always]
and $\Sigma R_x^2 = \Sigma R_y^2 = 1^2+2^2+\ldots+n^2 = n(n+1)(2n+1)/6$ [when there are ties in the ranks], it can be shown by algebra that the above formula can be shortened to
$$r_s = 1 - [6(\Sigma d^2)]/[n(n^2-1)] \quad \text{when there are no ties in the ranks.}$$
As ties typically make no difference in the first 3 decimals of r_s, this manual uses the shortened formula exclusively and notes when use of the longer formula gives a slightly different result.

4. The following table summarizes the calculations.

R_x	R_y	d	d^2
2	5	-3	9
6	2	4	16
3	3	0	0
5	8	-3	9
7	10	-3	9
10	9	1	1
9	1	8	64
8	7	1	1
4	6	-2	4
1	4	-3	9
55	55	0	122

$r_s = 1 - [6(\Sigma d^2)]/[n(n^2-1)]$
$\quad = 1 - [6(122)]/[10(99)]$
$\quad = 1 - .739$
$\quad = .261$

H_o: $\rho_s = 0$
H_1: $\rho_s \neq 0$
$\alpha = .05$
C.R. $r_s < -.648$
$\quad\;\; r_s > .648$
calculations:
$\quad r_s = .261$

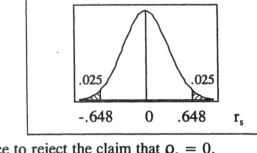

conclusion:
Do not reject H_o; there is not sufficient evidence to reject the claim that $\rho_s = 0$.
No; there does not appear to be a relationship between salary and physical demands.

5. The following table summarizes the calculations.

R_x	R_y	d	d^2
1	3	-2	4
2	5	-3	9
4	4	0	0
5	1	4	16
3	10	-7	49
6	7	-1	1
8	6	2	4
7	8	-1	1
10	2	8	64
9	9	0	0
55	55	0	148

$r_s = 1 - [6(\Sigma d^2)]/[n(n^2-1)]$
$\quad = 1 - [6(148)]/[10(99)]$
$\quad = 1 - .897$
$\quad = .103$

H_o: $\rho_s = 0$
H_1: $\rho_s \neq 0$
$\alpha = .05$
C.R. $r_s < -.648$
$\quad\;\; r_s > .648$
calculations:
$\quad r_s = .261$

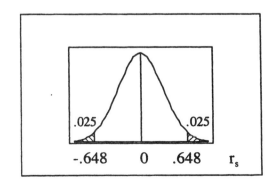

conclusion:
  Do not reject H_o; there is not sufficient evidence to reject the claim that $\rho_s = 0$.
  No; there does not appear to be a relationship between the two rankings.

6. The following table summarizes the calculations.

x	R_x	y	R_y	d	d^2
33.46	1	5.50	2	-1	1
50.68	2	5.00	2	1	1
87.92	4	8.08	3	1	1
98.84	5	17.00	6	-1	1
63.60	3	12.00	4	-1	1
107.34	6	16.00	5	1	1
	21		21	0	6

$r_s = 1 - [6(\Sigma d^2)]/[n(n^2-1)]$
  $= 1 - [6(6)]/[6(35)]$
  $= 1 - .171$
  $= .829$

$H_o: \rho_s = 0$
$H_1: \rho_s \neq 0$
$\alpha = .05$
C.R. $r_s < -.886$
    $r_s > .886$
calculations:
    $r_s = .829$

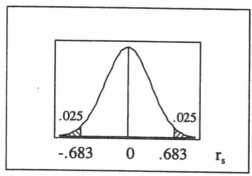

conclusion:
  Do not reject H_o; there is not sufficient evidence to conclude that $\rho_s \neq 0$.

7. The following table summarizes the calculations.

x	R_x	y	R_y	d	d^2
71	7.0	125	6	1	1
70.5	5	119	3.5	1.5	2.25
71	7.0	128	8.5	-1.5	2.25
72	9	128	8.5	.5	.25
70	3.0	119	3.5	-.5	.25
70	3.0	127	7	-4	16
66.5	1	105	1	0	0
70	3.0	123	5	-2	4
71.0	7	115	2	5	25
	45.0		45.0	0.0	51.00

$r_s = 1 - [6(\Sigma d^2)]/[n(n^2-1)]$
  $= 1 - [6(51.00)]/[9(80)]$
  $= 1 - .425$
  $= .575$

$H_o: \rho_s = 0$
$H_1: \rho_s \neq 0$
$\alpha = .05$
C.R. $r_s < -.683$
    $r_s > .683$
calculations:
    $r_s = .575$
conclusion:
  Do not reject H_o; there is not sufficient evidence to conclude that there is a correlation between heights and weights of supermodels.
  NOTE: Using formula 9-1 (since there are ties) yields $r_s = .557$. Compare this to the parametric hypothesis test using Pearson's correlation in section 9-2 exercise #10.

8. The following table summarizes the calculations.

x	R_x	y	R_y	d	d^2
100	8	7	5	3	9
14	5.5	4.4	3	2.5	6.25
14	5.5	5.9	4	1.5	2.25
35.2	7	1.6	1	6	36
12	4	10.4	8	-4	16
7	3	9.6	7	-4	16
5	2	8.9	6	-4	16
1	1	4.2	2	-1	1
	36.0		36.0	0.0	102.50

$r_s = 1 - [6(\Sigma d^2)]/[n(n^2-1)]$
  $= 1 - [6(102.5)]/[8(63)]$
  $= 1 - 1.220$
  $= -.220$

H_o: $\rho_s = 0$
H_1: $\rho_s \neq 0$
$\alpha = .05$
C.R. $r_s < -.738$
$r_s > .738$
calculations:
$r_s = -.220$
conclusion:
Do not reject H_o; there is not sufficient evidence to conclude that there is any correlation between salary and number of viewers.

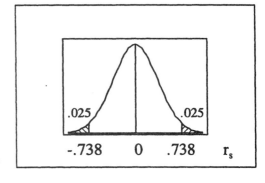

NOTE: Using formula 9-1 (since there are ties) yields $r_s = -.228$. Compare this to the parametric hypothesis test using Pearson's correlation in section 9-2 exercise #9.

9. The following table summarizes the calculations, where grams of fat is denoted x.

x	R_x	y	R_y
.07	16	3.7	7.5
.02	8.0	3.6	4.5
.01	6	3.6	4.5
.03	12.0	4.0	14.0
.03	12.0	4.0	14.0
.00	3.0	3.6	4.5
.03	12.0	3.8	9
.03	12.0	3.7	7.5
.06	15	4.1	16
.00	3.0	3.9	11.0
.02	8.0	3.9	11.0
.02	8.0	3.3	1
.00	3.0	3.5	2
.00	3.0	3.6	4.5
.00	3.0	3.9	11.0
.03	12.0	4.0	14.0
	136.0		136.0

Since there are so many ties, use formula 9-1 on the ranks.
n = 16
$\Sigma R_x = 136$
$\Sigma R_x^2 = 1474$
$\Sigma R_y = 136$
$\Sigma R_y^2 = 1486.5$
$\Sigma R_x R_y = 1320$

$n(\Sigma R_x^2) - (\Sigma R_x)^2 = 16(1474) - (136)^2$
$= 5088$
$n(\Sigma R_y^2) - (\Sigma R_y)^2 = 16(1486.5) - (136)^2$
$= 5288$
$n(\Sigma R_x R_y) - (\Sigma R_x)(\Sigma R_y) = 16(1320) - (136)(136)$
$= 2624$
$r_s = 2624/[\sqrt{5088}\sqrt{5288}]$
$= .5058$

H_o: $\rho_s = 0$
H_1: $\rho_s \neq 0$
$\alpha = .05$
C.R. $r_s < -.507$
$r_s > .507$
calculations:
$r_s = .506$
conclusion:

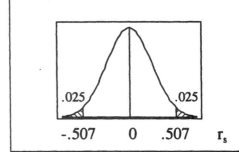

Do not reject H_o; there is not sufficient evidence to conclude that there is a correlation between grams of fat and calorie count.

NOTE: Compare this to the parametric hypothesis test using Pearson's correlation in section 9-2 exercise #9.

10. Refer to the data page for exercises #10, #11 and #12.
Since there are many ties, use formula 9-1 applied to the ranks.
The critical values are $\pm z/\sqrt{n-1} = \pm 1.96/\sqrt{39} = \pm.314$.
$n(\Sigma R_c^2) - (\Sigma R_c)^2 = 40(22138.50) - (820)^2 = 213140$
$n(\Sigma R_B^2) - (\Sigma R_B)^2 = 40(22138.00) - (820)^2 = 213120$
$n(\Sigma R_c R_B) - (\Sigma R_c)(\Sigma R_B) = 40(19581.50) - (820)(820) = 110860$

INSTRUCTOR'S SOLUTIONS Chapter 12

Below are the relevant values for exercises #10, #11 and #12.

	exercise #10				exercise #11						exercise #12					
#	cho	R_i	BMI	R_d	nic	R_n	tar	R_t	car	R_c	act	R_a	1dp	R_1	5dp	R_5
01	264	25	19.6	5.5	1.2	24.0	16	23.5	15	19.0	30	4.0	28	5.0	28	3.5
12	181	20	23.8	19.5	1.2	24.0	16	23.5	15	19.0	25	1	29	7.0	27	2
03	267	26	19.6	5.5	1.0	15.0	16	23.5	17	26.0	31	6.0	32	13.5	28	3.5
04	384	34	29.1	30	0.8	8.5	9	8.0	6	3	33	11.0	29	7.0	30	5.5
05	98	8.5	35.2	23	0.1	1	1	1	1	1	29	3	30	10.5	26	1
06	62	4.5	21.4	12	0.8	8.5	8	5.5	8	6	36	19.0	35	20	35	16.0
07	126	13	22.0	15	0.8	8.5	10	10	10	8.0	39	19.0	35	20.0	34	13.0
08	89	6	27.5	26	1.0	15.0	16	23.5	17	26.0	37	22.5	32	13.5	34	13.0
09	531	37	33.5	37	1.0	15.0	14	16.5	13	14.0	32	8	27	3.0	33	9.5
10	130	14	20.6	9	1.0	15.0	13	14.0	13	14.0	28	2	25	1	35	16.0
11	175	19	29.9	33	1.1	20.0	13	14.0	13	14.0	43	30	41	29.5	38	26.5
12	44	3	17.7	1	1.2	24.0	15	19.0	15	19.0	37	22.5	30	10.5	37	24.5
13	8	2	24.0	21	1.2	24.0	16	23.5	15	19.0	36	19.0	33	15.5	36	20.5
14	112	10	28.9	29	0.7	5.5	9	8.0	11	10.5	37	22.5	40	27.5	36	20.5
15	462	36	37.7	38	0.9	11	11	11	15	19.0	34	14	34	17.5	45	31
16	62	4.5	18.3	2	0.2	2	2	2	3	2	41	29	38	25.5	36	20.5
17	98	8.5	19.8	7	1.4	28.5	18	28.5	18	28.5	40	28	33	15.5	33	9.5
18	447	35	29.8	32	1.2	24.0	15	19.0	15	19.0	33	11.0	35	20.0	34	13.0
19	125	12	29.7	31	1.1	20.0	13	14.0	12	12	35	16.0	40	27.5	36	20.5
20	318	32	31.7	36	1.0	15.0	15	19.0	16	23.5	33	11.0	27	3.0	33	9.5
21	325	33	23.8	19.5	1.3	27	17	27	16	23.5	31	6.0	27	3.0	31	7
22	600	39	44.9	40	0.8	8.5	9	8.0	10	8.0	33	11.0	30	10.5	30	5.5
23	237	23	19.2	3	1.0	15.0	12	12	10	8.0	35	16.0	37	24	38	26.5
24	173	18	28.7	28	1.0	15.0	14	16.5	17	26.0	38	25.5	38	25.5	39	28
25	309	31	28.5	27	0.5	3	5	3	7	4.5	37	22.5	29	7.0	33	9.5
26	94	7	19.3	4	0.6	4	6	4	7	4.5	31	6.0	36	22.5	36	20.5
27	280	28	31.0	35	0.7	5.5	8	5.5	11	10.5	38	25.5	34	17.5	37	24.5
28	254	24	25.1	22	1.4	28.5	18	28.5	15	19.0	35	16.0	30	10.5	35	16.0
29	123	11	22.8	16.5	1.1	20.0	16	23.5	18	28.5	33	11.0	36	22.5	40	29.5
30	596	38	30.9	34		435.0		435.0		435.0	39	27	41	29.5	36	20.5
31	301	30	26.5	25							46	31	42	31	40	29.5
32	223	22	21.2	11								496.0		496.0		496.0
33	293	29	40.6	39												
34	146	15	21.9	13.5												
35	149	16.5	26.0	24												
36	149	16.5	23.5	18												
37	920	40	22.8	16.5												
38	271	27	20.7	10												
39	207	21	20.5	8												
40	2	1	21.9	13.5												
		820.0		820.0												

exercise #10

$\Sigma R_c^2 = 22138.50$

$\Sigma R_B^2 = 22138.00$

$\Sigma R_c R_B = 19581.50$

exercise #11

$\Sigma R_n^2 = 8509.00$

$\Sigma R_t^2 = 8530.00$

$\Sigma R_c^2 = 8519.00$

$\Sigma R_n R_t = 8355.50$

$\Sigma R_n R_c = 7994.25$

exercise #12

$\Sigma R_a^2 = 10394.50$

$\Sigma R_1^2 = 10401.50$

$\Sigma R_5^2 = 10887.00$

$\Sigma R_a R_1 = 9531.75$

$\Sigma R_a R_5 = 9369.50$

$H_o: \rho_s = 0$
$H_1: \rho_s \neq 0$
$\alpha = .05$
C.R. $r_s < -.314$
$\quad\quad r_s > .314$
calculations:
$\quad r_s = 110860/[\sqrt{213140}\sqrt{213120}\,]$
$\quad\quad = .5202$

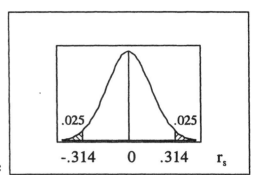

conclusion:
   Reject H_o; there is sufficient evidence to conclude that $\rho_s \neq 0$ (in fact, that $\rho_s > 0$).
Yes, there is a significant positive correlation between cholesterol level and body mass.
NOTE: Compare this to the parametric hypothesis test using Pearson's correlation in section 9-2 exercise #17.

11. Refer to the data page for exercises #10, #11 and #12.
Since there are many ties, use formula 9-1 applied to the ranks.
The critical values are $\pm z/\sqrt{n-1} = \pm 1.96/\sqrt{28} \pm .370$.
$\quad n(\Sigma R_n^2) - (\Sigma R_n)^2 = 29(8509) - (435)^2 = 57536$
$\quad n(\Sigma R_t^2) - (\Sigma R_t)^2 = 29(8530) - (435)^2 = 58145$
$\quad n(\Sigma R_c^2) - (\Sigma R_c)^2 = 29(8519) - (435)^2 = 57826$
$\quad n(\Sigma R_n R_t) - (\Sigma R_n)(\Sigma R_t) = 29(8355.50) - (435)(435) = 53084.50$
$\quad n(\Sigma R_n R_c) - (\Sigma R_n)(\Sigma R_c) = 29(7994.25) - (435)(435) = 42608.25$

a. Use the nicotine (n) & tar (t) values.
$H_o: \rho_s = 0$
$H_1: \rho_s \neq 0$
$\alpha = .05$
C.R. $r_s < -.370$
$\quad\quad r_s > .370$
calculations:
$\quad r_s = 53084.50/[\sqrt{57536}\sqrt{58145}]$
$\quad\quad = .9177$

conclusion:
   Reject H_o; there is sufficient evidence to conclude that the nicotine content is (positively) correlated with the amount of tar.

b. Use the nicotine (n) & carbon monoxide (c) values.
$H_o: \rho_s = 0$
$H_1: \rho_s \neq 0$
$\alpha = .05$
C.R. $r_s < -.370$
$\quad\quad r_s > .370$
calculations:
$\quad r_s = 42608.25/[\sqrt{57536}\sqrt{57826}]$
$\quad\quad = .7387$

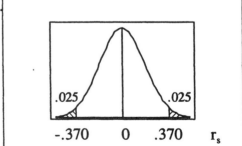

conclusion:
   Reject H_o; there is sufficient evidence to conclude that the nicotine content is (positively) correlated with the amount of carbon monoxide.

c. Tar is a better predictor than carbon monoxide, because it has a higher correlation with nicotine.

NOTE: Compare this to the parametric hypothesis test using Pearson's correlation in section 9-2 exercise #20.

12. Refer to the data page for exercises #10, #11 and #12.
    Since there are many ties, use formula 9-1 applied to the ranks.
    The critical values are $\pm z/\sqrt{n-1} = \pm 1.96/\sqrt{30} = \pm.358$.
    $n(\Sigma R_a^2) - (\Sigma R_a)^2 = 31(10394.50) - (496)^2 = 76213.5$
    $n(\Sigma R_1^2) - (\Sigma R_1)^2 = 31(10401.50) - (496)^2 = 76430.5$
    $n(\Sigma R_5^2) - (\Sigma R_5)^2 = 31(10887.00) - (496)^2 = 75981.0$
    $n(\Sigma R_a R_1) - (\Sigma R_a)(\Sigma R_1) = 31(9531.75) - (496)(496) = 49468.25$
    $n(\Sigma R_a R_5) - (\Sigma R_a)(\Sigma R_5) = 31(9369.50) - (496)(496) = 44438.50$

    a. Use the actual (a) & five-day predicted (5) values.
       $H_0: \rho_s = 0$
       $H_1: \rho_s \neq 0$
       $\alpha = .05$
       C.R. $r_s < -.358$
            $r_s > .358$
       calculations:
       $r_s = 44438.50/[\sqrt{76213.5}\sqrt{75981.0}]$
            $= .5840$
       conclusion:
       Reject H_0; there is sufficient evidence to conclude that $\rho_s \neq 0$ (in fact, that $\rho_s > 0$).
       Yes, there is a correlation between the actual and the 5-day predicted temperatures.
       No, correlation does not imply agreement.

    b. Use the actual (a) & one-day predicted (1) values.
       $H_0: \rho_s = 0$
       $H_1: \rho_s \neq 0$
       $\alpha = .05$
       C.R. $r_s < -.358$
            $r_s > .358$
       calculations:
       $r_s = 49468.25/[\sqrt{76213.5}\sqrt{76430.5}]$
            $= .6482$
       conclusion:
       Reject H_0; there is sufficient evidence to conclude that $\rho_s \neq 0$ (in fact, that $\rho_s > 0$).
       Yes, there is a correlation between the actual and the 1-day predicted temperatures.
       No, correlation does not imply agreement.

    c. One expects the 1-day prediction to be more highly correlated with the actual temperature, and the results in parts (a) and (b) support that expectation. A high correlation implies only that the temperatures are related, not that they are the same – if the predictions were always exactly 30 degrees too high, the correlation would be perfect despite the lack of agreement. In many correlation problems the two variables cannot agree (no matter how high the correlation) because they don't even have the same units (e.g., height and weight).

    NOTE: Compare this to the parametric hypothesis test using Pearson's correlation in section 9-2 exercise #21.

13. a. $t_{6,.025} = 2.365$; $r_s^2 = (2.447)^2/[(2.447)^2 + 6] = .499$, $r_s = \pm.707$
    b. $t_{13,.025} = 2.160$; $r_s^2 = (2.160)^2/[(2.160)^2 + 13] = .264$, $r_s = \pm.514$
    c. $t_{28,.025} = 2.048$; $r_s^2 = (2.048)^2/[(2.048)^2 + 28] = .130$, $r_s = \pm.361$
    d. $t_{28,.005} = 2.763$; $r_s^2 = (2.763)^2/[(2.763)^2 + 28] = .214$, $r_s = \pm.463$
    e. $t_{6,.005} = 3.707$; $r_s^2 = (3.707)^2/[(3.707)^2 + 6] = .696$, $r_s = \pm.834$

14. Let x = the tobacco times
        y = the alcohol times
    Ranking the scores produces the following summary statistics.

$\Sigma R_x = 1275$ $\Sigma R_y = 1275$ for $d = R_x - R_y$, $\Sigma d = 0$
$\Sigma R_x^2 = 42038.00$ $\Sigma R_y^2 = 41624.50$ $\Sigma d^2 = 10700.00$
$\Sigma R_x R_y = 36481.25$

Using the first formula,
$r_s = 1 - [6(\Sigma d^2)]/[n(n^2-1)]$
$= 1 - [6(10700.00)]/[50(2499)] = 1 - 64200/124950 \; 1 - .514 = .486$

Using the second formula,
$n(\Sigma R_x^2) - (\Sigma R_x)^2 = 50(42038.00) - (1275)^2 = 476275$
$n(\Sigma R_y^2) - (\Sigma R_y)^2 = 50(41624.50) - (1275)^2 = 455600$
$n(\Sigma R_x R_y) - (\Sigma R_x)(\Sigma R_y) = 50(36481.25) - (1275)(1275) = 198437.5$
$r_s = 198437.5/[\sqrt{476275} \sqrt{455600}] = .426$

The difference appears to be non-trivial. Since the first formula is exact only when there are no ties and the second formula is always exact, the result from the second formula is better. For n=50, both results are well into the critical region and would lead to the rejection of the hypothesis of no correlation – but in other cases, the use of the different formulas could result in different conclusions.

12-7 Runs Test for Randomness

NOTE: In each exercise, the item that appears first in the sequence is considered to be of the first type and its count is designated by n_1.

1. $n_1 = 11$ (# of H's) $G = 3$ (# of runs)
   $n_2 = 7$ (# of M's)   CV: 5,14 (from Table A-10)
   No; the sequence does not appear to be random – fewer runs than expected by chance.

2. $n_1 = 10$ (# of M's) $G = 7$ (# of runs)
   $n_2 = 6$ (# of F's)   CV: 4,13 (from Table A-10)
   Yes; the sequence appears to be random.

3. $n_1 = 10$ (# of A's) $G = 10$ (# of runs)
   $n_2 = 10$ (# of B's)   CV: 6,16 (from Table A-10)
   Yes; the sequence appears to be random.
   NOTE: This example illustrates an obvious shortcoming of this test – it considers only the number of runs, and not the pattern within the runs. Here there seems to be an obvious pattern of 2 A's followed by 2 B's – but the <u>number</u> of runs is not considered unusual.

4. $n_1 = 10$ (# of T's) $G = 4$ (# of runs)
   $n_2 = 10$ (# of F's)   CV: 6,16 (from Table A-10)
   No; the sequence does not appear to be random – fewer runs than expected by chance.

5. Since $n_1 = 12$ and $n_2 = 8$, use Table A-10.
   H_o: the sequence is random
   H_1: the sequence is not random
   $\alpha = .05$
   C.R. $G \leq 6$
   $G \geq 16$
   calculations:
   $G = 10$
   conclusion:
   Do not reject H_o; there is not sufficient evidence to reject the claim that the values occur in a random sequence.

   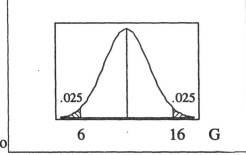

   A lack of randomness would mean there was a pattern. Recognition of that pattern would give the bettor an advantage and put the casino at a disadvantage.

6. Since $n_1 = 10$ and $n_2 = 11$, use Table A-10.
   H_o: the sequence is random
   H_1: the sequence is not random
   $\alpha = .05$
   C.R. $G \leq 6$
   $\qquad G \geq 17$
   calculations:
   $\qquad G = 10$
   conclusion:
   Do not reject H_o; there is not sufficient evidence to reject the claim that the genders occur in a random sequence.

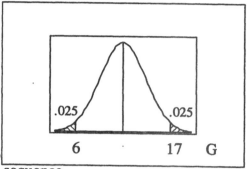

7. Since $n_1 = 19$ and $n_2 = 13$, use Table A-10.
   H_o: the sequence is random
   H_1: the sequence is not random
   $\alpha = .05$
   C.R. $G \leq 10$
   $\qquad G \geq 23$
   calculations:
   $\qquad G = 6$
   conclusion:
   Reject H_o; there is sufficient evidence to reject the claim that the values occur in a random sequence.

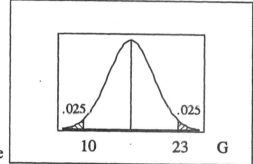

Fred should conclude that there is not randomness in the marital status of the women he selects. There are fewer runs than expected by chance alone – i.e., the married and single women tend to come in bunches. Perhaps his "random" selection process includes input from the present date in selecting the next date – and that women tend to associate with and recommend friends of their same marital status.

8. Since $n_1 = 19$ and $n_2 = 14$, use Table A-10.
   H_o: the sequence is random
   H_1: the sequence is not random
   $\alpha = .05$
   C.R. $G \leq 11$
   $\qquad G \geq 23$
   calculations:
   $\qquad G = 20$
   conclusion:
   Do not reject H_o; there is not sufficient evidence to reject the claim that the winners occur in a random sequence.

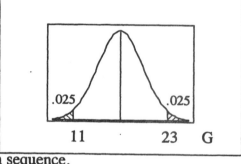

The result doesn't speak to the relative abilities of the two leagues (i.e., to whether or not there are an equal number of A's and N's). A low number of runs would indicate that one league tends to dominate for a while, and then the other. A high number of runs would indicate a suspicious tendency for the leagues to alternate winning.

9. Since $n_1 = 18$ and $n_2 = 14$, use Table A-10.
   H_o: the sequence is random
   H_1: the sequence is not random
   $\alpha = .05$
   C.R. $G \leq 10$
   $\qquad G \geq 23$
   calculations
   $\qquad G = 15$
   conclusion:
   Do not reject H_o; there is not sufficient evidence to reject the claim that the values occur in a random sequence.

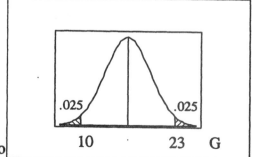

Yes; it appears that we elect Democrat and Republican candidates in a sequence that is random.

10. Define July 1 to be the middle of the year.
    A A A A A A B B B B B B
    Since $n_1 = 6$ and $n_2 = 6$, use Table A-10.
    H_o: A&B dates occur in a random sequence
    H_1: A&B dates do not occur in a random sequence
    $\alpha = .05$
    C.R. $G \leq 3$
         $G \geq 11$
    calculations:
      $G = 2$
    conclusion:
      Reject H_o; there is sufficient evidence to reject the claim that dates after and before the middle of the year occur in a random sequence and to conclude that the sequence is not random (in fact, that dates on the same side of the middle tend to occur in groups).

    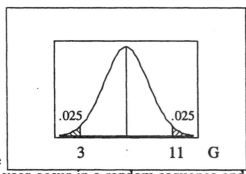

11. The median of the 21 values (arranged in numerical order from $x_1 = 1000$ to $x_{21} = 11568$) is $x_{11} = 3000$. Passing through the values chronologically and assigning A's and B's as directed (and ignoring values equal to the median), yields the sequence
    B B B B B B B B B B A A A A A A A A A A
    Since $n_1 = 10$ and $n_2 = 10$, use Table A-10.
    H_o: B&A values occur in a random sequence
    H_1: B&A values do not occur in a random sequence
    $\alpha = .05$
    C.R. $G \leq 6$
         $G \geq 16$
    calculations:
      $G = 2$
    conclusion:
      Reject H_o; there is sufficient evidence to reject the claim that values below and above the median occur in a random sequence and to conclude that the sequence is not random (in fact, that values on the same side of the median tend to occur in groups).

    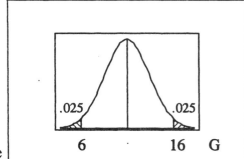

    The results suggest that the values in the stock market are steadily increasing and not randomly fluctuating around a long-term typical value.

12. The mean of the 21 values is 45418.1. The sequence of A&B values is as follows.
    A A A B A A A A A A A B B B B B B B B B B
    Since $n_1 = 10$ and $n_2 = 11$, use Table A-10.
    H_o: the A&B sequence is random
    H_1: the A&B sequence is not random
    $\alpha = .05$
    C.R. $G \leq 6$
         $G \geq 17$
    calculations:
      $G = 4$
    conclusion:
      Reject H_o; there is sufficient evidence to reject the claim that the values above and below the mean occur in a random sequence and to conclude that the sequence is not random (in fact, that values on the same side of the mean tend to occur in groups.

    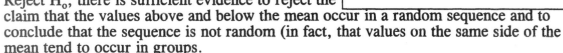

    There seems to be a downward trend. This could be attributed to safer highways and/or safer cars.

S-390 INSTRUCTOR'S SOLUTIONS Chapter 12

13. The sequence in O's and E's is as follows.
    O E O O O E E O O O E O O O O E O E E E
    E E E O O E O E O O O E E E E E O O O O
    E O O O O O O O O E O E E E O O E O E E
    O O E O E O E O E E E E E E E E E E O O
    E E E E E O E E E O O E E O O O E E O O
    Since $n_1 = 49$ and $n_2 = 51$, use the normal approximation.
    $\mu_G = 2n_1n_2/(n_1+n_2) + 1$
         $= 2(49)(51)/100 + 1 = 50.98$
    $\sigma_G^2 = [2n_1n_2(2n_1n_2-n_1-n_2)]/[(n_1+n_2)^2(n_1+n_2-1)]$
         $= [2(49)(51)(4898)]/[(100)^2(99)] = 24.727$
    H_0: the sequence is random
    H_1: the sequence is not random
    $\alpha = .05$ [assumed]
    C.R. $z < -z_{.025} = -1.96$
         $z > z_{.025} = 1.96$
    calculations:
    $G = 43$
    $z_G = (G - \mu_G)/\sigma_G$
         $= (43 - 50.98)/\sqrt{24.73}$
         $= -7.98/4.973 = -1.605$
    conclusion:
    Do not reject H_0; there is not sufficient evidence to conclude that the sequence is not random.

14. Since $n_1 = 57$ and $n_2 = 40$, use the normal approximation.
    $\mu_G = 2n_1n_2/(n_1+n_2) + 1$
         $= 2(57)(40)/97 + 1 = 48.010$
    $\sigma_G^2 = [2n_1n_2(2n_1n_2-n_1-n_2)]/[(n_1+n_2)^2(n_1+n_2-1)]$
         $= [2(57)(40)(4463)]/[(97)^2(96)] = 22.531$
    H_0: the sequence is random
    H_1: the sequence is not random
    $\alpha = .05$ [assumed]
    C.R. $z < -z_{.025} = -1.96$
         $z > z_{.025} = 1.96$
    calculations:
    $G = 54$
    $z_G = (G - \mu_G)/\sigma_G$
         $= (54 - 48.010)/\sqrt{22.531}$
         $= 5.990/4.747 = 1.262$
    conclusion:
    Do not reject H_0; there is not sufficient evidence to reject the claim that the sequence is random.

15. The 150 runners finished in the following order by gender.
    16M F 2M F 10M F   M F 4M F 7M 2F 11M F 9M 2F 5M 3F
     4M F 3M 2F   M F 4M 3F   M F 2M F   M F 4M 2F 4M F
      M 2F 4M 2F   M F 4M F   M F 2M F   3M F 2M 2F 4M 2F
    There are 111M's, 39F's and 54 runs.
    Since $n_1 = 111$ and $n_2 = 39$, use the normal approximation.
    $\mu_G = 2n_1n_2/(n_1+n_2) + 1$
         $= 2(111)(39)/150 + 1 = 58.72$
    $\sigma_G^2 = [2n_1n_2(2n_1n_2-n_1-n_2)]/[(n_1+n_2)^2(n_1+n_2-1)]$
         $= [2(111)(39)(8508)]/[(150)^2(149)] = 21.965$

H_o: the M&F sequence is random
H_1: the M&F sequence is not random
$\alpha = .05$ [assumed]
C.R. $z < -z_{.025} = -1.96$
$z > z_{.025} = 1.96$
calculations:
$G = 54$
$z_G = (G - \mu_G)/\sigma_G$
$= (54 - 58.72)/\sqrt{21.96}$
$= -4.72/4.687 = -1.007$

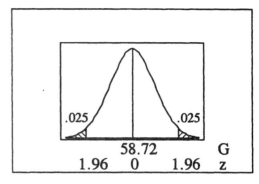

conclusion:
Do not reject H_o; there is not sufficient evidence to conclude that the sequence is not random.
No, there is not sufficient evidence to support the claim that male runners tend to finish before female runners – the hypothesis that finishing order was gender-random cannot be rejected.

16. The mean age of the 150 runners was 38.9 They finished in the following order by age, according to whether younger (Y) or older (E) than the mean.
```
 Y 2E 3Y 2E 2Y  E  Y  E  Y  E 3Y  E 2Y 4E  Y 3E 3Y
 E 2Y  E 2Y 2E  Y  E 2Y  E 6Y 4E 2Y  E  Y 5E 2Y  E
2Y 2E 7Y 3E 3Y 2E 4Y  E  Y 3E 3Y 6E 4Y  E  Y  E  Y
 E 5Y  E  Y  E 3Y 2E 2Y 4E 2Y 4E  Y 3E 3Y  E 2Y 2E
There are 80Y's, 70E's and 68 runs.
```
Since $n_1 = 80$ and $n_2 = 70$, use the normal approximation.
$\mu_G = 2n_1 n_2/(n_1+n_2) + 1$
$= 2(80)(70)/150 + 1 = 75.667$
$\sigma_G^2 = [2n_1 n_2(2n_1 n_2 - n_1 - n_2)]/[(n_1+n_2)^2(n_1+n_2-1)]$
$= [2(80)(70)(11050)]/[(150)^2(149)] = 36.916$

H_o: the Y&E sequence is random
H_1: the Y&E sequence is not random
$\alpha = .05$ [assumed]
C.R. $z < -z_{.025} = -1.96$
$z > z_{.025} = 1.96$
calculations:
$G = 68$
$z_G = (G - \mu_G)/\sigma_G$
$= (68 - 75.667/\sqrt{36.916}$
$= -7.667/6.076 = -1.262$

conclusion:
Do not reject H_o; there is not sufficient evidence to reject the claim that the sequence is random.
No, there is not sufficient evidence to support the claim that younger runners tend to finish before older runners – the hypothesis that finishing order was age-random cannot be rejected.

17. The minimum possible number of runs is $G = 2$ and occurs when all the A's are together and all the B's are together (e.g., A A B B).
The maximum possible number of runs is $G = 4$ and occurs when the A's and B's alternate (e.g., A B A B).
Because the critical region for $n_1 = n_2 = 2$ is
C.R. $G \leq 1$
$G \geq 6$
the null hypothesis of the sequence being random can never be rejected at the .05 level. Very simply, this means that it is not possible for such a small sample to provide 95% certainty that a non-random phenomenon is occurring.

18. a. and b. The 84 sequences and the number of runs in each are as follows.

```
AAABBBBBB-2  BAAABBBBB-3  BBAAABBBB-3  BBBAAABBB-3  BBBBAAABB-3  BBBBBAAAB-3
AABABBBBB-4  BAABABBBB-5  BBAABABBB-5  BBBAABABB-5  BBBBAABAB-5  BBBBBAABA-4
AABBABBBB-4  BAABBABBB-5  BBAABBABB-5  BBBAABBAB-5  BBBBAABBA-4              BBBBBBAAA-2
AABBBABBB-4  BAABBBABB-5  BBAABBBAB-5  BBBAABBBA-4
AABBBBABB-4  BAABBBBAB-5  BBAABBBBA-4
AABBBBBAB-4  BAABBBBBA-4
AABBBBBBA-3

ABAABBBBB-4  BABAABBBB-5  BBABAABBB-5  BBBABAABB-5  BBBBABAAB-5  BBBBBABAA-4
ABABABBBB-6  BABABABBB-7  BBABABABB-7  BBBABABAB-7  BBBBABABA-6
ABABBABBB-6  BABABBABB-7  BBABABBAB-7  BBBABABBA-6
ABABBBABB-6  BABABBBAB-7  BBABABBBA-6
ABABBBBAB-6  BABABBBBA-6
ABABBBBBA-5

ABBAABBBB-4  BABBAABBB-5  BBABBAABB-5  BBBABBAAB-5  BBBBABBAA-4
ABBABABBB-6  BABBABABB-7  BBABBABAB-7  BBBABBABA-6
ABBABBABB-6  BABBABBAB-7  BBABBABBA-6
ABBABBBAB-6  BABBABBBA-6
ABBABBBBA-5

ABBBAABBB-4  BABBBAABB-5  BBABBBAAB-5  BBBABBBAA-4
ABBBABABB-6  BABBBABAB-7  BBABBBABA-6
ABBBABBAB-6  BABBBABBA-6
ABBBABBBA-5

ABBBBAABB-4  BABBBBAAB-5  BBABBBBAA-4
ABBBBABAB-6  BABBBBABA-6
ABBBBABBA-4

ABBBBBAAB-4  BABBBBBAA-4
ABBBBBABA-5

ABBBBBBAA-3
```

c. At the right is the distribution for G, the number of runs, found from the above sequences. Based on this distribution, a two-tailed test at the .05 level (that places .025 or less in each tail) has
    C.R. G ≤ 2
         G ≥ 8
(for which it will never be possible to reject by being in the upper tail).

G	P(G)
2	2/84 = .023
3	7/84 = .083
4	21/84 = .250
5	24/84 = .286
6	20/84 = .238
7	10/84 = .229
	84/84 = 1.000

d. The critical region in part (c) agrees exactly with Table A-10.

Review Exercises

1. Let the after-course scores be group 1.
claim: median difference ≠ 0

subj	A	B	C	D	E	F	G	H	I	J
R-M	+	0	−	+	+	+	−	+	+	+

n = 9: 7+'s and 2−'s
H_o: median difference = 0
H_1: median difference ≠ 0
α = .05
C.R. x ≤ $x_{L,9,.025}$ = 1
     x ≥ $x_{U,9,.025}$ = 9−1 = 8
calculations:
    x = 2 (using less frequent count)
    x = 7 (using + count)

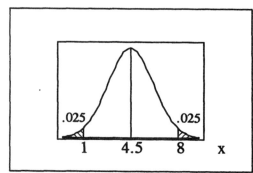

conclusion:
    Do not reject H_o; there is not sufficient evidence to conclude that the median difference is different from 0 – i.e., there is not sufficient evidence to reject the claim that the course has no effect.

2. Let the after-course scores be group 1.
    claim: the populations have the same distribution

    ```
    A-B |  20   0  -10   40   30   10  -30   20   20   10
    R   |   5   -   -2    9  7.5    2 -7.5    5    5    2
    ΣR- =  9.5
    ΣR+ = 35.5
    ΣR  = 45.0
    ```
    check: $\Sigma R = n(n+1)/2 = 9(10)/2 = 45$
    H_0: the populations have the same distribution
    H_1: the populations have different distributions
    $\alpha = .05$
    C.R. $T \leq T_{L,9,.025} = 6$
    $T \geq T_{U,9,.025} = 45-6 = 39$
    calculations
    $T = 9.5$ (using the smaller ranks)
    $T = 35.5$ (using the positive ranks)

    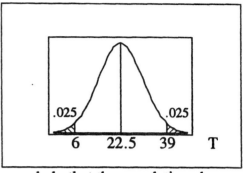

    conclusion:
    Do not reject H_0; there is not sufficient evidence to conclude that the populations have different distributions – i.e., there is not sufficient evidence to reject the claim that the course has no effect.

3. Let x = the number of women hired.
    claim: $p < .5$
    22 +'s (women)   44 -'s (men)
    n = 66 +'s or -'s
    Since n > 25, use z with
    $\mu_x = n/2 = 66/2 = 33$
    $\sigma_x = \sqrt{n}/2 = \sqrt{66}/2 = 4.062$
    H_0: $p = .5$
    H_1: $p < .5$
    $\alpha = .01$
    C.R. $z < -z_{.01} = -2.326$
    calculations:
    x = 22
    $z_x = [(x+.5)-\mu_x]/\sigma_x$
    $= [22.5 - 33]/4.062$
    $= -10.5/4.062 = -2.585$
    conclusion:
    Reject H_0; there is sufficient evidence to conclude that $p < .5$ (i.e., that there is bias against women in the hiring practices).

4. Below are the ordered scores for each group.
    claim: the populations have the same distribution

    ```
    beer    R    liquor    R
    .129    1    .182      9           n₁ = 12        ΣR₁ =  89.5
    .146    2    .185     10           n₂ = 14        ΣR₂ = 261.5
    .148    3    .190     12.5
    .152    4    .205     15           n = Σn = 26    ΣR = 351.0
    .154    5    .220     17
    .155    6    .224     18           check: ΣR = n(n+1)/2
    .164    7    .225     19                       = 26(27)/2
    .165    8    .226     20                       = 351
    .187   11    .227     21           R = ΣR₁ = 89.5
    .190   12.5  .234     22
    .203   14    .241     23           μ_R = n₁(n+1)/2
    .212   16    .247     24               = 12(27)/2
           89.5  .253     25               = 162
                .257      26           σ²_R = n₁n₂(n+1)/12
                         261.5              = (12)(14)(27)/12 = 378
    ```

H_0: the populations have the same distribution
H_1: the populations have different distributions
$\alpha = .05$
C.R. $z < -z_{.025} = -1.96$
$z > z_{.025} = 1.96$
calculations:
$z_R = (R - \mu_R)/\sigma_R$
$= (89.5 - 162)/\sqrt{378}$
$= -72.5/19.442 = -3.729$

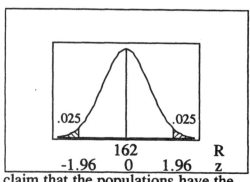

conclusion:
Reject H_0; there is sufficient evidence to reject the claim that the populations have the same distribution and to conclude that the distributions are different (in fact, that the beer scores are lower).
It appears that the liquor drinkers are more dangerous.

5. The following table summarizes the calculations.

x	R_x	y	R_y	d	d^2
29	5	31	8.5	-3.5	12.25
35	9	27	2	7	49
28	3.5	29	6,5	-3	9
44	10	25	1	9	81
25	2	31	8.5	-6.5	42.25
34	8	29	6.5	1.5	2.25
30	6	28	4.0	2	4
33	7	28	4.0	3	9
28	3.5	28	4.0	-0.5	0.25
24	1	33	10	-9	81
	55.0		55.0	0.0	290.00

$r_s = 1 - [6(\Sigma d^2)]/[n(n^2-1)]$
$= 1 - [6(290)]/[10(99)]$
$= 1 - 1.758$
$= -.758$

H_0: $\rho_s = 0$
H_1: $\rho_s \neq 0$
$\alpha = .05$
C.R. $r_s < -.648$
$r_s > .648$
calculations:
$r_s = -.758$
conclusion:
Reject H_0; there is sufficient evidence to reject the claim that $\rho_s=0$ conclude that $\rho_s \neq 0$ (in fact, that $\rho_s < 0$).

Yes; based on these results you can expect to pay more for gas if you buy a heavier car. Since the calculations are based entirely on ranks, and the ranks would not change whether the weights are given in pounds or 100-pounds, such a modification would not change the results at all.
NOTE: Using formula 9-1 (since there are ties) yields $r_s = -.796$.

6. The sequence in O's and E's is as follows.
O O E O O O O O E E E O E E O E O E E O
O O O E E E E O O E O O O O O E O E E E
Since $n_1 = 22$ and $n_2 = 18$, use the normal approximation.
$\mu_G = 2n_1n_2/(n_1+n_2) + 1$
$= 2(22)(18)/40 + 1 = 20.8$
$\sigma_G^2 = [2n_1n_2(2n_1n_2-n_1-n_2)]/[(n_1+n_2)^2(n_1+n_2-1)]$
$= [2(22)(18)(752)]/[(40)^2(39)] = 9.545$
H_0: the O&E sequence is random
H_1: the O&E sequence is not random
$\alpha = .05$
C.R. $z < -z_{.025} = -1.96$
$z > z_{.025} = 1.96$

calculations:
G = 18
$z_G = (G - \mu_G)/\sigma_G$
    $= (18 - 20.8)/\sqrt{9.545}$
    $= -2.8/3.089 = -.906$

conclusion:
Do not reject H_o; there is not sufficient evidence to conclude that the sequence is not random.

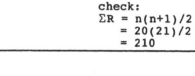

7. Below are the scores for each group.
   The group listed first is group 1, etc.

subco	R	compa	R	midsi	R	fulls	R
595	5	1051	15	629	6	1985	17
1063	16	1193	18	1686	20	971	12
885	10	946	11	880	9	996	14
519	3	984	13	181	1	804	8
422	2	584	4	645	7	1376	19
	36		61		43		70

$n_1 = 5 \quad R_1 = 36$
$n_2 = 5 \quad R_2 = 61$
$n_3 = 5 \quad R_3 = 43$
$n_4 = 5 \quad R_4 = 70$

$n = \Sigma n = 20 \quad \Sigma R = 210$

check:
$\Sigma R = n(n+1)/2$
      $= 20(21)/2$
      $= 210$

H_o: the populations have the same distribution
H_1: the populations have different distributions
$\alpha = .05$
C.R. $H > \chi^2_{3,.05} = 7.815$
calculations:
$H = [12/n(n+1)] \cdot [\Sigma(R_i^2/n_i)] - 3(n+1)$
  $= [12/20(21)] \cdot [(36)^2/5 + (61)^2/5 + (43)^2/5 + (70)^2/5] - 3(21)$
  $= [.0286] \cdot [2353.2] - 63 = 4.234$

conclusion:
Do not reject H_o; there is not sufficient evidence to conclude that leg injury measurements for the four weight categories are not all the same.

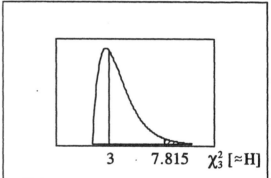

No; for this particular measurement, the data do not show that the heavier cars are safer in a crash.

8. The following table summarizes the calculations.

x	R_x	y	R_y	d	d^2
91	5	4.56	2	3	9
92	6	6.48	6	0	0
82	2	5.99	5	-3	9
85	3	7.92	8	-5	25
87	4	5.36	4	0	0
80	1	3.32	1	0	0
94	7	7.32	7	0	0
97	8	5.27	3	5	25
	36		36	0	68

$r_s = 1 - [6(\Sigma d^2)]/[n(n^2-1)]$
   $= 1 - [6(68)]/[8(63)]$
   $= 1 - .810$
   $= .190$

H_o: $\rho_s = 0$
H_1: $\rho_s \neq 0$
$\alpha = .05$
C.R. $r_s < -.738$
     $r_s > .738$
calculations:
$r_s = .190$

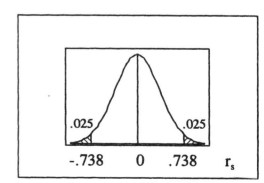

conclusion:
Do not reject H_o; there is not sufficient evidence to conclude that $\rho_s \neq 0$.
The results suggest that the higher priced tapes do not necessarily have higher performance ratings. When buying tapes, choose less expensive ones.

Cumulative Review Exercises

1. a. Since $n_1 = 20$ and $n_2 = 7$, use Table A-10.
    H_0: the M&F sequence is random
    H_1: the M&F sequence is not random
    $\alpha = .05$
    C.R. $G \le 6$
    $\quad\quad\; G \ge 16$
    calculations:
    $\quad G = 15$
    conclusion:
    $\quad$ Do not reject H_0; there is not sufficient
    evidence to reject the claim that the sequence is random with respect to gender

    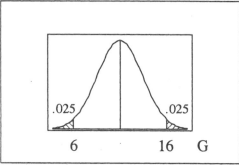

    b. Let p be the proportion of women: $\hat{p} = x/n = 7/27 = .259$
    claim: $p \ne .50$
    H_0: $p = .50$
    H_1: $p \ne .50$
    $\alpha = .05$
    C.R. $z < -z_{.025} = -1.96$
    $\quad\quad z > z_{.025} = 1.96$
    calculations:
    $z_{\hat{p}} = (\hat{p} - \mu_{\hat{p}})/\sigma_{\hat{p}}$
    $\quad = (.259 - .50)/\sqrt{(.50)(.50)/27}$
    $\quad = -.241/.0962 = -2.502$
    P-value $= 2 \cdot P(z < -2.50) = 2 \cdot (.0062) = .0124$
    conclusion:
    $\quad$ Reject H_0; there is sufficient evidence to conclude that $p \ne .50$ (in fact, that $p < .50$).

    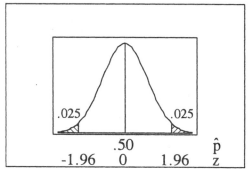

    c. Let x = the number of women.
    claim: $p \ne .50$
    $\quad$ 7 +'s (women) $\quad$ 20 -'s (men)
    $\quad$ n = 27 +'s or -'s
    Since n > 25, use z with
    $\mu_x = n/2 = 27/2 = 13.5$
    $\sigma_x = \sqrt{n}/2 = \sqrt{27}/2 = 2.598$
    H_0: $p = .50$
    H_1: $p \ne .50$
    $\alpha = .05$
    C.R. $z < -z_{.025} = -1.96$
    $\quad\quad z > z_{.025} = 1.96$
    calculations:
    $x = 7$
    $z_x = [(x+.5) - \mu_x]/\sigma_x$
    $\quad = [7.5 - 13.5]/2.598 = -6/2.598 = -2.309$
    P-value $= 2 \cdot P(z < -2.31) = 2 \cdot (.0104) = .0208$
    conclusion:
    $\quad$ Reject H_0; there is sufficient evidence to conclude that $p \ne .50$ (in fact, that $p < .50$).
    NOTE: The smaller P-value in part (b) [.0124 < .0208] indicates that the parametric tests are generally better than their non-parametric counterparts.

    d. Let p be the proportion of women: $\hat{p} = x/n = 7/27 = .259$
    $\hat{p} \pm z_{.025} \cdot \sqrt{\hat{p}\hat{q}/n}$
    $.259 \pm 1.96 \cdot \sqrt{(.259)(.741)/27}$
    $.259 \pm .165$
    $.094 < p < .425$

    e. There is no evidence that the passengers were not sampled in random sequence according to gender. But the above results cannot address whether or not the sample is biased against either gender without knowing the proportion of each gender in the population of passengers. If the sample contains significantly more of one gender than another, that might reflect the gender distribution of the passengers and not gender bias by the pollster. Otherwise, there are no problems with the survey in these areas.

2. The following table summarizes the preliminary calculations necessary for all parts of this exercise. Since there are ties, the shortcut formula for r_s using $d=R_x-R_y$ will not be used.
   Let x = height of the winner.
   Let y = height of the runner-up.

```
           raw scores              sign  signed     ranks
        x        y      d=x-y       d    rank d    R_x    R_y
       76       64       12         +      7        8      1
       66       71       -5         -     -5        1      4.5
       70       72       -2         -     -2.5      2.5    6.5
       70       72       -2         -     -2.5      2.5    6.5
       74       68        6         +      6        6.5    2
       71.5     71        0.5       +      1        4      4.5
       73       69.5      3.5       +      4        5      3
       74       74        0         0      -        6.5    8
Σv    574.5    561.5     13.0                      36      36
Σv²  41325.25 39476.25  225.50                    203.00  203.00
Σuv           40288.00                             143.00
```

a. $n(\Sigma xy) - (\Sigma x)(\Sigma y) = 8(40288.00) - (574.1)(561.5) = -277.75$
   $n(\Sigma x^2) - (\Sigma x)^2 = 8(41325.25) - (574.5)^2 = 551.75$
   $n(\Sigma y^2) - (\Sigma y)^2 = 8(39476.25) - (561.5)^2 = 527.75$
   $r = [n(\Sigma xy) - (\Sigma x)(\Sigma y)]/[\sqrt{n(\Sigma x^2) - (\Sigma x)^2} \cdot \sqrt{n(\Sigma y^2) - (\Sigma y)^2}]$
     $= -277.75/[\sqrt{551.75} \cdot \sqrt{527.75}] = -.5147$

   $H_0: \rho = 0$
   $H_1: \rho \neq 0$
   $\alpha = .05$
   C.R. $r < -.707$ OR   C.R. $t < -t_{6,.025} = -2.447$
        $r > .707$ $t > t_{6,.025} = 2.447$

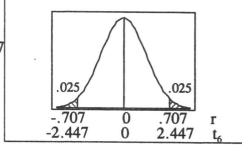

   calculations: calculations:
   $r = -.515$ $t_r = (r - \mu_r)/s_r$
                              $= (-.515 - 0)/\sqrt{(1-(-.515)^2)/6}$
   conclusion: $= -.515/.3500 = -1.471$
   Do not reject H_0; there is not sufficient evidence to reject the claim that $\rho = 0$.
   No; there is no significant correlation between the heights.

b. Refer to the summary calculations at the beginning of the exercise.
   Since there are ties, use R_x for x and R_y for y in formula 9-1.
   $n(\Sigma xy) - (\Sigma x)(\Sigma y) = 8(143.00) - (36)(36) = -152.00$
   $n(\Sigma x^2) - (\Sigma x)^2 = 8(203.00) - (36)^2 = 328.00$
   $n(\Sigma y^2) - (\Sigma y)^2 = 8(203.00) - (36)^2 = 328.00$
   $r = -152.00/[\sqrt{328.00} \cdot \sqrt{328.00}] = -.4634$
   $H_0: \rho_s = 0$
   $H_1: \rho_s \neq 0$
   $\alpha = .05$
   C.R. $r_s < -.738$
        $r_s > .738$
   calculations:
   $r_s = -.463$

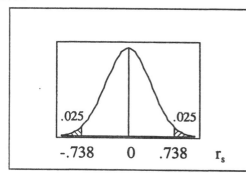

conclusion:
   Do not reject H_o; there is not sufficient evidence to conclude that there is any correlation between the heights.
   NOTE: The smaller (in absolute value) r_s in part (b) [$|-.463| < |-.515|$] indicates that non-parametric tests are generally weaker than their parametric counterparts.

c. Refer to the summary calculations at the beginning of the exercise.
   claim: median difference ≠ 0
       n = 7: 4+'s and 3-'s
   H_o: median difference = 0
   H_1: median difference ≠ 0
   $\alpha = .05+$
   C.R. $x \le x_{L,7,.025} = 0$
       $x \ge x_{U,7,.025} = 7-0 = 7$
   calculations:
       x = 3 (using less frequent count)
       x = 4 (using + count)
   conclusion:
   Do not reject H_o; there is not sufficient evidence to conclude that the median difference is different from 0.

d. Refer to the summary calculations at the beginning of the exercise.
   claim: the populations have different distributions
       ΣR− = 10   n = 7 non-zero ranks
       ΣR+ = 18
       ΣR = 28
   check: ΣR = n(n+1)/2 = 7(8)/2 = 28
   H_o: the populations have the same distribution
   H_1: the populations have different distributions
   $\alpha = .05$
   C.R. $T \le T_{L,7,.025} = 2$
       $T \ge T_{U,7,.025} = 28-2 = 26$
   calculations:
       T = 10 (using the smaller ranks)
       T = 18 (using the positive ranks)
   conclusion:
   Do not reject H_o; there is not sufficient evidence to conclude there is a difference between the heights of the winning and losing candidates.

e. Refer to the summary calculations at the beginning of the exercise.
     n = 8   Σd = 13.0 $\bar{d} = 1.625$
             Σd² = 225.50 $s_d = 5.403$
   original claim: $\mu_d \ne 0$
   H_o: $\mu_d = 0$
   H_1: $\mu_d \ne 0$
   $\alpha = .05$
   C.R. $t < -t_{7,.025} = -2.365$
       $t > t_{7,.025} = 2.365$
   calculations:
       $t_{\bar{d}} = (\bar{d} - \mu_{\bar{d}})/s_{\bar{d}}$
          = $(1.625 - 0)/(5.403/\sqrt{8})$
          = $1.625/1.910 = .851$
   P-value = $2 \cdot P(t_7 > .851) > .20$
   conclusion:
   Do not reject H_o; there is not sufficient evidence to conclude that $\mu_d \ne 0$.

f. The is not sufficient evidence to conclude there is any correlation between the heights of the winning and losing candidates, nor that there is any difference between the heights.

Chapter 13

Statistical Process Control

13-2 Control Charts for Variation and Mean

NOTE: In this section, k = number of sample subgroups
n = number of observations per sample subgroup

Although the $\bar{x}$ line is not specified in the text as a necessary part of the run chart, such a line is included in the Minitab examples and is included in this manual.

1. a. Process data are data arranged in sequence according to time.
    b. Process data are out of statistical control when they exhibit more than natural variation – i.e., when the data include patterns, cycles, or unusual points.
    c. A process is out of statistical control whenever any of the following 3 criteria are met:
       (1) The data exhibit a non-random trend, pattern, or cycle.
       (2) The is a point beyond the upper or lower control limit.
       (3) The are 8 or more consecutive points on the same side of the center line.
    d. Random variation is due to chance; assignable variation is the result of identifiable causes.
    e. An R chart monitors the ranges of consecutive samples on size n and is used to determine whether the process variation is within statistical control. An $\bar{x}$ chart monitors the means of consecutive sample of size n and is used to determine whether the process average is within statistical control.

The following chart summarizes the information necessary for exercises #2, #3 and #4.

sample	consumption			$\bar{x}$	R
1a	4762	3875	2657	3764.7	2105
1b	4358	2201	3187	3248.7	2157
2a	4504	3237	2198	3313.0	2306
2b	2511	3020	2857	2796.0	509
3a	3952	2785	2118	2951.7	1834
3b	2658	2139	3071	2622.7	932
4a	3863	3013	2023	2966.3	1840
4b	2953	3456	2647	3018.7	809

2. There are k·n = (8 samples)·(3 observations/sample) = 24 observations on the run chart.
    $\bar{\bar{x}} = \Sigma\bar{x}/k = \Sigma x/(k \cdot n) = 74045/24 = 3085.2$

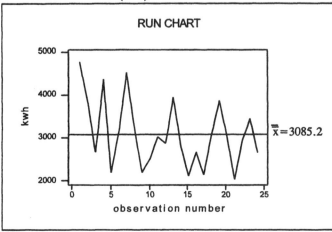

No; there does not appear to be a pattern suggesting that the process is not within statistical control. Some of the variation might be explained as follows: (1) a downward trend reflecting changes in the home (e.g., improved insulation) or home usage (e.g., children leaving home) or a long term climate shift; (2) a fairly consistent cycle of ups and downs reflecting repeating seasonal patterns.

3. $\bar{R} = \Sigma R/k = 12492/8 = 1561.5$ $\quad$ LCL $= D_3\bar{\bar{R}} = 0.000(1561.5) = 0.0$
$\quad$ UCL $= D_4\bar{R} = 2.574(1561.5) = 4019.3$

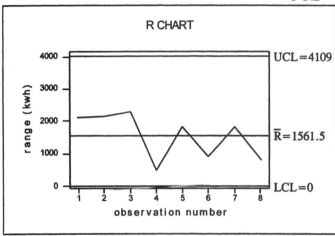

The process variation is within statistical control.

4. $\bar{\bar{x}} = \Sigma\bar{x}/k = 24681.8/8 = 3085.2$
$\bar{R} = \Sigma R/k = 12492/8 = 1561.5$
LCL $= \bar{\bar{x}} - A_2\bar{R} = 3085.3 - (1.023)(1561.5) = 3085.3 - 1597.4 = 1487.9$
UCL $= \bar{\bar{x}} + A_2\bar{R} = 3085.3 + (1.023)(1561.5) = 3085.3 + 1597.4 = 4682.7$

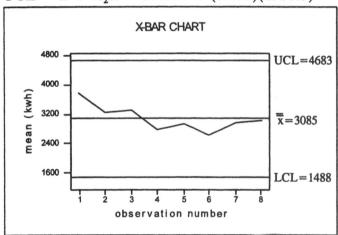

The process mean is within statistical control. A process is not in statistical control when there is a pattern and/or unusual data point that cannot be attributed to reasonable random effects. For household energy data, any of the following factors could make the process go out of statistical control: unusual weather, family adjustments (increase or decrease in size, adoption of conservation habits), house adjustments (new insulation).

5. $\bar{R} = \Sigma R/k = 1374/25 = 54.96$ $LCL = D_3\bar{R} = 0.076(54.96) = 4.18$
   $UCL = D_4\bar{R} = 1.924(54.96) = 105.74$

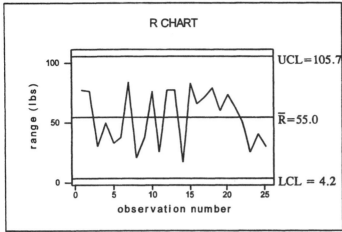

The process variation appears to be within statistical control.

6. $\bar{\bar{x}} = \Sigma\bar{x}/k = 6677.9/25 = 267.116$
   $\bar{R} = \Sigma R/k = 1374/25 = 54.96$
   $LCL = \bar{\bar{x}} - A_2\bar{R} = 267.116 - (.419)(54.96) = 267.116 - 23.028 = 244.088$
   $UCL = \bar{\bar{x}} + A_2\bar{R} = 267.116 + (.419)(54.96) = 267.116 + 23.028 = 290.144$

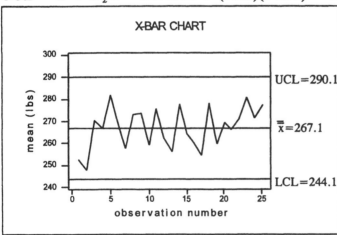

The process mean appears to be within statistical control.

7. There are $k \cdot n = (20 \text{ samples}) \cdot (5 \text{ obs/sample}) = 100$ observations on the run chart.
   For convenience, the problem is worked in milligrams (i.e., the original weights x 1000).
   $\bar{\bar{x}} = \Sigma\bar{x}/k = \Sigma x/(k \cdot n) = 570980/100 = 5709.8$

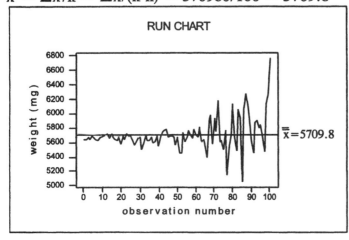

Yes, there appears to be a pattern that suggests the process is not within statistical control – viz., the variability is increasing. In practical terms, this suggests that the equipment is not holding its tolerance and that the process needs to be shut down for an adjustment.

8. For convenience, the problem is worked in milligrams (i.e., the original weights x 1000).
$\bar{R} = \Sigma R/k = 7292/20 = 364.6$ $LCL = D_3\bar{R} = 0.000(364.6) = 0.0$
                                              $UCL = D_4\bar{R} = 2.114(364.6) = 770.8$

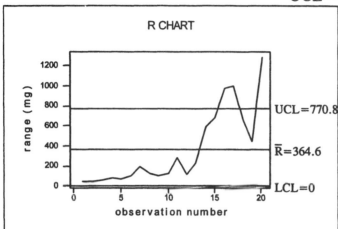

The process variation is not within statistical control. To be out of control, the process must meet at least one of the three out-of-control criteria. This process actually meets all three. (1) There is a pattern to the R values – in this case, an upward drift. (2) There is a point beyond the upper or lower control limit – in this case, there are 3 points above the UCL. (3) There are at least eight consecutive points on the same side of the center line – in this case, the first 13 points are below the center line.

9. For convenience, the problem is worked in milligrams (i.e., the original weights x 1000).
$\bar{\bar{x}} = \Sigma\bar{x}/k = 114196.0/20 = 5709.8$
$\bar{R} = \Sigma R/k = 7292/20 = 364.6$
$LCL = \bar{\bar{x}} - A_2\bar{R} = 5709.8 - (.577)(364.6) = 5709.8 - 210.4 = 5499.4$
$UCL = \bar{\bar{x}} + A_2\bar{R} = 5709.8 + (.577)(364.6) = 5709.8 + 210.4 = 5920.2$

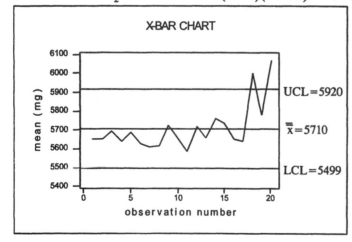

The process mean is not within statistical control. To be out of control, the process must meet at least one of the three out-of-control criteria. This process actually meets all three. (1) There is a pattern to the $\bar{x}$ values – in this case, an upward drift. (2) There is a point beyond the upper or lower control limit – in this case, there are 2 points above the UCL. (3) There are at least eight consecutive points on the same side of the center line – in this case, the first 8 points are below the center line. Yes, corrective action is needed.

10. There are 52 observations, one for each Monday, on the run chart.
$\bar{\bar{x}} = \Sigma x/52 = 5.22/52 = .1004$

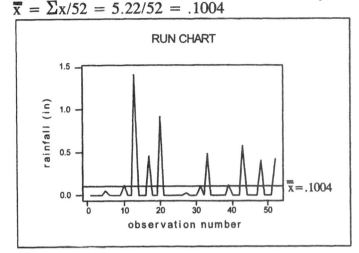

No, the process does not appear to be within statistical control. There are occasional exceptionally large (i.e., relative to the others) values that seem to be outliers. And there are two runs of at least eight consecutive observations on the same side of the center line – the values for weeks 1-9 are below $\bar{\bar{x}}$ and the values for weeks 21-30 are below $\bar{\bar{x}}$.

The following chart summarizes the information necessary for exercises #11 and #12.

week	$\bar{x}$	R	week	$\bar{x}$	R	week	$\bar{x}$	R	week	$\bar{x}$	R
1	.019	.05	14	.000	.00	27	.036	.14	40	.067	.24
2	.027	.10	15	.097	.40	28	.027	.11	41	.003	.02
3	.101	.71	16	.237	.87	29	.010	.05	42	.003	.02
4	.186	.64	17	.067	.47	30	.017	.12	43	.181	.68
5	.016	.05	18	.054	.24	31	.082	.44	44	.243	1.48
6	.091	.64	19	.036	.14	32	.044	.18	45	.290	1.28
7	.051	.30	20	.237	.92	33	.164	.64	46	.139	.96
8	.001	.01	21	.049	.27	34	.134	.85	47	.116	.79
9	.039	.16	22	.003	.01	35	.009	.03	48	.109	.41
10	.121	.39	23	.000	.00	36	.017	.12	49	.017	.08
11	.181	.78	24	.101	.71	37	.050	.26	50	.000	.00
12	.029	.17	25	.097	.33	38	.057	.40	51	.106	.74
13	.416	1.41	26	.000	.00	39	.017	.12	52	.146	.43

11. $\bar{R} = \Sigma R/k = 20.36/52 = .3915$ $\quad$ LCL $= D_3\bar{R} = 0.076(.3915) = .030$
$\quad$ UCL $= D_4\bar{R} = 1.924(.3915) = .753$

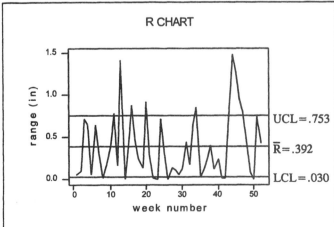

The process variation is not within statistical control. Since there are points above the upper control limit and points below the lower control limit, reject the claim that rainfall amounts exhibit stable variation. While only one point above or below the appropriate

S-404 INSTRUCTOR'S SOLUTIONS Chapter 13

control limits is sufficient evidence of unstable variation, these data have multiple points beyond each limit.

12. $\bar{\bar{x}} = \Sigma\bar{x}/k = 4.3414/52 = .0835$
    $\bar{R} = \Sigma R/k = 20.36/52 = .3915$
    LCL $= \bar{\bar{x}} - A_2\bar{R} = .0835 - (.419)(.3915) = .0835 - .1641 = -.0806$, truncated at 0
    UCL $= \bar{\bar{x}} + A_2\bar{R} = .0835 + (.419)(.3915) = .0835 + .1641 = .2475$

    NOTE: The LCL is truncated at 0 because of the story problem – since x is the number of inches of rainfall, which cannot be negative. For variables such as temperature, for which x can assume negative values, such truncation would not be appropriate.

    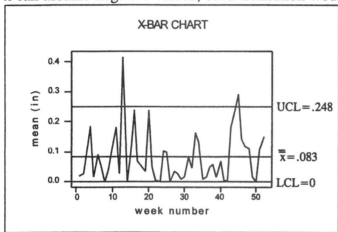

    The process mean is not within statistical control, since there are points above the upper control limit.

13. $\bar{s} = \Sigma s/k = 193.85/20 = 9.6925$ LCL $= B_3\bar{s} = 0(9.6925) = 0$
                                                   UCL $= B_4\bar{s} = 2.266(9.6925) = 21.963$

    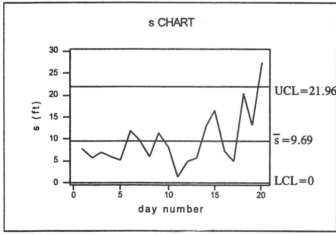

    This is very similar to the R chart given in this section, and both charts indicate the process variation is not within statistical control. For small n's in the sub-samples (n=4 in this example), there will be a very high correlation between the values of R and s – and so the two charts will be almost identical, but with different labels on the vertical axis.

14. $\bar{\bar{x}} = \Sigma\bar{x}/k = 129.00/20 = 6.45$
    $\bar{s} = \Sigma s/k = 193.85/20 = 9.6925$
    LCL $= \bar{\bar{x}} - A_3\bar{s} = 6.45 - (1.628)(9.6925) = 6.45 - 15.779 = -9.329$
    UCL $= \bar{\bar{x}} + A_3\bar{s} = 6.45 + (1.628)(9.6925) = 6.45 + 15.779 = 22.229$

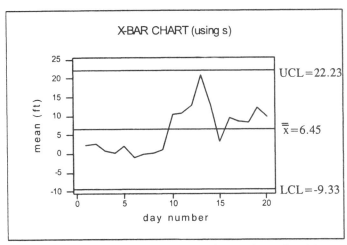

This is almost identical to the chart given in this section based on R. The plot of the points does not change, nor does the line for $\bar{\bar{x}}$ – only the UCL and LCL lines changed by a fraction of a unit. Considering the extra work and more decimal places involved working with s, the chart based on R should be preferred.

13-3 Control Charts for Attributes

1. This process is within statistical control. Since the first third of the sample means are generally less than the overall mean, the middle third are generally more than the overall mean, and the final third are generally less than the overall mean, however, one may wish to check future analyses to see whether such a pattern tends to repeat itself.

2. This process is within statistical control.

3. This process is out of statistical control. There is an upward trend, and there is a point above the upper control limit.

4. This process is out of statistical control. There are points lying beyond the control limits and there appears to be an upward trend.

5. $\bar{p} = (\Sigma x)/(\Sigma n) = (30+29+\ldots+23)/(13)(100,000) = 332/1,300,000 = .000255$
    $\sqrt{\bar{p}\cdot\bar{q}/n} = \sqrt{(.000255)(.999745)/100000} = .0000505$
    $LCL = \bar{p} - 3\sqrt{\bar{p}\cdot\bar{q}/n} = .000255 - 3(.0000505) = .000255 - .000152 = .000104$
    $UCL = \bar{p} + 3\sqrt{\bar{p}\cdot\bar{q}/n} = .000255 + 3(.0000505) = .000255 + .000152 = .000407$

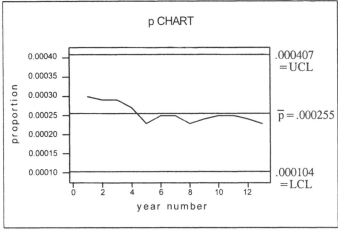

NOTE: The 13 sample proportions are .00030, .00029, .00029,..., .00023.
There are three out-of-control criteria, and meeting any one of them means that the process is not within statistical control. This process meets two of those criteria. (1) There is a pattern – in this case, a downward trend. (3) The run of 8 rule applies – in this case, the last 9 points are all below the center line. By definition, therefore, process is statistically unstable. In this instance, however, that is good and means that the age 0-4 death rate for infectious diseases is decreasing.

6. $\bar{p} = (\Sigma x)/(\Sigma n) = (29+33+...+24)/(20)(1000) = 572/20000 = .0286$
$\sqrt{\bar{p}\cdot\bar{q}/n} = \sqrt{(.0286)(.9714)/1000} = .00527$
LCL $= \bar{p} - 3\sqrt{\bar{p}\cdot\bar{q}/n} = .0286 - 3(.00527) = .0286 - .0158 = .0128$
UCL $= \bar{p} + 3\sqrt{\bar{p}\cdot\bar{q}/n} = .0286 + 3(.00527) = .0286 + .0158 = .0444$

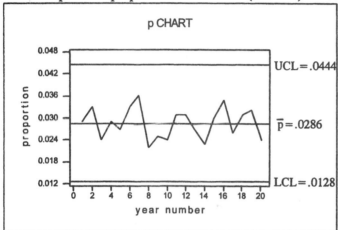

NOTE: The 20 sample proportions are: .029 .033 .024 .029 ... .024
The process is within statistical control. Viewing the data as a statistical process, they do not suggest a problem that should be corrected. Viewing the data as crime statistics, they indicate that there has been no reduction in the crime rate -- and that could be considered a problem that should be corrected. On the positive side, however, the data indicate that there has been no increase in the crime rate either.

7. $\bar{p} = (\Sigma x)/(\Sigma n) = (3+3+1...+5)/(52)(7) = 127/364 = .3489$
$\sqrt{\bar{p}\cdot\bar{q}/n} = \sqrt{(.3489)(.6511)/7} = .1801$
LCL $= \bar{p} - 3\sqrt{\bar{p}\cdot\bar{q}/n} = .3489 - 3(.1801) = .3489 - .5404 = -.192$ [truncated at 0.000]
UCL $= \bar{p} + 3\sqrt{\bar{p}\cdot\bar{q}/n} = .3489 + 3(.1801) = .3489 + .5404 = .889$

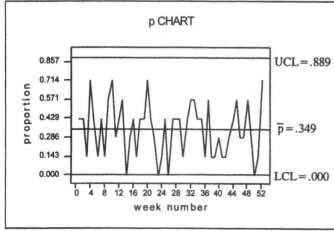

NOTE: The 52 sample proportions are: 3/7, 3/7, 1/7,...
The vertical labels correspond to 0/7=.000, 1/7=.143, 2/7=.286, 3/7=.429,...
The process is within statistical control. Viewing the proportion of days each week with

measurable rainfall as a statistical process, it does not need correction. Whether the proportions are too high or too low for individual preferences or other needs is another issue.

8. a. The results for Japan are given below.
$\bar{p} = (\Sigma x)/(\Sigma n) = (58+60+...+63)/(8)(10,000) = 496/80,000 = .00620$
$\sqrt{\bar{p}\cdot\bar{q}/n} = \sqrt{(.00620)(.99380)/10000} = .0007850$
LCL $= \bar{p} - 3\sqrt{\bar{p}\cdot\bar{q}/n} = .00620 - 3(.0007850) = .00620 - .00235 = .00385$
UCL $= \bar{p} + 3\sqrt{\bar{p}\cdot\bar{q}/n} = .00620 + 3(.0007850) = .00620 + .00235 = .00855$

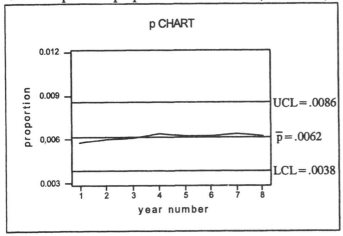

NOTE: The 13 sample proportions are: .0058 .0060 .0061 .0064 ... .0063..

b. The results for the United States are given below.
$\bar{p} = (\Sigma x)/(\Sigma n) = (98+94+...+87)/(8)(10,000) = 729/80,000 = .00911$
$\sqrt{\bar{p}\cdot\bar{q}/n} = \sqrt{(.00911)(.99089)/10000} = .0009502$
LCL $= \bar{p} - 3\sqrt{\bar{p}\cdot\bar{q}/n} = .00911 - 3(.0009502) = .00911 - .00285 = .00626$
UCL $= \bar{p} + 3\sqrt{\bar{p}\cdot\bar{q}/n} = .00911 + 3(.0009502) = .00911 + .00285 = .01196$

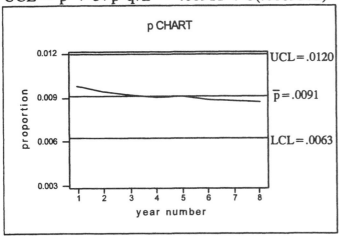

NOTE: The 13 sample proportions are: .0098 .0094 .0092 .0090 ... .0087...

c. The two p-charts are given above on the same scale. Both rates are within statistical control. The rates in Japan are lower and more consistent, and they may exhibit a very slight upward trend. The rates in the United States are higher and more variable, and they appear to exhibit a slight downward trend.

9. $\bar{p} = (\Sigma x)/(\Sigma n) = (25+24+...+31)/(13)(100,000) = 375/1,300,000 = .000288$
$n\cdot\bar{p} = (100,000)(.000288) = 28.846 \ [= \bar{x}]$
$\sqrt{n\cdot\bar{p}\cdot\bar{q}} = \sqrt{(100000)(.000288)(.999712)} = 5.370$

$$LCL = n\cdot\bar{p} - 3\sqrt{n\cdot\bar{p}\cdot\bar{q}} = 28.846 - 3(5.370) = 28.846 - 16.110 = 12.736$$
$$UCL = n\cdot\bar{p} + 3\sqrt{n\cdot\bar{p}\cdot\bar{q}} = 28.846 + 3(5.370) = 28.846 + 16.110 = 44.956$$

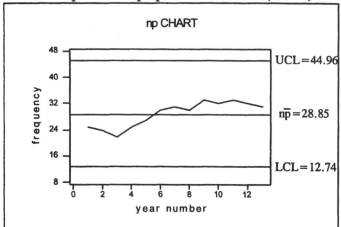

The charts are identical except for the labels. Since the labels on the np chart are n=100,000 times the labels on the chart for p, the center line is 100,000(.000288) = 28.486 and the upper control limit is 100,000(.000450) = 44.956 and the lower control limit is 100,000(.000127) = 12.736. The placement of the points does not change.

10. $\bar{p} = (\Sigma x)/(\Sigma n) = .05$
In both parts (a) and (b), the center line occurs at .05.
 a. n = 100
  $\sqrt{\bar{p}\cdot\bar{q}/n} = \sqrt{(.05)(.95)/100} = .0218$
  $LCL = \bar{p} - 3\sqrt{\bar{p}\cdot\bar{q}/n} = .05 - 3(.0218) = .05 - .0654 = 0$ [since it cannot be negative]
  $UCL = \bar{p} + 3\sqrt{\bar{p}\cdot\bar{q}/n} = .05 + 3(.0218) = .05 + .0654 = .1154$
 b. n = 300
  $\sqrt{\bar{p}\cdot\bar{q}/n} = \sqrt{(.05)(.95)/300} = .0126$
  $LCL = \bar{p} - 3\sqrt{\bar{p}\cdot\bar{q}/n} = .05 - 3(.0126) = .05 - .0378 = .0123$
  $UCL = \bar{p} + 3\sqrt{\bar{p}\cdot\bar{q}/n} = .05 + 3(.0126) = .05 + .0378 = .0878$
 c. The lower and upper control limits are closer to the center line in part (b). This has the advantage of being better able (i.e., on the basis of less deviance from the long run average) to detect when the process is out of statistical control, but it has the disadvantage of requiring the examination of a larger sample size. In addition, the chart in part (a) has LCL = 0.000 and cannot determine if the observed proportion is significantly less than .05. The chart in part (b) would be better able to detect a shift from 5% to 10% because the larger sample size would cause less fluctuation about the 5% or 10% long run averages and make the shift more noticeable.

Review Exercises

The chart at the right summarizes the information necessary for exercises #1, #2 and #3.

$\bar{\bar{x}} = \Sigma\bar{x}/k$
$= 126.922/11$
$= 11.538$

$\bar{R} = \Sigma R/k$
$= 55.86/11$
$= 5.078$

year	$\bar{x}$	R
1980	12.480	9.33
1981	12.106	2.99
1982	10.707	6.26
1983	9.430	5.11
1984	13.372	7.09
1985	10.468	4.32
1986	15.084	6.34
1987	11.684	4.50
1988	10.240	3.99
1989	10.734	4.50
1990	10.706	1.34

1.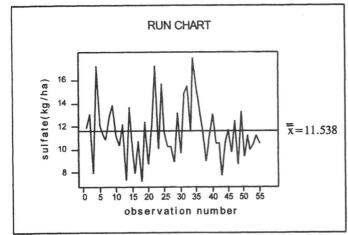

No; there does not appear to be a pattern suggesting that the process is not within statistical control.

2. $\bar{R} = \Sigma R/k = 55.86/11 = 5.078$ $LCL = D_3\bar{R} = 0(5.078) = 0$
   $UCL = D_4\bar{R} = 2.114(5.078) = 10.73$

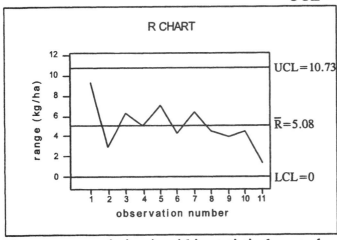

The process variation is within statistical control.

3. $\bar{\bar{x}} = \Sigma \bar{x}/k = 126.922/11 = 11.538$
   $\bar{R} = \Sigma R/k = 55.86/11 = 5.078$
   $LCL = \bar{\bar{x}} - A_2\bar{R} = 11.538 - (.577)(5.078) = 11.538 - 2.930 = 8.608$
   $UCL = \bar{\bar{x}} + A_2\bar{R} = 11.538 + (.577)(5.078) = 11.538 + 2.930 = 14.468$

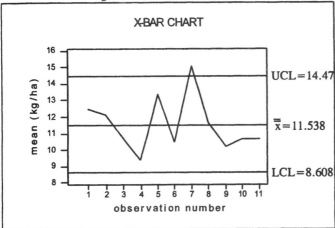

The process mean is not within statistical control because there is a point above the UCL.

4. $\bar{p} = (\Sigma x)/(\Sigma n) = (270+264+\ldots+343)/(13)(100{,}000) = 4183/1{,}300{,}000 = .00322$
$\sqrt{\bar{p}\cdot\bar{q}/n} = \sqrt{(.00322)(.99678)/100000} = .000179$
LCL $= \bar{p} - 3\sqrt{\bar{p}\cdot\bar{q}/n} = .00322 - 3(.000179) = .00322 - .00053 = .00268$
UCL $= \bar{p} + 3\sqrt{\bar{p}\cdot\bar{q}/n} = .00322 + 3(.000179) = .00322 + .00053 = .00375$

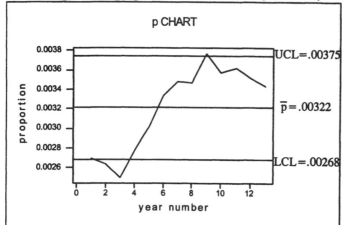

NOTE: The 13 sample proportions are: .00270 .00264 .00250 .00278 ... .00343
There are three out-of-control criteria, and meeting any one of them means that the process is not within statistical control. This process meets all three criteria. (1) There is a pattern – in this case, an upward trend. (2) There is a point outside the lower or upper control limits – in this case, there are two points below the LCL and one point above the UCL. (3) The run of 8 rule applies – in this case, the last 8 points are all above the center line. By definition, therefore, process is statistically unstable. In this instance, that means that the age 65+ death rate for infectious diseases is increasing.

5. $\bar{p} = (\Sigma x)/(\Sigma n) = (608+466+\ldots+491)/(15)(1{,}000) = 7031/15000 = .4687$
$\sqrt{\bar{p}\cdot\bar{q}/n} = \sqrt{(.4687)(.5313)/1000} = .01578$
LCL $= \bar{p} - 3\sqrt{\bar{p}\cdot\bar{q}/n} = .4687 - 3(.01578) = .4687 - .0473 = .4214$
UCL $= \bar{p} + 3\sqrt{\bar{p}\cdot\bar{q}/n} = .4687 + 3(.01578) = .4687 + .0473 = .5161$

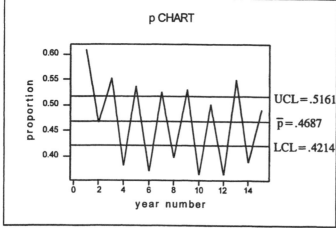

NOTE: The 15 sample proportions are: .608 .466 .552 .382 ... .491
There are three out-of-control criteria, and meeting any one of them means that the process is not within statistical control. This process meets the first two criteria. (1) There is a pattern – in this case, an alternating trend. (2) There is a point outside the lower or upper control limits – in this case, most of the points are beyond the control limits. By definition, therefore, process is statistically unstable. The pattern is caused by the fact that the national elections occur every two years, every other national election involving a presidential election and bring out the voters in larger numbers.

Cumulative Review Exercises

1. a. $\bar{p} = (\Sigma x)/(\Sigma n) = (10+8+...+11)/(20)(400) = 150/8000 = .01875$
   $\sqrt{\bar{p}\cdot\bar{q}/n} = \sqrt{(.01875)(.98125)/400} = .006782$
   LCL $= \bar{p} - 3\sqrt{\bar{p}\cdot\bar{q}/n} = .01875 - 3(.006782) = .01875 - .02035 = 0$ [truncate at 0]
   UCL $= \bar{p} + 3\sqrt{\bar{p}\cdot\bar{q}/n} = .01875 + 3(.006782) = .01875 + .02035 = .0391\ 5161$
   NOTE: The LCL is truncated at 0 because of the story problem – since $\bar{p}$ is the proportion of defectives in the sample, which cannot be negative.

   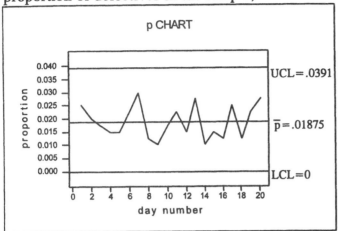

   NOTE: The 20 sample proportions are: $10/400 = .025$, $8/400 = .020$,...,$11/400 = .0275$
   The process is within statistical control, and so the data can be treated as coming from a binomial population with a fixed mean and variance.

   b. Using all the data combined, $\hat{p} = x/n = 150/8000$
   $\hat{p} \pm z_{.025}\cdot\sqrt{\hat{p}\hat{q}/n}$
   $.01875 \pm 1.960\cdot\sqrt{(.01875)(.98125)/8000}$
   $.01875 \pm .00297$
   $.0158 < p < .0217$

   c. original claim $p > .01$
   $H_o: p = .01$
   $H_1: p > .01$
   $\alpha = .05$
   C.R. $z > 1.645$
   calculations:
   $z_{\hat{p}} = (\hat{p} - \mu_{\hat{p}})/\sigma_{\hat{p}}$
   $= (.01875 - .01)/\sqrt{(.01)(.99)/8000}$
   $= .00875/.001112$
   $= 7.866$
   P-value $= P(z>7.87) = 1 - P(z<7.87) = 1 - .9999 = .0001$
   conclusion:
   Reject H_o; there is sufficient evidence to conclude that $p > .01$.

2. a. let $A_i =$ the ith point is above the center line
   $P(A_i) = .5$ for each value of i
   $P(A) = P(\text{all 8 above}) = P(A_1)\cdot P(A_2)\cdot...\cdot P(A_8) = (.5)^8 = .00391$

   b. let $B_i =$ the ith point is below the center line
   $P(B_i) = .5$ for each value of i
   $P(B) = P(\text{all 8 below}) = P(B_1)\cdot P(B_2)\cdot...\cdot P(B_8) = (.5)^8 = .00391$

   c. Notice that the events in parts (a) and (b) are mutually exclusive.
   $P(A \text{ or } B) = P(A) + P(B) - P(A \text{ and } B)$
   $= .00391 + .00391 - 0 = .00781$

3. The following chart summarizes the information in the same manner used for the kwh consumption data in section 13-2 exercises #2, #3 and #4. The analysis also follows the format of those exercises.

sample	temperature			x	R
1a	32	35	59	42.0	27
1b	76	66	42	61.3	34
2a	22	33	56	37.0	34
2b	70	63	42	58.3	28
3a	30	38	55	41.0	25
3b	71	61	38	56.7	33
4a	32	40	57	43.0	25
4b	72	65	45	60.7	27

a. Run Chart: There are k·n = (8 samples)·(3 observations/sample) = 24 observations.
$\bar{X} = \Sigma\bar{x}/k = 400/8 = 50.00$

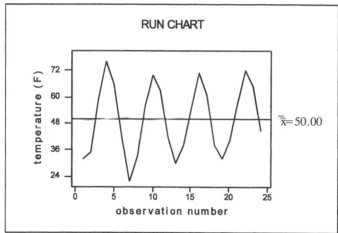

The process is not within statistical control because there is a pattern of cycles indicating something other than random variation is at work.

b. R Chart: $\bar{R} = \Sigma R/k = 233/8 = 29.125$ $LCL = D_3\bar{R} = 0(29.125) = 0$
$UCL = D_4\bar{R} = 2.544(29.125) = 74.97$

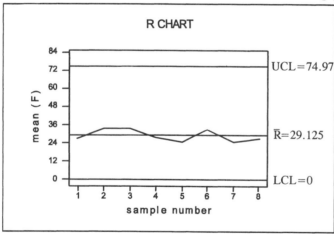

The process variation is within statistical control.

c. $\bar{x}$ Chart: $\bar{\bar{x}} = \Sigma\bar{x}/k = 400/8 = 50.00$
$\bar{R} = \Sigma R/k = 233/8 = 29.125$
$LCL = \bar{\bar{x}} - A_2\bar{R} = 50.00 - (1.023)(29.125) = 50.00 - 29.79 = 20.21$
$UCL = \bar{\bar{x}} + A_2\bar{R} = 50.00 + (1.023)(29.125) = 50.00 + 29.79 = 79.79$

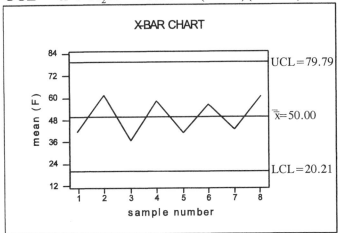

The process mean is not within statistical control because there is an alternating pattern of points above and below the overall mean.

4. Letting x represent temperature (°F) and y represent consumption (kwh), the following numerical summary contains all the necessary information for the n=24 paired values.
$\Sigma x = 1200$ $\Sigma y = 74045$
$\Sigma x^2 = 65850$ $\Sigma y^2 = 242624810$ $\Sigma xy = 3562714$
$n(\Sigma x^2)-(\Sigma x)^2 = 140400$ $n(\Sigma y^2)-(\Sigma y)^2 = 340333415$ $n(\Sigma xy)-(\Sigma x)(\Sigma y) = -3348864$

a. $r = -3348864/[\sqrt{140400}\sqrt{340333415}] = -.484$
The critical values being $r = \pm.396$ (for n=25 at the $\alpha=.05$ level), we conclude there is a significant negative linear correlation between outside temperature and energy consumption.

b. $b_1 = -3348864/140400 = -23.852$
$b_0 = (74045/24) - (-23.852)(1200/24) = 4277.824$
The regression line is $\hat{y} = 4277.824 - 23.852x$

c. $\hat{y}_{60} = 4277.824 - 23.852(60) = 2847$ kwh

FINAL NOTE: Congratulations! You have completed statistics – the course that everybody likes to hate. I hope that this manual has helped to make the course a little more understandable for you – and that you leave the course with an appreciation of broad principles, and not memories of merely manipulating formulas. I wish you well in your continued studies, and that you achieve your full potential wherever your journey of life may lead.

Suggested Course Syllabus

This book contains much more material than can be covered in the typical one-semester introductory statistics course. Sample syllabi are provided on the following pages as an aid to instructors in planning their individual courses. The "class hour" referred to in these outlines consists of approximately one hour. In addition to the sample syllabi that follow, there are many other configurations that successfully satisfy individual needs and preferences.

When selecting the sections to be included, consider assigning some sections as out-of-class independent work. For example, after students understand the basic concepts of confidence intervals and hypothesis testing, consider assigning exercises from Chapter 8 (Inferences From Two Samples) to be done using a computer software package or a TI-83 Plus calculator. This is an excellent way to include important and useful concepts while encouraging students to extend their basic knowledge to different circumstances.

In planning a new course syllabus for the Ninth Edition of *Elementary Statistics*, users of the Eighth Edition should consider these major changes:

1. In Chapters 6, 7, and, 8, proportions are now covered first, before means and standard deviations. Reasons for this change: Students see proportions frequently in the media, and they tend to be more interested in data that are expressed as proportions. Also, proportions are generally easier to work with than means, so students can better focus on the important principles of estimating parameters and testing hypotheses.

2. In Chapter 5, there is a new section discussing "Sampling Distributions and Estimators," so students can better understand some of the reasons behind the methods introduced in later chapters.

3. In Chapter 5, Sections 5-3 and 5-4 from the Eighth Edition (nonstandard normal distributions) are now combined in Section 5-3. This change is motivated by the new format of Table A-2, which makes it easier for students to work with normal distributions.

4. Section 6-4 from the Eighth Edition (Sample Size Required to Estimate μ) is now included in Section 6-3 (Estimating a Population Mean: σ Known) along with confidence intervals for estimating μ.

5. Because the topic of statistical process control is included less frequently than the topic of nonparametric statistics, those two topics have been switched so that Chapter 12 now covers methods of nonparametric statistics and Chapter 13 covers statistical process control.

Syllabus Version A

Class Hour	Text Section	Topic
1	1-1, 1-2, 1-3	Overview; Nature of Data; Critical Thinking
2	1-4	Design of Experiments
3	2-1, 2-2, 2-3	Frequency Distributions and Visualizing Data
4	2-4	Measures of Center
5	2-5	Measures of Variation
6	2-5	Measures of Variation
7	2-6	Measures of Relative Standing
8	2-7	Exploratory Data Analysis
9	Ch. 1, 2	Review
10	Ch. 1, 2	Test 1
11	3-1, 3-2, 3-3	Fundamentals of Probability and Addition Rule
12	3-4	Multiplication Rule: Basics
13	4-1, 4-2	Random Variables
14	4-3	Binomial Probability Distributions
15	4-4	Mean, Variance, and Standard Deviation for the Binomial Distribution
16	Ch. 3, 4	Review
17	Ch. 3, 4	Test 2
18	5-1, 5-2	The Standard Normal Distribution
19	5-3	Applications of Normal Distributions
20	5-4	Sampling Distributions and Estimators
21	5-5	The Central Limit Theorem
22	6-1, 6-2	Estimating a Population Proportion
23	6-3	Estimating a Population Mean: σ Known
24	6-4	Estimating a Population Mean: σ Not Known
25	6-5	Estimating a Population Variance
26	Ch. 5, 6	Review
27	Ch. 5, 6	Test 3
28	7-1, 7-2	Basics of Hypothesis Testing
29	7-2, 7-3	Testing a Claim about a Proportion
30	7-3	Testing a Claim about a Proportion
31	7-4	Testing a Claim about a Mean: σ Known
32	7-5	Testing a Claim about a Mean: σ Not Known
33	7-6	Testing a Claim about Standard Deviation or Variance
34	Ch. 7	Review
35	Ch. 7	Test 4
36	9-1, 9-2	Correlation
37	9-2, 9-3	Correlation and Regression
38	9-3	Regression
39	10-3	Contingency Tables
40		Group Project Presentations
41		Group Project Presentations
42	Ch. 9, 10	Review
43	Ch. 9, 10	Test 5
44		Review for Comprehensive Final Exam
45		Review for Comprehensive Final Exam

Syllabus Version B: Relaxed Pace

Class Hour	Text Section	Topic
1	1-1, 1-2	Overview; Nature of Data
2	1-3	Critical Thinking
3	2-1, 2-2, 2-3	Frequency Distributions and Visualizing Data
4	2-4	Measures of Center
5	2-5	Measures of Variation
6	2-5	Measures of Variation
7	2-6	Measures of Relative Standing
8	2-7	Exploratory Data Analysis
9	Ch. 1, 2	Review
10	Ch. 1, 2	Test 1
11	3-1, 3-2	Fundamentals of Probability
12	3-2	Fundamentals of Probability
13	4-1, 4-2	Random Variables
14	4-3	Binomial Probability Distributions
15	4-4	Mean, Variance, and Standard Deviation for the Binomial Distribution
16	Ch. 3, 4	Review
17	Ch. 3, 4	Test 2
18	5-1, 5-2	The Standard Normal Distribution
19	5-3	Applications of Normal Distributions
20	5-3	Applications of Normal Distributions
21	5-5	The Central Limit Theorem
22	6-1, 6-2	Estimating a Population Proportion
23	6-2	Estimating a Population Proportion
24	6-3	Estimating a Population Mean: σ Known
25	6-4	Estimating a Population Mean: σ Not Known
26	Ch. 5, 6	Review
27	Ch. 5, 6	Test 3
28	7-1, 7-2	Basics of Hypothesis Testing
29	7-2, 7-3	Testing a Claim about a Proportion
30	7-3	Testing a Claim about a Proportion
31	7-4	Testing a Claim about a Mean: σ Known
32	7-5	Testing a Claim about a Mean: σ Not Known
33	7-5	Testing a Claim about a Mean: σ Not Known
34	Ch. 7	Review
35	Ch. 7	Test 4
36	9-1, 9-2	Correlation
37	9-2, 9-3	Correlation and Regression
38	9-3	Regression
39		Group Project Presentations
40		Group Project Presentations
41	Ch. 9	Review
42	Ch. 9	Test 5
43		Review for Comprehensive Final Exam
44		Review for Comprehensive Final Exam

Syllabus Version C: EARLY COVERAGE OF CORRELATION AND REGRESSION

Class Hour	Text Section	Topic
1	1-1, 1-2, 1-3	Overview; Nature of Data; Critical Thinking
2	1-4	Design of Experiments
3	2-1, 2-2, 2-3	Frequency Distributions and Visualizing Data
4	2-4	Measures of Center
5	2-5	Measures of Variation
6	2-5	Measures of Variation
7	2-6	Measures of Relative Standing
8	2-7	Exploratory Data Analysis
9	**9-1, 9-2**	**Correlation**
10	**9-3**	**Regression**
11	**9-3**	**Regression**
12	Ch. 1, 2, 9	Review
13	Ch. 1, 2, 9	Test 1
14	3-1, 3-2, 3-3	Fundamentals of Probability and Addition Rule
15	3-4	Multiplication Rule: Basics
16	4-1, 4-2	Random Variables
17	4-3	Binomial Probability Distributions
18	4-4	Mean, Variance, and Standard Deviation for the Binomial Distribution
19	5-1, 5-2	The Standard Normal Distribution
20	5-3	Applications of Normal Distributions
21	5-4	Sampling Distributions and Estimators
22	5-5	The Central Limit Theorem
23	5-6	Normal as Approximation to Binomial
24	Ch. 3, 4, 5	Review
25	Ch. 3, 4, 5	Test 2
26	6-1, 6-2	Estimating a Population Proportion
27	6-3	Estimating a Population Mean: σ Known
28	6-4	Estimating a Population Mean: σ Not Known
29	6-5	Estimating a Population Variance
30	7-1, 7-2	Basics of Hypothesis Testing
31	7-2, 7-3	Testing a Claim about a Proportion
32	7-3	Testing a Claim about a Proportion
33	7-4	Testing a Claim about a Mean: σ Known
34	7-5	Testing a Claim about a Mean: σ Not Known
35	7-6	Testing a Claim about Standard Deviation or Variance
36	10-3	Contingency Tables
37	Ch. 6, 7, 10	Review
38	Ch. 6, 7, 10	Test 3
39		Group Project Presentations
40		Group Project Presentations
41		Review for Comprehensive Final Exam
42		Review for Comprehensive Final Exam

Syllabus Version D: MINIMUM COVERAGE OF PROBABILITY

Class Hour	Text Section	Topic
1	1-1, 1-2, 1-3	Overview; Nature of Data; Critical Thinking
2	1-4	Design of Experiments
3	2-1, 2-2, 2-3	Frequency Distributions and Visualizing Data
4	2-4	Measures of Center
5	2-5	Measures of Variation
6	2-5	Measures of Variation
7	2-6	Measures of Relative Standing
8	2-7	Exploratory Data Analysis
9	Ch. 1, 2	Review
10	Ch. 1, 2	Test 1
11	**3-1, 3-2**	**Fundamentals of Probability**
12	4-1, 4-2	Random Variables
13	4-3	Binomial Probability Distributions
14	4-4	Mean, Variance, and Standard Deviation for the Binomial Distribution
15	5-1, 5-2	The Standard Normal Distribution
16	5-3	Applications of Normal Distributions
17	5-4	Sampling Distributions and Estimators
18	5-5	The Central Limit Theorem
19	Ch. 3, 4, 5	Review
20	Ch. 3, 4, 5	Test 2
21	6-1, 6-2	Estimating a Population Proportion
22	6-3	Estimating a Population Mean: σ Known
23	6-4	Estimating a Population Mean: σ Not Known
24	6-5	Estimating a Population Variance
25	7-1, 7-2	Basics of Hypothesis Testing
26	7-2, 7-3	Testing a Claim about a Proportion
27	7-3	Testing a Claim about a Proportion
28	7-4	Testing a Claim about a Mean: σ Known
29	7-5	Testing a Claim about a Mean: σ Not Known
30	7-6	Testing a Claim about Standard Deviation or Variance
31	Ch. 6, 7	Review
32	Ch. 6, 7	Test 3
33	9-1, 9-2	Correlation
34	9-2, 9-3	Correlation and Regression
35	9-3	Regression
36	10-1, 10-2	Multinomial Experiments
37	10-3	Contingency Tables
38	Ch. 9, 10	Review
39	Ch. 9, 10	Test 4
40		Group Project Presentations
41		Group Project Presentations
42		Review for Comprehensive Final Exam
43		Review for Comprehensive Final Exam

Syllabus Version E: INTEGRATION OF SOME NONPARAMETRIC STATISTICS

Class Hour	Text Section	Topic
1	1-1, 1-2, 1-3	Overview; Nature of Data; Critical Thinking
2	1-4	Design of Experiments
3	12-7	**Runs Test for Randomness**
4	2-1, 2-2, 2-3	Frequency Distributions and Visualizing Data
5	2-4	Measures of Center
6	2-5	Measures of Variation
7	2-5	Measures of Variation
8	2-6	Measures of Position
9	2-7	Exploratory Data Analysis
10	Ch. 1, 2, 12	Review
11	Ch. 1, 2, 12	Test 1
12	3-1, 3-2, 3-3	Fundamentals of Probability and Addition Rule
13	3-4	Multiplication Rule: Basics
14	4-1, 4-2	Random Variables
15	4-3	Binomial Probability Distributions
16	4-4	Mean, Variance, and Standard Deviation for the Binomial Distribution
17	5-1, 5-2	The Standard Normal Distribution
18	5-3	Applications of Normal Distributions
19	5-4	Sampling Distributions and Estimators
20	5-5	The Central Limit Theorem
21	5-6	Normal as Approximation to Binomial
22	Ch. 3, 4, 5	Review
23	Ch. 3, 4, 5	Test 2
24	6-1, 6-2	Estimating a Population Proportion
25	6-3	Estimating a Population Mean: σ Known
26	6-4	Estimating a Population Mean: σ Not Known
27	6-5	Estimating a Population Variance
28	7-1, 7-2	Basics of Hypothesis Testing
29	7-2, 7-3	Testing a Claim about a Proportion
30	7-3	Testing a Claim about a Proportion
31	7-4	Testing a Claim about a Mean: σ Known
32	7-5	Testing a Claim about a Mean: σ Not Known
33	7-6	Testing a Claim about Standard Deviation or Variance
34	Ch. 6, 7	Review
35	Ch. 6, 7	Test 3
36	9-1, 9-2	Correlation and Regression
37	9-3	Regression
38	**12-6**	**Rank Correlation**
39	**12-5**	**Kruskal-Wallis Test**
40	10-3	Contingency Tables
41	Ch. 9, 10, 12	Review
42	Ch. 9, 10, 12	Test 4
43		Group Project Presentations
44		Group Project Presentations
45		Review for Comprehensive Final Exam

Syllabus Version F: Two Semester Course

A two-semester sequence allows for more time to better incorporate computer usage, cooperative group activities, simulation techniques in probability, and bootstrap resampling techniques in estimating parameters and testing hypotheses.

First Semester:

- Chapter 1 (Sections 1-1 through 1-4):   4 classes
- Chapter 2 (Sections 2-1 through 2-7):   7 classes
- Chapter 3 (Sections 3-1 through 3-7):   6 classes
- Chapter 4 (Sections 4-1 through 4-5):   5 classes
- Chapter 5 (Sections 5-1 through 5-7):   7 classes
- Chapter 6 (Sections 6-1 through 6-5):   7 classes
- Chapter 7 (Sections 7-1 through 7-6):   8 classes

Second Semester:

- Chapter 8 (Sections 8-1 through 8-5):   8 classes
- Chapter 9 (Sections 9-1 through 9-6):   9 classes
- Chapter 10 (Sections 10-1 through 10-3): 4 classes
- Chapter 11 (Sections 11-1 through 11-3): 6 classes
- Chapter 12 (Sections 12-1 through 12-7): 4 classes
- Chapter 13 (Sections 13-1 through 13-3): 9 classes
- Chapter 14 (Section 14-1):   4 classes

Video Index
Elementary Statistics, 9e and *Elementary Statistics Using Excel, 2e*
Triola

VHS Tape #	Run Time	Section #	Title
		Chapter 1	**Introduction to Statistics**
1	08:50	1-2	Types of Data
1	13:05	1-3	Critical Thinking
1	10:22	1-4	Design of Experiments
		Chapter 2	**Describing, Exploring, and Comparing Data**
1	22:36	2-2	Frequency Distributions
		2-3	Visualizing Data
1	13:47	2-4	Measures of Center
1	16:37	2-5	Measures of Variation
1	14:50	2-6	Measures of Relative Standing
1	10:36	2-7	Exploratory Data Analysis (EDA)
		Chapter 3	**Probability**
2	19:42	3-2	Fundamentals
2	15:19	3-3	Addition Rule
2	16:35	3-4	Multiplication Rule: Basics
		3-5	Multiplication Rule: Complements and Conditional Probability
2	07:19	3-6	Probabilities Through Simulations
2	17:36	3-7	Counting
		Chapter 4	**Probability Distributions**
3	15:00	4-2	Random Variables
3	17:10	4-3	Binomial Probability Distributions
		4-4	Mean, Variance, and Standard Deviation for the Binomial Distribution
3	11:13	4-5	The Poisson Distribution
		Chapter 5	**Normal Probability Distributions**
3	15:30	5-2	The Standard Normal Distribution
3	14:17	5-3	Applications of Normal Distributions
3	12:26	5-4	Sampling Distributions and Estimators
3	14:44	5-5	The Central Limit Theorem
3	12:29	5-6	Normal as Approximation to Binomial
3	07:23	5-7	Determining Normality
		Chapter 6	**Estimates and Sample Sizes**
4	13:47	6-2	Estimating a Population Proportion
4	10:15	6-3	Estimating a Population Mean: σ Known
4	10:00	6-4	Estimating a Population Mean: σ Not Known
4	08:59	6-5	Estimating a Population Variance
		Chapter 7	**Hypothesis Testing**
4	12:46	7-2	Basics of Hypothesis Testing
4	11:42	7-3	Testing a Claim about a Proportion
4	07:12	7-4	Testing a Claim about a Mean: σ Known
4	08:12	7-5	Testing a Claim about a Mean: σ Not Known
4	15:06	7-6	Testing a Claim about Variation

Triola
Elementary Statistics, 9e and *Elementary Statistics Using Excel, 2e*
Video Index

VHS Tape #	Run Time	Section #	Title
		Chapter 8	**Inferences from Two Samples**
4	14:51	8-2	Inferences about Two Proportions
4	13:56	8-3	Inferences about Two Means: Independent Samples
4	13:41	8-4	Inferences from Matched Pairs
4	11:47	8-5	Comparing Variation in Two Samples
		Chapter 9	**Correlation and Regression**
5	07:34	9-2	Correlation
5	08:54	9-3	Regression
5	08:21	9-4	Variation and Prediction Intervals
5	07:20	9-5	Multiple Regression
5	17:28	9-6	Modeling
		Chapter 10	**Multinomial Experiments and Contingency Tables**
5	20:00	10-2 10-3	Multinomial Experiments: Goodness-of-Fit Contingency Tables: Independence and Homogeneity
		Chapter 11	**Analysis of Variance**
5	08:02	11-2	One-Way ANOVA
5	11:33	11-3	Two-Way ANOVA
		Chapter 12	**Nonparametric Statistics**
5	14:55	12-2 12-3	Sign Test Wilcoxon Signed-Ranks Test
		Chapter 13	**Statistical Process Control**
5	15:22	13-2 13-3	Control Charts for Variation and Mean Control Charts for Attributes